Phytochemicals and Phytopharmaceuticals

Phytochemicals and Phytopharmaceuticals

Editors

Fereidoon Shahidi

Chi-Tang Ho

AOCS PRESS

Champaign, Illinois

AOCS Mission Statement
To be a forum for the exchange of ideas, information, and experience among those with a professional interest in the science and technology of fats, oils, and related substances in ways that promote personal excellence and provide high standards of quality.

The paper used in this book is acid-free and falls within the guidelines established to ensure permanence and durability.

Library of Congress Cataloging-in-Publication Data
Phytochemicals and phytopharmaceuticals / editors, Fereidoon Shahidi, Chi-Tang Ho.
p. cm.
Includes bibliographical references and index.
ISBN 1-893997-05-7 (alk. paper)
1. Materia medica, Vegetable. 2. Botanical chemistry. 3. Pharmacognosy. I. Shahidi, Fereidoon, 1951– II. Ho, Chi-Tang, 1944–

RS164.P534 1999
615'.32—dc21 99-056987
CIP

Printed in the United States of America with vegetable oil-based inks.
00 99 98 97 5 4 3 2 1

Preface

Plants provide a rich source of phytochemicals and phytopharmaceuticals and have been used since antiquity to treat, manage, or possibly cure different diseases. Many of the active components of plants are phenolic in nature and serve as a defense mechanism to protect plants from photosynthetic stress, reactive oxygen species, herbivores, and wounds and act as filter for ultraviolet light. A myriad of compounds, together with their precursors and reactions products, provide a cocktail, or a soup, which collectively participates in the survival of plants. The oldest Chinese review of medicinal materials, *Shennong Bencao Jing*, published in 100–200 AD, recorded 365 herbs. In 1596, the first *Compendium of Materia Medica* (Ben Cao Gang Mu) was published in China and included 4,773 materials from plants. Similarly, some religious teachings refer to the role of plant materials in disease prevention and treatment. Garlic, a popular medicinal plant, has been recorded in early medical writings, such as the *Papyrus Ebers* scrolls, as having a remedial power.

Although synthetic drugs have dominated modern Western medicine for the past several decades, the leads for synthetic drugs have usually been derived from a plant origin and some 20% of the drug supply is from plant sources. In recent years, due to increasing demands of consumers for alternative and preventive health management, there has been a renewed interest in plant materials. Thus, the market for supplements and nutraceuticals has expanded exponentially, and the public interest in phytochemicals and phytopharmaceuticals has been considerable. This stimulated us to develop this volume from contributions made by an international group of experts in their respective research areas related to plant bioactives. We are grateful to the participating authors for their outstanding contributions, which have made the publication of this state-of-the-art monograph possible.

Fereidoon Shahidi
Chi-Tang Ho

Contents

Preface v

Chapter 1 Plants as Medicine: The Role of Phytochemicals in Optimal Health 1
Clare M. Hasler

Chapter 2 Functional Foods and Phytochemicals Research Incentives: U.S. Policy and Industry Preference 13
Nancy M. Childs and Rhond Scott Witwer

Chapter 3 New Opportunities to Bring Nutritionally Superior Food Products to the Market 22
J.W. Finley

Chapter 4 Conceptualization of the Prooxidant and Antioxidant Actions of Plant Food Chemicals 32
Okezie I. Aruoma

Chapter 5 A Novel Approach for Discovering Phytopharmaceuticals 47
Ramesh C. Pandey, Renuka Misra, Luben K. Yankov, Bohos Mesrob, and Amlan Dutta

Chapter 6 Variables Affecting the Phytochemical Contents of Garlic and Its Health Benefits 62
Harunobu Amagase, Hiromichi Matsuura, and Yoichi Itakura

Chapter 7 Citrus Limonoids: Biochemistry and Possible Importance to Human Nutrition 79
Shin Hasegawa, Luke K.T. Lam, and Edward G. Miller

Chapter 8 Limonoid Glucosides: Systemic Effects on Oral Carcinogenesis . 95
E.G. Miller, M.T. Record, W.H. Binnie, and S. Hasegawa

Chapter 9 Bioactive Components of Flaxseed: Occurrence and Health Benefits 106
B. Dave Oomah and G. Mazza

Chapter 10 Medicinal Lignans from Flaxseed: Isolation and Purification .. 122
Neil D. Westcott, and Alister D. Muir

Chapter 11 Reaction of α-Tocopherol During the Peroxidation of Phospholipids in Liposomes 132
Ryo Yamauchi

Chapter 12 Serum Antioxidant Capacity Assessed by an Iron-Dependent Assay in Carotid Stenotics Given Tocotrienols and Tocopherols 146
T.R. Watkins, D.K. Kooyenga, and M.L. Bierenbaum

Chapter 13 Role of Flavonols and Anthocyanins from Fruits and Vegetables in Cancer Prevention 154
Barbara Raab, Gisela Jacobasch, Lutz Heinevetter, and Stefanie Maurer

Chapter 14 Dietary Intake of Flavonoids and Isoflavonoids by Japanese and Their Pharmacokinetics and Bioactivities 164
Shaw Watanabe, Yusuke Arai, Ryota Haba, Mariko Uehara, Herman Adlercreutz, Kayako Shimoi, and Naohide Kinae

Chapter 15 Membrane Localizations of Carotenoids and α-Tocopherol and Their Antioxidant Properties for Singlet Oxygen 175
Kenji Fukuzawa, Yasuki Inokami, and Junji Terao

Chapter 16 Carotenoid Regeneration of Phenolic Compounds from Phenoxyl Radicals in Relation to Oxidative Defense 185
Alan Mortensen, Steven A. Everett, and Leif H. Skibsted

Chapter 17 Inhibition of Carcinogenesis by Tea Constituents and the Mechanisms Involved 192
Chung S. Yang, Janelle M. Landau, Guang-yu Yang, Sungbin Kim, and Jie Liao

Chapter 18 Antioxidative Activity of (-)-Epicatechin in Rat Plasma: Intragastric Administration and Its Metabolites 202
Mariusz K. Piskula, Edson L. da Silva, and Junji Terao

Chapter 19 Antibacterial Actions of Tea Polyphenols and Their Practical Applications in Humans 214
Yukihiko Hara

Chapter 20 Free Radical–Scavenging Action of Catechin and Related Compounds in Homogeneous and Micellar Solutions 222
Kazuo Mukai, Yu Kanesaki, Yoshifumi Egawa, and Shin-ichi Nagaoka

Chapter 21 Modulating Prooxidative Mitotic Signaling as Action Mechanisms of Cancer Chemopreventive Agents Including Curcumin, Epigallocatechin Gallate, and Other Phytopholyphenols 239
Jen-Kun Lin, Yu-Li Lin, Yu-Chih Liang, Yen-Chou Chen, and Shoei-Yn Lin-Shiau

Chapter 22 Radical-Scavenging Activity of Maillard Reaction Substances . . 252
M. Kawane, Y. Yoshiki, K. Okubo, and M. Tsunakawa

Chapter 23 Reactive Oxygen–Scavenging Compound Isolated from Adzuki Beans 261
T. Iida, Y. Yoshiki, M. Kawane and K. Okubo

Chapter 24 Antioxidative and Cytotoxic Components of Highbush Blueberry (*Vaccinium corymbosum* L.) 271
Mingfu Wang, Jiangang Li, Yu Shao, Tzou-Chi Huang, Mou-Tuan Huang, Chee-Kok Chin, Robert T. Rosen, and Chi-Tang Ho

Chapter 25 Antioxidants of Evening Primrose 278
F. Shahidi, M. Wettasinghe, R. Amarowicz, and M.A. Khan

Chapter 26 Antioxidant and Antitumor Activity of Rosemary Leaves 296
Chi-Tang Ho, Mou-Tuan Huang, You-Rong Lou, Wei Ma, Yu Shao, Guor-Jien Wei, Mingfu Wang, and Chee-Kok Chin

Chapter 27 Antitumor Activity of an Extract of the Gum Exudate from the Tree *Boswellia serrata* 308
Mou-Tan Huang, Wei Ma, Vladimir Badmaev, Yu Shao, *Chee-Kok Chin, and Chi-Tang Ho*

Chapter 28 Inhibitory Effect of Resveratrol and Related Compounds on the Macromolecular Synthesis in HL-60 Cells and the Metabolism of 7,12-Dimethylbenz[a]anthracene by 314
Mouse Liver Microsomes
Benaifer D. Dubash, Bo Lin Zheng, Calvin Hyungchan Kim, Kan He, Yu Shao, Qun Yi Zheng, Chi-Tang Ho, and Mou-Tuan Huang

Chapter 29 Chemiluminescence of a Hydrogen Peroxide/Gallic Acid/ Horseradish Peroxidase System 321
Y. Yoshiki, H. Yuan, M. Kawane, T. Iida, K. Okubo, H. Ohrui, T. Ishizawa, and S. Kawabata

Chapter 30 Bioactive Metabolites from Yucatecan Medicinal Plants 332
Rocío de L. Borges-Argáez, Fabiola Escalante-Erosa, Filogonio May-Pat, Leticia Medina-Baizabál, Yazmín del C. Ojeda-Uc, Sergio R. Peraza-Sánchez, Matilde Pérez-Rodríguez, Norma E. Salazar-Aguilar, Paulino Simá-Polanco,and Luis M. Peña-Rodríguez

Chapter 31 Antioxidative Fractions from the Leaves of Betel (*Piper betle* L.) 342
N. Andarwulan, T.J. Susanto, and R. Syarief

Chapter 32 Affinities of Dietary Phenolic Antioxidants for Lipid Bilayers . 349
Tsutomu Nakayama

Chapter 33 CimiPure (*Cimicifuga racemosa*): A Standardized Black Cohosh Extract with Novel Triterpene Glycoside for Menopausal Women 360
Qun Y. Zheng, Kan He, Lynn Pilkington, Yu Shao, and Bolin Zheng

Chapter 34 Natural Products of Agrochemical Interest from Mexican Labiatae 371
Baldomero Esquivel, Ana Adela Sánchez, and Eduardo Aranda

Chapter 35 Synthetic Study of a Highly Antitumorigenic Marine Phytochemical, Halichondrin B 386
Kiyoshi Horita, Sansei Nishibe, and Osamu Yonemitsu

Chapter 36 Phenolics from Deciduous Leaves and Coniferous Needles as Sources of Novel Control Agents for Lepidopteran Forest Pests 398
Mamdouh M. Abou-Zaid, Gary G. Grant, Blair V. Helson, Clifford W. Beninger, and Peter de Groot

Chapter 37 Pharmacologic Aspects and Analysis of Potato Glycoalkaloids . . 417
Peter Sporns, Darcy C. Abell, and Darcy R. Driedger

Index 425

Chapter 1

Plants as Medicine: The Role of Phytochemicals in Optimal Health

Clare M. Hasler

Department of Food Science and Human Nutrition and Functional Foods for Health Program, University of Illinois, Urbana, IL 61801

Introduction

Overwhelming epidemiologic evidence indicates that a plant-based diet affords protection against the development of certain chronic diseases, particularly cancer (Steinmetz and Potter 1991a). In a 1992 review of ~200 epidemiologic studies (Block *et al.* 1992), a significant protective effect of fruit and vegetable consumption against a variety of cancer types was found in 128 of 156 dietary studies. More importantly, for most cancer sites, individuals whose intake of fruits and vegetables ranked in the lower one fourth of the population experienced twice the risk of cancer vs. those with a high intake. More recently, a report from the World Cancer Research Fund and the American Institute for Cancer Research (1997) recommended year-round consumption of a variety of vegetables and fruits (400–800 g/d) as well as a variety of starchy or protein-rich foods of plant origin to reduce cancer risk.

In addition to the macro- and micronutrients (e.g., protein, fat, carbohydrates, energy, vitamins, or minerals) necessary for normal metabolism, a plant-based diet contains numerous nonnutritive components known as "phytochemicals," which may also play an important role in health enhancement (American Dietetics Association 1995). Because many phytochemicals are not well characterized and their mode(s) of action not well established, research on these physiologically active plant components is currently an area of intense effort. This chapter will provide a brief overview of the role of phytochemicals in disease prevention, with a focus on bioactive components from soy, garlic, flaxseed, tomatoes, tea, and citrus fruits.

Diet in Cancer Chemoprevention

Cancer is a leading cause of mortality in most developed countries. In the United States, for example, >1,300,000 new cancer cases were diagnosed in 1997, with 560,000 lives lost. One of every four deaths in the United States is related to cancer (American Cancer Society 1997), with approximately one third of those linked directly to diet (Doll and Peto 1981). Thus, dietary modification—in particular, the consumption of fruits and vegetables—may be an effective mechanism to reduce the morbidity and mortality associated with this devastating disease (Gillman 1996).

Fruits, vegetables and other plants contain numerous low-molecular-weight secondary metabolites collectively known as "phytochemicals" (Rhodes 1996), which may play a crucial role in cancer prevention (Murakami *et al.* 1996). Plant chemists have long been aware that one particular class of phytochemicals, phytoalexins, is produced by plants as a defense mechanism in lieu of an immune system (Smith and Banks 1986). However, it has been only relatively recently that these same phytochemicals have been studied for their essential role in human disease prevention and health promotion (Begley 1994). Phytochemicals have clearly become the nutrition buzzword of the 1990s, and many food companies are looking for ways to enhance the levels of phytonutrients in traditional products (Sloan 1996).

More than a dozen classes of anticarcinogenic compounds that can alter nearly every stage of the cancer process have been identified in plants. This cancer chemopreventive effect is thought to be achieved by a variety of mechanisms, including the following: (i) inducing detoxification enzymes, (ii) inhibiting nitrosamine formation, (iii) providing substrates for the formation of antineoplastic agents, (iv) diluting and binding carcinogens in the digestive tract, (v) altering hormone metabolism, or (vi) acting as antioxidants (Steinmetz and Potter 1991b). Some of the physiologically important phytochemicals (and the foods that contain them) receiving increased research attention at present include soy isoflavones, citrus limonoids, tomato lycopene, flaxseed lignans, tea catechins, and garlic allylic sulfides. Although there is emerging evidence concerning the importance of certain phytochemicals in the prevention of cardiovascular disease (Howard and Kritchevsky 1997) and stroke (Gillman *et al.* 1995), overwhelmingly, research efforts have focused on the cancer chemoprotective factors in fruits and vegetables. According to Potter, "Vegetables and fruit contain the anticarcinogenic cocktail to which we are adapted. We abandon it at our peril" (Potter 1991, Potter and Graves 1991).

Soy Isoflavones

Soy has clearly been a plant food in the spotlight in the 1990s (Kuhn 1996). In addition to being recognized as an excellent source of protein, equivalent in quality to animal protein (Young 1991), soy is being investigated intensely for its ability to treat and prevent a variety of chronic diseases, including cancer (Messina and Erdman 1995; Messina and Erdman 1998).

In 1990, an important workshop at the United States National Cancer Institute examined the relationship between soy consumption and cancer risk. In this workshop, several classes of anticarcinogenic phytochemicals in soybeans were identified, i.e., protease inhibitors, phytosterols, saponins, phenolic acids, phytic acid, and isoflavones (Messina and Barnes 1991). Of these, the isoflavones are unique because soybeans are the only significant dietary source of these biologically active compounds (Price and Fenwick 1985).

The primary isoflavones in soy, genistein and daidzein, are structurally similar to the estrogenic steroids (e.g., estradiol) and have been shown to possess both estrogenic and antiestrogenic activity (Verdeal and Ryan 1979). Because they are weak estrogens (0.001 times the activity of estradiol), isoflavones may act as antiestrogens by competing with the more potent, naturally occurring endogenous estrogens for binding to the estrogen receptor (Coward *et al.* 1991). This may, in part, provide a biological basis for observations that soy consumption may reduce the risk of estrogen-dependent cancers (Messina *et al.* 1994). Southeast Asian populations, such as those of Japan, Korea, and Taiwan, who consume 20–80 mg of genistein/d (derived almost entirely from soy), have a significantly lower incidence of breast and prostate cancer. In comparison, dietary intake of genistein in the United States is only 1–3 mg/d (Barnes *et al.* 1995).

In addition to their estrogenic and antiestrogenic effects, soy isoflavones can also modulate a number of other cellular mechanisms that may contribute to the cancer chemopreventive benefits of these isoflavones (Barnes 1997). In particular, genistein has been shown to act as a potent and specific inhibitor of protein tyrosine kinases (Akiyama *et al.* 1987). In addition, genistein inhibits DNA topoisomerase II activity (Constantinou *et al.* 1990), alters cell cycle–specific events (Matsukawa *et al.* 1993), induces apoptosis (Bergamaschi *et al.* 1993), and inhibits angiogenesis, a process known to be essential for tumor growth (Fotsis *et al.* 1993).

The potential for reducing cancer risk through the consumption of soy products is significant, as evidenced by epidemiologic, *in vitro*, and *in vivo* (animal) data (Fournier *et al.* 1998). However, these observations must be confirmed in well-controlled human clinical intervention trials. Such trials are currently in progress to assess the role of soy in reducing the risk of breast, colon, and prostate cancer.

Tomato Lycopene

More than 600 naturally occurring pigments known as carotenoids are widely dispersed throughout the plant and animal kingdom (Olsen and Krinsky 1995). The most widely studied of the carotenoids has been β-carotene, not only because of its ability to serve as a precursor to vitamin A, but because of increasing evidence that this compound may reduce cancer risk (Peto *et al.* 1981). Unfortunately, recent clinical trials investigating β-carotene supplements in subjects at risk for lung cancer have yielded negative results, and even suggest that β-carotene in isolated form may exacerbate this disease (Greenberg and Sporn 1996).

Recent research has been emerging on the cancer chemopreventive potential of another physiologically active carotenoid that has been relatively ignored, namely, lycopene (Clinton 1998). In contrast to other carotenoids, which are typically widely distributed through many fruits and vegetables, lycopene occurs in very high concentrations in only a few plant foods, i.e., tomatoes (and processed tomato products), pink grapefruit, guava, and watermelon (Stahl and Sies 1996). Lycopene is also the most abundant carotenoid in plasma (Ascherio *et al.* 1992).

Interest in the role of lycopene in cancer risk reduction was stimulated significantly by findings from a recent prospective cohort study of >47,000 men. In that study, men who consumed processed tomato products (e.g., tomatoes, tomato sauce, pizza) ≥10 times per week had less than one half the risk of developing advanced prostate cancer (Giovannucci *et al.* 1995). This finding supported a previous observation that tomato intake was significantly related to a lower risk of prostate cancer in Seventh-Day Adventists (Mills *et al.* 1989).

The proposed mechanism by which lycopene reduces prostate cancer risk is likely related to its antioxidant function because lycopene has been shown to be the most efficient quencher of singlet oxygen in biological systems (Di Mascio *et al.* 1989). Interestingly, lycopene is the most abundant carotenoid in the prostate gland in which it occurs at physiologically relevant concentrations (Clinton *et al.* 1996). An additional mechanism by which lycopene may exert a cancer chemoprotective effect is the enhancement of gap junction–mediated intercellular communication (Zhang *et al.* 1991).

In addition to prostate cancer, other types of cancer have also been shown to be inversely related to tomato intake, including those of the digestive tract, pancreas, cervix, bladder, and skin (reviewed recently by Gerster 1997). More recent evidence also shows that low plasma lycopene levels may be an independent risk factor for lung cancer, particularly in smokers (Li *et al.* 1997). It is anticipated that there will be an increase in the sale of tomato products in response to continued research findings on the health benefits of lycopene (Best 1997).

Citrus Limonoids

Citrus fruit consumption has been shown to protect against a variety of human cancers (Block 1991). Although oranges, lemons, limes, and grapefruits are a principal source of a number of important nutrients, including ascorbic acid, folate, fiber, and vitamin E, it has been suggested that other component(s) are responsible for the anticancer activity of these foods. Citrus fruits are particularly high in a class of monoterpenes known as the limonoids (Hasegawa *et al.* 1980).

d-Limonene is the predominant monocyclic monoterpene found in the essential oils of citrus fruits and is responsible for the bitter taste of these foods. Over the last decade, significant evidence has been accumulating in support of the cancer chemopreventive effect of limonene in several animal model systems. This compound has been shown to be effective against both spontaneous and chemically induced rodent mammary tumors, as well as skin, liver, lung, and forestomach tumors (Crowell and Gould 1994). Possible chemopreventive mechanisms of action of limonene include the induction of hepatic detoxification enzymes glutathione-S-transferase and uridine diphosphoglucuronosyl transferase (Elegbede *et al.* 1993), the inhibition of post-translational isoprenylation of small G proteins, including $p21_{ras}$ (Crowell *et al.* 1991), and increased levels of transforming growth factor-1 and mannose-6-phosphate/insulin-like growth factor-II (Jirtle *et al.* 1993). The antitumor activity of limonene may be derived from its metabolites (Gould 1995); five metabolites have been characterized in human subjects as follows: dihydroperillic acid and perillic acid, their methyl esters, and limonene-1,2-diol (Crowell *et al.* 1994). On the basis of these observations and because it has little or no toxicity in humans (Igimi *et al.* 1991), limonene has been suggested as a good candidate for human clinical chemoprevention trial evaluation (Kelloff *et al.* 1994).

Flaxseed Lignans

Flaxseed has been used for food in Europe and Asia since 5000–8000 B.C. and continues to be used in baked goods and other foods in those areas (Carter 1993). The primary nutritional attributes of flaxseed are its relatively high level of the n-3 fatty acid, α-linolenic acid, which is higher in flaxseed (57%) than any other major seed oil (Carter 1993), and the high content of both soluble and insoluble fiber (Hadley *et al.* 1992). More recently, however, research has focused on a class of estrogenic compounds known as the lignans (Harris and Haggerty 1993).

Lignans are diphenolic compounds containing a 2,3-dibenzylbutane structure (Setchell *et al.* 1980). Although it has long been known that >100 lignans exist in plants (Rao 1978), it was not until 1981 that lignans were identified in humans (Setchell *et al.* 1981a). The two primary mammalian lignans, enterodiol and enterolactone, are formed in the intestinal tract by bacterial action on the plant lignan precursors, secoisolariciresinol diglycoside and matairesinol, respectively (Setchell *et al.* 1981b). Flaxseed meal and flour are by far the most significant sources of precursors to the mammalian lignans, producing 75–800 times more than any other plant food (Thompson *et al.* 1991).

Lignans exhibit a number of biological activities that may contribute to their cancer preventive properties (Thompson 1995). Most significantly, however, enterodiol and enterolactone are structurally similar to both naturally occurring and synthetic estrogens and have been shown to possess weakly estrogenic and antiestrogenic activities (Kurzer *et al.* 1995), suggesting that they may play a role in the prevention of estrogen-dependent cancers. In rodents, flaxseed has been shown to decrease the number of aberrant colonic crypt foci (Serraino and Thompson 1992a), reduce mammary gland proliferation and nuclear aberration (Serraino and Thompson 1991), and reduce mammary tumor size when fed during the promotion stage of carcinogenesis (Serraino and Thompson 1992b).

Few studies have evaluated the effects of flaxseed feeding on risk markers for cancer in humans. Phipps *et al.* (1993) demonstrated that the ingestion of 10 g of flaxseed per day was associated with several hormonal changes representative of decreased ovarian dysfunction in premenopausal women, i.e., significantly longer luteal phase lengths, significantly higher luteal phase progesterone/estradiol ratios, and fewer anovulatory cycles. Given that ovarian dysfunction has been linked to increased risk for breast cancer, these findings may explain why diets low in fiber, and hence, lignans, may be related to increased risk. Although no epidemiologic studies have yet been conducted on flaxseed intake and cancer risk, Adlercreutz and co-workers found that the urinary lignan excretion was significantly lower in postmenopausal breast cancer patients compared with postmenopausal controls eating a normal mixed or a lactovegetarian diet (Adlercreutz *et al.* 1982). Urinary excretion of lignans is also highly correlated with dietary intake of flaxseed or other grain products (Axelson *et al.* 1982). Clearly, much more work has to be done to determine the role of flaxseed in the prevention of breast and other hormone-dependent cancers.

Garlic Organosulfur Compounds

Garlic has been utilized by practitioners of traditional medicine for >4000 years (Bolton *et al.* 1982). The characteristic flavor and pungency of garlic (hence its nickname "the stinking rose") are due to an abundance of naturally occurring oil- and water-soluble sulfur-containing components. These are also likely the bioactive substances responsible for the various medicinal effects ascribed to this plant (Agarwal 1996).

The cancer-preventative effect of garlic has been observed in a number of epidemiologic studies (Reuter *et al.* 1996), with stomach cancer appearing to be the type of neoplasia whose risk is most clearly reduced by garlic consumption (Cipriani *et al.* 1991, You *et al.* 1988 and 1989). In studies conducted in China, the incidence of stomach cancer in Cangshan County, where fresh garlic consumption averages 20 g/d, was only 8% that of the stomach cancer incidence in Qixia County, where garlic consumption averages <1 g/d (Han 1993). The proposed mechanism by which garlic consumption reduces stomach cancer may be related to the destruction of carcinogenic nitroso compounds in the stomach by garlic constituents.

Garlic consumption has also been shown to reduce the risk of colon cancer. In a study of 41,387 postmenopausal women in Iowa, garlic was the only food (of 127 assessed) that showed a significant association with decreased colon cancer risk (Steinmetz *et al.* 1994). Consumption of one or more servings of fresh or powdered garlic per week resulted in a 50% lower risk of cancer of the distal colon and a 35% lower risk of cancers anywhere in the colon.

Unfortunately, not all epidemiologic studies have shown that garlic can reduce cancer risk. A higher incidence of lung carcinoma was found to be associated with the use of garlic supplements in The Netherlands (Dorant *et al.* 1994). Garlic supplement use was also ineffective in reducing the risk of breast (Dorant *et al.* 1995) and colorectal carcinoma (Dorant *et al.* 1996). Similarly, garlic has not been consistently shown to reduce cancer risk in animals (Dorant *et al.* 1993). Inconclusive results are likely due to variations in the type of garlic compounds or preparations used by various investigators; there is considerable variation in the quantity of organosulfur compounds available in fresh and commercially available garlic products (Lawson *et al.* 1991).

Tea Catechins

Tea is second only to water as the most widely consumed beverage in the world (Weisburger 1997), with a per capita consumption of ~120 mL/d (Stoner and Mukhtar 1995). Of the ~2.5 million metric tons of dried tea manufactured annually, the majority (78%) is black tea; the remainder is green tea (20%) with a small percentage (2%) of oolong tea.

Approximately 30% of the total dry weight of fresh tea leaves is composed of polyphenols, including flavandiols, flavonoids, phenolic acids, and flavanols (more commonly referred to as catechins) (Harbowy and Balentine 1997). Catechins are the predominant and most significant of all tea components (Graham 1992). The four major green tea catechins are (-)-epigallocatechin-3-gallate (EGCG), (-)-epigallocatechin (EGC), (-)-epicatechin-3-gallate (ECG), and (-)-epicatechin (EC). In recent years, there has been intense interest in the various pharmacologic actions of tea (American Health Foundation 1992).

Three recent reviews have summarized the epidemiologic studies on the cancer chemopreventive effects of tea, with inconclusive results. A report by the Working Group of the International Agency for Research on Cancer (1991) concluded that there was inadequate evidence for the anticarcinogenic effect of tea in humans. A similar conclusion was reached in a review by Yang and Wang (1993). In that review, 100 epidemiologic studies were reviewed, examining the relationship of tea consumption to cancers of the bladder and urinary tract, breast, colon and rectum, esophagus, kidney, liver, lung, nasopharynx, pancreas, stomach, and uterus. Approximately two thirds of the studies found no relationship between tea consumption and cancer risk; 20 studies found a positive relationship and only 14 studies found that tea consumption reduced cancer risk. A more recent review by Katiyar and Mukhtar (1996) also concluded that population studies were inconclusive. However, these authors stated that tea consumption "is likely to have preventive effects in reducing cancer risk" and suggested that additional case-controlled studies be undertaken.

In contrast to inconclusive results obtained from epidemiologic studies, research findings in laboratory animals clearly support a chemopreventive effect of tea components (Clydesdale 1997). Tea has been shown to afford protection against cancers of the skin, lung, forestomach, esophagus, duodenum, pancreas, liver, breast, and colon (Katiyar and Mukhtar 1996). Although additional well-designed case-control studies, as well as clinical intervention trials, are required to assess whether tea consumption can reduce cancer risk in humans, authors of a recent review state, "No other agent tested for possible chemoprevention effects in animal models has elicited such strong activity as tea and its components at the concentrations usually consumed by humans" (Dreosti *et al.* 1997).

Conclusion

Evidence is rapidly emerging that nonnutrient components in plant foods may play a critical role in the prevention of chronic diseases, particularly cancer. Most of this evidence has been derived from epidemiologic, animal, or *in vitro* studies. To determine the optimal intake levels of phytochemicals, well-controlled human clinical intervention trials will be required. Until that time, the best advice is still to consume a plant-based diet that is high in fruits, vegetables, grains, and legumes.

References

Adlercreutz, H., T. Fotsis, R. Heikkinen, J.T. Dwyer, M. Woods, B.R Goldin, and S.L. Gorbach, Excretion of the Lignans Enterolactone and Enterodiol and of Equol in

Omnivorous and Vegetarian Postmenopausal Women and in Women with Breast Cancer, *Lancet ii*: 1295–1299 (1982).

Agarwal, K.C., Therapeutic Actions of Garlic Constituents, *Med. Res. Rev. 16*: 111–124 (1996).

Akiyama, T., J. Ishida, S. Nakagawa, H. Ogawara, S. Watanabe, N.M. Itoh, M. Shibuya, and Y. Fukami, Genistein, a Specific Inhibitor of Tyrosine-Specific Protein Kinases, *J. Biol. Chem. 262*: 5592–5595 (1987).

American Cancer Society, *Cancer Facts and Figures—1997*, Atlanta, Georgia, 1997.

American Dietetics Association Position Statement, Phytochemicals and Functional Foods, *J. Am. Diet. Assoc. 95*: 493–496 (1995).

American Health Foundation, Physiological and Pharmacological Effects of *Camellia sinensis* (Tea): Implications for Cardiovascular Disease, Cancer, and Public Health, *Prev. Med. 21*: 329–391 and 503–553 (1992).

Ascherio, A., M.J. Stampfer, G.A. Colditz, E.B. Rimm, L. Litin, and W.C. Willett, Correlations of Vitamin A and E Intakes with the Plasma Concentrations of Carotenoids and Tocopherols Among American Men and Women, *J. Nutr. 122*: 1792–1801 (1992).

Axelson, M., J. Sjovall, B.E. Gustafsson, and K.D.R. Setchell, Origin of Lignans in Mammals and Identification of a Precursor from Plants, *Nature 298*: 659–660 (1982).

Barnes, S., Evolution of the Health Benefits of Soy Isoflavones, *Proc. Soc. Exp. Biol. Med. 217*: 386–392 (1997).

Barnes, S., G. Peterson, and L. Coward, Rationale for the Use of Genistein-Containing Soy Matrices in Chemoprevention Trials for Breast and Prostate Cancer, *J. Cell. Biochem. (Suppl.) 22*: 181–187 (1995).

Begley, S., Beyond Vitamins, *Newsweek, April 25*: 45–49 (1994).

Bergamaschi, G., V. Rosti, M. Danova, L. Ponchio, C. Lucotti, and M. Cazzola, Inhibitors of Tyrosine Phosphorylation Induce Apoptosis in Human Leukemic Cell Lines, *Leukemia 7*: 2012–2018 (1993).

Best, D., All Natural and Nutraceutical, *Prep. Foods 166*: 32–38 (1997).

Block, G., Vitamin C and Cancer Prevention: The Epidemiologic Evidence, *Am. J. Clin. Nutr. 53*: 270S–282S (1991).

Block, G., B. Patterson, and A. Subar, Fruit, Vegetables and Cancer Prevention: A Review of the Epidemiological Evidence, *Nutr. Cancer 18*: 1–29 (1992).

Bolton, S., G. Null, and W. Troetel, The Medical Uses of Garlic—Fact and Fiction, *Am. Pharm. 22*: 448–451 (1982).

Carter, J., Flaxseed as a Source of Alpha Linolenic Acid, *J. Am. Coll. Nutr. 12*: 551 (1993).

Cipriani, F., E. Buiatti, and D. Palli, Gastric Cancer in Italy, *Ital. J. Gastroenterol. 23*: 429–435 (1991).

Clinton, S.K., Lycopene: Chemistry, Biology, and Implications for Human Health and Disease, *Nutr. Rev. 56*: 35–51 (1998).

Clinton, S.K., C. Emenhiser, S.J. Schwartz, D.G. Bostwick, A.W. Williams, B.J. Moore, and J.W. Erdman, *cis-trans* Lycopene Isomers, Carotenoids, and Retinol in the Human Prostate, *Cancer Epidemiol. Biomark. Prev. 5*: 823–833 (1996).

Clydesdale, F.M. (ed.), Tea and Health, *Crit. Rev. Food Sci. Nutr. 37*: 691–785 (1997).

Constantinou, A., K. Kiguchi, and E. Humberman, Induction of Differentiation and DNA Strand Breakage in Human HL-60 and K562 Leukemia Cells by Genistein, *Cancer Res. 50*: 2618–2624 (1990).

Coward, L., N. Barnes, K.D.R. Setchell, and S. Barnes, Genistein, Diadzein and Their 8-Glycoside Conjugates: Antitumor Isoflavones in Soybean Foods from American and Asian Diets, *J. Agric. Food Chem. 41*: 1961–1967 (1991).

Crowell, P.L., R.R. Chang, Z. Ren, C.E. Elson, and M.N. Gould, Selective Inhibition of Isoprenylation of 21–26-kDa Proteins by the Anticarcinogen *d*-Limonene and Its Metabolites, *J. Biol. Chem. 266*: 17679–17685 (1991).

Crowell, P.L., C.E. Elson, H.H. Bailey, A. Elegbede, J.D. Haag, and M.N. Gould, Human Metabolism of the Experimental Cancer Therapeutic Agent *d*-Limonene, *Cancer Chemother. Pharmacol. 35*: 31–37 (1994).

Crowell, P.L., and M.N. Gould, Chemoprevention and Therapy of Cancer by *d*-Limonene, *Crit. Rev. Oncog. 5*: 1–22 (1994).

Di Mascio, P., S. Kaiser, and H. Sies, Lycopene as the Most Efficient Biological Carotenoid Singlet Oxygen Quencher, *Arch. Biochem. Biophys. 274*: 532–538 (1989).

Doll, R., and R. Peto, The Causes of Cancer: Quantitative Estimates of Avoidable Risks of Cancer in the United States Today, *J. Natl. Cancer Inst. 66*: 1192–1308 (1981).

Dorant, E., P.A. van den Brandt, and R.A. Goldbohm, A Prospective Cohort Study on Allium Vegetable Consumption, Garlic Supplement Use and the Risk of Lung Carcinoma in The Netherlands, *Cancer Res. 54*: 6148–6153 (1994).

Dorant, E., P.A. van den Brandt, and R.A. Goldbohm, Allium Vegetable Consumption, Garlic Supplement Intake, and Female Breast Carcinoma Incidence, *Breast Cancer Res. Treat. 33*: 163–170 (1995).

Dorant, E., P.A. van den Brandt, and R.A. Goldbohm, A Prospective Cohort Study on the Relationship Between Onion and Leek Consumption, Garlic Supplement Use and the Risk of Colorectal Carcinoma in The Netherlands, *Carcinogenesis 17*: 477–484 (1996).

Dorant, E., P.A. van den Brandt, R.A. Goldbohm, R.J.J. Hermus, and F. Sturmans, Garlic and Its Significance for the Prevention of Cancer in Humans: A Critical Review, *Br. J. Cancer 67*: 424–429 (1993).

Dreosti, I.E., M.J. Wargovich, and C.S. Yang, Inhibition of Carcinogenesis by Tea: The Evidence from Experimental Studies, *Crit. Rev. Food Sci. Nutr. 37*: 761–770 (1997).

Elegbede, J.A., T.H. Maltzman, C.E. Elson, and M.N. Gould, Effects of Anticarcinogenic Monoterpenes on Phase II Hepatic Metabolizing Enzymes, *Carcinogenesis 14*: 1221–1223 (1993).

Fotsis, T., M. Pepper, H. Adlercreutz, G. Fleischmann, G. Hase, R. Montesano, and L. Schweigerer, Genistein, a Dietary-Derived Inhibitor of *in vitro* Angiogenesis, *Proc. Natl. Acad. Sci. USA 90*: 2690–2694 (1993).

Fournier, D.B., J.W. Erdman, Jr., and G.B. Gordon, Soy, Its Components, and Cancer Prevention: A Review of the *In Vitro*, Animal, and Human Data. *Cancer Epidemiology, Biomarkers & Prevention 7:* 1055–1065 (1998).

Gerster, H., The Potential Role of Lycopene for Human Health, *J. Am. Coll. Nutr 16*: 109–126 (1997).

Gillman, M.W., Enjoy Your Fruits and Vegetables. Eating Fruit and Vegetables Protects Against the Common Chronic Diseases of Adulthood, *Br. Med. J. 313*: 765–766 (1996).

Gillman, M.W., L.A. Cupples, D. Gagnon, B.M. Posner, R.C. Ellison, W.P. Castelli, and P.A. Wolf, Protective Effect of Fruits and Vegetables on Development of Stroke in Men, *J. Am. Med. Assoc. 273*: 1113–1117 (1995).

Giovannucci, E., A. Ascherio, E.B. Rimm, M.J. Stampfer, G.A. Colditz, and W.C. Willett, Intake of Carotenoids and Retinol in Relation to Risk of Prostate Cancer, *J. Natl. Cancer Inst. 87*: 1767–1776 (1995).

Gould, M.N., Prevention and Therapy of Mammary Cancer by Monoterpenes, *J. Cell. Biochem. (Suppl.) 22*: 139–144 (1995).

Graham, H.N., Green Tea Composition, Consumption and Polyphenol Chemistry, *Prev. Med. 21*: 334–350 (1992).

Greenberg, R.E., and M.B. Sporn, Antioxidant Vitamins, Cancer, and Cardiovascular Disease, *N. Engl. J. Med. 334*: 1189–1190 (1996).

Hadley, M., C. Lacher, and J. Mitchell-Fetch, Fiber in Flaxseed, *Proc. Flax Inst. 54*: 79–83 (1992).

Han, J., Highlights of the Cancer Chemoprevention Studies in China, *Prev. Med. 22*: 712–722 (1993).

Harbowy, M.E., and D.A. Balentine, Tea Chemistry, *Crit. Rev. Plant Sci. 16*: 415–480 (1997).

Harris, R.K., and W.J. Haggerty, Assays from Potentially Anticarcinogenic Phytochemicals in Flaxseed, *Cereal Foods World 38*: 147–151 (1993).

Hasegawa, S., R.D. Bennett, and C.P. Verdon, Limonoids in Citrus Seeds: Origin and Relative Concentration, *J. Agric. Food Chem. 28*: 922–925 (1980).

Howard, B.V., and D. Kritchevsky, Phytochemicals and Cardiovascular Disease. A Statement for Health Care Professionals from the American Heart Association, *Circulation 95*: 2591–2593 (1997).

Igimi, H., R. Tamura, K. Toraishi, F. Yamamoto, A. Kataoka, Y. Ikejiri, T. Hisatsugu, and H. Shimura, Medical Dissolution of Gallstones: Clinical Experience of *d*-Limonene as a Simple, Safe, and Effective Solvent, *Dig. Dis. Sci. 36*: 200–208 (1991).

International Agency for Research on Cancer, Coffee, Tea, Mate, Methylxanthines, and Methylgloxyal, *IARC Monogr. Eval. Carcinog. Risks Hum. 51*: 1–513 (1991).

Jirtle, R.L., J.D. Haag, E.A. Ariazi, and M.N. Gould, Increased Mannose 6-Phosphate/Insulin-Like Growth Factor II Receptor and Transforming Growth Factor 1 Levels During Monoterpene-Induced Regression of Mammary Tumors, *Cancer Res. 53*: 3849–3852 (1993).

Katiyar, S.K., and H. Mukhtar, Tea in Chemoprevention of Cancer: Epidemiologic and Experimental Studies (Review), *Int. J. Oncol. 8*: 221–238 (1996).

Kelloff, G.J., C.W. Boone, V.E. Steele, J.A. Crowell, R. Lubet, and C.C. Sigman, Progress in Cancer Chemoprevention: Perspectives on Agent Selection and Short-Term Clinical Intervention Trials, *Cancer Res. (Suppl.) 54*: 2015S–2024S (1994).

Kuhn, M.E., Soy in the Spotlight. Disease-Fighting Benefits May Change the Image of the Once-Lowly Bean, *Food Proc. May*: 52–58 (1996).

Kurzer, M.S., J.L. Slavin, and H. Adlercreutz, Flaxseed, Lignans, and Sex Hormones, in *Flaxseed in Human Nutrition*, edited by S. Cunnane, and L.U. Thompson, AOCS Press, Champaign, IL, 1995, pp. 136–144.

Lawson, L.D., Z.-Y.J. Wang, and B.G. Hughes, Identification and HPLC Quantitation of the Sulfides and Dialk(en)yl Thiosulfinates in Commercial Garlic Products, *Planta Med. 57*: 363–370 (1991).

Li, Y., M. Elie, W.S. Blaner, P. Brandt-Rauf, and J. Ford, Lycopene, Smoking and Lung Cancer, *Proc. Am. Assoc. Cancer Res. 38*: #758 (abstr.) (1997).

Matsukawa, Y., N. Marui, T. Sakai, Y. Satomi, M. Yoshida, K. Matsumoto, H. Nishino, and A. Aoike, Genistein Arrests Cell Cycle Progression at G2-M, *Cancer Res. 53*: 1328–1331 (1993).

Messina, M., and S. Barnes, The Role of Soy Products in Reducing Risk of Cancer, *J. Natl. Cancer Inst. 83*: 541–546 (1991).

Messina, M.J., and J.W. Erdman, Jr., Proceedings from The First International Symposium on the Role of Soy in Preventing and Treating Chronic Disease, *Am. J. Clin. Nutr. J. Nutr. 125 (Suppl.)*: S567–S808 (1995).

Messina, M.J., and J.W. Erdman, Jr., Proceedings from The Second International Symposium on the Role of Soy in Preventing and Treating Chronic Disease, *Am. J. Clin. Nutr. 68 (Suppl.)*: 1329S–1544S (1998).

Messina, M., and V. Messina, Increasing Use of Soyfoods and Their Potential Role in Cancer Prevention, *J. Am. Diet. Assoc. 91*: 836–840 (1991).

Messina, M.J., V. Persky, K.D.R. Setchell, and S. Barnes, Soy Intake and Cancer Risk: A Review of the *in vitro* and *in vivo* Data, *Nutr. Cancer 21*: 113–131 (1994).

Mills, P.K., W.L. Beeson, R.L. Phillips, and G.E. Fraser, Cohort Study of Diet, Lifestyle, and Prostate Cancer in Adventist Men, *Cancer 64*: 598–604 (1989).

Mukhtar, A., H. Ohigashi, and K. Koshimizu, Anti-Tumor Promotion with Food Phytochemcials: A Strategy for Cancer Chemoprevention, *Biosci. Biotechnol. Biochem. 60*: 1–8 (1996).

Olson, J.A., and N.I. Krinsky, Introduction: The Colorful, Fascinating World of the Carotenoids: Important Physiologic Modulators, *FASEB J. 9:* 1547–1550 (1995).

Peto, R., R. Doll, J.D. Buckley, and M.B. Sporn, Can Dietary Beta-Carotene Materially Reduce Cancer Rates? *Nature 290*: 201–208 (1981).

Phipps, W.R., M.C. Martini, J.W. Lampe, J.L. Slavin, and M.S. Kurzer, Effect of Flax Seed Ingestion on the Menstrual Cycle, *J. Clin. Endocrinol. Metab. 77*: 1215–1219 (1993).

Potter, J.D., The Epidemiology of Diet and Cancer: Evidence of Human Maladaptation, in *Nutrition and Cancer Prevention. Investigating the Role of Macronutrients,* edited by T.C. Moon, and M.S. Micozzi, Dekker, New York, 1991.

Potter, J.D., and K.D. Graves, Diet and Cancer: Evidence and Mechanisms—An Adaptation Argument, in *Nutrition, Toxicology and Cancer,* edited by I. Rowland, CRC Press, Boca Raton, 1991.

Price, K.R., and G.R. Fenwick, Naturally Occurring Estrogens in Foods: a Review, *Food Addit. Contam. 2*: 73–106 (1985).

Rao, C.B.S., *The Chemistry of Lignans,* Andhra University Press, Waltair, India, 1978.

Reuter, H.D., H.P. Koch, and L.D. Lawson, Therapeutic Effects and Applications of Garlic and Its Preparations, in *Garlic. The Science and Therapeutic Application of* Allium sativum *L. and Related Species*, edited by H.P. Koch, and L.D. Lawson, 2nd edn., Williams and Wilkins, Baltimore, 1996, pp. 136–148.

Rhodes, M.J.C., Physiologically-Active Compounds in Plant Foods: An Overview, *Proc. Nutr. Soc. 55*: 371–384 (1996).

Serraino, M.R., and L.U. Thompson, The Effect of Flaxseed in Early Risk Markers for Mammary Carcinogenesis, *Cancer Lett. 60*: 135–142 (1991).

Serraino, M.R., and L.U. Thompson, The Effect of Flaxseed Supplementation on the Initiation and Promotional Stages of Mammary Tumorigenesis, *Nutr. Cancer 17*: 153–159 (1992a).

Serraino, M.R., and L.U. Thompson, Flaxseed Supplementation and Early Markers of Colon Carcinogenesis, *Cancer Lett. 63*: 159–165 (1992b).

Setchell, K.D.R., A.M. Lawson, S.P. Borriello, R. Harkness, H. Gordon, D.M.L. Morgan, D.N. Kiek, H. Adlercreutz, L.C. Anderson, and M. Axelson, Lignan Formation in Man—Microbial Involvement and Possible Roles in Relation to Cancer, *The Lancet ii*: 4–7 (1981a).

Setchell, K.D.R., A.M. Lawson, E. Conway, N.F. Taylor, D.N. Kirk, G. Cooley, R.D. Farrant, S. Wynn, and M. Axelson, The Definitive Identification of the Lignans *trans*-2,3-bis(3- Hydroxybenzyl)—butyrolactone and 2,3-bis(3-Hydroxybenzyl)butane-1,4-diol in Human and Animal Urine, *Biochem. J. 197*: 447–458 (1981b).

Setchell, K.D.R., A.M. Lawson, F.L. Mitchell, H. Adlercreutz, D.N. Kirk, and M. Axelson, Lignans in Man and in Animal Species, *Nature 287*: 740–742 (1980).

Sloan, A.E., The Top 10 Trends to Watch and Work On, *Food Technol. 50*: 55–71 (1996).

Smith, D.A., and S.W. Banks, Biosynthesis, Elicitation and Biological Activity of Isoflavanoid Phytoalexins, *Phytochemistry 25*: 979–995 (1986).

Stahl, W., and H. Sies, Lycopene: A Biologically Important Carotenoid for Humans? *Arch. Biochem. Biophys. 336*: 1–9 (1996).

Steinmetz, K.A., H. Kushi, R.M. Bostick, A.R. Folsom, and J.D. Potter, Vegetables, Fruit, and Colon Cancer in the Iowa Women's Health Study, *Am. J. Epidemiol. 139*: 1–15 (1994).

Steinmetz, K., and J. Potter, Vegetables, Fruit and Cancer. I. Epidemiology, *Cancer Causes Control 2 (Suppl.)*: 325–357 (1991a).

Steinmetz, K., and J. Potter, Vegetables, Fruit and Cancer. II. Mechanisms, *Cancer Causes Control 2 (Suppl.)*: 427–442 (1991b).

Stoner, G.D., and H. Mukhtar, Polyphenols as Cancer Chemopreventive Agents, *J. Cell. Biochem. (Suppl.) 22*: 169–180 (1995).

Thompson, L.U., Flaxseed, Lignans, and Cancer, in *Flaxseed in Human Nutrition*, edited by S. Cunnane, and L.U. Thompson, AOCS Press, Champaign, IL, 1995, pp. 219–236.

Thompson, L.U., P. Robb, M. Serraino, and F. Cheung, Mammalian Lignan Production from Various Foods, *Nutr. Cancer 16*: 43–52 (1991).

Verdeal, K., and D.S. Ryan, Naturally-Occurring Estrogens in Plant Foodstuffs: A Review, *J. Food Prot. 42*: 577–583 (1979).

Weisburger, J.H., Tea and Health: A Historical Perspective, *Cancer Lett. 114*: 315–317 (1997).

World Cancer Research Fund and American Institute for Cancer Research, *Food, Nutrition and the Prevention of Cancer: A Global Perspective,* 1997. Banta Book Group, Menasha, WI.

Yang, C.S., and Z.-Y. Wang, Tea and Cancer, *J. Natl. Cancer Inst. 85*: 1038–1049 (1993).

You, W.-C., W.J. Blot, Y.-S. Chang, A.G., Ershow, Z.T. Yang, Q. An, B.E. Henderson, J.R. Fraumeni, Jr., and T.G. Wang, Allium Vegetables and Reduced Risk of Stomach Cancer, *J. Natl. Cancer Inst. 81*: 162–164 (1989).

You, W.-C., W.J. Blot, Y.-S. Chang, A.G. Ershow, Z.T. Yang, Q. An, B.E. Henderson, G.W. Xu, J.F. Fraumeni, Jr., and T.G. Wang, Diet and High Risk of Stomach Cancer in Shandong, China, *Cancer Res. 48*: 3518–3523 (1988).

Young, V.R., Soy Protein in Relation to Human Protein and Amino Acid Nutrition, *J. Am. Diet. Assoc. 91*: 828–835 (1991).

Zhang, L.-X., R.V. Cooney, and J.S. Bertram, Carotenoids Enhance Gap Junctional Communication and Inhibit Lipid Peroxidation in C3H/10T1/2 Cells: Relationship to Their Cancer Chemopreventive Action, *Carcinogenesis 12*: 2109–2114 (1991).

Chapter 2

Functional Foods and Phytochemicals Research Incentives: U.S. Policy and Industry Preference

Nancy M. Childs[a] and Rhonda Scott Witwer[b]

[a]Department of Food Marketing, Saint Joseph's University, Philadelphia, PA 19131, and [b]Monsanto Company, St. Louis, MO

Introduction

The conclusion of the 20th century has witnessed an extraordinary amount of discovery, and subsequent attention, on the link between dietary nutrients and disease prevention. Consumer interest in this area has been very strong and has encouraged market response. The food, pharmaceutical, and dietary supplement industries are beginning to deliver these active ingredients, as whole foods, individual nutrients, food ingredients, phytochemicals, or extracts, within a regulated environment.

The Nutrition Labeling and Education Act (NLEA) of 1990 (Public Law 1994a) and the Dietary Supplement Health and Education Act (DSHEA) of 1994 (Public Law 1994b) control the labeling criteria for food and dietary supplements that make a statement on their product labels about the nutrient-health link. The legislation also specifies the need for nutrition research to substantiate the petitioned health claims. At this juncture in the decade, a considerable amount of comment has been made regarding the lack of new health claim petitions under the NLEA health claim protocol (Anonymous 1996a and 1997a, Pappalardo 1996, Petrucelli 1996). By mid-1997, only nine health claims were approved, with seven of these approved in the initial NLEA legislation.

Why Nutrition Research Incentives Are Proposed

Much speculation exists concerning the lack of private sector petitioning for new health claim review and approval. Foremost is a concern that the high hurdles for the quantity and quality of research required for Food and Drug Administration (FDA) approval, termed the "significant scientific agreement," are too daunting for the food industry to pursue. The food industry traditionally reports a low research and development (R & D) investment/sales ratio of 0.5% (Anonymous 1997c, Kevin 1997) and does not approach the pharmaceutical industry's vigorous R & D investment ratio, which reported 19% in 1995 (Anonymous 1997c, Kevin 1997). There is also a concern that the generic nature of the approved health claim, which is available for use by all qualifying products, does not protect the investment of the firm engaged in securing the health claim. Additionally, all research completed in pursuit of the health claim becomes publicly available for review and comment and cannot be protected as proprietary research. The current health claim approval system is thus both generic and public (Childs 1997).

This seemingly structural discouragement for investing in private sector nutrition research runs counter to an accelerating need for solid nutrition research and discovery to provide substantiation of the health preventive benefits comparable to the design and rigor of medical research. This is necessary for the category to mature beyond its present stage and to engage health professionals in encouraging functional food and phytochemical consumption to optimize personal and public health.

Nutrition research and product documentation are mitigated by the presence of a biomarker for monitoring the progress (and subsequent successful intervention) of a chronic disease state. Cholesterol levels serve as biomarkers for heart disease as does blood pressure for hypertension. Discovering and documenting appropriate biomarkers is a delicate undertaking, which itself requires substantial investment in research. Additionally, there is a concern that international markets are encouraging more nutrition research through government-industry partnerships such as the European Food and Industry Research (FAIR) programs (Pitts 1997) and through proprietary claim approval as awarded by the Japanese Food of Specified Health Use (FOSHU) designation (Furikawa 1993). Many of these regulatory environments include publicly funded health care for the aging populations. To moderate their health care costs, these governments are amenable to permitting and promoting functional food contributions to achieve disease prevention. These developments open the possibility for the industry's competitive advantage to be captured by international competitors.

The Keystone National Policy Dialogue on Food, Nutrition and Health discussed the need for research incentives and presented several models for consideration. As incentive formats, they range on a continuum from providing full exclusivity to providing none (Anonymous 1996a). The FDA has continued the dialogue by establishing a working committee on research incentives (FDA 1997) and the academic program at the University of Illinois' Functional Foods for Health Program has conducted a closed workshop on this topic (Developing Incentives for New Functional Foods 1997).

Survey Objectives

The research survey had three basic objectives as follows:

1. The assessment of industry attitudes concerning the need for research incentives.
2. The determination of industry preferences among the six research formats presented in the Keystone National Policy Dialogue on Food, Nutrition and Health.
3. The identification of the differences in preference, if they exist, between type and size of company, functional responsibility of respondent, and industry.

The investigated research incentives as defined in the survey are presented in Table 2.1. The various incentives are not uniquely exclusive. Their overlapping

TABLE 2.1
Definitions of Proposed Research Incentives

The final report of the Keystone Dialogue identified several formats for encouraging private investment in research. Implementing legislation may be required to establish some of these economic incentives.

Lead Time: Confidential advance notice of approval given to the petitioner of a health claim, so that a marketing plan and market launch can be prepared in advance of the public notification.

Exclusivity: The organization that invests in the research supporting an approved health claim would be awarded exclusive use of that health claim in the market for a specified period of time.

Tax Credit: A portion of the research and development investment supporting the health claim would be awarded as a tax credit.

Compulsory Health Claim Royalties: Approved health claims would continue to be available to all companies, but the organization that performs the research necessary for the approval of the health claim would be awarded a royalty that would be assessed against products that were similarly qualified or used the new health claim.

Orphan Drug Model: A combination of the following: (i) 7 years of market exclusivity; (ii) 50% of the research development costs treated as tax credit (a maximum limit provided); (iii) federal grant money (maximum of $100,000/yr) for clinical testing expenses; and (iv) assistance on research protocol.

Research Pools: The creation of research pools for functional foods may encourage research. Traditionally, agricultural commodity boards and cooperatives have successfully promoted research *via* research pools.

nature is demonstrated in Figure 2.1, which illustrates that exclusivity is not a unique concept, but is present in the Orphan Drug, Health Claim Royalties, and Lead Time to Market options as well.

The mailing list of the Functional Foods for Health Program at the University of Illinois comprised the list of original respondents. The survey's target audience was defined as individuals from companies selling or intending to sell functional foods. The final unadjusted response rate was 16%, yielding 184 qualified respondents. An adjusted response rate accommodating duplicate mailings, multiple mailings per corporation, and unqualified nonrespondents would be considerably higher.

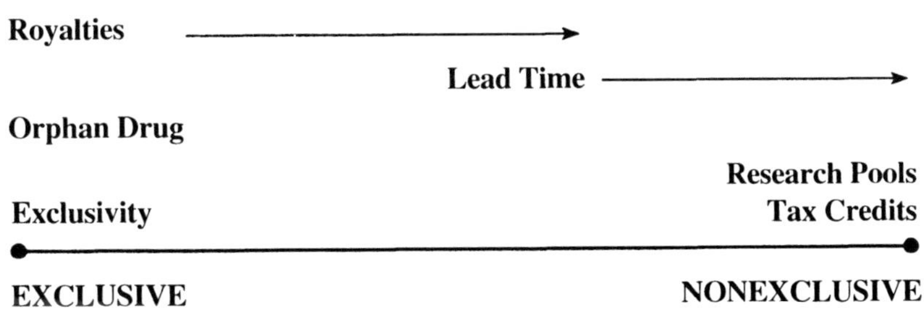

Fig. 2.1. Research incentives classification.

Sample Characteristics

By industry, the sample of qualified respondents was represented by the food industry (55%), the pharmaceutical industry (13%), and the dietary supplement industry (32%). The sizes of responding corporations in the sample included 26% with ≤$50 million in sales, 62% with sales >$50 million, and 12% unknown. The sample split 60/40 between respondents who were science and technical in their job function and those for whom marketing and commercialization were their corporate responsibility.

Preferences for Research Incentives

The overall preference expressed was for an exclusive research incentive (Table 2.2); this is not surprising, in review, given the prevalence of exclusivity within the other options (see Fig. 2.1). The two least preferred incentive options were Tax Credits and Research Pools, both being nonexclusive in nature.

Attempting to examine more than just the incentive preference, the survey delved deeper to determine the rationale for incentive preference by the manner in which the incentive was thought to reward the research investor. Key areas probed were the firm's direct likelihood to increase profits, to increase its market share, and to grow the size of the overall market category, depending on the research incentive offered.

All three industry segments strongly believed that the exclusivity incentive would yield higher functional food profits, more so than any of the other formats proposed. Similarly, all three industries believed that the exclusivity format would offer the most potential for growing the firm's individual market share. The former suggests that value capture through price premium or larger volume is obtainable *via* exclusivity, thus best growing profits. The latter suggests that volume gains are expected *via* exclusivity, best driving market share growth. Both are private gains recovered by the investing firm.

TABLE 2.2
Responses to Survey on Preferred Incentive Format to Grow the Overall Market Category[a]

Preferred Research Incentive[b]	Mean
Exclusivity	2.0
Orphan drug model	3.1
Lead time to market	3.6
Compulsory royalties	3.6
Tax credits	4.1
Research pools	4.3

[a]n = 184
[b]A response of 1 indicated most preferred.

The last inquiry regarding which incentive format would grow the overall market category, a more public gain, resulted in more diverse responses. The nonexclusive options of the present generic claim format and the existing research pools incentive emerged as the two leading responses. This indicates a belief that such nonexclusive formats promote fullest access to the market and would generate both products and marketing that would best grow the category. It is deduced, by the earlier responses, that although these nonexclusive incentives grow the category, they are not perceived to do so in a manner that would increase individual corporate profits or market share. In fact, the nonexclusivity would likely generate copycat products based on price competition and not product differentiation. Price competition and generic products would likely decrease profits. Details of responses to questions concerning research incentive formats are presented in Tables 2.3–2.5. A careful perusal notes the pharmaceutical industry's stronger preference for exclusive formats, reflecting its current successful experience with exclusivity provided through patents in its ethical drug industry.

Mechanisms for Exclusivity

There was strong preference expressed, representing nearly two thirds of the sample, for a government agency to award exclusivity. Over 40% of the sample felt that the Department of Health and Human Services (DHHS)/FDA should be the awarding agency. Responses concerning the suggested fixed number of years for

TABLE 2.3
Research Incentive Formats Chosen Most Likely to Yield Higher Profits

Industry	Food n = 99	Pharmaceutical n = 23	Dietary supplement n = 54
Exclusivity	52%	61%	48%
Orphan drug model	17	22	22
Compulsory royalties	9	13	9
Existing generic claim	8	0	7

TABLE 2.4
Research Incentive Formats Chosen Most Likely to Increase Market Share

Industry	Food n = 99	Pharmaceutical n = 23	Dietary supplement n = 54
Exclusivity	54%	57%	48%
Orphan drug model	15	4	13
Lead time	10	9	4
Research pools	7	13	6
Existing generic claim	5	9	13

TABLE 2.5
Research Incentive Formats Chosen Most Likely to Grow the Market Catgory

Industry	Food n = 96	Pharmaceutical n = 23	Dietary supplement n = 52
Research pools	24%	13%	17%
Existing generic claim	22	30	25
Exclusivity	15	9	19
Compulsory royalties	14	13	8
Lead time	13	0	4
Orphan drug model	7	22	3

which exclusivity should be awarded ranged from 0 to beyond the patent expiration period of 17 years. Nearly 20% of the sample did agree that 5 years would be appropriate, which was the dominant sample mode displayed in Figure 2.2.

Desire for Formal Government Recognition of Functional Foods

There was strong affirmation among respondents for a government-recognized definition for functional foods. Similarly, there was a strong preference expressed for a defined regulatory pathway for the approval of functional foods. In both cases, as Tables 2.6 and 2.7 indicate, the pharmaceutical industry expressed the strongest desire for regulatory clarity, again reflecting their familiarity with the more regulated nature of their own industry. The dietary supplement industry was less concerned about regulatory pathways than were the food and pharmaceutical industries. This suggests their more entrepreneurial industry approach and the less confining regulatory environment of the Dietary Supplement Health and Education Act of 1994. The dietary supplement industry yielded a highly significant reverse opinion on whether a government-provided regulatory pathway would increase their likelihood to invest in research, thus underscoring this latter observation (Table 2.8).

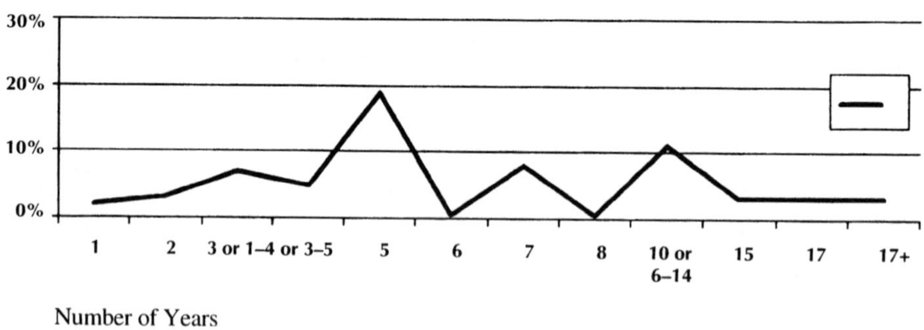

Fig. 2.2. Fixed number of years of exclusivity that should be granted.

TABLE 2.6
Would You Like a Govermental Recognized Definition of Functional Foods?

Industry	Food	Pharmaceutical	Dietary supplement
Yes	73%	66%	57%
No	27	33	43

TABLE 2.7
Should a Defined Regulatory Pathway Exist for Functional Foods?

Industry	Food	Pharmaceutical	Dietary supplement
Yes	87%	75%	69%
No	13	25	31

TABLE 2.8
Would Defined Regulatory Pathway Increase Likelihood of Investment?

Industry	Food	Pharmaceutical	Dietary supplement
Yes	74%	60%	48%
No	26	40	52

Preliminary Findings

Among many potential findings, the survey showed a strong preference for proprietary exclusive claims across the sample. The pharmaceutical and dietary supplement industries particularly preferred exclusivity. The nonexclusive formats evidenced by the existing generic claims and research pools were found to be best for growing the market category. There was an association of exclusivity with higher profits (52%), with share growth (49%), but not with market category growth (16%).

Industry indicated that exclusivity would maximize privately funded research and discovery for functional foods, would maximize the approval rate for health claims, and would increase functional food profitability. It also indicated that research pools would provide the most consumer education on diet and health and would maximize opportunities for improving public health. Certainly the current campaigns by the Florida Citrus Board utilizing the FDA approved health claims and the Fluid Milk Processor Board's milk moustache campaign promoting nutrition information with celebrities are examples of marketing efforts funded by grower pools.

There also was strong demand shown for government-provided regulations for functional foods, particularly for a recognized definition (65%) and approval pathways (75%). In the current generic claim environment, industry is anxious for a risk-reduced approach to research investments in functional foods.

Additionally, the survey research documented differences by industry suggest that consensus in the functional food industry will be hard to reach. As long as the functional food industry participants commercialize products under different regulatory constricts, NLEA and DSHEA, it will be difficult for them to reach common ground on commercialization paths and appropriate levels of research.

In conclusion, a strong, credible, vibrant functional food industry is dependent on the quantity and quality of research available to isolate and document the safety and efficacy of functional properties. Additionally, the discovery of viable biomarkers to document endpoints and monitor individual needs and progress is necessary to move the field forward in an efficient manner for consumers and health providers. Accordingly, there is a strong need for consumer and health provider education to accompany the maturation of this field. As the commercial marketplace grows in value and prominence for the consumer, it is essential that there be structural private sector market incentives to encourage research investment for continued discovery and documentation.

References

Anonymous, Keystone National Policy Dialogue on Food, Nutrition and Health Final Report, The Keystone Center, Washington, DC, 1996a.

Anonymous, *The Need for a Research-Intensive Nutraceutical Industry: What Can Congress Do?* The Claims-Research Connection, a Proposal by the Foundation for Innovation in Medicine, NJ, 1996b.

Anonymous, Research Incentives: What Does Industry Want? *Nutr. Bus. J.*, August 5 (1997a).

Anonymous, *Standard and Poor's Industry Surveys*, Vol. 1 (1997b).

Anonymous, U.S. Bureau of the Census, Statistical Abstract of the United States: 1997, 117th edition, Washington, DC, 1997c.

Childs, N.M., Industry Incentives for Nutrition Research: What Does it Mean for Consumers? *American Dietetic Association Functional Food Session: Therapies for the Future?* Boston, MA, October 1997.

Developing Incentives for New Functional Foods, closed associates meeting held by the University of Illinois Functional Foods for Health Program, Chicago IL, November 1997.

FDA Working Committee on Research Incentives, reporting to FDA Food Advisory Committee, Alexandria, VA, September 1997.

Furikawa, T., *A Comparison of the U.S., European and Japanese Nutraceutical Health and Medical Claim Rules*, The Foundation for Innovation in Medicine, New York, NY, January 1993.

Kevin, K., The 1997 Top 100 R and D Survey, *Food Process. 58*: 65–70 (1997).

Pappalardo, J.K., Evaluating the NLEA: Where's the Beef? *J. Public Policy Market. 15*: 153–156 (1996).

Petrucelli, P.J., Consumer and Marketing Implications of Information Provision: The Case of the Nutrition Labeling and Education Act of 1990, *J. Public Policy Market. 15*: 150–153 (1996).

Pitts, E., The Food and Drink Industry Advisory Council, *European Union Emerging Technologies Programme: Functional Foods*, Belfast, Ireland, April 1997.

Public Law No. 101-535, 104 Stat 2353 [codified at 21 U.S.C. § 343 (1994a)].

Public Law No. 103-417, 108 Stat 4325 (1994b enacted S. 784).

Chapter 3

New Opportunities to Bring Nutritionally Superior Food Products to the Market

J.W. Finley

Kraft Foods, Glenview, IL 60025

Introduction

Technology now affords opportunities to introduce unique products to the market; these products can provide health benefits or enhance the quality of life of the consumer. Modern biotechnology has developed and successfully introduced genetically improved crops with extended shelf life, improved disease resistance, and enhanced agronomic characteristics. Enhancements to food processing and packaging have helped the rapid expansion of the fresh or near-fresh produce business. This is also driven by the consumer's growing interest in health-promoting foods. In this chapter, the challenges to the introduction of nutritionally superior food products will be discussed. Bringing these new products to the market is not without barriers, but the rewards are likely equal to or greater than the challenges.

Consumer interest and concern about nutrition is a growing issue, particularly in North America. Two factors are driving this interest. First, the population is aging: The baby boomers are now approaching the age of 50 and discovering that they are no longer invincible. They are concerned not only about the aging process itself but also with the assurance of a high quality of life during the extended aging process.

A second cause for greater consumer attention to health and disease prevention is driven by the escalating costs of health care. The emphasis has increased interest in disease prevention and general well-being compared with the conventional approach in medicine of disease treatment. The prevention concerns have fueled a growing market in vitamins, minerals, supplements, and food supplements that prevent or delay the onset of chronic disease. From a business perspective, this trend is very tempting, considering that the food industry generally grows at <5%/y, whereas the health supplement business is growing at >20%/y.

Consumer Interest and Concerns

Consumer interest and concerns about nutrition represent growing issues, particularly in North America. In the last 10 years, emphasis has shifted from avoiding the bad to searching for foods that contain health-promoting ingredients. Consumers express these concerns by seeking foods that they perceive to be healthy and avoiding foods or food components that are perceived as unhealthy. Examples include the increased consumption of fruits and vegetables and the success of low- and no-fat products in the marketplace. In a recent publication by the United Soybean Board, it was stated that 84% of consumers avoid foods they perceive as bad and 89% are proactively seeking and consuming products thought to be healthy (Anonymous 1996). This concern can be seen graphically in Figure 3.1. From the graph, it is clear that the public is concerned about nutrition. The same survey did report that nearly three fourths of the population felt that there was too much conflicting information in the press. The same group also felt that it did not adversely affect their ability to make choices. Where does the public acquire nutrition information? According to a recent series of focus groups, most of the public and a substantial portion of the health community receive their nutrition information from the media and particularly TV news magazines (Schmidt 1997). Most health professionals, not to mention the public, do not have the time or inclination to follow significant scientific controversies about nutrition in the scientific literature. The result is that the information comes from hearsay and the popular press. It is important, therefore, that the information provided to the press be clear and understandable. As scientists, we should also try to be objective and factual. In the early 1980s, the scientific community managed to convey a fear of eating to the public, due in part to the constant emphasis on potentially toxic materials in foods. Although some of this was substantiated, much of it was overstated and caused undue concern. We must ensure that the press receives accurate and objective information about health- and nutrition-related issues.

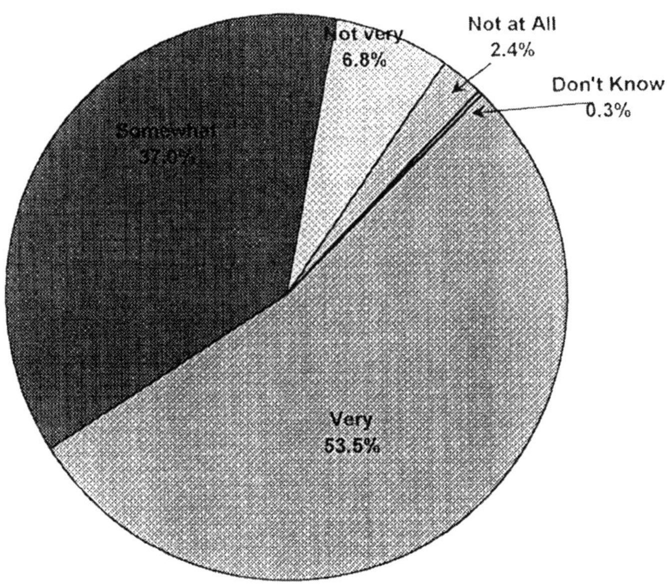

Fig 3.1. Consumer concern about nutrition.

What is the consumer looking for in food products? Health, vitality, and gratification! In other words, the ideal food will help prevent the onset of chronic disease, help the consumer perform tasks better, and taste as good or better than currently available foods. For health, consumers want products that will enhance their quality of life through prevention or delay of chronic disease, particularly cardiovascular disease and cancer. For vitality, the 50-y-old baby boomer wants to feel and look like he or she is 25. Thus, products that enhance performance belong with or alongside the health-promoting product. This can be seen in the market expansion of performance bars and performance drinks directed at the athletic community as well as the weekend warriors. For gratification, flavor and quality are the key factors. We must always remember that the consumer wants foods that taste good. All nutritionally enhanced products must deliver organoleptic satisfaction equal to or greater than that of conventional foods. This is particularly critical if it is a product such as soy, which delivers the phytoestrogen genestein. Soy is not the favorite flavor of the Western diet; thus, to deliver the soy in foods, we must find ways that are not only appealing but taste good as well. To gain the benefit, the foods must be consumed regularly; thus, quality and flavor become major objectives. The foods must be appealing if they are to be consumed on a regular basis, which translates into repeat sales.

Health-Promoting Foods

Examining the health issues a little more closely, we see that health-promoting foods fall into three categories, i.e., essential nutrients, sustenance, and recovery, as shown in Figure 3.2. Essential nutrients are vitamins, minerals, lipids, and proteins required for health. These nutrients are the basics for energy and health as defined by classical nutrition. With foods in this category, we can encourage greater consumption particularly of fruits and vegetables, which are good sources

HEALTH

ESSENTIAL NUTRIENT	SUSTENANCE	RECOVERY
Balanced calorie needs Nutrients to prevent disease	Nutrients for health maintenance	Foods to improve recovery External nutrition–Dietetic Foods
Fruits Vegetables Starch Dairy Meat	Antioxidants Balanced calories Healthy Profile Ready-to-eat Foods	External Nutrition Meal Supplements Dietetic Products

Fig 3.2. Health-promoting foods falling into three categories, their role and source.

of vitamins and minerals. Clearly, this group of foods also provides sources of phytochemicals, which are of emerging interest for delay of onset of chronic disease. The phytochemicals, antioxidants, and balanced calories that maintain health are considered sustenance. Recovery focuses on foods that are designed to help in the prevention of further deterioration or the recovery from specific diseases.

The sustenance area may be the area of greatest opportunity because it extends from supplements to ready-to-eat foods These are the foods that improve health and performance and prevent or delay the onset of chronic disease. Foods in this category help ameliorate disease, for example, foods that meet special needs for diabetics. Commercially, bars with controlled release of carbohydrates are being sold to help diabetics maintain moderate levels of blood glucose for prolonged periods of time.

Of most interest here is the category that includes the health promotion foods, i.e., those foods whose components will help prevent the onset of chronic disease. As health-promoting ingredients are identified and their efficacy documented, our next challenge is to find ways to deliver them in formats that the consumer will accept and use repeatedly. The bioactive components may be extracted and formulated into a food product or delivered as an integral part of food. Such foods may contain new or unique food ingredients from natural sources or plant materials enriched in bioactive materials through genetic manipulation. Regardless of the delivery system, we now find ourselves in the fuzzy areas between foods and drugs. It is in this confusing and controversial area that nutrition, food processing, and pharmaceuticals intersect. All are converging around the central theme of wellness and seeking profitable means to exploit the business opportunities. Emerging from this area, we are seeing a number of supplements and foods that have been called nutraceuticals, functional foods or hypernutritious foods. Although none of these terms are particularly appealing, they do help in their own way to define the fuzzy area between pharmaceuticals and food. The result is a new industry, largely populated at present by small companies and a few large players, that is starting to stick its toe in the water. Neither a food nor a pharmaceutical company can afford to ignore a business that is growing in excess of 20%/y.

A wide range of opportunities exists in the functional foods and nutraceutical category. Figure 3.3 illustrates the current segmentation of the market. It can be seen that the meal/snack/meal replacement segment is the largest single portion of the market. The beverage business, however, accounts for another major slice. Assuming that one has identified a new ingredient or component, how does one identify the best opportunities and nature of the market segment the product should enter? Foods occupy a larger segment, but the supplement industry is growing much more quickly. The identification of the new material and demonstration of benefit comprise one of the critical first steps, followed by a careful analysis of the market size.

Many of the most interesting nutraceuticals are phytochemicals found either in plants that are not normally consumed in high levels or in portions of the plant normally

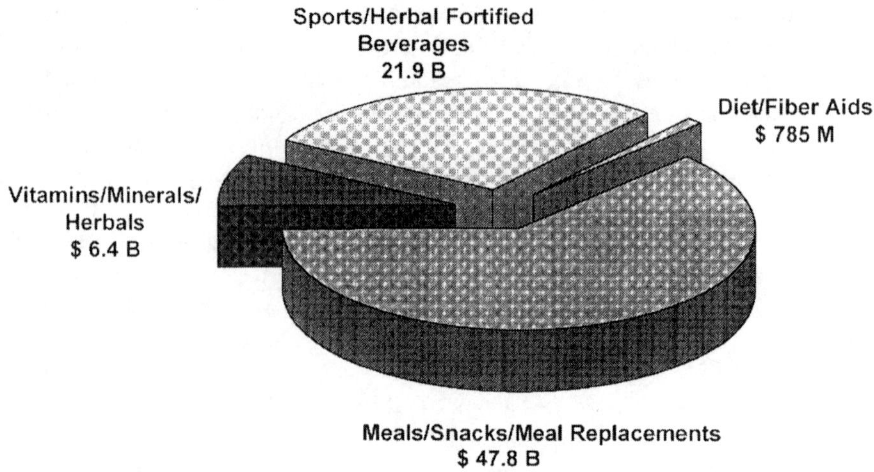

Fig 3.3. Current U.S. nutraceutical market; M, million; B, billion.

considered inedible. For example, soy provides genestein, a mild phytoestrogen that appears to have health benefits in the prevention of osteoperosis. Soy isolates, concentrates, and flours that are rich in genestein can be prepared. Although the application of soy isolates and flours is growing, their wide acceptance is limited. The best traditional sources are soy milk or tofu, which are not considered highly palatable by consumers of Western diets.

Another example is the flavanols found in grape seeds and skins. The flavanols are antioxidants that would protect food against oxidation and the consumer against cardiovascular disease (CVD). Because of their taste, most consumers do not consider these portions of the grape and the compounds themselves very palatable.

What are the Opportunities?

In the short term, we can enhance foods with naturally occurring components that protect against disease and ensure that the foods contain optimum levels of vitamins. In conjunction with providing foods with efficacious levels of vitamins and phytochemicals, we should also encourage the consumer to eat more fruits and vegetables. One trip to the supermarket will tell you that this is happening. The produce section is becoming the showplace of the market. Controlled atmosphere packaging has also resulted in a whole new business of ready-to-use vegetables. However, the overall consumption of fruits and vegetables is not increasing at the desired level. The USDA food pyramid suggests consumption of five fruits and vegetables per day. Most of the public does not approach this level of intake. A significant opportunity exists in finding improved delivery systems that provide fruits or vegetables at efficacious levels in convenient and tasty formats. In other words, we need to make them taste better so the public will eat them.

In the medium term, we can look for foods to be enriched with new and unique nutrients from unconventional sources. Examples might include flavanols from grape seeds or soy products with enhanced levels of antioxidants and/or phytoestrogens. Many of these products can be developed through enhanced food technology and conventional plant breeding. We also can expect to see many more candidates for low-calorie ingredients such as Olestra® and Sorbestrin®. Carbohydrate-based bulking agents are likely to be the next wave of low-calorie ingredients.

In the longer term, we believe that biotechnology will provide us with new and unique foods that taste good and are enhanced or designed to deliver health-promoting ingredients. These longer-term approaches imply a more complex development project, and the time lines can become very complex, as illustrated in Figure 3.2.

In Figure 3.4, we can see some of the major programs involved in the evolution of a new ingredient. The overlap of programs is essential timely delivery of new materials—the greater the overlap, the greater the risk. The risk, however, can be overcome by being the first to market with the new product.

Understanding the hurdles that will be encountered along the way will help in the planning and evaluation of the project before major investment occurs. Focusing on a product with an up-front understanding of the risks and rewards helps raise the comfort level of all involved, particularly the decision makers. What are the major hurdles that can impede the ultimate success of a new development project? First, we need the concept, i.e., a valid idea with a reasonable hypothesis and definition of the final product. We do not want to simply change a fatty acid or an amino acid in a food unless we have a product concept in mind from which a benefit can be realized. It is important to try to develop new platforms that result in families of ingredients rather than single ingredients.

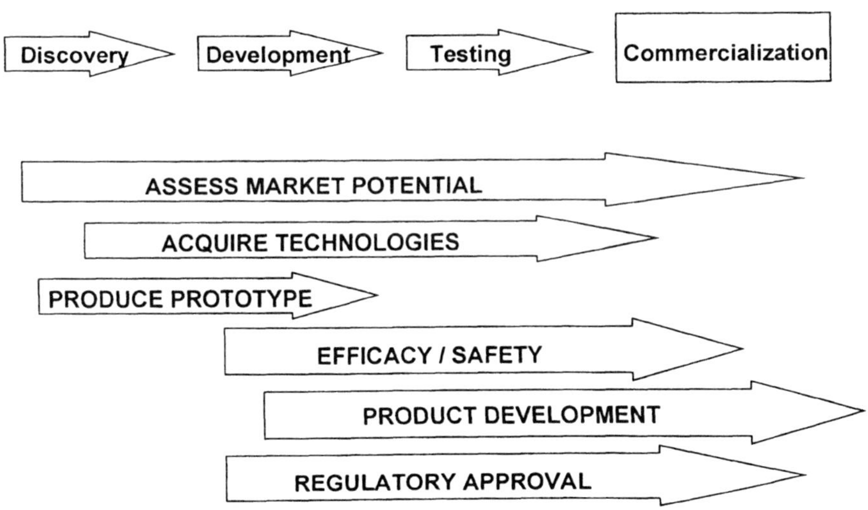

Fig 3.4. Nutraceutical development market.

What are the barriers to bringing a new biotechnology-based product to market? The resulting intellectual property must be protectable. The product must be safe, and a reasonable safety testing program must be considered to demonstrate that there are no significant risks with the new product. Connected to this, the regulatory barriers, both domestic and international, must be considered. The concept must be a valid business proposition. If successful, can added value be captured? The bottom line is whether it will make money for the corporation. Consumer concerns are always an issue, particularly with technology-based products. How will the consumer react? Will they accept the new product? Why should the consumer pay the price of a new product? Conversely, is there a way to capture the added value of the product?

First, consider the intellectual property hurdles. Be sure you own the technology; with genetically engineered products, in particular, the patenting and ownership of a gene or a construct are critical. One must stay on top of the field and know who owns what! When a valuable gene is identified for a particular application, it must be determined whether it has been isolated for the first time or whether it can be licensed. What technologies could threaten the patent and can these be captured to protect the investment? With new ingredients or food components, the application patents in foods are generally as valuable as the composition of matter patents; the latter are generally much more difficult to obtain in natural products.

Assuming that a desirable technology that can operate without infringement and be fully protected has been identified, what are the technical barriers to be considered?

Technical feasibility for the application would seem to be the most difficult barrier, but generally it is the most clearly defined. We establish and develop the required technology to meet a specific goal. For example, with the rapidly advancing genomics and molecular biology seen today, the deployment of nutritionally improved products is technically feasible in many if not most agricultural products. Cost of goods for production can be a huge advantage if a particularly desirable component can be overexpressed in a plant. For example, overexpression of genestein in soy would provide soy products that provide the health benefit but may be used at lower concentrations, thus overcoming flavor issues. One consideration is how long it will take. Soy plants are easy to deal with, because multiple crops can be grown in 1 y by growing successive generations in different parts of the world. Tree crops on the other hand are much more difficult because of the long generation times. With any new development, the fit with current or desired business ventures must be considered. In Figure 3.5, looking at the progression of the technology from test tube to dinner plate, we see that once the transgenic plants are developed and ready for field trials, we are ready to start on the next hurdles in the development process. The first is regulatory, i.e., the product must be shown to be safe to the regulators. Generally, both FDA and USDA approvals will be required. To make the introduction of a new crop attractive, there must be some benefit to

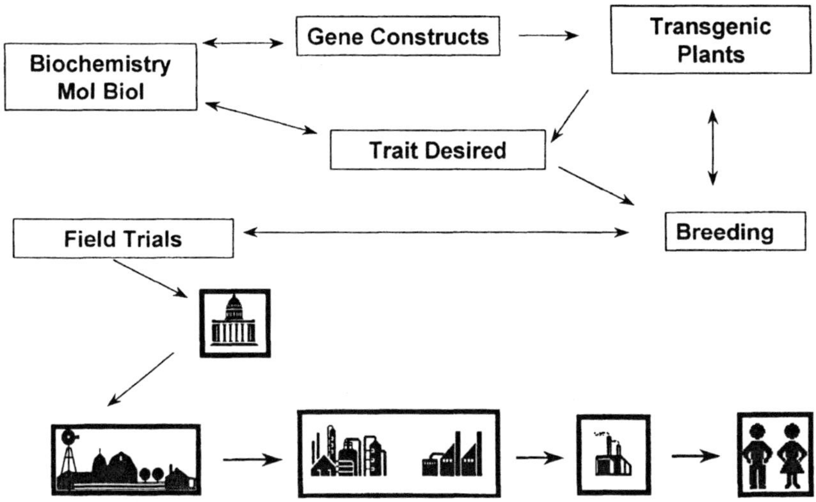

Fig 3.5. Plant biotechnology from test tube to dinner plate.

the farmer such as better yield or less herbicide or pesticide required for production. Two examples are the Roundup Ready® products and products with improved insect resistance through addition of genes that produce microbial insecticides. In addition, there should be benefits to the elevator operator and the processor. For example, are the soy products that are higher in oil less likely to break? The point is that the best products will produce value throughout the value chain. Importantly, the value must be clear to the consumer who will eventually pay for the added value.

The food manufacturer must see benefit in the new product because it is likely he will have to pay a price premium for the ingredient, and he must find a way to recoup the cost with a better label claim or cost-saving improvement in product functionality. Two cases in which improved label claims might be realized are grains with higher levels of β-glucan or higher levels of genestein in soy when the efficacy is proven and the labeling is approved.

Ultimately all products must benefit the consumer, i.e., lower cost if the farmer achieves better yields or perceived benefits from better quality, flavor, fewer defects (insect-resistant potatoes) or perhaps components that provide a health benefit.

Potential Consumer Benefits of Genetically Modified Food Ingredients

Quality

- Insect resistance provides better-looking products
- Improved flavor and color
- Delayed ripening makes products available for a greater portion of the year

- Better shelf stability *via* delayed lipid oxidation
- Foods with fewer gastrointestinal side effects, e.g., low raffinose/stachyose beans
- Foods with fewer allergens

Health Benefits

- More essential nutrients
- Higher concentration or new sources of beneficial phytochemicals
- "Naturally" produced low-calorie fats
- Products with more "good-for-you" fiber

Many of these benefits can be realized through the application of modern biotechnology. The time required to realize commercialization of products produced through biotechnology is rapidly diminishing; however, the reality is that even with the greatest success, the time from initiation to significant production can be 4–8 y. Because the costs are very significant, we must consider the major business hurdle, i.e., will the concept ultimately produce a profitable business that is large enough to justify the investment? How do we, as scientists, convince management and marketing that the new ingredient or product is worthy of investment? One powerful tool is to estimate the net present value (NPV) based on projected market data. For a major corporation to embark on a major research program, the NPV of the targeted product may have to be between $75 and $100 million. This is driven by the high cost of the research and testing of the products, which requires the use of high technology, before they can appear on the market. In addition to the basic development costs of biotechnology, which can be considerable, the safety testing can also represent a major effort and enormous cost. When estimating the NPV, the risk factors for the testing must be included in the model.

With this information, one can begin to develop a focused business case. We would all like to have low-cost/high-reward projects. Unfortunately, very few projects fall into this category. Most biotechnology projects will fall into the category of high cost and high likelihood of success. That is not to say high spending guarantees success. It simply means that the cost of testing is high, but as the product moves through the testing, the greater the likelihood of success. The challenge is to constantly evaluate the research and the market potential. We should ask ourselves every day whether we are headed in the right direction. Clearly, we need to find ways to control cost, but we must do a complete and thorough business analysis as well as safety analysis. Both are ongoing throughout the life of the project.

Conclusions

New products and ingredients offer large growth potential for the food industry. Products of particular interest are foods that enhance the quality of life through delay of the onset of chronic disease. Many of these products can be realized through biotechnology. Although there are a number of significant hurdles to be considered, large markets are possible, and there will be significant rewards for those who choose to take the risk. The degree of risk must be evaluated in terms of the real market potential. This market potential should be regularly reassessed throughout the life of the project.

References

Anonymous, *National Report on Consumer Attitudes About Nutrition,* United Soybean Board, 1996.

Schmidt, D.B., *Communicating the Benefits of Functional Foods—Insights from Consumer and Health Professional Focus Groups*, Abstract, A213th American Chemical Society Meeting, San Francisco, CA, April 13–17, 1997.

Chapter 4

Conceptualization of the Prooxidant and Antioxidant Actions of Plant Food Chemicals

Okezie I. Aruoma

Department of Biology, Imperial College of Science, Technology and Medicine, Sir Alexander Fleming Building, Imperial College Road, London SW7 2AZ, UK

Introduction

Lipids may be considered from the standpoint of the following four distinctive interfacial groups: bulk food lipids (e.g., oils), dispersed food lipids (e.g., membranes or emulsions such as salad dressings), dispersed lipids in living organisms (membranes and organelles), and lipids found in watery fluids in organisms (e.g., cytoplasm or plasma) (Aruoma 1996, Frankel 1980, Porter 1993). When considering the free radical reaction of lipid oxidation, the prooxidant and antioxidant considerations may have different implications for each of the groups, e.g., they may be totally irrelevant for the bulk food lipids but highly relevant for the dispersed food lipids and biological fluids in organisms (Aruoma 1996).

It is perfectly possible for an antioxidant to protect in one system but fail to protect, or even sometimes cause damage, in others. Antioxidant inhibitors of lipid peroxidation may not protect other targets (such as DNA and protein) against damage, and sometimes can even aggravate such damage (see below). This may not matter much in the food matrix because damage to DNA and proteins, unless very excessive, will not alter the taste or texture of food (unlike lipid peroxidation). However, it is of the greatest importance *in vivo*. For example, butylated hydroxyanisole (BHA) is a powerful inhibitor of lipid peroxidation, and yet huge dietary doses of it can induce cancer of the rat forestomach, and it has been suggested that oxidative DNA damage could be involved (Schildermann *et al.* 1995).

Plant extracts have antioxidant indications. Typical examples include extracts and active principles from rosemary, sage, cocoa shells, oats, tea, olives, garlic, ginger, red onion skin, grapes, apple cuticle, wheat gliadin, korum rind, licorice, nutmeg, clove, oregano, thyme, mustard leaf seed, chia seed, peanut seed coat, birch bark, carob pod, tempeli, yam, mango, mangosteen, and vanilla (Aruoma *et al.* 1992, Löliger 1991, Pratt 1993, Shahidi and Wanasundara 1992, Shahidi, 1997). Vitamin E, vitamin C, β-carotene, flavonoids, and other polyphenols found in some of these extracts are widely discussed as potential antioxidant prophylactics. Flavonoids are naturally occurring benzo-γ-pyrone derivatives, which are ubiquitous in plant cells and are therefore accessible to consumers through diets of plant origin.

Antioxidant Characterization

A compound might exert antioxidant actions *in vivo* or in food by inhibiting generation of reactive oxygen species (ROS) or by directly scavenging free radicals. Additionally, *in vivo,* an antioxidant might act by raising the levels of endogenous antioxidant defenses [e.g., by up-regulating expression of the genes encoding superoxide dismutase (SOD), catalase, or glutathione peroxidase].

Simple experiments can be performed to examine direct antioxidant ability *in vitro* and to test for possible prooxidant effects on different molecular targets. This "screening" approach can be used to rule out direct antioxidant activity *in vivo* (Fig. 4.1). For example, a compound that is poorly effective *in vitro* will not be any better *in vivo* (Aruoma 1997). The interrelationship between ROS and antioxidants in humans is very complex indeed. Humans possess a variety of radical-scavenging mechanisms. These include the antioxidant enzymes superoxide dismutase, catalase, and glutathione peroxidase. The low-molecular-weight antioxidants include glutathione (GSH), uric acid, α-tocopherol (vitamin E), and ascorbic acid (vitamin C). Scavenging enzymes and antioxidants can inhibit free radical production by chelating the transition metal catalysts, breaking chain reactions, reducing concentrations of ROS, and by scavenging initiating radicals. Cytoprotective enzymes are located within both the hydrophilic and hydrophobic compartments of the cell. Alpha-Tocopherol delays lipid peroxidation by reacting with chain-propagating peroxyl radicals faster than these radicals can react with proteins or fatty acid side-chains (Dix and Aikens 1993, Duval and Poelman 1995, Liebler 1993). In theory, beta-carotene has remarkable antioxidant chemistry, but this has been difficult to demonstrate in a beneficial manner in biological systems (Byers and Perry 1992, Sies *et al.* 1992). Beta-Carotene can interact with a free radical to form a β-carotene–derived radical, which, in the presence of oxygen, forms a peroxyl

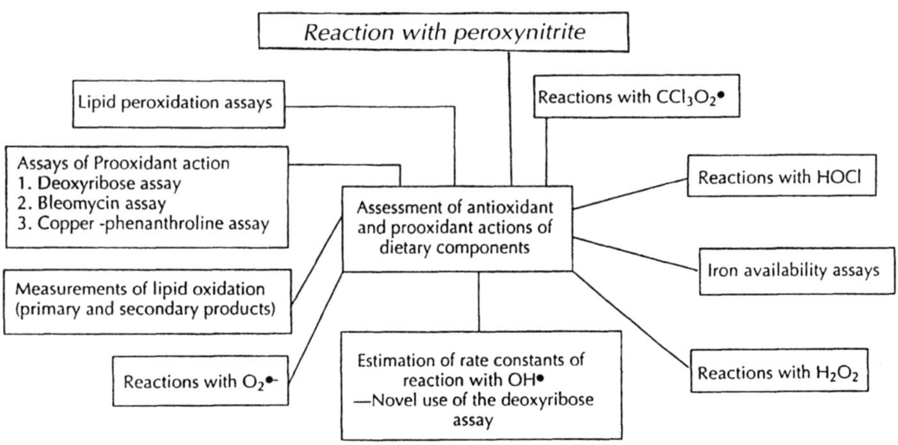

Fig. 4.1. Experimental approaches for the evaluation of antioxidant and potential prooxidant actions *in vitro.*

radical. The abilities to recycle vitamin E– and β-carotene–derived radicals may vary *in vivo*. Thus, the accumulation of the β-carotene radicals may present a prooxidant status *in vivo*. Ascorbate (vitamin C) is often claimed to be an important antioxidant *in vivo*. Its antioxidant property is related to the fact that the dehydroascorbate radical is much less reactive than are many of the radicals that can be scavenged by ascorbate (Rose and Bode 1993, Seib and Tolbert 1982). Intracellular enzymic systems exist *in vivo* to reduce the dehydroascorbate back to ascorbate using NADH (the NADH-semi-dehydro-ascorbate reductase enzyme) or GSH (the dehydroascorbate reductase enzyme) as sources of reducing power. Understanding the biopharmacy of ascorbic acid is of great importance (see Rumsey and Levine 1998).

Different Approaches to the Characterization of Antioxidant and Prooxidant Actions of Dietary Antioxidants

Antioxidants act at different levels in the oxidative sequence involving lipids. The extent to which oxidation of fatty acids and their esters occurs depends on the chemical structure of the fatty acid. Measurements of lipid peroxidation using rat liver or cardiac microsomes, ox-brain phospholipid liposomes, arachidonic acid, and other lipid model systems (e.g., bulk oil and emulsified oil systems) should be the first line of tests to establish the potential antioxidant action of dietary antioxidant compounds (Aruoma *et al.* 1992 and 1993, Brand-Williams *et al.* 1995, Frankel 1993, Pryor *et al.* 1988 and 1993). An antioxidant index based on ability to scavenge peroxyl radicals (Akaike *et al.* 1995, Aruoma *et al.* 1995) provides support for antioxidant efficacy in *in vitro* systems. The trichloromethylperoxyl radical ($CCl_3O_2^{\bullet}$) is a reactive peroxyl radical that is frequently used as a model to assess the ability of a compound to react with peroxyl radicals. $CCl_3O_2^{\bullet}$ is a "clean" lipid-soluble radical for *in vitro* studies of scavenging ability (Aruoma *et al.* 1995, Packer *et al.* 1978). The $CCl_3O_2^{\bullet}$ can be generated by radiolysis of an aqueous mixture of propan-2-ol and CCl_4. The values arising from these studies can be regarded only as indicative of antioxidant propensities *in vitro* (Table 4.1).

Unfortunately, the process described above is expensive and requires a high degree of proficiency. Simpler but equally informative methods for assessing antioxidant ability of nutrient components have been described in the literature (Aruoma *et al.*1990, Cao *et al.* 1993, Laranjinha *et al.* 1994, Lissi *et al.* 1995, Miller and Rice-Evans 1996, Rice Evans *et al.* 1995, Pryor *et al.* 1993). One example is the spectrophotometric technique, total antioxidant activity (TAA) or the Trolox equivalent antioxidant activity (TEAC). The method involves the generation of the long-lived specific radical cation chromophore of 2,2′-azinobis-(3-ethylbenzothiazoline-6-sulfonic acid) (ABTS) by controlled chemical oxidation. The $ABTS^{\bullet+}$ radical cation has absorption maxima in the near-infrared region at 645, 734, and 815 nm. The TEAC reflects the ability of hydrogen- or electron-donating antioxidants to scavenge the $ABTS^{\bullet+}$ radical cation compared with that of Trolox. The antioxidant suppresses the A_{734} to an extent and on a time scale dependent on

TABLE 4.1
Reactions of "Natural" Antioxidants in Assays Involving Reactive Oxygen Species (*In Vitro*)

Dietary components	Rate constant for reaction with $CCl_3O_2^{\bullet}$ $[m^{-1}s^{-1}]$	Dietary source of the components
Vanillic acid	3.97×10^6	Classic vanilla flavor
Vanillin	9.54×10^7	Classic vanilla flavor
Thymol	3.82×10^6	Essential oil of *Thymus vulgaris*
Carvacrol	3.92×10^5	In oil of thyme
6-Gingerol	4.67×10^6	In ginger oils
Zingerone	5.63×10^6	Isolated from ginger root
Ferulic acid	7.5×10^6	From chia, Mexican plant
(+)Catechin	6.1×10^6	Green tea
(±)Catechin	6.1×10^6	Green tea
(–)Epicatechin	7.3×10^6	Green tea
Hydroxytyrosol	8.37×10^6	Olive oil and olive fruits
Carnosic acid	2.7×10^7	Rosemary extract
Carnosol	$1\text{–}3 \times 10^6$	Rosemary extract
α-Tocopherol (Vitamin E)	4.89×10^8	Various forms naturally occurring
Vitamin C	1.3×10^8	Synthetic, naturally occurring
Trolox C	2.23×10^8	Synthetic vitamin E analog
Propyl gallate	1.67×10^7	Synthetic compound
Quercetin	3.9×10^7	Fruits and vegetables
Morin	3.01×10^8	Fruits and vegetables
Chrysin	9.86×10^7	Fruits and vegetables
Fisetin	4.09×10^8	Fruits and vegetables

the antioxidant activity. Table 4.2 (Rice-Evans and Miller 1996) shows the TEAC of flavonoids, phenyl propanoids, and carotenoids.

Under certain conditions, antioxidants that protect lipids against free radical damage may actually accelerate damage to other molecules such as DNA, carbohydrates, and proteins. The deoxyribose assay (Aruoma 1994, Halliwell *et al.* 1988) allows determination of rate constants of reactions with $OH^{\bullet}$ radicals, assessment of abilities to exert prooxidant action, and assessment of abilities to chelate metal ions.

In tests for the ability of a compound to exert prooxidant action *in vitro* by using the so-called deoxyribose assay, it is the ability of the compound to mediate a reaction similar to that of ascorbate that constitutes the basis of the evaluation. The Fe(III)-EDTA (ethylendiaminetetraacetic acid) complex has a tested propensity to be reduced by the prooxidant (if indeed it is able to do so). The redox potentials of other metal complexes (which may be physiologically relevant) would vary. In the absence of EDTA, iron ions are equally available to both the deoxyribose and the compound under test. Thus, compounds that are able to preferentially

TABLE 4.2
Trolox Equivalent Antioxidant Capacity (TEAC) for a Variety of Naturally Occurring Flavonoids, Phenylpropanoids, Carotenoids and Vitamins

Antioxidant	Sources	TEAC (mmol/L)[a]
Vitamins		
Vitamin C	Fruit and vegetables	1.0 ± 0.02
Vitamin E	Grains, nuts, oils	1.0 ± 0.03
Flavonoids		
Anthocyanidins		
Oenin	Black grapes/red wine	1.8 ± 0.02
Cyanidin	Grapes, raspberry, strawberry	4.4 ± 0.02
Delphinidin	Aubergine skin	4.4 ± 0.11
Flavonols		
Quercetin	Onion, apple skin, berries, black grapes, tea, broccoli	4.7 ± 0.10
Kaempferol	Endive, leek, broccoli, grapefruit, tea	1.3 ± 0.08
Flavone		
Rutin	(Rutinoside of quercetin)	2.4 ± 0.12
Luteolin	Lemon, olive, celery, red pepper	2.1 ± 0.05
Chrysin	Fruit skin	1.4 ± 0.07
Apigenin	Celery, parsley	1.5 ± 0.08
Flavanols		
(Epi)catechin	Black grapes/red wine	2.4 ± 0.02
Epigallocatechin	Teas	3.8 ± 0.06
Epigallocatechin gallate		4.8 ± 0.06
Epicatechin gallate		4.9 ± 0.02
Flavanone		
Naringin	Peel of citrus fruit	0.24 ± 0.02
Taxifolin	Citrus fruit	1.9 ± 0.03
Hydroxycinnamates		
Ferulic acid	Grains, tomatoes, spinach, cabbage, asparagus	1.9 ± 0.02
Caffeic acid	White grapes, olives, spinach, cabbage, asparagus	1.3 ± 0.01
Chlorogenic acid	Apples, pears, cherries, plums, tomatoes, peaches	1.3 ± 0.02
p-Coumaric acid	White grapes, tomatoes, spinach, cabbage, asparagus	2.2 ± 0.06
Carotenoids		
Carotenes		
Lycopene	Tomatoes	2.9 ± 0.15
β-Carotene	Carrots, sweet potato, tomatoes, paprika, green vegetables	1.9 ± 0.10
α-Carotene	Tomatoes, carrots, green vegetables	1.3 ± 0.04

Continued

TABLE 4.2
(Continued)

Antioxidant	Sources	TEAC (mmol/L)[a]
α-Carotene	Tomatoes, carrots, green vegetables	1.3 ± 0.04
Xanthophylls		
β-Cryptoxanthin	Mango, papaya, peaches, paprika, oranges	2.0 ± 0.02
Lutein	Banana, satsuma peel, egg yolk, green vegetables	1.5 ± 0.10
Zeaxanthin	Paprika, satsuma peel	1.4 ± 0.04
Astaxanthin	Salmon, food colorants, crab	0.03 ± 0.03
Canthaxanthin	Carrots, kale, food colorants, red peppers	0.02 ± 0.02

[a]*Source*: Rice-Evans and Miller (1996). The values of TEAC varied by 2–11% in replicative measurements.

chelate iron and present the resulting metal complex in a less redox-active form compared with EDTA-metal complex will protect deoxyribose against damage in the presence of ascorbate and H_2O_2. Substances that inhibit in the assay are also those that are able to bind iron ions strongly enough to remove them from deoxyribose. Thus, EDTA removes iron ions from deoxyribose, but iron-EDTA chelates are very effective in generating $OH^•$ so that the deoxyribose is still degraded, this time by $OH^•$ in "free" solution, rather than by $OH^•$ formed on the deoxyribose molecule, for example (Aruoma *et al.* 1987, Halliwell *et al.* 1988).

Studies using the deoxyribose assay can provide useful information on the likelihood that molecules could chelate metal ions in a way that prevents them from catalyzing $OH^•$ formation. Thus, when iron is added to the assay mixture as ferric chloride instead of as ferric-EDTA, some of the Fe^{3+} ions bind to deoxyribose, and damage to the sugar becomes site specific such that the $OH^•$ formed by bound iron ions immediately attacks the deoxyribose. The ability of a substance to inhibit deoxyribose degradation under these reaction conditions is a measure of its ability to interfere with site-specific hydroxyl radical generation.

When ascorbate is omitted from the deoxyribose reaction mixture, the ability of added compounds to reduce the Fe^{3+}-EDTA complex can be tested. This gives an index of prooxidant action, allowing assessment of antioxidant molecules in non-lipid systems. "The Fe^{3+}-EDTA complex has a tested propensity to be reduced by the prooxidant (if, indeed, it is able to do so). It follows that the redox potentials of other metal complexes (which may be physiologically relevant) would vary." This simple idea led to the proposal to use assays involving DNA damage to specifically test for the abilities of dietary antioxidants to exert prooxidant actions, different from their intended abilities to minimize oxidation of lipids. The tests involve measurement of DNA damage in the presence of bleomycin-iron and copper-1,10-phenanthroline complexes (reviewed in Aruoma 1996).

Oxidative DNA Damage and Its Use in Assessing Prooxidant and Antioxidant Actions

Incubation of calf-thymus DNA with a system producing $OH^{\bullet}$ radicals gives rise to extensive chemical modification of the DNA bases in a way that appears to be diagnostic for $OH^{\bullet}$ (Aruoma *et al.* 1989). Incubation of DNA with Fe^{3+}-EDTA, ascorbate and H_2O_2 leads to significant rises in the amounts of several oxidatively modified bases; this is characteristic of attack by $OH^{\bullet}$. Addition of antioxidants will modulate the extent of the DNA damage measured. Omission of ascorbate from the reaction mixture greatly decreases the DNA base modification. This provides a window to test the ability of dietary antioxidants to exert prooxidant properties *in vitro*.

The results of gas chromatography-mass spectrometry (GC-MS) analysis of modified bases in DNA are often expressed as nanomoles of modified bases per milligram of DNA (equivalent to pmol/μg DNA). However, it is easy to convert these data into the actual number of bases modified. Dividing the amount of nmol bases/mg DNA by 3.14 (or multiplying by 0.318) gives the number of modified bases per 10^3 bases in DNA, i.e., 1 nmol/mg DNA corresponds to ~318 modified bases per 10^6 DNA bases.

Use of the Bleomycin Iron–Dependent DNA Damage to Assess Prooxidant and Antioxidant Actions

Bleomycin, an antitumor antibiotic, binds to DNA by using its bithiazole and terminal amine residues; it also complexes with metals (such as iron) using the β-amino-alaninepyrimidine-β-hydroxy histidine portion of the molecule. Bleomycin binds iron ions, and the bleomycin-iron complex will degrade DNA in the presence of O_2 and a reducing agent such as ascorbic acid.

The bleomycin-Fe^{3+} complex by itself is inactive in inducing damage in DNA. Oxygen and a reducing agent or hydrogen peroxide are required for the damage to DNA to occur. DNA cleavage by bleomycin releases some free bases and base propenals in amounts that are stoichiometric with strand cleavage (Burger *et al.* 1981, Giloni *et al.* 1981, Petering *et al.* 1990, Sigiura *et al.* 1982). A positive test is obtained when the compound is able to reduce the bleomycin-Fe^{3+}-DNA complex to the more active bleomycin-Fe^{2+}-DNA complex (in the presence of oxygen) in the absence of added ascorbate in the reaction mixture, resulting in DNA damage (Aruoma 1996).

Use of the Copper-1,10-Phenanthroline–Dependent DNA Damage to Assess Prooxidant and Antioxidant Actions

The copper-1,10-phenanthroline (OP) complex has nuclease activity and has been used for structural studies on DNA. The copper-phenanthroline complex can be used to assess prooxidant actions. Hydrogen peroxide is implicated in the mechanism of the DNA damage by the copper-phenanthroline system. Hydroxyl radicals are involved in the damage to DNA caused by the copper-phenanthroline system (Dizdaroglu *et al.* 1990, Que *et al.* 1980). Unlike the bleomycin-iron–mediated damage to DNA, damage in the copper-phenanthroline system is confined mainly to the DNA bases. The small amount of DNA sugar damage is what the copper-phenanthroline assay measures. When a reducing agent is omitted from the reaction mixture, no damage to deoxysugar in DNA occurs. Increasing the concentrations of the reducing agents such as ascorbate and/or mercaptoethanol leads to increased deoxysugar damage. This is in agreement with earlier observations that the nuclease activity of the copper-phenanthroline complex is potentiated by thiols, a superoxide generating system, xanthine-xanthine oxidase, and NADH in the presence of hydrogen peroxide (Reich *et al.* 1981).

The deoxyribose assay allows determination of rate constants of reactions with $OH^{\bullet}$ radicals, assessment of abilities to exert prooxidant action, and assessment of abilities to chelate metal ions. The net outcome can be further investigated by using assays involving DNA damage to assess prooxidant actions. The positive prooxidant actions in the deoxyribose system rely on the ability of the compounds to promote reduction of Fe^{3+} to Fe^{2+} chelates and hence $OH^{\bullet}$ formation in the presence of H_2O_2. Assays involving DNA rely on ability to reduce either the iron-bleomycin-DNA or copper-1,10,phenanthroline-DNA complex. If the compound under test promotes the two reactions described, it possesses a prooxidant property and must be subjected to a more careful evaluation. Compounds that have antioxidant indications will reduce the extent of damage however. A compound might be prooxidant in the deoxyribose system and/or the DNA systems but sometimes not in both. There are often solubility problems. Organic solvents do not affect the outcome of DNA-dependent assays. Thus, in cases in which the deoxyribose assay cannot be performed because of solubility restriction, the copper-phenanthroline assay would suffice. Circumventing potential prooxidant action could contribute to increased protective ability of dietary antioxidants toward susceptible substrates (Aruoma 1996).

Scavenging of Hypochlorous Acid

Hypochlorous acid, HOCl, is produced by the neutrophil-derived enzyme myeloperoxidase at sites of inflammation and when activated neutrophils infiltrate reoxygenated tissue (Weiss 1989). The enzyme oxidizes chloride ions, $Cl^{\bullet}$, in the presence of H_2O_2. HOCl can attack many biological molecules. For example, the proteolytic inhibitor α_1-antiproteinase (α_1-AP) is the major inhibitor in human plasma of proteolytic enzymes such as elastase. α_1-AP accounts for ~90% of the elastase-inhibitory capacity of human serum (Travis and Salvesen 1983). Thus, its inactivation by HOCl might greatly potentiate tissue damage because elastase is also released from activated neutrophils.

What is most significant, perhaps, is the observation that phenolic compounds can react effectively with HOCl, hence protecting susceptible targets against oxidation (Aruoma *et al.* 1992 and 1993). This may have physiologic significance, given the interest in the use of dietary natural antioxidants to manipulate disease states.

Assays for hypochlorous acid that could be performed with ease to test the ability of an antioxidant to react with the molecule include the elastase assay (Aruoma *et al.* 1992, Wasil *et al.* 1987), assay with catalase (Aruoma and Halliwell 1987), inhibition of taurine-chloramine formation (Aruoma *et al.* 1991, Thomas *et al.* 1985, Weiss *et al.* 1983), and an assay based on the oxidation of 5-thio-2-nitrobenzoic acid (TNB) (Ching *et al.* 1994). In the last-mentioned, the ability of the antioxidant molecule to scavenge HOCl is determined by measuring the oxidation of TNB to 5,5′ dithiobis (2-nitrobenzoic acid) (DTNB) at 412 nm.

Hypochlorous acid may not have great relevance to food materials, although it is widely used in bleaches (as the sodium salt, NaOCl) and could be generated during irradiation of foods that contain chloride ions under anoxic conditions. In the context of food handling, HOCl is a major active constituent of chlorine-based bleaches often used to disinfect equipment with which foods will come into contact.

An antioxidant protecting against damage by HOCl might do so not only by scavenging HOCl but also by inhibiting myeloperoxidase. Thus, several thiols are not only good scavengers of HOCl but also act as competing substrates for myeloperoxidase and therefore slow down HOCl formation (Cuperus *et al.* 1985, Svensson and Lindvall 1988). Ascorbic acid is a good HOCl scavenger and a substrate for myeloperoxidase, but its effects on HOCl production are complex (Marquez and Dunford 1990). Ergothioneine (2-mercaptohistidinetrimethylbetaine) is formed *via* hercynine from histidine, methionine, and cysteine in microorganisms and is not biosynthesized in animals. Concentrations up to 1–2 mmol/L have been reported in human and other mammalian tissues. Ergothioniene is an important natural antioxidant, capable of scavenging HOCl (see Fig. 4.2) (Aruoma *et al.* 1997) at physiologic concentrations. Several phenolic compounds including flavonoids react quickly with HOCl and can protect α_1-AP and other susceptible targets against damage *in vitro*. This may have physiologic significance, given the interest in the use of natural phenolic antioxidants as therapeutic agents.

Reactions with Peroxynitrite: Novel Method for Assessing Antioxidant Capacity of Nutrient Components

One suggested toxicity of $NO^\bullet$ involves its reaction with the superoxide radical ($O_2^{\bullet-}$) to give peroxynitrite ($ONOO^-$), a reaction that occurs with a rate constant of 6.7×10^9 mol/(L · s). Peroxynitrite induces peroxidation of lipids, oxidizes methionine and -SH residues in proteins, depletes antioxidants, and causes DNA damage (reviewed in Beckman 1996a). Addition of peroxynitrite to biological fluids leads to nitration of tyrosine residues, whose presence appears to be a "marker" of peroxynitrite-dependent damage *in vivo* (Beckman 1996b). Peroxynitrite also inactivates

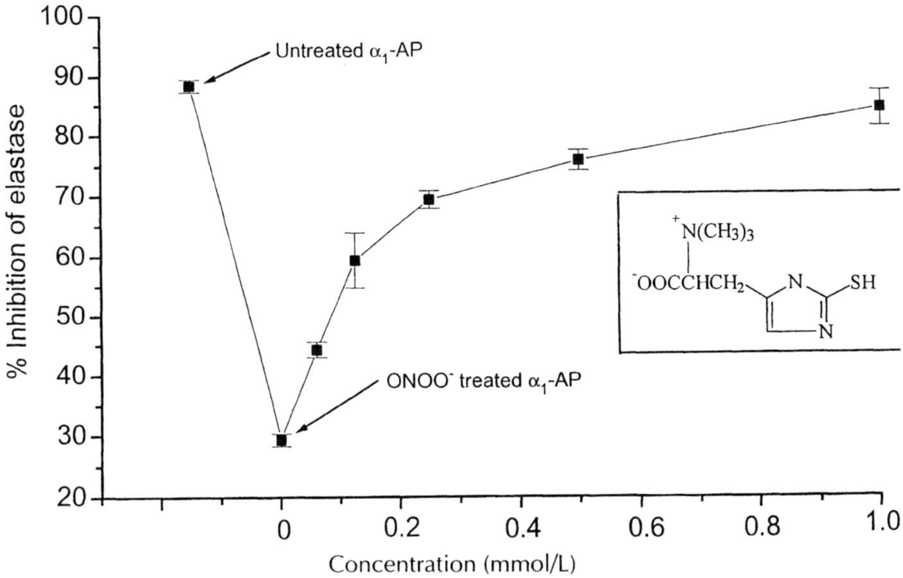

Fig. 4.2. Prevention of α_1-antiproteinase (α_1-AP) inactivation. The assay can be conducted as described by Whiteman *et al.* (1996a). Elastase, α_1-AP and ergothioneine (10 mmol/L stock) are dissolved in water. α_1-AP may also be dissolved in phosphate-buffered saline, pH 7.4 (140 mmol/L NaCl, 2.7 mmol/L KCl, 16 mmol/L Na_2HPO_4, 2.9 mmol/L KH_2PO_4) to a concentration of 4 mg/mL and elastase in the same buffer to 5 mg/mL. The volume of α_1-AP required to inhibit elastase 80–90% (typically 60–70 µL) is added to buffer (500 mmol/L K_2HPO_4/KH_2PO_4, pH 7.4) with or without 0.1 mL of compound to be tested to give a volume of 0.945 mL and incubated in a water bath at 37°C for 15 min, after the addition of peroxynitrite (typically 5 µL) to a final concentration of 0.5 mmol/L. The sample is vortexed for 10 s and incubated for 5 min. Elastase (typically 50 µL) is added and the sample is incubated at 37°C for a further 15 min, followed by addition of 2.0 mL of buffer. After 15 min, 0.1 mL of elastase substrate (SANA) is added, and the rate of reaction is followed at 410 nm for 30 s. *Source*: Aruoma *et al.* 1997.

α_1-AP, the major inhibitor of serine proteases such as elastase in human body fluids, supporting the suggestion that peroxynitrite generation *in vivo* can facilitate both oxidative and proteolytic damage.

The reaction of peroxynitrite with tyrosine (in proteins) and phenolic "antioxidant" compounds and its inactivation of the proteolytic inhibitor α_1-AP have been adopted as assays for antioxidant activity (Aruoma *et al.* 1997, Beckman *et al.* 1992, Ramezanian *et al.* 1996, Whiteman *et al.* 1996a and 1996b). This is illustrated by the action of ergothioneine, a natural antioxidant (see Fig. 4.3). Scavenging of $ONOO^-$ by antioxidants, based on assays involving tyrosine nitration and inactivation of α_1-AP, is suggested as a useful research tool that provides additional valuable information on the antioxidant profile of the biomolecules.

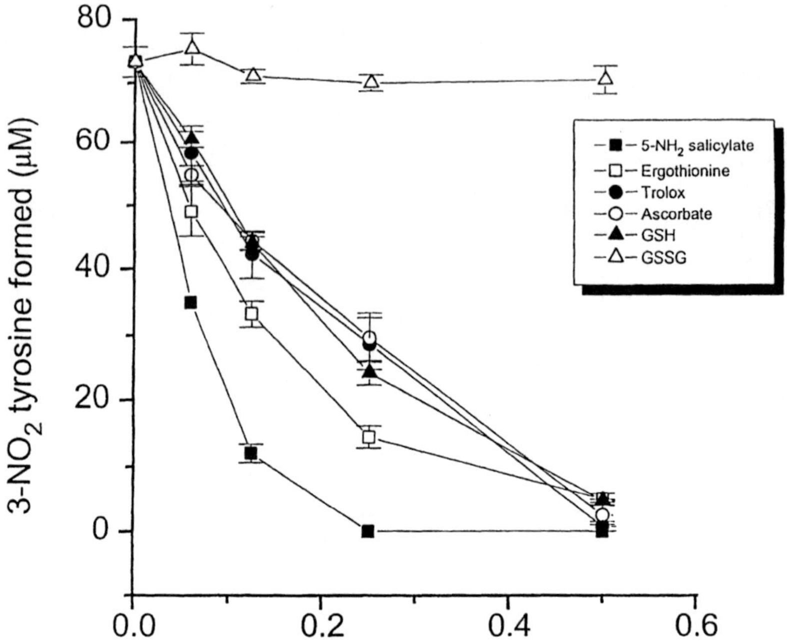

Fig. 4.3. Measurement of tyrosine nitration. This is illustrated with ergothioneine. The experiment may be conducted essentially as described. A stock concentration (10 mmol/L) of DL-tyrosine was prepared in 10 mL by adding 8 mL of water to 250 μL of 10% (wt/vol) KOH followed by 250 μL of 5% phosphoric acid with 1.5 mL of water. Tyrosine solution (0.1 mL) together with 0.1 mL of a solution of the compound to be tested was added to a plastic test tube containing 0.795 mL of buffer (500 mmol/L K_2HPO_4/KH_2PO_4, pH 7.4) and incubated in a water bath at 37°C for 15 min. After this time, peroxynitrite (typically 5 μL) was added to a final concentration of 1 mmol/L, the tubes vortexed for 15 s and incubated. Measurement of 3-nitrotyrosine can be performed using a Spherisorb 5-μm ODS2 C_{18} column with a guard column C_{18} cartridge. The suggested eluent is 500 mmol/L KH_2PO_4/H_3PO_4, pH 3.01, with 20% methanol (vol/vol) at a flow rate of 1 mL/min; the ultraviolet (UV) detector should be set at 274 nm (sensitivity 0.02) with a detection limit of 0.5 μmol/L. The 3-nitrotyrosine detected was confirmed by spiking with standards. Peak heights of tyrosine and 3-nitrotyrosine can then be measured and concentrations calculated from a standard curve. Abbreviations: GSH, glutathione; GSSG, oxidized glutathione. *Source*: Aruoma *et al.* 1997.

Indeed ergothioneine has been shown to protect cells against oxidative DNA damage by $ONOO^-$ (Aruoma *et al.* 1999).

References

Akaike, T., S. Ijiri, K. Sato, T. Katsuki, and H. Maeda, Determination of Peroxyl Radical Scavenging Activity in Food by Using Bactericidal Action of Alkyl Peroxyl Radical, *J. Agric. Food Chem. 43*: 1864–1870 (1995).

Aruoma, O.I., Deoxyribose Assay for Detecting Hydroxyl Radicals, *Methods Enzymol. 233*: 57–66 (1994).

Aruoma, O.I., Assessment of Potential Pro-Oxidant and Antioxidant Actions, *J. Am. Oil Chem. Soc. 73*: 1617–1625 (1996).

Aruoma, O.I., Extracts as Antioxidant Prophylactic Agents, *INFORM 8:* 1236–1242 (1997).

Aruoma, O.I., D. Akanmu, R. Cecchini, and B. Halliwell, Evaluation of the Ability of the Angiotensin-Converting Enzyme Inhibitor Captopril to Scavenge Reactive Oxygen Species, *Chem. Biol. Interact. 77*: 303–314 (1991).

Aruoma, O.I., P.J. Evans, H. Kaur, L. Sutcliffe, and B. Halliwell, An Evaluation of the Antioxidant and Potential Pro-Oxidant Properties of Food Additives and Trolox c, Vitamin E and Probucol, *Free Radic. Res. Commun. 10*: 143–157 (1990).

Aruoma, O.I., and B. Halliwell, Action of Hypochlorous Acid on the Antioxidant Protective Enzymes Superoxide Dismutase, Catalase and Glutathione Peroxidase, *Biochem. J. 248*: 973–976 (1987).

Aruoma, O.I., B. Halliwell, R. Aeschbach, and J. Löliger, Antioxidant and Pro-Oxidant Properties of Active Rosemary Constituents: Carnosol and Carnosic Acid, *Xenobiotica 22*: 257–268 (1992).

Aruoma, O.I., B. Halliwell, and M. Dizdaroglu, Iron Ion-Dependent Modification of Bases in DNA by the Superoxide Radical Generating System Hypoxanthine/Xanthine Oxidase, *J. Biol. Chem. 264*: 20509–20512 (1989).

Aruoma, O.I., A. Murcia, J. Butler, and B. Halliwell, Evaluation of the Antioxidant and Pro-Oxidant Action of Gallic Acid and Its Derivatives *J. Agric. Food. Chem. 41*: 1880–1885 (1993).

Aruoma, O.I., J.P.E. Spencer, J. Butler, and B. Halliwell, Reaction of Plant-Derived and Synthetic Antioxidants with Trichloromethylperoxyl Radicals, *Free Radic. Res. 22*: 187–190 (1995).

Aruoma, O.I., J.P.E. Spencer, and N. Mahmood, Protection Against Oxidative Damage and Cell Death by the Natural Antioxidant Ergothioneine, *Food Chem. Toxicol. 37:* 1043–1053 (1999).

Aruoma, O.I., M. Whiteman, T.G. England, and B. Halliwell, Scavenging of Peroxynitrite by Ergothioneine, a Non-Cytotoxic Antioxidant, *Biochem. Biophys. Res. Commun. 231*: 389–391 (1997).

Beckman, J.S., The Physiological and Pathological Chemistry of Nitric Oxide, in *The Biological Chemistry of Nitric Oxide*, edited by J. Lancaster, Academic Press, New York, 1996a, pp. 1– 83.

Beckman, J.S., Oxidative Damage and Tyrosine Nitration from Peroxynitrite, *Chem. Res. Toxicol. 9*: 836–844 (1996b).

Beckman, J.S., H. Ischiropoulos, L. Zhu, M. van der Woerd, C.D. Smith, J. Harrison, J.C. Martin, and J.H. Tsai, Kinetics of Superoxide Dismutase- and Iron-Catalysed Nitration of Phenolics by Peroxynitrite, *Arch. Biochem. Biophys. 298*: 438–445 (1992).

Brand-Williams, W., M.-E. Cuvelier, and C. Berset, Use of a Free Radical Method to Evaluate Antioxidant Activity, *Lebensm.-Wiss. Technol. 28*: 25–30 (1995).

Burger, R.M., J. Peisach, and S.B. Horwitz, Mechanism of Bleomycin Action: *in vitro* Studies, *Life Sci. 28*: 715–727 (1981).

Byers, T., and G. Perry, Dietary Carotenes, Vitamin C and Vitamin E as Protective Antioxidants in Human Cancers, *Annu. Rev. Nutr. 12*: 139–159 (1992).

Cao, G., H.M. Alessio, and R.G. Cutler, Oxygen Radical Absorbance Capacity Assay for Antioxidants, *Free Radic. Biol. Med. 14*: 303–311 (1993).

Ching, T.L., J. de Jong, and A. Bast, A Method for Screening Hypochlorous Acid Scavengers by Inhibition of the Oxidation of 5-Thio-2-nitrobenzoic Acid: Application to Anti-Asthmatic Drugs, *Anal. Biochem. 218*: 377–381 (1994).

Cuperus, R.A., A.O. Muijsers, and R. Wever, Antiarthritic Drugs Containing Thiol Groups Scavenge Hypochlorite and Inhibit Its Formation by Myeloperoxidase from Human Leukocytes, *Arthritis Rheum. 28*: 1228–1233 (1985).

Dix, T.A., and J. Aikens, Mechanism and Biological Relevance of Lipid Peroxidation, *Chem. Res. Toxicol. 6*: 2–18 (1993).

Dizdaroglu, M., O.I. Aruoma, and B. Halliwell, Modification of Bases in DNA by Copper-Ion-1,10-phenanthroline Complexes, *Biochemistry 29:* 8447–8451 (1990).

Duval, C., and M.C. Poelman, Scavenger Effect of Vitamin E and Derivatives on Free Radicals Generated by Photo-Irradiated Pheomelanin, *J. Pharm. Sci. 84*: 107–110 (1995).

Frankel, E.N., Lipid Oxidation, *Prog. Lipid Res. 19*: 1–22 (1980).

Frankel, E.N., In Search of Better Methods to Evaluate Natural Antioxidants and Oxidative Stability in Food Lipids, *Trends Food Sci. Technol. 4*: 220–225 (1993).

Giloni, L., M. Takeshita, F. Johnson, C. Iden, and A.P. Grollman, Bleomycin-Induced Strand Scission of DNA: Mechanism of Deoxyribose Cleavage, *J. Biol. Chem. 256*: 8608–8615 (1981).

Halliwell, B., M. Grootveld, and J.M.C. Gutteridge, Methods for the Measurement of Hydroxyl Radicals in Biochemical Systems: Deoxyribose Degradation and Aromatic Hydroxylation, *Methods Biochem. Anal. 33:* 59–90 (1988).

Laranjinha, J.A.N., L.M. Almeida, and V.M.C. Madeira, Reactivity of Dietary Phenolic Acids with Peroxyl Radicals: Antioxidant Activity upon Low Density Lipoprotein Oxidation, *Biochem. Pharmacol.* 48: 487–494 (1994).

Liebler, D., The Role of Metabolism in the Antioxidant Function of Vitamin E, *Crit. Rev. Toxicol. 23*: 147–169 (1993).

Lissi, E., M. Salim-Hanna, C. Pascual, and M.D. Del Castillo, Evaluation of Total Antioxidant Potential (TRAP) and Total Antioxidant Reactivity from Luminol Enhanced Chemiluminescence Measurements, *Free Radic. Biol. Med. 18*: 153–158 (1995).

Löliger, J., The Use of Antioxidants in Food, in *Free Radicals and Food Additives*, edited by O.I. Aruoma, and B. Halliwell, Taylor & Francis, London, 1991, pp. 121–150.

Marquez, L.A., and H.B. Dunford, Reaction of Compound III of Myeloperoxidase with Ascorbic Acid, *J. Biol. Chem.* 265: 6074–6078 (1990).

Miller, N.J., and C.A. Rice-Evans, Spectrophotometric Determination of Antioxidant Activity, *Redox Rep. 2*: 161–171 (1996).

Packer, J.E., T.F. Slater, and R.L. Wilson, Reactions of the CCl_4-Related Peroxy Free Radical ($CCl_3O_2^{\bullet}$) with Amino Acids: Pulse Radiolysis Evidence, *Life Sci. 23*: 2617–2620 (1978).

Petering, D.H., R.W. Byrnes, and W.E. Antholine, The Role of Redox-Active Metals in the Mechanism of Action of Bleomycin, *Chem. Biol. Interact. 73*: 133–182 (1990).

Porter, W.L., Paradoxical Behavior of Antioxidants in Food and Biological Systems, *Toxicol. Ind. Health 9*: 93–122 (1993).

Pratt, D.E., Antioxidants Indigenous to Foods, *Toxicol. Ind. Health 9*: 63–75 (1993).

Pryor, W.A., J.A. Cornicelli, L.J. Devall, B. Tait, B.K. Trivedi, D.T. Witak, and M. Wu, A Rapid Screening Test to Determine the Antioxidant Potencies of Natural and Synthetic Antioxidants, *J. Org. Chem. 58*: 3521–3532 (1993).

Pryor, W.A., T. Strickland, and D.F. Church, Comparison of the Efficiencies of Several Natural and Synthetic Antioxidants in Aqueous Sodium Dodecyl Sulfate Micelle Solution, *J. Am. Chem. Soc. 110*: 2224–2229 (1988).

Que, B.G., K.M. Downey, and A.G. So, Degradation of Deoxyribonucleic Acid by a 1,10-Phenanthroline-copper Complex: The Role of Hydroxyl Radicals, *Biochemistry 19*: 5987–5991 (1980).

Ramezanian, M.S., S. Padmaja, and W.H. Koppenol, Nitration and Hydroxylation of Phenolic Compounds by Peroxynitrite, *Chem. Res. Toxicol. 9*: 232–240 (1996).

Reich, K.A., L.E. Marshall, D.R. Graham, and D.S. Sigman, Cleavage of DNA by the Phenanthroline-Copper Ion Complex. Superoxide Mediates the Reaction Dependent on NADH and Hydrogen Peroxide, *J. Am. Chem. Soc. 103*: 3582–3584 (1981).

Rice-Evans, C., N.J. Miller, G.P. Boldwell, P.M. Bramley, and J.B. Pridham, The Relative Antioxidant Activities of Plant-Derived Polyphenolic Flavonoids, *Free Radic. Res. 22*: 375–383 (1995).

Rice-Evans, C.A., and N.J. Miller, Antioxidant Activities of Flavonoids as Bioactive Components of Food, *Biochem. Soc. Trans. 24*: 790–795 (1996).

Rose, R.C., and A.M. Bode, Biology of Free Radical Scavengers: An Evaluation of Ascorbate, *FASEB J. 7*: 1135–1142 (1993).

Rumsey, S.C., and M. Levine, Absorption, transport, and disposition of ascorbic acid in humans. *Nutr. Biochem. 9:* 116–130 (1998).

Schildermann, P.A.E.L., F. Ten Hoor, and J.C.S. Kleinjas, Induction of Oxidative DNA Damage and Early Lesions in Rat Gastro-Intestinal Epithelium in Relation to Prostaglandin H Synthase-Mediated Metabolism of Butylated Hydroxyanisole, *Food Chem. Toxicol. 33*: 99–109 (1995).

Seib, P.A., and B.M. Tolbert, *Ascorbic Acid, Metabolism, and Uses,* American Chemical Society, Washington, DC, 1982.

Shahidi, F., *Natural Antioxidants. Chemistry, Health Effects and Applications*, AOCS Press, Champaign, IL, 1997.

Shahidi, F., and J.P.K.P.D. Wanasundara, Phenolic Antioxidants, *Crit. Rev. Food Sci. Nutr. 32*: 67– 103 (1992).

Sies, H., W. Stahl, and A.R. Sundquist, Antioxidant Functions of Vitamins: Vitamin E and C, Beta-Carotene and Other Carotenoids, *Ann. N.Y. Acad. Sci. 669*: 7–20 (1992).

Sigiura, Y., T. Suzuki, J. Kuwahara, and H. Tanaka, On the Mechanism of Hydrogen Peroxide-, Superoxide-, and Ultraviolet Light-Induced DNA Cleavages of Inactive Bleomycin Iron-(III) Complex, *Biochem. Biophys. Res. Commun. 105*: 1511–1518 (1982).

Svensson, B.E., and S. Lindvall, Myeloperoxidase Oxidation of Cysteamine, *Biochem. J. 249*: 521–530 (1988).

Thomas, E.L., M.B. Grisham, D.F. Melton, and M.M. Jefferson, Evidence for a Role of Taurine in the *in vitro* Oxidative Toxicity of Neutrophils Towards Erythrocytes, *J. Biol. Chem. 260*: 3321–3329 (1985).

Travis, J., and G.S. Salvesen, Human Plasma Proteinase Inhibitor, *Annu. Rev. Biochem. 52*: 655–709 (1983).

Wasil, M., B. Halliwell, D.C.S. Hutchinson, and H. Baum, The Action of Human Extracellular Fluids: Effect of Human Serum and Its Protein Components on the

Inactivation of α_1-Antiproteinase by Hypochlorous Acid and by Hydrogen Peroxide, *Biochem. J. 243*: 219–223 (1987).

Weiss, S.J., Tissue Destruction by Neutrophils, *N. Engl. J. Med. 320*: 365–376 (1989).

Weiss, S.J., M.D. Lampert, and S.T. Test, Long-Lived Oxidants Generated by Human Neutrophils Characterization and Bioactivity, *Science 222*: 625–628 (1983).

Whiteman, M., H. Kaur, and B. Halliwell, Protection Against Peroxynitrite Dependent Tyrosine Nitration and α_1-Antiproteinase Inactivation by Some Anti-Inflammatory Drugs and by the Antibiotic Tetracycline, *Ann. Rheum. Dis. 55*: 383–387 (1996a).

Whiteman, M., H. Tritschler, and B. Halliwell, Protection Against Peroxynitrite-Dependent Tyrosine Nitration and α_1-Antiproteinase Inactivation by Oxidized and Reduced Lipoic Acid, *FEBS Lett. 379*: 74–79 (1996b).

Chapter 5

A Novel Approach for Discovering Phytopharmaceuticals

Ramesh C. Pandey, Renuka Misra, Luben K. Yankov, Bohos Mesrob, and Amlan Dutta

Xechem, Incorporated, New Brunswick, NJ 08901

Introduction

Xechem is a fully integrated pharmaceutical company dedicated to discovering new drug leads from natural sources. Screening and search for natural sources of anti-cancer compounds, antiviral agents, antibiotics active against drug-resistant bacteria, antifungal and antiprotozoal agents, immunomodulatory and antiaging/memory-enhancing agents are ongoing activities. The company is fully equipped for the extraction, fractionation, and isolation of pure compounds from medicinal plants, microbial fermentation, marine organisms, or semisynthetic processes. It also collaborates with research organizations in Europe, India, and the People's Republic of China to investigate phytopharmaceuticals from natural sources that have been used in these countries for centuries (Table 5.1).

Drug Development Rationale

Drugs derived from natural sources have a long historical track record. Ground breaking drugs from natural products that are still in use today include plant-derived drugs such as digitalin, morphine, quinine, vincristine, atropine, and paclitaxel, and drugs of

TABLE 5.1
Collaborations for Screening of Natural Products

Collaborating institute	Collection type
International Institute of Ayurveda, India	Herbal plants used in Indian Ayurvedic system of medicine
Beijing Medical University, People's Republic of China	Traditional Chinese herbal medicine
Kunming Institute of Botany, People's Republic of China	Traditional Chinese herbal medicine
Institute of Genetics, Academy of Sciences, Moldova	Isolated chemical compounds
Center for Deep-Sea Ecology and Biotechnology, Rutgers University, New Brunswick, NJ	Marine organisms

microbial origin such as penicillin, vancomycin, streptomycin, gentamycin, doxorubicin, bleomycin, amphotericin B, and cyclosporin A. Three notable anticancer drugs that have entered the market in the last few years, Taxol, Taxotere, and Topotecan, are of botanical origin. Recently, several anticancer lead compounds such as bryostatin I and didemnin B have been isolated from marine organisms.

Natural sources provide a structural diversity of compounds that cannot be matched in the laboratory. Nor is structural information required regarding a natural product to perform screening. In addition, although early assays using animal models were time consuming, labor intensive, and empirical, the advent of reliable *in vitro* screens and sensitive chemical and biochemical assay methods has allowed screening to be performed rapidly and effectively.

Although natural products research is a proven way to generate lead compounds, it has certain disadvantages compared with newer sample generation technologies such as combinatorial chemistry. The collection of novel compounds and extracts is labor intensive and requires collaborations with institutions and companies from other countries. Also, persons at the collection site must have a certain degree of knowledge about the ethnomedicinal uses of the samples and the chemical taxonomy, which are used to increase the probability of a "hit" during the screening process. In addition, even if the ethnomedicinal use is known, it is difficult to translate that knowledge into the various assay classifications such as antibacterial, antiviral, or anticancer. Usually, the samples are complex mixtures that are less amenable to direct assaying and require some sort of sample preparation. However, we believe that the advantages of natural sources outweigh their problems.

Screening Process

Figure 5.1 illustrates Xechem's screening process to identify a drug candidate of natural origin. In the next section, each of the steps will be described in more detail. Several iterations through the process may be required to identify a drug candidate.

Selection, Collection, and Initial Screening of Samples

The selection and collection of samples require a certain initial, or "rational" screening. For example, Xechem screens for classes of microbes, plants, and chemical compounds known to be sources of diverse secondary metabolites and bioactivity.

Another means of rational screening is ethnomedicinal knowledge. Ethnomedicinal practices predate modern medical practice by thousands of years. An enormous amount of information about the medicinal properties of various plant and animal extracts exists in various ethnic cultures and can be utilized in setting up successful rational screens. In a study conducted at the National Cancer Institute (NCI), Spjut and Perdue (1976) compared random screening with screening according to traditional medicinal practices. Using various literature references, they divided the species tested in NCI assays into categories based on ethnomedicinal use. They found that the frequency of finding the primary activity of plants selected according to tradi-

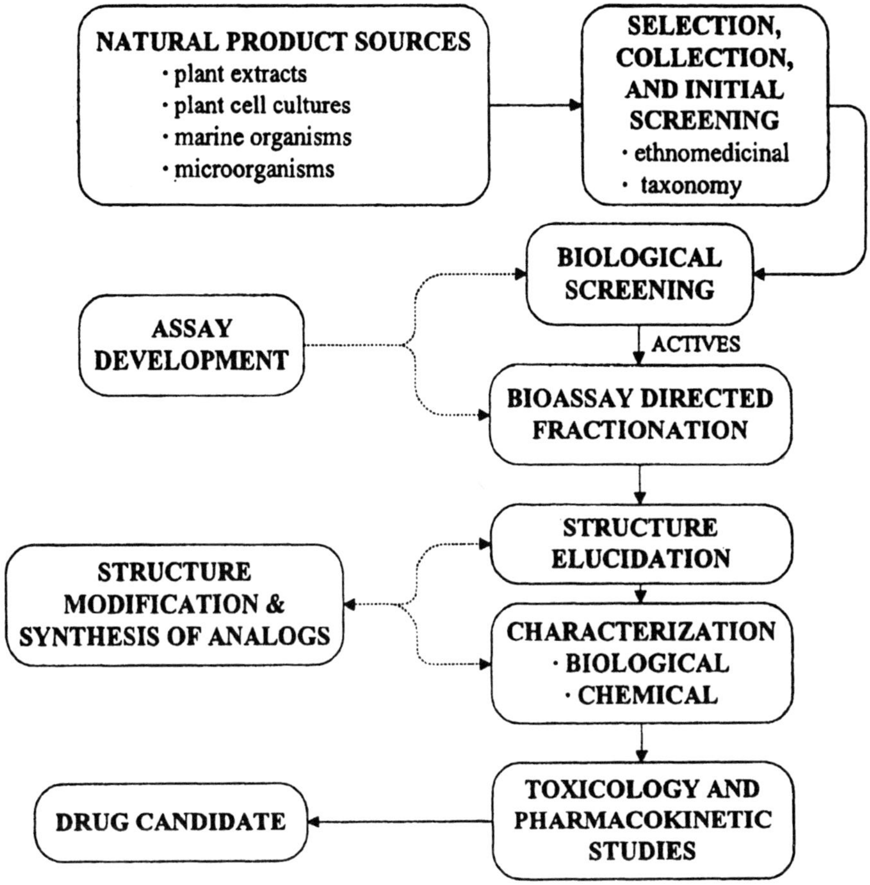

Fig. 5.1. Flowchart that shows the procedure followed at Xechem to generate a drug candidate.

tional medicinal use was 2–5 times greater than in plants selected by random screening. Figure 5.2 shows the steps followed in ethnomedicinal screening.

The general procedure for the preparation of crude plant extracts is shown in Figure 5.3. Methanol is usually used as the extractant. The methanolic extract is dried and assayed in *in vitro* assays. The active extracts are fractionated by solvent/solvent partition and/or chromatographed to identify the active fraction or compounds in the extract.

Collections set up based on taxonomic information are also being utilized for screenings. We have collaborated with The Center for Deep-Sea Ecology and Biotechnology, at Rutgers University for the screening of newly discovered and taxonomically classified marine organisms for pharmaceutical activity. We therefore have access to an enormous collection of deep-sea organisms including bacte-

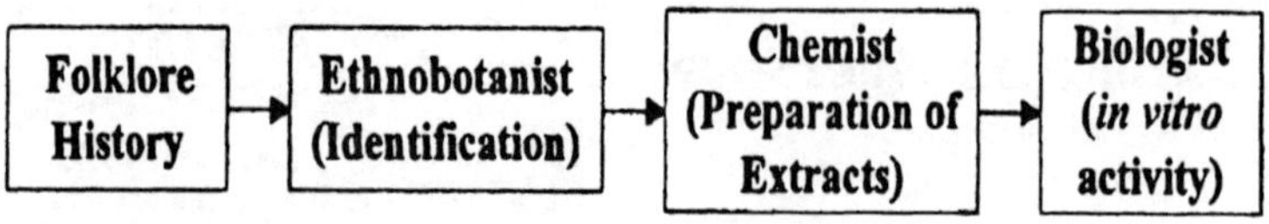

Fig. 5.2. Selection and testing of bioactive plant material.

ria, algae, vestimentaferan tubeworms, polychaete annelids, bivalve, gastropod mollusks and other lower invertebrates. Several of these organisms have been collected from Juan de Fuca Ridge (depth, 2580 m) and Alvin Dive (East Pacific Rise 9–10 N). The specific sites included deep-sea hydrothermal vents and sulfide/methane seeps in extreme conditions of temperature, pressure, and high concentrations of toxic chemicals and radionuclides. Currently, a total of 1,769 frozen samples are available from the Rutgers collection. Many of these have been classified using 16S ribosomal RNA phylogeny. Others will be classified using alternate techniques, and their nutritional requirements determined.

Those that can be grown under controlled conditions will be grown on suitable media. Because these organisms flourish at temperatures >80°C and pressure >2000 psi and are exposed to highly carcinogenic polycyclic aromatic compounds, they may have unique mechanisms by which they degrade, detoxify, and tolerate this environment. Screening this unique biota, therefore, may lead to new classes of compounds that have no counterparts in terrestrial living beings. We are also testing a collection of compounds isolated at the Institute of Genetics, Academy of Sciences, Moldova, in various bioactivity screens.

Our rational screening strategy has yielded several promising leads. The "hit ratio" was higher than expected, especially for samples of ethnomedicinal origin. Table 5.2 shows a summary of primary screening activities on a section of the samples tested.

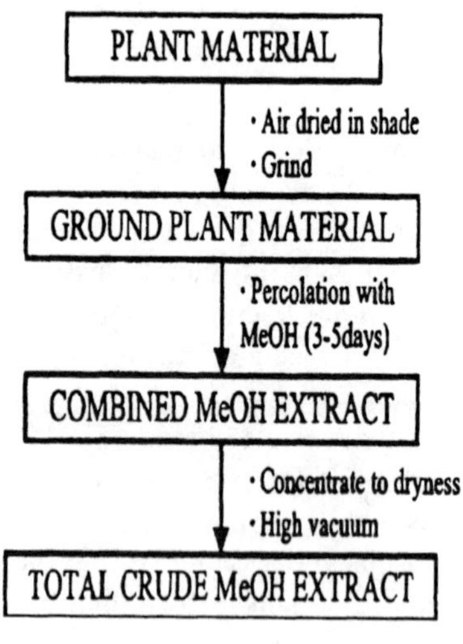

Fig. 5.3. The general extraction scheme followed to prepare crude extracts from plant material.

TABLE 5.2
Results of Antimicrobial and Antitumor Screening

Source	Number samples		Numer with promising activity		
	Crude extract	Pure compounds	Anti-bacterial	Anti-fungal	Anti-tumor
Medicinal plants					
China	100	85	23	19 (11)[a]	3
India	600[b]	45	11	9 (6)[a]	12
Isolated chemical compounds from Moldova	—	88	Under investigation		
Taxane compounds	5	30	0[c]	—	9
Marine organisms	100[b,d]	—	3	2	1

[a]Numbers in parentheses indicate compounds with antibacterial and antifungal activity.
[b]Only a small number of the samples were screened.
[c]Only anti-*Pseudomonas cepacia* activity.
[d]Of 1769 frozen organisms collected.

We also utilize plant cell cultures for screening studies. Plant cell culture–based screening provides us with a tool to investigate plants that are rare or difficult to acquire. Because the cell lines originate from a single plant organ, it is easier to investigate intraspecies and organ-specific variations in the phytochemical composition. For example, we have created a large number of plant cell lines from the *Taxus* species to search for taxanes with paclitaxel-like properties. These cell lines were screened using a proprietary technology, **E**nhanced **X**echem **I**ntegrated **S**creening **T**echnique (EXIST), a novel screening method that makes use of plant cell cultures and immiscible liquid second phases. The concept behind EXIST is shown in Figure 5.4. Phytochemicals produced by the cells are extracted *in situ* into the second phase. The

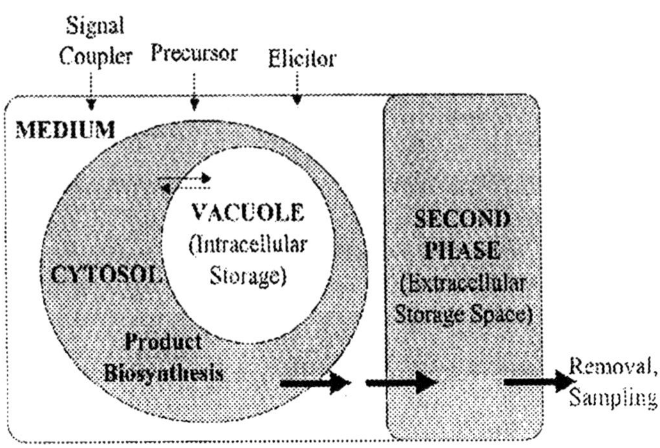

Fig. 5.4. Overall concept of **E**nhanced **X**echem **I**ntegrated **S**creening **T**echnique (EXIST).

second phase acts as an artificial accumulation site, preventing product degradation and relieving feedback inhibition in the cells, thereby enhancing the yield from the cells. Different biosynthesis inducers such as methyl jasmonate, and elicitors such as arachidonic acid, are added individually or together to generate different distributions of the phytochemicals in the second phase. The metabolites are then back-extracted from the second phase to generate crude extracts, which are assayed in various screens. The data are analyzed by a simple deconvolution program that assumes that the compounds have linear, independent effects on the assay to predict which compounds in the crude extract are responsible for the bioactivity. These compounds are targeted for purification, structure determination, and structure-activity relationship studies. Advantages of this technique are as follows:

1. Minor components with finite potency are enhanced with two-phase culture and various enhancement techniques compared with standard plant cell cultures.
2. The second phase is sampled; therefore, this is a nondestructive screening method in which the cells need not be killed to analyze the compounds produced, making the process feasible for time-course experiments using the same culture.
3. The use of production-enhancement techniques during screening provides a good basis for process optimization during scale-up, which requires large quantities of a specific compound.

Biological Screening

We have targeted a few therapeutic areas for the discovery of new lead compounds. These include screening for anticancer compounds, antiviral agents [including acquired immunodeficiency syndrome (AIDS)], antibiotics active against drug-resistant bacteria, antifungal and anti-inflammatory agents, cholesterol-lowering drugs, and antiaging/memory-enhancing agents (see Fig. 5.5).

Screening for Anticancer Activity. The crude extracts that suggest anticancer potential are assayed on a human tumor cell cytotoxicity test panel. The cytotoxicity of the extracts is determined using a microculture tetrazolium (MTT) assay. The cell line panel consists of human cancer cell lines that represent ovarian, breast, prostate, renal, and lung cancers. The basic *in vitro* system measures the ability of the test panel of various cell lines to replicate in the presence of the test compound. Controls consisting of normal diploid cells are used to determine the selective cytotoxicity of the test substance for the neoplastic cells. Samples with a median effective dose ($ED_{50)}$ $\leq$ 10 mg/mL are considered active and are subjected to further tests. We also have an ongoing cooperative program with the NCI for *in vitro* testing. Purified and well- characterized compounds are sent to the NCI for their standard 60 human tumor cell line screening (Boyd and Paull 1995).

In addition, samples are subjected to a tubulin assembly assay to identify paclitaxel-like compounds (see Fig. 5.7 for structures). The antitumor property of

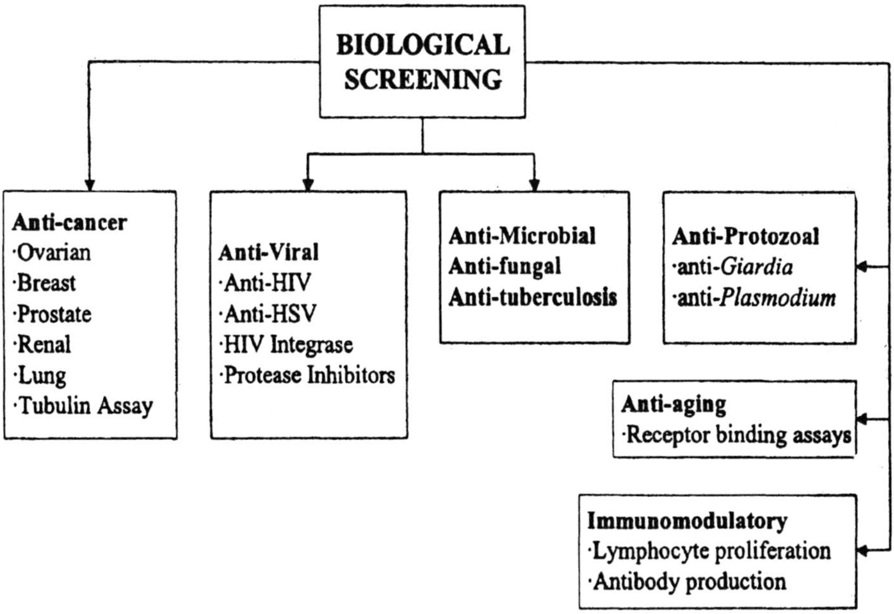

Fig. 5.5. List of therapeutic targets for the initial screening of natural products. Abbreviations: HIV, human immunodeficiency virus; HSV, herpes simplex virus.

paclitaxel is known to be due to its ability to promote tubulin assembly into microtubules. In the presence of paclitaxel, microtubules resist depolymerization, thus interfering with the G_2 and M phases of the cell cycle. The tubulin assembly assay used at Xechem is based on this property of paclitaxel (Shelanski *et al.* 1973).

Screening for Antibacterial Compounds. Crude extracts that indicate antibacterial activity are tested using the agar-well diffusion method against a wide spectrum of organism-specific and drug-resistant pathogens (listed in Table 5.3). The minimum inhibitory concentration (MIC) of the active extracts is determined using the *in vitro* Alamar Blue microplate assay. Compounds that appear promising in the

TABLE 5.3
Organism-Specific and Drug-Resistant Bacteria Targeted in Antibacterial Screens

Methicillin-resistant *Staphylococcus aureus*
Vancomycin-resistant *Enterococcus faecalis*
Penicillin-resistant *Streptococcus pneumoniae*
Pseudomonas aeruginosa
Burkholderia cepacia
Entercoccus faecalis
Escherichia coli
Multiple drug–resistant *Mycobacterium tuberculosis*

early screening are tested against a battery of fresh clinical isolates with an emphasis on the drug-resistant strains.

Screening for Antifungal Compounds. Xechem uses the National Committee for Clinical Laboratory Standards (NCCLS) method for antifungal compounds. The primary screening panel includes reference cultures of *Candida albicans, Candida glabrata,* and *Aspergillus niger.* Compounds showing activities are reexamined using reference cultures and azole-resistant strains of the fungi as shown in Table 5.4.

Screening for Central Nervous System (CNS) Active Compounds (Antiaging and Memory-Enhancing Agents). The primary screen for antiaging and memory-enhancing agents is comprised of the 40 *in vitro* receptor-binding assays of the National Institute of Mental Health (NIMH), which use primarily rat brain homogenates. The scheme used to isolate the CNS active compounds is shown in Figure 5.6.

Bioassay Directed Fractionation of Active Extracts

Once the initial screening discovers a promising active extract, we devote an intense interdisciplinary effort to identify and isolate the active compounds. If the sample is of microbial or plant cell culture origin, the fermentation group improves the strain and manipulates media and environmental parameters to produce the compounds in large amounts for determination of activity. If the sample originates from another institution, Xechem contracts with the originator for a sufficient supply for isolation and bioactivity studies. In either case, the natural products group works with the assay specialist to separate the compounds, guided at each step by bioassay data.

Structure Elucidation

After an active compound is isolated, the structure of the compound is then elucidated and fully characterized. The structural identity of the active compounds is established by using classical and modern physicochemical methods, including

TABLE 5.4
Fungal Pathogens Targeted for Screening

Molds	Yeast
Aspergillus fumigatus	*Candida albicans*
Aspergillus flavus	*Candida tropicalis*
Aspergillus niger	*Candida parapsilosis*
Cladosporium species	*Candida glabrata*
Fusarium species	*Cryptococcus neoformans*
Microsporum canis	*Histoplasma capsulatum* (dimorphic)
Trichophyton rubrum	

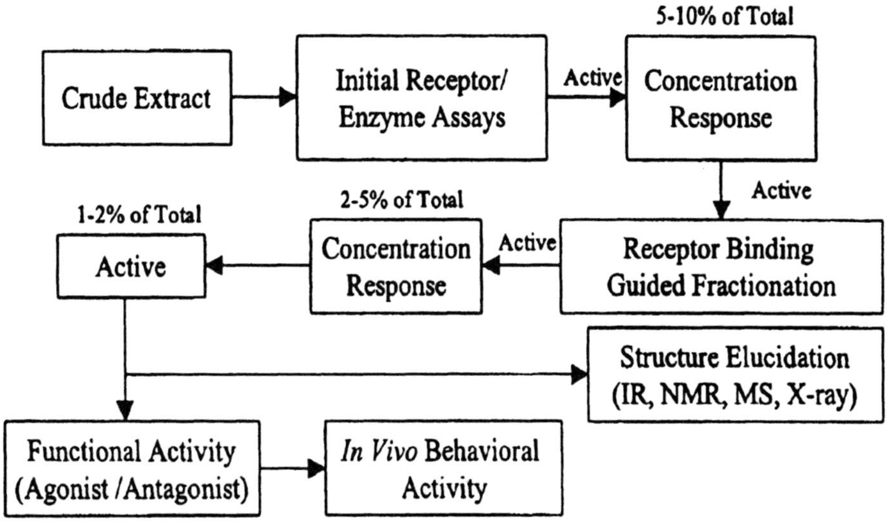

Fig. 5.6. Flowchart for screening central nervous system (CNS) active compounds from natural products. Abbreviations: IR, infrared; NMR, nuclear magnetic resonance; MS, mass spectrometry.

infrared (IR), ultraviolet (UV)-Vis, ^{1}H and ^{13}C nuclear magnetic resonance (NMR), and various mass spectral (MS) techniques, as necessary. In addition, we develop a detailed compilation of other physical and chromatographic data including melting point, elemental analysis, solubility data, thin-layer chromatography (TLC) parameters, high-performance liquid chromatography (HPLC) data, thermo gravimetric analysis (TGA), and differential scanning calorimetry (DSC). When necessary, the preliminary stability, solubility, and analytical studies of materials produced, as well as preparation of analytical reference standards, are conducted in-house.

Biological and Chemical Characterization

The bioactivity of the compound is tested in a variety of *in vitro* and *in vivo* biological assays. The potency and the specificity toward a particular disease marker are determined. The characterization process can also provide a good idea of the toxicity profile of the active compound.

In vivo testing is conducted on the purified compounds to confirm the therapeutic potential shown in the *in vitro* studies. We collaborate with the Southern Research Institute in Birmingham, Alabama to conduct the *in vivo* studies in animal models.

Enmeshed with the characterization process are chemical activities such as generating an impurity profile in which the minor constituents of the purified chemical compound are characterized and the structure-activity relationship of the compound is studied. This is done by modifying the structure of the compound and studying its bioactivity. Structure modification can be accomplished by synthesis or, if the compounds are of cell culture origin, biologically by modifying the media or changing the precursors. The specific methods to be used for the preparation of various analogs depend upon the given target compound. Such modifications of structure are important because, although the lead compound may have a critical characteristic, it usually will not fit the ideal profile of a new drug. The drug candidate will result from a rational redesign of the lead compound.

Toxicology and Pharmacokinetic Studies

Ideally, a drug candidate should exert a therapeutic effect in a dose range that has no toxic effect on the body. A critical element of drug development studies is to assess the safety of the drug candidate in a series of acute and subacute toxicity studies. Simultaneously, studies on the metabolic fate of the drug and the mechanism of action are conducted. The pharmacokinetic profile of the drug is generated with the determination of various parameters that indicate the drug absorption, distribution, excretion, and the concentration of the drug in various organs.

After completion of safety assessment studies, all of the data are evaluated to determine whether the drug is promising enough to merit clinical studies. If it is, then an Investigational New Drug Application (INDA) is prepared and filed with the Food and Drug Administration (FDA). The study then moves from the test tubes and animal studies to the clinic. Various phases of clinical studies are conducted to ascertain the safety and efficacy of the drug in humans, resulting finally in the preparation and filing with the FDA of a New Drug Application (NDA). Due to the great time and expense associated with developing a drug candidate, a continuous economic evaluation is required to assess whether continued development of the drug candidate is economically justifiable.

Specific Illustration

An example of the drug development process is given to illustrate some of the aspects of the drug development process at Xechem. A primary goal of Xechem is to develop a generic version of paclitaxel [(**1**), Fig. 5.7], an anticancer drug for the treatment of breast and ovarian cancers. In our search for a second generation of potent taxoid

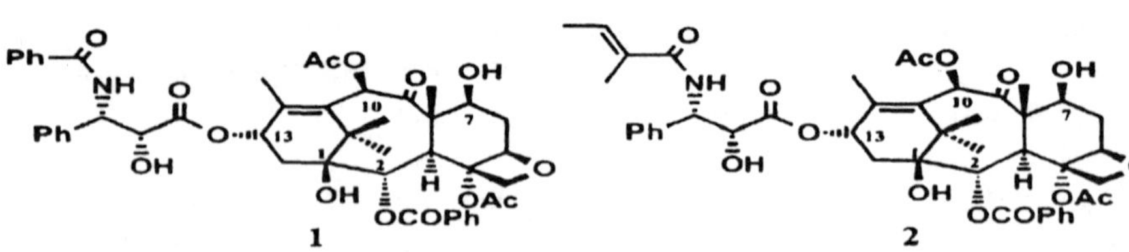

1 **2**

Fig. 5.7. The chemical structures of paclitaxel (**1**) and cephalomannine (**2**).

drugs with improved solubility characteristics, higher potency, and enhanced ability to overcome or escape drug resistance, we have found that new drug candidates can be generated by modifying cephalomannine (**2**), which is structurally similar to paclitaxel with some of the same cytotoxic properties. The raw material for paclitaxel also contains large quantities of cephalomannine. The taxane ring of cephalomannine is identical to that of paclitaxel. The difference between the two is in the side chain. In the case of paclitaxel, the amino group in the phenyl isoserine side chain is benzoylated, whereas in cephalomannine, the amino group is condensed with (*E*)-2-methyl-2-butenoic acid (tiglated). Epimerized paclitaxel and cephalomannine also exist in the crude raw material because paclitaxel and cephalomannine can be epimerized at carbon-7, either thermally or in the presence of a catalyst (aprotonic, acid or base).

During structure modification studies, we were able to synthesize a number of promising anticancer agents when we halogenated the side chain double bond of cephalomannine. The dichloro- and dibromo-analogs of cephalomannine and 7-*epi*-cephalomannine showed significant paclitaxel-like activity in the *in vitro* tests conducted in the NCI 60-human tumor cell line panel. Further *in vitro* and *in vivo* evaluation of the analogs was conducted at the Southern Research Institute to substantiate the cytotoxic activities of the analogs. The structures of some of the active analogs are shown in Figure 5.8. The compounds were prepared by selective chlorination or bromination of the side chain double bond of cephalomannine and 7-*epi*-cephalomannine in a controlled environment. The individual stereomers were separated and their structures elucidated by ^{13}C and ^{1}H NMR and electron ionization (EI) and fast atom bombardment (FAB^+) mass spectrometry (Pandey *et al.* 1998, Yankov *et al.* 1996).

The *in vitro* and *in vivo* evaluation of the halogenated analogs indicated that they exhibited a pattern of multilog differential cytotoxicity similar to that of paclitaxel. As an example, the 50% growth inhibition (GI_{50}) and the total growth inhibition (TGI) data from the NCI 60 human tumor cell line panel of one of the brominated analogs, XCLY 401759, is correlated with the assay results of other anticancer drugs with known mechanisms of action. The correlation data are shown in Tables 5.5 and 5.6. Note the high correlation of the XCLY 401759 GI_{50} and TGI data with the paclitaxel data, implying that the differential cytotoxic effect of the analog is similar to paclitaxel, and that XCLY 401759 probably has the same mechanism of action as that of paclitaxel. Additional *in vitro* studies indicated that the dependence of cytotoxicity on the duration of exposure was similar to that of paclitaxel. The *in vitro* tests also indicated a wider therapeutic range of the analogs in comparison to paclitaxel. Other advantages included higher solubility in 50% Cremophor and 50% ethanol vehicle, the current delivery system for paclitaxel. XCLY-401759 was tested *in vivo* in animal models at the Southern Research Institute. The anticellular efficacy of the compound was demonstrated on several neoplastic tumor cell lines that had proved to be most sensitive in *in vitro* testing, MX-1 (breast), NCI-H522 (non-small cell lung), RXF 393 (renal), and OVCAR-3 (ovarian), which were subcutaneously implanted in mice.

The *in vivo* data are summarized in Table 5.7. The $\log_{10}$ cell kill by the XCLY 401759 was >1 on the MX-1, NCI-H522, and OVCAR-3 subcutaneous tumors.

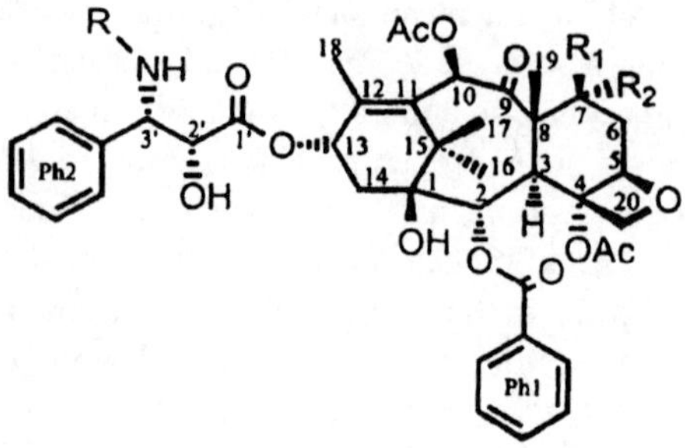

	R	R_1	R_2		R	R_1	R_2
3	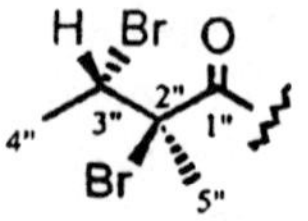	H	OH	**7**	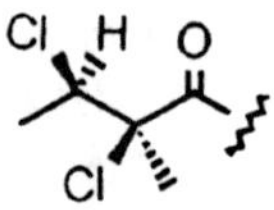	OH	H
4	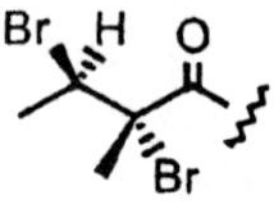	H	OH	**8**	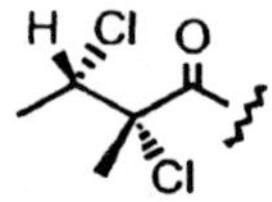	OH	H
5	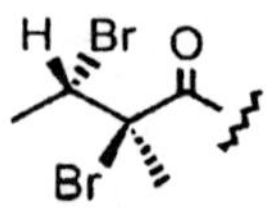	OH	H	**9**	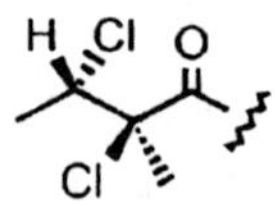	OH	H
6	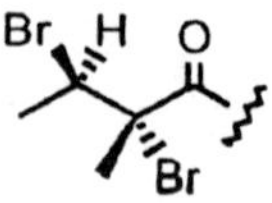	OH	H	**10**	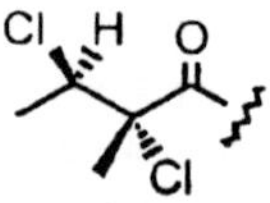	OH	H

Fig. 5.8. Chemical structures of some of the active taxane analogs developed at Xechem.

The percentage of growth delay was substantial in all three cases. It was >200% in all three cases and 400% for the MX-1 breast cancer tumors.

Several U.S. and International Patents are currently pending with five U.S. patents issued which are directed to the halogenated analogs for use as anticancer agents (Pandey and Yankov 1997). Current research at Xechem is targeted toward filing for an IND on the analogs in the near future.

Conclusions

In summary, we are actively developing drug candidates based on available biodiversity to develop natural products as drugs. Natural products screening is being carried out in collaboration with various international and national institutions. The

TABLE 5.5
Correlation of the GI_{50} Data of XCLY 401759 in the NCI 60 Cell Line Assay with the Assay Results of Other Anticancer Drugs with Known Mechanisms of Action[a,b]

	Chemical name	NSC test number	LCONC	Max X	Correlation coefficient	n
1	Paclitaxel	125973	–4.60	21	0.825	60
2	MDR Rhod30	999991	0.00	1	0.811	10
3	Vinblastine sulfate	49842	–5.60	127	0.755	60
4	Actinomycin D	3053	–6.60	71	0.713	60
5	Phyllanthoside	328426	–5.60	19	0.699	60
6	Bisantrene HCL	337766	–3.60	10	0.686	60
7	Macbecin II	330500	–3.30	12	0.663	59
8	Bruceantin	165563	–3.70	14	0.618	60
9	Chromomycin A3	58514	–4.00	8	0.604	60
10	Deoxydoxorubicin	267469	–3.70	13	0.590	60
11	*S*-Trityl-L-cysteine	83265	–3.90	15	0.586	60

[a]Abbreviations: GI_{50}, 50% growth inhibition; NCI, National Cancer Institute; NSC, National Service Center; LCONC, log of highest concentration tested; Max X, number of tests; n = number of cell lines tested.
[b]The log of the highest concentration of XCLY 401759 tested is –4.0 M.

natural product collections fall into three broad-based categories, i.e., plants, microbes, and marine organisms. These categories are mutually complementary, i.e., products isolated from any one category are unlikely to be found in the others. Ethnomedicinal and taxonomic information is utilized primarily as a rational basis for screening. We are interested in searching for anticancer compounds, antiviral

TABLE 5.6
Correlation of the TGI data of XCLY 401759 in the NCI 60 Cell Line Assay with the Assay Results of Other Anticancer Drugs with Known Mechanisms of Action[a,b]

	Chemical name	NSC test number	LCONC	Max X	Correlation coefficient	n
1	Paclitaxel	125973	–4.60	20	0.830	59
2	Vinblastine sulfate	49842	–5.60	128	0.727	59
3	Rhizoxin	332598	–9.00	9	0.605	59
4	Maytansine	153858	–4.00	15	0.598	59
5	Vincristine sulfate	67574	–3.00	62	0.527	59
6	Macbecin II	330500	–3.30	12	0.501	59
7	Phyllanthoside	328426	–5.60	19	0.493	59
8	*S*-Trityl-L-cysteine	83265	–3.90	15	0.484	59
9	Bactobolin	325014	–3.65	11	0.451	59
10	Ccnu	79037	–3.30	58	0.430	59
11	Pancratiastatin	349156	–3.65	11	0.422	59

[a]Abbreviations: TGI, total growth inhibition; NCI, National Cancer Institute; NSC, National Service Center; LCONC, log of highest concentration tested; Max X, number of tests; n = number of cell lines tested.
[b]The log of the highest concentration of XCLY 401759 tested is –4.0 M.

TABLE 5.7
Effect of XCLY-401759 Administered Intravenously Against Subcutaneously Implanted Human Tumor Xenografts[a]

Tumor	Dosage[b] (mg/kg)	Drug deaths	Growth delay[c] %(T – C)/C	Log $_{10}$ cell kill[d]	Day 55 tumor-free survivors
MX-1	160.0	2	403	1.7	1
(Breast)	106.7	0	277	1.1	1
	71.1	0	245	0.9	0
NCI-H522	120.0	0	188	1.3	1
(Non-small	80.0	0	179	1.2	0
cell lung)	53.3	0	>229	>1.6	0
RXF 393	160.0	0	63	0.0	0
(Renal)	106.7	0	32	–0.2	0
	71.1	0	22	–0.3	0
OVCAR-3	160.0	1	>350	>2.1	1
(Ovarian)	106.7	0	255	1.5	0
	71.1	0	329	2.0	0

[a]Female athymic mice were implanted with tumor fragments on d 0; n = 10 for treated controls and n = 5 for every other group.
[b]Treatment was initiated on d 6, 7, 6, and postimplant for the MX-1, NCI-H522, RXF 393, and OVCAR-3 models, respectively, when all mice in each study had small but measurable tumors.
[c]An evaluation size of two mass doublings was used for all experiments. Control tumors attained that size in 6.2, 14.9, 6.0, and 12.2 d for the MX-1, NCI-H522, RXF 393, and OVCAR-3 models, respectively.
[d]The initial doubling time for the control tumors was 3.7, 5.6, 2.7, and 5.5 d for the MX-1, NCI- H522, RXF 393, and OVCAR-3 models, respectively.

agents (including AIDS), antibiotics active against drug-resistant bacteria, antifungal and antiprotozoal agents, immunomodulatory agents, and antiaging/memory-enhancing agents. The drug discovery process is an interdisciplinary effort, which is facilitated by a flexible and technologically innovative organization.

Acknowledgments

The authors thank Professor Sun Han-Dong of the Kunming Institute of Botany, Professor Zhang Li-Hie of the Beijing Medical University, Professor Sukh Dev, Professors Richard A. Lutz and Robert C. Vrijenhoek of The Center for Deep-Sea Ecology and Biotechnology, Rutgers University, the International Institute of Ayurveda, and the Institute of Genetics, Academy of Sciences, Moldova, for providing collections of natural products to the screening program and the National Cancer Institute, the National Institute of Mental Health, and the Southern Research Institute for assistance in *in vitro* and *in vivo* testing of compounds. We are also grateful to Dr. Saroj Mishra for helpful suggestions and critical review of the manuscript.

References

Boyd, M.R., and K.D. Paull, Some Practical Considerations and Applications of the National Cancer Institute *in vitro* Anticancer Drug Discovery Screen, *Drug Dev. Res. 34*: 91 (1995).

Pandey, R.C., and L.K. Yankov, Isolation and Purification of Paclitaxel from Organic Matter Containing Paclitaxel, Cephalomannine and Other Related Taxanes, U.S. Patent 5,654,448 (1997).

Pandey, R.C., L.K. Yankov, A. Poulev, R. Nair, and S. Caccamese, Synthesis and Separation of Potential Anticancer Active Dihalocephalomannine Diastereomers from Extracts of *Taxus yunnanensis*, *J. Nat. Prod. 61*: 56–63 (1998)

Shelanski, M.L., F. Gaskin, and C.R. Cantor, Microtubule Assembly in the Absence of Added Nucleotides, *Proc. Natl. Acad. Sci. USA 70*: 765–770 (1973).

Spjut, R.W., and R. E. Perdue, Plant Folklore: A Tool for Predicting Sources of Antitumor Activity, *Cancer Treat. Rep. 60*: 979–985 (1976).

Yankov, L.K., R. Nair, A. Poulev, A. Dutta, and R.C. Pandey, Novel Efficient Process for the Purification of Paclitaxel and Preparation of Its Analogues with Anticancer Activity, The Monroe E. Wall Symposium, Harnessing Biodiversity for Therapeutic Drugs and Foods: Developing Products for the 21st Century, New Brunswick, NJ, June 2–5, 1996.

Chapter 6

Variables Affecting the Phytochemical Contents of Garlic and Its Health Benefits

Harunobu Amagase[a], Hiromichi Matsuura[b], and Yoichi Itakura[b]

[a]Department of Research and Development, Wakunaga of America Company, Mission Viejo, CA 92691

[b]Institute for OTC Research, Wakunaga Pharmaceutical Company, Hiroshima 739-11, Japan

Introduction

It has long been known that extraction increases the potency and bioavailability of various crude herbs, and eliminates their harsh and toxic characteristics. The irritating, acidic, and oxidizing compounds in raw garlic can also be eliminated by extraction. Because many adverse reactions to garlic ingestion can be attributed to its harsh and irritating oil-soluble constituents derived from allicin, in some cultures in fact, garlic is soaked or extracted with alcohol, wine, milk, or vinegar and used therapeutically.

Although there are many types of garlic preparations available commercially, they fall into only three categories, i.e., dehydrated garlic powder, garlic oil, or garlic extract. Because garlic contains various naturally occurring enzymes, once it has been ruptured, a cascade of chemical reaction occurs. Under the cooking or manufacturing processes, the highly unstable and reactive allylic thiosulfinates, including allicin, are transformed into various compounds in a short time period. Other chemical constituents are also stabilized during the process. However, because the structure of chemical constituents in garlic is so complex, the final number of chemical constituents in each garlic preparation is quite different and is dependent upon the process (Block 1992, Iberl *et al.* 1990).

Garlic has been shown to exhibit cardioprotective hypolipidemic and antiplatelet activity (Kleijnen *et al.* 1989, Lau 1987, Randerson 1993, Steiner *et al.* 1996a and 1996b) and to improve peripheral circulation (Kikuchi *et al.* 1994, Okuhara 1994). Beyond these attributes, aged garlic extract (AGE) has been reported to possess hepatoprotective (Dion and Milner 1996, Horie *et al.* 1989 and 1992, Ide *et al.* 1996a and 1996b, Lau *et al.* 1991, Nagai and Yamawaki 1974, Nakagawa *et al.* 1985 and 1988, Tadi *et al.* 1991, Yamasaki *et al.* 1991), immune-enhancing (Abdullah *et al.* 1989, Geng *et al.* 1997, Hirao *et al.* 1987, Kandil *et al.* 1987, Lau 1989, Lau *et al.* 1991, Morioka *et al.* 1993, Nagai 1973, Reeve *et al.* 1993, Tadi *et al.* 1990), anticancer, and chemopreventive activities (Amagase and Milner 1993, Amagase *et al.* 1996, Dorant 1993, Lau *et al.* 1986, Nishino *et al.* 1989, Reeve *et al.* 1993, Riggs *et al.* 1997). Furthermore, AGE exhibits antioxidative activities, whereas other garlic preparations have been shown to stimulate oxidation (Imai *et al.* 1994). Preparations such as AGE, which are uniquely designed to contain small amounts of the oil-soluble constituents and a rich array of the water-soluble constituents such as *S*-allyl cysteine (SAC), are much less toxic than other forms of garlic (Nakagawa *et al.* 1980). AGE and its related preparations have been confirmed to be safe by various toxicologic studies (Imada 1990, Kanezawa *et al.* 1984, Nakagawa *et al.* 1980, 1984a and 1984b, Sumiyoshi *et al.* 1984, Yoshida *et al.* 1984) and by a long history of human consumption. These pharmacologic effects of AGE may be due to the conversion products that are formed during the aging process.

In this paper, we investigate and review the effects of several kinds of garlic preparations on physiologic and psychological stress in mice; we examine the antioxidative and free radical–scavenging effects of various preparations with the use of the chemiluminescence and 1,1-diphenyl-2-picrylhydrazyl (DPPH) assays *in vitro* and in mice, immunological factors *in vitro* and in humans, pentane release into the mice breath, and peripheral microcirculation in healthy human subjects by thermography.

Samples

Garlic preparations used in this study were raw garlic juice (RGJ), heated garlic juice (HGJ), processed (dehydrated) garlic powder (PGP), and aged garlic extract (AGE). Heating deactivates various enzymes in the intact garlic bulb. One of the enzymes in garlic, i.e., alliinase, cannot convert alliin into various compounds including allicin if the intact garlic bulb is heated. When garlic is processed without heat, enzymatic conversion takes place during the process and yields greater quantities of SAC and stable compounds than when garlic is heated.

RGJ was prepared from skinned raw garlic cloves that were crushed in a Waring blender (Waring Products Division, Dynamics Corporation of America, New Hartford, CT, USA) for 1 min together with an equal weight of water. The mixture was then allowed to stand for 30 min at 25°C. After filtration through cheese cloth, the raw garlic sample was obtained. The contents of allicin and alliin in RGJ, analyzed by the method of Imai *et al.* (1994), were 0.21 and 0%, respectively. The extract content of RGJ was 17.2%.

HGJ was prepared from skinned raw garlic cloves that were soaked in hot water at 95°C for 30 min. The preparation methods for HGJ were identical to those used in preparing RGJ. HGJ samples contained 18.4% extract content, 0.266% alliin, and 0.0005% allicin.

AGE, manufactured under a license issued by the Ministry of Health Welfare of Japan, is formulated by Wakunaga Pharmaceutical Company (Osaka, Japan) by soaking sliced garlic in aqueous ethanol and extracting over a 10-mo period at room temperature. The resulting end product contains 15.1% extract content and 0.0258% alliin, but no allicin. Although HGJ contains ~10 times more alliin than AGE, SAC concentration in HGJ is ~20 times less than that of AGE. Allicin was not detected in either preparation. The purchased PGP powder contained 0.418% alliin, but no allicin.

Antistress Effects

Four kinds of stress models in mice, including a forced swimming test (Tadano *et al.* 1992), mechanical treadmill running test, an immobilization stress test, and a cooling/rewarming test (Karasawa *et al.* 1984) were used to assess physiologic and psychological stress (Ushijima *et al.* 1997).

Animals. Male ddY or ICR mice were purchased from Japan SLC (Hamamatsu, Japan). They were kept in an air-conditioned room (23 ± 1°C in temperature and 55 ± 5% humidity) and given a conventional diet (CE-2, Kurea Japan, Tokyo, Japan) and water *ad libitum*.

Forced Swimming Test. The forced swimming test was conducted in accordance with the method of Tadano *et al.* (1992). Five-week-old ddY mice were plunged individually into a vertical glass cylinder (height 30 cm, diameter 15 cm) filled to a height of 15 cm with water maintained at 20°C for 5 min, and the duration of immobility was measured. Immediately after the mice were removed from the water, garlic samples were administered orally using a stomach tube and the mice were allowed to dry for 60 min at room temperature (23 ± 1°C). Then, the mice were again placed into the cylinder for 5 min and the extent of immobility was quantified. A mouse was judged to be immobile when it ceased struggling and remained floating motionless in the water, making only those movements necessary to keep its head above water.

Each active swimming period was calculated by subtracting the time of immobility from the total time (5 min) in the cylinder. The antistress effect was expressed by the ratio of the active swimming period after garlic administration to the active swimming period before garlic administration.

RGJ, HGJ, and AGE were administered at dosages of 2 and 4 mL/kg; PGP was dissolved in distilled water and administered at dosages of 300 and 600 mg/kg.

Mechanical Treadmill Running. Seven-week-old ddY mice were made to run on a mechanical treadmill apparatus (model KN-73, Natsume Seisakusho, Tokyo, Japan). The apparatus consisted of a wide, endless belt, riding on metal rollers. An acrylic box, partitioned into 10 individual compartments of 10 cm × 57 cm, was suspended to cover the belt, providing a limited running area for each mouse. Motivation was provided by an electric shock needle plate at the rear of each compartment. The belt was set at a 5% incline, running at a constant speed of 30 m/min (mice were pretrained by running for 2 d, 7 and 4 min each day). The mice were deprived of food overnight, and garlic samples were administered orally on the next day with the use of a stomach tube. RGJ, HGJ, and AGE were administered at 5 mL/kg; PGP was dissolved in distilled water and administered at 750 mg/kg.

Thirty minutes after the administration of samples, running times were measured under the conditions described above.

Immobilization Stress. Nine-week-old ICR mice were individually held in a polystyrene tube (length 12 cm, diameter 3 cm, FALCON 2070, Becton Dickinson, Franklin Lakes, NJ) with an air hole for 16 h per day (6:00 P.M.–10:00 A.M.) and placed in home cages for the remaining 8 h with food and water *ad libitum.* Mice were exposed to immobilization stress for 2 d. With the use of a stomach tube, garlic samples were administered to mice for 10 consecutive days before the mice were placed back into polystyrene tubes for a second time. RGJ, HGJ, and AGE were administered at dosages of 2 and 10 mL/kg; PGP was dissolved in distilled water and administered at dosages of 300 and 1500 mg/kg. After the second stress exposure, the animals were killed and both adrenal glands were dissected and weighed.

Cooling/Rewarming Test. The cooling/rewarming test was performed according to the method of Karasawa *et al.* (1984). After overnight food deprivation, 5-wk-old ddY mice were placed at constant room temperature (20°C) for 30 min to stabilize their rectal temperature. After the rectal temperature of the mice was measured with a digital thermometer (model DTM-900, M.G. Giken, Tokyo, Japan), garlic samples were administered orally using a stomach tube. RGJ, HGJ, and AGE were administered at dosages of 2 and 10 mL/kg; PGP was dissolved in distilled water and administered at dosages of 300 and 1500 mg/kg.

After administration of garlic samples, mice were placed in a restraint cage (90 × 28 mm inner dimension) and the lower part of their bodies was immersed in 15°C water for 10 min. The mice were then gently wiped with a dry towel, and rectal temperature was measured every 10 min for 60 min.

Antioxidation

Oxidation, especially lipid peroxidation associated with the oxygen radical, has been considered a major cause of cancer, atherosclerosis, and the aging process (Greenstock 1993, Rice-Evans and Diplock 1993). Garlic and its constituents have demonstrated antioxidant and radial-scavenging properties, suggesting prevention of these diseases (Kitahara *et al.* 1974, Horie *et al.* 1989 and 1992, Yamasaki *et al.* 1994, Numagami *et al.* 1996, Geng *et al.* 1997, Ide *et al.* 1997, Ide and Lau 1997a and 1997b). Two kinds of *in vitro* studies, low-level chemiluminescence and hydrogen peroxide (Ide *et al.* 1997, Ide and Lau 1997a and 1997b), and an *in vivo* study (Ide *et al.* 1996a and 1996b), pentane release into the mice breath, were used to evaluate the antioxidant effects of AGE and its constituents.

Low-Level Chemiluminescence (LLC). The emission of LLC from the liver microsomal fraction initiated by *t*-butyl hydroperoxide was utilized to measure the oxidative condition. *S*-Alk(en)yl-L-cysteines were prepared by alk(en)ylation of L-cysteine with alk(en)yl bromide. Standard allicin was prepared and purified by silica gel column chromatography [stepwise elution of *n*-hexane-ethyl acetate (9:1) and (4:1)] followed by preparative high-performance liquid chromatography (HPLC) [column: TSK-GEL ODS80T_M (Toso, Tokyo, Japan) (21 mm × 30 cm), mobile phase: 30% MeCN] and was kept at –80°C until analysis. *S*-Allylmercapto-cysteine (SAMC) was synthesized as previously described (Block *et al.* 1984, Stoll and Seebeck 1951, Yoshimura 1958). Other chemicals were of analytical grade and purchased commercially.

Determinations were carried out on a Shimadzu LC-6A HPLC system (Shimadzu, Kyoto, Japan). The contents of allicin and ajoene were measured by an HPLC method [column: TSK-GEL ODS80T_M (4.6 mm × 15 cm); mobile phase, 40% CH_3CN (0.8mL/min); analytical wavelength, 254 nm] without pretreatment of the garlic preparations. Alliin and *S*-allylcysteine (SAC) were determined by ion-exchange chromatography [column, TSK-GEL ODS80T_M (4.6 mm × 15 cm); mobile phase, 21:5 mmol/L sodium citrate buffer (pH 3.25) for alliin; 53.7 mmol/L sodium citrate buffer (pH 4.0) for SAC; column temperature, 40°C] followed by postderivatization using o-phthalaldehyde (OPA) (Wako Pure Chemical Industries, Osaka, Japan) (fluorescent detection, excitation 340 nm; emission 455 nm). The test sample for the analysis of SAC was prepared by the addition of 5% trichloroacetic acid.

Liver microsomes were prepared from male SD-strain rats (6- to 10-wk-old). After the liver microsomal fraction was washed in 1.15% KCl solution, it was resuspended in the same solution and then stored at –80°C until use. The protein concentration of the microsomal fraction was determined by the Bio-Rad (Bio-Rad Laboratories, Hercules, CA, USA) protein assay kit and adjusted to a concentration of 0.8 mg/mL.

The emission intensity of LLC was measured by a chemiluminescence analyzer equipped with a CL-10 counter and a CL-100 detector (Tohoku Electronic Industries, Sendai, Japan). The chemiluminescence circular cell (diameter: 50 mm, height: 10 mm) was kept at 37°C and the chemiluminescence intensity was measured and recorded every 10 s. The procedures were carried out in an air-saturated condition. The mixture of 3 mL of 0.1 mol/L potassium phosphate buffer (pH 7.4), containing the microsomal fraction (0.8 mg as protein), and 0.2 mL of test sample solution [final concentration of sample: 0.15% (wt/vol) of garlic preparations and 5 and 10 mmol/L of pure compounds] was incubated for 5 min; the reaction was initiated by the addition of 0.8 mL of *t*-butyl hydroperoxide (final concentration: 2 mmol/L). The emission intensity was measured for a total of 15 min including 5 min of preincubation. The test samples and *t*-butyl hydroperoxide were dissolved in 0.1 mol/L phosphate buffer, pH 7.4.

Hydrogen Peroxide. The scavenging activity of AGE was observed in the fraction eluted with ammonia through cation-exchange chromatography. Malondialdehyde that was produced during lipid peroxidation was measured by that method.

Pentane Release. An injection of carbon tetrachloride, a potent prooxidant in the liver, causes pentane in the breath of mice. AGE was administered to mice orally before a carbon tetrachloride injection to evaluate the efficacy of its antioxidant activity. Pentane in the breath of mice was determined by gas chromatography (GC).

Peripheral Microcirculation

Healthy male subjects (n = 12) were divided randomly into two groups aged 22 through 42 y (average 31.5 y) and 34 through 43 y (average 39.8 y) (Okuhara 1994). Each group was crossed over after an interval of at least 2 d. Blood flow to the hands and feet was assessed by thermogram both before and after administration of 1.6 mL of either AGE or HGJ.

The six subjects in the continuous administration test were divided randomly into two groups, the AGE group and the HGJ group, which were crossed over after an interval of at least 2 d, becoming the HGJ group and the AGE group, respectively; 0.8 mL of AGE or HGJ was administered in gelatin capsules twice a day after meals, over a period of 14 d.

On the basis of the report (Nasu 1984) that the skin temperatures of healthy subjects are stable at room temperatures of 24–26°C, the skin temperature measurements were taken at a room temperature of 24–25°C. All skin temperature measurements were taken in the morning, using a Nihon Denki Sanei Thermotracer TH 1100 (NEC Sanei Instruments, Ltd., Tokyo, Japan); the subjects were acclimated beforehand by exposing their limbs for 30 min in a room at 24–25°C so that skin temperatures were stabilized. Measurements were taken in the morning before administration and at 30, 60, 90, and 120 min in the case of single administration; in the case of continuous administration, measurements were made before administration and after 14 d.

Measurements were taken, and average skin temperatures were calculated within the areas of the backs of the hands and the backs of the feet. The testing period was from January through May, 1994.

Results and Discussion

Organosulfur compounds in garlic cloves have been shown to transform into a number of different compounds during processing (Iberl *et al.* 1990). Furthermore, different methods of processing result in different chemical constituents in garlic preparations. The garlic preparations used in these studies contained different organosulfur compounds. The water extract of raw garlic contains mainly allicin for a short time period. The water extract of heat-treated garlic contains mainly alliin. Heating deactivates many characteristics such as various enzyme activities in the intact garlic bulb, including that of alliinase, which is an essential enzyme to convert alliin into various compounds through allicin. Many stable water-soluble organosulfur compounds, including *S*-alk(en)ylcysteines, their sulfoxides, and γ-glutamyl derivatives in AGE were found. In addition, a variety of oil-soluble organosulfur compounds have been identified and quantified in AGE by Weinberg *et al.* (1993). AGE was developed to be superior in benefits to raw garlic and without any odor or irritating side effects. A natural aging process that not only maximizes the benefits of garlic, but eliminates garlic's pungent odor and harsh side effects was used for this process. The manufacturing process, comparative studies, and clinical studies indicate the superior benefits and differences of AGE over other garlic products. Because AGE is processed without heating, enzymatic conversion takes place during the process, yielding greater quantities of SAC and stable compounds than HGJ. Although HGJ contains ~10 times more alliin than AGE, SAC concentration in HGJ is ~20 times less than in AGE. Allicin was not detected in either preparation. In these studies, many beneficial effects were found in the allicin-free garlic preparation. Therefore, allicin is not necessary or essential for garlic to exhibit its benefits in various areas. According to analytical results, all types of garlic products available today, namely, garlic oil, powder, and aged garlic/aged garlic extract products, each produced by a different manufacturing process, contain different beneficial ingredients and some similar components, yet in different amounts. There is also variation within the groups, i.e., the constituents of one brand of garlic powder product may not be the same as those of the others. The resulting compounds in a garlic product depend on how the garlic is processed. Therefore, all manufacturers should have both efficacy and safety data on their specific products, not on garlic in general. The high-quality manufacturing process, as well as numerous comparative studies from scientific journals, has confirmed the superiority of the AGE over other forms of garlic in terms of both safety and efficacy.

Davies *et al.* (1982) reported that the free radical concentration was elevated in muscles after treadmill running, and Brady *et al.* (1979) showed that forced swimming induced an increase of lipid peroxide in both muscle and liver. HGJ and PGP showed insignificant effects on both swimming, treadmill running, and the cooling/rewarming times. The second active swimming period was reduced by 64% from the initial dunking in the control mice. In mice given 2 mL/kg of RGJ, the active swimming period was prolonged by 84.1%, whereas a dose of 4 mL/kg of RGJ produced no measurable effect. The active swimming period of mice given HGJ was not influenced at a dose of 2 mL/kg, and was slightly prolonged at a dose of 4 mL/kg, although it was not significantly different. The active swimming period for mice given AGE was prolonged by 82.3 and 90.0% at a dosage of 2 and 4 mL/kg, respectively. No significant effect on the active swimming period was demonstrated in mice given PGP at either dose. Although these two forms of garlic do not supply oxygen to the muscles, AGE significantly prolonged the active swimming period, the treadmill running time, improved distal circulation, and inhibited reduction of rectal temperature. The running time of the control mice was 929 s. The running time of the mice given RGJ was not influenced compared with the controls. The running time of the mice given HGJ and PGP was slightly, but not significantly prolonged. The running time of the mice given AGE was prolonged to 1611 s.

The rectal temperatures of the mice immersed in water at 15°C for 10 min dropped from 35.0 to 19.7°C. After removal from water, the rectal temperature of the mice gradually recovered to 27.4°C over 60 min. The recovery of rectal temperatures in the mice given RGJ was delayed significantly compared with that of the control mice. The recoveries of rectal temperatures in mice given HGJ and PGP were not influenced. However, the recovery of rectal temperatures in mice given AGE was significantly accelerated in a dose-dependent manner. Furthermore, the drop in rectal temperatures by cool water immersion in the mice given the AGE at 10 mL/kg was significantly inhibited. RGJ and HGJ enhanced the emission, whereas AGE demonstrated an inhibitory effect. This result suggests that RGJ and HGJ cause oxidative damage to cell membranes. The damaging effect of RGJ might be due to one or several oxidative components, such as allicin. AGE was found to be effective for both swimming and running. Its effects were likely to be due mainly to its antioxidant properties.

Among the above garlic preparations, only AGE has antioxidant effects, such as decreasing the light emission of the liver microsomal fraction induced by *t*-butyl hydroperoxide. AGE exhibited a significant decrease in the emission of LLC from the liver microsomal fraction initiated by *t*-butyl hydroperoxide, whereas it was stimulated by RGJ, HGJ, and PGP. Both tables 6.1 and 6.2 show the percentage of increase/decrease of light intensity evaluated at the time of the maximum emission. The percentage for RGJ and HGJ could not be calculated due to the continuous increase of chemiluminescence. AGE decreased the maximum emission by 30% at the concentration of 0.15% (wt/vol). Among a variety of organosulfur compounds, *S*-allylcysteine (SAC) and *S*-allylmercapto-L-cysteine (SAMC) also decreased the emission of LLC (Table 6.2). AGE was found to inhibit the emission of LLC and the early formation of thiobarbituric acid-reactive substances (TBARS). The fact that AGE inhibited emissions indicates that it inhibited the formation of free radicals. RGJ and HGJ, on the other hand, enhanced such emissions, demonstrating the formation of free radicals or oxidative activity. Further, AGE is rich in SAC and SAMC, which have demonstrated potent antioxidant and radical-scavenging activities of their own. A laboratory analysis using the same method as above has shown that eight of the top garlic products on the market enhance oxidative activity, whereas AGE shows antioxidant activity. The formation of TBARS at the time of maximum emission was decreased to 51% in the presence of AGE compared with that in phosphate buffer as a control. However, no significant suppression was observed at the end point (after 15 min incubation). Of

TABLE 6.1
Inhibition of Low-Level Chemiluminescence by Garlic Preparations

Sample	Inhibition (%)[a]
Aged garlic extract	30.7 ± 8.7
Water extract of raw garlic	−110.7 ± 22.7

[a]Inhibition percentage was calculated by [1− (Etm − Etp) (Ecm − Ecp)] × 100. Etm and Etp are the emission intensities of test samples at maximum emission and preincubation. Ecm and Ecp are the emission intensities of the control at maximum emission and preincubation. Sample concentration: 0.15% (wt/vol); n = 5.

TABLE 6.2
Inhibition of Low-Level Chemiluminescence by Water-Soluble Organosulfur Compounds Derived from Garlic

	Inhibition (%)			
Sample	5 mmol/L	n	10 mmol/L	n
S-Methyl-L-cysteine	36.8 ± 9.8	4	46.0 ± 11.9	4
S-Propyl-L-cysteine	28.0 ± 4.7	4	32.8 ± 6.9	4
S-(1-Propenyl)-L-cysteine	34.1 ± 9.5	4	48.5 ± 5.2	4
S-Allyl-L-cysteine	33.4 ± 10.2	8	45.4 ± 7.2	11
γ-Glutamyl-*S*-methyl-L-cysteine	23.2 ± 6.0	4	31.9 ± 4.9	4
γ-Glutamyl-*S*-propyl-L-cysteine	21.5 ± 6.6	4	29.0 ± 6.2	4
γ-Glutamyl-*S*-allyl-L-cysteine	25.4 ± 5.1	4	36.5 ± 10.0	6
(+) Alliin	36.9 ± 8.6	4	50.4 ± 10.8	4
Methiin	14.9 ± 1.4	3	28.3 ± 6.9	3
S-Allylmercapto-L-cysteine	60.0 ± 5.0	3	82.3 ± 2.3	3
Glutathione	70.3 ± 4.0	3	80.7 ± 1.8	3

samples in this study, the inhibition of SAMC was the strongest and nearly the same as that of glutathione (used as a reference compound) at the time of maximum emission. *S*-Alk(en)ylcysteines showed greater inhibition compared with their substituted γ-glutamyl derivatives. AGE, SAC, and SAMC were shown to scavenge hydrogen peroxide *in vivo*, and AGE had a stronger scavenging effect than RGJ. When the corresponding fraction of AGE was compared with the water extracts of raw garlic, the scavenging activity of the AGE fraction was found to be significantly stronger than that of the water extract.

Pentane evolution in the breath, which is caused by lipid peroxidation and was determined by GC, was also diminished by oral treatment with AGE. Oral consumption of AGE before carbon tetrachloride injection significantly reduced the amount of pentane in the breath by 80% compared with that of the control mice 2 h after injection. Because pentane is derived from oxidized fatty acids, these data indicate that AGE was effective in suppressing lipid peroxidation caused by carbon tetrachloride in mice. Imai *et al.* (1994) also reported the antioxidant effects of AGE in comparison with RGJ, and HGJ using the LLC method accompanied by lipid peroxidation. When there is an acute exercise load, energy is produced aerobically or anaerobically, depending on the intensity of the exercise. In the forced swimming test, an aerobic exercise, the supply of oxygen was sufficient to produce energy without lactic acid formation in the muscle (Nagai *et al.* 1992). In contrast, hard exercise such as treadmill running is aerobic at first, but subsequently becomes anaerobic and lactic acid formation increases (Hill 1914). Under anaerobic conditions, blood circulation becomes important for the supply of oxygen to and excretion of lactic acid from the muscles. Although RGJ demonstrated an antistress effect under aerobic conditions at a low dosage of 2 mL/kg, this effect may be reduced at 4 mL/kg, and no effect was observed at 5 mL/kg under the anaerobic condition of treadmill running. RGJ may reduce the antistress effect under aerobic conditions by its inhibitory effect on the supply of oxygen. It is well known that raw garlic induces anemia (Katsunuma 1932, Kuzutani 1934, Miyamoto 1935, Nakagawa *et al.* 1980) and gastrointestinal dysfunction (Kodera 1994, Nakagawa *et al.*1980, Shashikanth *et al.* 1986, Subrahmanyan *et al.* 1958). The lack of effect of RGJ under anaerobic conditions could be due to its deteriorating effect on the distal circulation, which would hinder the delivery of oxygen to and the removal of lactic acid from the muscles, causing them to reach threshold faster and become fatigued.

In addition, the efficacy of AGE may also benefit from its ability to stimulate energy metabolism as suggested earlier. Immobilization stress is generally used to evaluate a stress effect. This stress induces hypertrophy of the adrenal gland, which is one of the triad of changes that occur in a stressed body (Selye 1936). This change is induced by stimulation of the pituitary-adrenocortical axis functions. Of the four garlic preparations used in this study, only AGE prevented hypertrophy of the adrenal gland. The average adrenal weight of the mice exposed to the immobilization stress was increased to 5.93 mg in comparison with that of unstressed mice, 4.63 mg. No significant effect was observed in the mice given RGJ, HGJ, or PGP, whereas the stress-induced adrenal hypertrophy was significantly inhibited in the mice given AGE at 10 mL/kg. This finding is probably closely related to the effects found in the other three tests. It has been shown that, of the four garlic preparations tested, only AGE exhibited effectiveness on both physiologic and psychological stresses. It was concluded that AGE possesses a superior "nutritional tonic effect," capable of maintaining homeostasis of the body.

AGE functions as an antioxidant that quenches free radicals, suggested to be the cause of aging, heart/cardiovascular disease, cancer, and other diseases (Greenstock 1993, Rice-Evans and Diplock 1993). AGE also suppresses the formation of TBARS at the maximum emission point, the early stage; the chemiluminescence is generated from the self-reaction of *t*-BuOO•. Because it is generated from the self-reaction of *t*-BuOO• at the late stage and breaks down free radicals and lipid peroxides, the result is the formation of TBARS. Thus, it suggests AGE could mainly scavenge *t*-BuOO•. AGE also possess an antirancidity ability to protect cell membranes and an overall ability to quench free radicals, whereas other garlic preparations, including raw garlic, stimulate the oxidative condition (Yamasaki *et al.* 1994). It also protects lymphocytes from irradiation damage, whereas fresh garlic worsened conditions to a greater degree than in an irradiated control (Lau 1989). AGE also protects liver microsomal membranes from chemical (lipid peroxide formation) and physical (decrease of membrane fluidity) changes induced by lipid peroxidation and protects vascular endothelial cells from oxidant injury upon exposure to hydrogen peroxide. The antioxidant activity of AGE may be responsible for its cardioprotective effect against the anticarcinogenic drug, Doxorubicin, which has been shown to induce strong oxidative damage to the cardiac muscle (Kojima *et al.* 1994). Both AGE and SAC have been shown to protect cells that line blood vessels from oxidant injury caused by hydrogen peroxide (Yamasaki *et al.* 1994). The antioxidant activity may be indirectly useful for retardation of the aging process and for prevention of cancer and atherosclerosis (Greenstock 1993, Moriguchi 1993, Rice-Evans and Diplock 1993). The two major organosulfur compounds in AGE, i.e., SAC and SAMC, showed radical-scavenging activity in both chemiluminescence and DPPH assays. These findings are closely related to the previously reported antihepatotoxic activity of these compounds, i.e., SAC and SAMC prevented CCl_4-induced cytotoxicity in isolated liver cells and in the mouse liver (Nakagawa *et al.* 1985 and 1988).

Furthermore, because oral administration of SAC and its analogs makes it bioavailable (Nagae *et al.* 1994), it could be effective for scavenging radicals in the body; thus, these compounds may play an important role in the antioxidative activity of AGE (Nagae *et al.* 1994). It has been reported that the maximum blood concentration of SAC was reached 60 min after oral administration in animals (Nagae *et al.* 1994). Because superior bioavailability of SAC after oral administration has been reported, both AGE and SAC may be effective for prevention of diseases induced by oxidative damage through their antioxidant and radical-scavenging activities.

AGE has been found to be more beneficial to blood than other forms of garlic. Test tube samples (Freeman and Kodera 1995) have shown that raw garlic and dehydrated raw garlic powders oxidize red blood cells, causing them to irreversibly loose their oxygen-carrying capacity, whereas AGE causes no such oxidation.

AGE and other antioxidative vitamins and minerals, such as vitamin A and selenium, may have a synergistic anticarcinogenic effect on carcinogen-induced mammary cancer (Amagase *et al.* 1996).

AGE provides nutrients and a protein fraction, which may be conducive to the growth of beneficial bacteria. These bacteria (*Bifidobacterium bifidum* and *Lactobacillus acidophilus*) produce both acid and antibacterial factors, which have been suggested to decrease the growth of pathogens and their production of toxic and carcinogenic compounds.

Because garlic has been commonly used for centuries as a tonic, and in folk medicine for the treatment of lumbago, arteriosclerosis, and sensitivity to cold, it is thought that it may be effective in improving blood circulation and in inhibiting platelet aggregation. AGE has been confirmed to raise skin temperatures clinically and to improve such symptoms as numbness of the limbs and chilling of the limbs and lumbar area (Kikuchi *et al.* 1994). When garlic was tested against serum lipids (Lau 1987, Steiner *et al.* 1996a and 1996b) and fatigue among athletic students (Kawashima *et al.*1986), the results suggested that it promotes blood circulation and the discharge of fatigue products from the blood. Those who seem to be medically normal nonetheless suffer from indefinable discomfort or chills in some specific part of their bodies, especially during the winter. Moreover, a majority of these people suffer chills in more than one location, and, not surprisingly, skin temperatures are lower in those locations. Skin temperature is affected by a variety of factors, but the most important factor in determining skin temperature is the blood flow directly below the surface of the skin. In particular, skin blood flow in the peripheral extremities of the limbs is said to operate only with the object of thermoregulation. The skin temperatures of the AGE group were 28.63°C before administration and 29.68°C after 30 min, 29.90°C (peak temperature) after 60 min, 29.52°C after 90 min, and 28.75°C after 120 min. Thus, temperatures rose by 1.05°C after 30 min and by 1.27°C after 60 min; these results were significant ($P < 0.01$ and $P < 0.05$, respectively) compared with temperatures before administration. The skin temperatures of the HGJ group were 28.83°C before administration and 29.50°C (peak temperature) after 30 min, 29.40°C after 60 min, 28.73°C after 90 min, and 28.72°C after 120 min. Thus, temperatures were higher by 0.67°C after 30 min and by 0.57°C after 60 min, but these differences were not significant. The skin temperatures of the AGE group were 22.62°C before administration and 23.13°C after 30 min, 23.25°C after 60 min, 23.77°C (peak temperature) after 90 min, and 22.88°C after 120 min. Thus, temperatures were higher by 0.51, 0.63, and 1.15°C after 30, 60, and 90 min, respectively.

The skin temperatures of the HGJ group were 23.12°C before administration and 23.32°C (peak temperature) after 30 min, 23.28°C after 60 min, 22.93°C after 90 min, and 23.00°C after 120 min. Thus, temperatures were higher by 0.20 and 0.16°C after 30 and 60 min, respectively. At no time after administration were the temperatures of either group significantly different from those before administration. Finally, there was no significant difference between the two groups in terms of changes in skin temperature after a single administration. Skin temperatures of the AGE group were 30.52°C before administration but 31.05°C after 14 d of administration, a rise of 0.53°C; the difference between the values before and after administration was significant ($P < 0.05$). The skin temperatures of the HGJ group were 30.92°C before administration and 30.62°C after 14 d of administration, a drop of 0.30°C. The skin temperatures of the AGE group were 26.62°C before administration and 28.98°C after 14 d of administration, a rise of 2.36°C; the difference between the values before and after administration was significant ($P < 0.01$). The skin temperatures of the HGJ group were 26.32°C before administration and 26.72°C after 14 d of administration, a rise of 0.40°C, but the difference was not significant. Differences were found between the two groups in the skin temperatures of the hands and feet after 14 d of administration ($P < 0.1$ for both). In the single administration test, the skin temperatures of the backs of the hands in the AGE group were higher 30 min after administration and reached a peak after 60 min, an increase in temperature that was significant at each stage compared with the values before administration. The temperatures then decreased gradually after that. Skin temperatures in the HGJ group peaked quickly at 30 min after administration and subsequently decreased gradually (after 60 min). The skin temperatures on the dorsal side of the feet in the AGE group rose gradually, peaked at 90 min after administration, and had decreased by the next measurement. Skin temperatures in the HGJ group rose slightly by 30 and 60 min after administration, and then declined slightly. In the continuous administration test, the skin temperatures on the dorsal side of the hands were significantly higher after 14 d of administration in the GE group, but were slightly lower in the GEC group. The skin temperatures on the dorsal side of the feet were clearly higher in the AGE group after 14 d of administration and also somewhat higher in the HGJ group. Test subjects in both groups also reported no particular feelings of cold in their feet during administration of either medicine. The HGJ used in the tests is a liquid preparation of heat-treated garlic, which apparently differs somewhat from garlic in its constituents (Ichikawa 1993, Matsuda 1991). It is not yet known what connection, if any, exists between skin temperatures and any active component of AGE. Above all, however, the peak in skin temperatures of the backs of the hands which occurred 60 min after administration in the AGE group is thought to have resulted because the active component, such as SAC of the AGE, reached its maximum blood concentration after 60 min; thus, there was a clinical improvement in blood flow resulting mainly from the effects of the AGE.

AGE is extracted from garlic through a long-term aging process. Its constituents include proteins, sugars, amino acids, sulfur compounds, lipids, vitamins, and enzymes; it is used as an ingredient in tonics and thus can be used for a period of time. Judging also by the results of the single administration and the clinical results reported earlier, the beneficial effect that continuous administration of AGE had on skin temperatures in this study is thought to have occurred because it continuously improved blood flow in the peripheral blood vessels that densely populate the dorsal side of the hands and feet; it also enhances microcirculatory function because it affects homeostasis. It is presumed that AGE and HGJ had different effects because of the difference in constituents between the two caused by the different methods of manufacture.

In conclusion, AGE is the best garlic preparation available, providing all of the benefits of garlic without undesirable side effects. These studies demonstrate that the conversion products (SAC or other constituents) formed during the process may be responsible for the various pharmacologic activities. In today's society, people suffer from many types of stresses and degenerative diseases, such as cardiovascular disease. AGE may prevent many of these conditions and symptoms.

References

Abdullah, T.H., D.V. Kirkpatrick, and J. Carter, Enhancement of Natural Killer Cell Activity in AIDS with Garlic, *J. Oncol. 21*: 52–53 (1989).

Amagase, H., and J. Milner, Impact of Various Sources of Garlic and Their Constituents on 7,12-Dimethylbenz[a]anthracene Binding to Mammary Cell DNA, *Carcinogenesis 14:* 1627–1631 (1993).

Amagase, H., E.M. Schaffer, and J.A. Milner, Dietary Components Modify the Ability of Garlic to Suppress 7,12-Dimethylbenz[a]anthracene-induced Mammary DNA Adducts, *J. Nutr. 126:* 817–824 (1996).

Block, E., The Organosulfur Chemistry of the Genus *Allium*. C. Implications for the Organic Chemistry of Sulfur, *Angew. Chem. Int. Ed. Engl. 31*: 1135–1178 (1992).

Block, E., S. Ahmad, M.K. Jain, R.W. Crecely, R. Apitz-Castro, and R. Cruz, (*E,Z*)-Ajoene: A Potent Antithrombic Agent from Garlic, *J. Am. Chem. Soc. 106*: 8295–8296 (1984).

Brady, P.S., L.J. Brady, and D.E. Ullrey, Selenium, Vitamin E and the Response to Swimming Stress in the Rat, *J. Nutr. 109*: 1103–1109 (1979).

Davies, K.J.A., A.T. Quintanihilha, G.A. Brooks, and L. Packer, Free Radicals and Tissue Damage Produced by Exercise, *Biochem. Biophys. Res. Commun. 107*: 1198–1205 (1982).

Dion, M.E., and J.A. Milner, Formation and Bioactivation of Nitrosomorpholine Is Inhibited by *S*-Allyl Cysteine, *FASEB J. 10:* A498 (1996).

Dorant, E., Garlic and Its Significance for the Prevention of Cancers in Humans: A Critical View, *Fr. J. Cancer 67*: 424–429 (1993).

Freeman, F., and Y. Kodera, Garlic Chemistry: Stability of *S*-(2-Propenyl)-2-propene-1-sulfinothioate (Allicin) in Blood, Solvents, and Simulated Physiological Fluids, *J. Agric. Food Chem. 43*: 2332– 2338 (1995).

Geng, Z., Y. Rong, and B.H.S. Lau, *S*-Allyl Cysteine Inhibits Activation of Nuclear Factor Kappa B in Human T Cells, *Free Radic. Biol Med. 23*: 345–350 (1997).

Greenstock, C.L., Radiation and Aging: Free Radical Damage, Biological Response and Possible Antioxidant Intervention, *Med. Hypotheses 41*: 473–482 (1993).

Hill, A.V., The Oxidative Removal of Lactic Acid, *J. Physiol. 48*: x–xi (1914).

Hirao, Y., I. Sumioka, S. Nakagami, M. Yamamoto, S. Hatono, S. Yoshida, T. Fuwa, and S. Nakagawa, Activation of Immunoresponder Cells by the Protein Fraction from Aged Garlic Extract, *Phytother. Res. 1*: 161–164 (1987).

Horie, T., S. Awazu, Y. Itakura, and T. Fuwa, Identified Diallyl Polysulfides from an Aged Garlic Extract Which Protects the Membranes from Lipid Peroxidation, *Planta Med. 58*: 468–469 (1992).

Horie, T., T. Murayama, T. Mishima, F. Itoh, Y. Minamide, T. Fuwa, and S. Awazu, Protection of Liver Microsomal Membranes from Lipid Peroxidation by Garlic Extract, *Planta Med. 55*: 506–508 (1989).

Iberl, E., G. Winkler, and K. Knobloch, Products of Allicin Transformation: Ajoenes and Dithiins, Characterization and Their Determination by HPLC, *Planta Med. 56*: 202–211 (1990).

Ichikawa, M., Wakunaga Pharmaceutical Co., *Materials 225* (1993).

Ide, N., and B.H.S. Lau, Garlic Compounds Inhibit Low Density Lipoprotein (LDL) Oxidation and Protect Endothelial Cells from Oxidized LDL-Induced Injury, *FASAB J. 11*: A122 (1997a).

Ide, N., and B.H.S. Lau, Garlic Compounds Protect Vascular Endothelial Cells from Oxidized Low Density Lipoprotein-Induced Injury, *J. Pharm. Pharmacol. 49*: 908–911 (1997b).

Ide, N., H. Matsuura, and Y. Itakura, Scavenging Effects of Aged Garlic Extract and Its Constituents on Active Oxygen Species, *Phytother. Res. 10*: 340–341 (1996a).

Ide, N., A.B. Nelson, and B.H.S. Lau, Aged Garlic Extract and Its Constituents Inhibit Cu^{+2}-Induced Oxidative Modification of Low Density Lipoprotein, *Planta Med. 63*: 263–264 (1997).

Ide, N., K. Ryu, J. Imai, T. Moriguchi, S. Nagae, H. Sumiyoshi, H. Matsuura, and Y. Itakura, Antioxidant and Radical Scavenging Effects of Aged Garlic Extract and Its Constituents, 2nd International Congress on Phytomedicine, September 11–14, 1996, Munich, Germany, 1996b.

Imada, O., Toxicity Aspects of Garlic, in *First World Congress on the Health Significance of Garlic and Garlic Constituents*, Nutrition International, Irvine, CA 92619–0632, 1990, p. 47.

Imai, J., N. Ide, S. Nagae, T. Moriguchi, H. Matsuura, and Y. Itakura, Antioxidant and Radical Scavenging Effects of Aged Garlic Extract and Its Constituents, *Planta Med. 60*: 417–420 (1994).

Kandil, O.M., T.H. Abdellah, and A. Elkadi, Garlic and the Immune System in Humans: Its Effects on Natural Killer Cells, *Fed. Proc. 46*: 441 (abstr.) (1987).

Kanezawa, A., S. Nakagawa, H. Sumiyoshi, K. Masamoto, H. Harada, S. Nakagami, S. Date, A. Yokota, M. Nishikawa, and T. Fuwa, General Toxicity Tests of Garlic Extract Preparation (KYOLEOPIN) Containing Vitamins, *Oyo Yakuri (Appl. Pharmacol.) 27*: 909–929 (1984).

Karasawa, H., S. Sakai, T. Sato, and M. Kato, Effects of the Medical Alcoholic Beverage (Yomeishu) on Rectal and Skin Temperature of Experimental Animals Subjected to Cooling Rewarming Test, *Oyo Yakuri-Pharmacometrics (Appl. Pharmacol.) 281*: 155–161 (1984).

Katsunuma, S., Effect of Garlic on Anaemia, *Exp. Med. 18*: 442–444 (1932).

Kawashima, Y., Y. Ochiai, and H. Shuzenji, Anti-Fatigue Effect of Aged Garlic Extract in Athletic Club Students, *Kiso To Rinsho (Preclin. Clin. Rep.) 20*: 8229–8245 (1986).

Kikuchi, N., Y. Nishimura, C. Tsukamoto, Y. Kawashima, H. Ochiai, H.Y. Ayashi, and I. Fujisaki, Clinical Effects of Leopin-5 (LE-5) and Garlic Extract (GE) on Peripheral Microcirculation, *Shinyaku To Rinsho (Jpn. J. New Remedies Clin.) 43*: 146–158 (1994).

Kitahara, S., S. Takahara, and K. Futamura, Peroxidase Activity in Garlic Product (Kyoleopin), *Kiso To Rinsho (Preclin. Clin. Rep.) 8*: 115–116 (1974).

Kleijnen, J., P. Knipschild, and G. Ter Riet, Garlic, Onions, and Cardiovascular Risk Factors. A Review of the Evidence from Human Experiments with Emphasis on Commercially Available Preparations, *Br. J. Clin. Pharmacol. 28*: 535–544 (1989).

Kodera, Y., Dietary Tolerance/Absorption/Metabolism of Garlic Phytochemicals, presented at *Designer Foods III Symposium: Research Update on Phytochemicals in Garlic, Soy and Licorice,* Washington, DC, May 1994.

Kojima, R., Y. Toyama, and S.T. Ohnishi, Protective Effects of an Aged Garlic Extract on Doxorubicin-Induced Cardiotoxicity in the Mouse, *Nutr. Cancer 22*: 163–173 (1994).

Kuzatani, S., Effects of Garlic (*Allium scorodoprasum* L.) on Anemia, *Clin. Pathol. Hematol. 3*: 1175–1233 (1934).

Lau, B.H.S., Effect of an Odor-Modified Garlic Preparation on Blood Lipids, *Nutr. Res. 7*: 139–149 (1987).

Lau, B.H.S., Detoxifying, Radio-Protective and Phagocyte-Enhancing Effects of Garlic, *Int. Clin. Nutr. Rev. 9*: 27–31 (1989).

Lau, B.H.S., J.L. Woolley, C.L. Marsh, G.R. Barker, D.H. Koobs, and R.R. Torrey, Superiority of Intralesional Immunotherapy with *Corynebacterium parvum* and *Allium sativum* in Control of Murine Transitional Cell Carcinoma, *J. Urol. 136*: 701–705 (1986).

Lau, B.H.S., T. Yamasaki, and D.S. Gridley, Garlic Compounds Modulate Macrophage and T-Lymphocyte Functions, *Mol. Biother. 3*: 103–107 (1991).

Matsuda, K., An Investigation of the Value of Scordinine as an Indicator in Processed Garlic, and a Method of Measuring Processed Garlic Using Aliin as the Indicator, *Tomita Drug Research Yearbook 18*: 117–127 (1991).

Miyamoto, T., Effects of Garlic on Hemograms, *J. Manchurian Med. 22*: 379–386 (1935).

Moriguchi, T., K. Takashina, P.J. Chu, and H. Saito, The Abstract of the Third China-Japan Joint Meeting on Pharmacology, Beijing, China, 1993, p. 242.

Morioka, N., L.L. Sze, D.L. Morton, and R.F. Irie, A Protein Fraction from Aged Garlic Extract Enhances Cytotoxicity and Proliferation of Human Lymphocytes Mediated by Interleukin-2 and Concanavalin A, *Cancer Immunol. Immunother. 37*: 316–322 (1993).

Nagae, S., M. Ushijima, S. Hatono, J. Imai, S. Kasuga, H. Matsuura, Y. Itakura, and Y. Higashi, Pharmacokinetics of the Garlic Compound *S*-Allyl Cysteine, *Planta Med. 60*: 214–217 (1994).

Nagai, K., Experimental Studies on the Preventative Effect of Garlic Extract Against Infection with Influenza Virus, *Kansenshogaku-Zasshi (Jpn. J. Infect. Dis.) 47*: 321–325 (1973).

Nagai, K., M. Hanazuka, S. Hizume, Anti-Fatigue Effects of Ginseng Extract and the Mixture of Ginseng and Astragali Extract in Mice, *Jpn. Pharmacol. Ther. 20*: 3535–3541 (1992).

Nagai, K., and M. Yamawaki, Preventive Effect of Aged Garlic Extract (Kyoleopin) on Liver Damage Induced by Carbon Tetrachloride in Rats, *Yakuri To Chiryo (Jpn. Pharmacol. Ther.) 2*: 1622–1635 (1974).

Nakagawa, S., S. Kasuga, and H. Matsuura, Prevention of Liver Damage by Aged Garlic Extract and Its Components in Mice, *Phytother. Res. 1*: 1–4 (1988).

Nakagawa, S., K. Masamoto, H. Sumiyoshi, and H. Harada, Acute Toxicity of Garlic Extract, *J. Toxicol. Sci. 9:* 57–60 (1984a).

Nakagawa, S., K. Masamoto, H. Sumiyoshi, K. Kunihiro, and T. Fuwa, Effects of Raw and Extracted-Aged Garlic Juice on Growth of Young Rats and Their Organs after Peroral Administration, *J. Toxicol. Sci. 5*: 91–112 (1980).

Nakagawa S., H. Sumiyoshi, K. Masamoto, A. Kanezawa, H. Harada, S. Nakagami, S. Date, A. Yokota, M. Nishikawa, and T. Fuwa, Acute and Subacute Toxicity Tests of a Ginseng and Garlic Preparation Containing Vitamin B1 (LEOPIN-FIVE), *Oyo Yakuri (Appl. Pharmacol.) 27*: 1133–1150 (1984b).

Nakagawa, S., S. Yoshida, Y. Hirao, S. Kusaga, and T. Fuwa, Cytoprotective Activity of Components of Garlic, Ginseng and Cuiwjia on Hepatocyte Injury Induced by Carbon Tetra-Chloride *in vitro*, *Hiroshima J. Med. Sci. 34*: 303–309 (1985).

Nasu, K., The Relationship Between Digital Blood Flow and Skin Temperature According to Strain Gauge Plethysmography, *Biomed. Therm. 4*: 85–87 (1984).

Nishino, H., A. Iwashima, Y. Itakura, H. Matsuura, and T. Fuwa, Antitumor-Promoting Activity of Garlic Extract, *Oncology 46*: 277–280 (1989).

Numagami, Y., S. Sato, and T. Ohnishi, Attenuation of Rat Ischemic Brain Damage by Aged Garlic Extracts: A Possible Protecting Mechanism as Antioxidants, *Neurochem. Int. 29*: 135–143 (1996).

Okuhara, T., Clinical Study of Garlic Extract on Peripheral Circulation, *Jpn. Pharmacol. Ther. 22*: 3695–3701 (1994).

Randerson, K., Cardiology Update. Garlic and the Healthy Heart, *Nurs. Stand. 7*: 51 (1993).

Reeve, V.E., M. Bosnic, E. Rozinova, and C. Boehm-Wilcox, A Garlic Extract Protects from Ultraviolet B (280–320 nm) Radiation-Induced Suppression of Contact Hypersensitivity *Photochem. Photobiol. 58*: 813–817 (1993).

Rice-Evans, C.A., and A.T. Diplock, Current Status of Antioxidant Therapy, *Free Radic. Biol. Med. 15*: 77–96 (1993).

Riggs, D.R., J.I. DeHaven, and D.L. Lamm, *Allium sativum* (Garlic) Treatment for Murine Transitional Cell Carcinoma, *Cancer 79*: 1987–1994 (1997).

Selye, H., A Syndrome Produced by Diverse Nocuous Agents, *Nature 138*: 32 (1936).

Shashikanth, K.N., S.C. Basappa, and J.S. Murthy, Effect of Feeding Raw and Boiled Garlic (*Allium sativum* L.) Extracts on the Growth, Caecal Microflora and Serum Proteins of Albino Rats, *Nutr. Rep. Int. 33:* 313–319 (1986).

Steiner, M., A.H. Kham, D. Holbert, and R.I.S. Lin, A Double-Blind Crossover Study in Moderately Hypercholesteremic Men That Compared the Effect of Aged Garlic Extract and Placebo Administration on Blood Lipids, *Am. J. Clin. Nutr. 64*: 866–870 (1996a).

Steiner, M., A.H. Khan, and R.I.S. Lin, A Double-Blind Crossover Study in Moderately Hypercholesterolemic Men Comparing the Effect of Aged Garlic Extract and Placebo Administration on Blood Lipids and Platelet Function, *Shinyaku To Rinsho (New Drug Clin.) 45*: 456–466 (1996b).

Stoll, A., and E. Seebeck, Chemical Investigations on Alliin, the Specific Principle of Garlic, *Helv. Chim. Acta 56*: 481–487 (1951).

Subrahmanyan, V., K. Krishanmurthy, and V. Sreenivasamurthy, The Effect of Incorporation of Garlic in the Diet on the Intestinal Microflora of Rats, *Ann. Biochem. Exp. Med. 18*: 85–89 (1958).

Sumiyoshi, H., A. Kanezawa, K. Masamoto, A. Kanezawa, K. Masamoto, H. Harada, S. Nakagami, A. Yokota, M. Nishikawa, and S. Nakagawa, Chronic Toxicity Test of Garlic Extract in Rats, *J. Toxicol. Sci. 9*: 61–75 (1984).

Tadano, T., T. Aizawa, T. Asao, M. Hozumi, and K. Kisara, Pharmacological Studies of Nutritive and Tonic Crude Drugs on Fatigue in Mice, *Folia Pharmacol. Japon (Folia Pharmacol. Jpn.) 100*: 423–431 (in Japanese) (1992).

Tadi, P.P., R.W. Teel, and B.H.S. Lau, Anticandidal and Anticarcinogenic Potentials of Garlic, *Int. Clin. Nutr. Rev. 10*: 423–429 (1990).

Tadi, P.P., R.W. Teel, and B.H.S. Lau, Organosulfur Compounds of Garlic Modulate Mutagenesis, Metabolism, and DNA Binding of Aflatoxin B1, *Nutr. Cancer 15*: 87–95 (1991).

Ushijima, M., I. Sumioka, M. Kakimoto, K. Yokoyama, N. Uda, H. Matsuura, E. Kyo, A. Suzuki, S. Kasuga, Y. Itakura, B. Petesch, and H. Amagase, Effect of Garlic and Garlic Preparations on Physiological and Psychological Stress in Mice, *Phytother. Res. 22*: 226–230 (1997).

Weinberg, D.S., M.L. Manier, M.D. Richardson, and F.G. Haibach, Identification and Quantification of Organosulfur Compliance Markers in Garlic Extract, *J. Agric. Food Chem. 41*: 37–41 (1993).

Yamasaki, T., L. Lin, and B.H.S. Lau, Garlic Compounds Protect Vascular Endothelial Cells from Hydrogen Peroxide-Induced Oxidant Injury, *Phytother. Res. 8*: 408–412 (1994).

Yamasaki, T., R.W. Teel, and B.H.S. Lau, Effect of Allixin, a Phytoalexin Produced by Garlic, on Mutagenesis, DNA-Binding and Metabolism of Aflatoxin B1, *Cancer Lett. 59*: 89–94 (1991).

Yoshida, S., Y. Hirao, and S. Nakagawa, Mutagenicity and Cytotoxicity Tests of Garlic, *J. Toxicol. Sci. 9*: 77–86 (1984).

Yoshimura, M., Studies on the Nutritional Value of Allium Plants (XI). Investigation of Specially Smelling Ingredients in Allium Plants, *Vitamins (Kyoto) 14*: 633–640 (1958).

Chapter 7

Citrus Limonoids: Biochemistry and Possible Importance to Human Nutrition

Shin Hasegawa[a], Luke K.T. Lam[b], and Edward G. Miller[c]

[a]Western Regional Research Center, U.S. Department of Agriculture, Agricultural Research Service, Albany, CA 94710

[b]LKT Laboratories, St. Paul, MN 55114

[c]Department of Biomedical Sciences, Baylor College of Dentistry, Texas A&M University System, Dallas, TX 75246

Introduction

Limonoids are a group of chemically related, highly oxygenated triterpenoids present in the Rutaceae and Meliaceae plant families. Citrus limonoids are very unique; with a few exceptions, they are present only in species related to the *Citrus* genus. Thirty-six limonoid aglycones and 17 limonoid glucosides have been isolated from *Citrus* and its hybrids (Hasegawa and Miyake 1996). Substantial progress has been made in the understanding of the biochemistry of citrus limonoids. It is now known how, when, and where these compounds are biosynthesized and accumulated in various tissues of *Citrus* (Hasegawa and Herman 1992, Hasegawa and Miyake 1996).

Limonoids are important quality constituents of citrus fruit. They are one of two bitter principles in *Citrus* (Emerson 1948). Limonoid bitterness lowers the quality of a variety of citrus juices. It has recently been shown, however, that these compounds may have important physiologic properties. Limonoids possess biological functions such as anticarcinogenic activity in laboratory animals (Lam and Hasegawa 1989, Lam *et al.* 1989, 1994a and 1994b, Miller *et al.* 1989, 1992, 1994a, 1994b, 1996 and 1997) and in human breast cancer cells (Guthrie *et al.* 1997), and antifeedant activity against insects and termites (Alford and Bentley 1986, Alford *et al.* 1987, Klocke and Kubo 1982).

Limonoids are accumulated mainly as glucoside derivatives in mature fruit tissues and are one of the major secondary metabolites in citrus fruit. One glass of commercial orange juice, for example, contains 65 mg of total limonoid glucosides (Fong *et al.* 1989). Limonoid glucosides may play an important role in overall fruit quality and may be important in human nutrition. Epidemiologic studies suggest that citrus fruit contains many chemopreventative agents that act to prevent several cancers (Block 1991). Limonoid glucosides may be contributing to this anticancer activity. Limonoid glucosides inhibit significantly the formation of chemically induced tumorigenesis in the oral cavity of hamsters when applied topically to the pouches (Miller *et al.* 1992) or reduce the number of chemically induced tumors when administered in the diet (Miller *et al.* 1997). They also effectively inhibit the proliferation and reduce the viability of hormone-independent human breast cancer cells in culture (Guthrie *et al.* 1997).

Anticarcinogenic activity of limonoid aglycones has been investigated more intensely than that of the limonoid glucosides. They induce a detoxifying enzyme system, glutathione S-transferase activity (Lam and Hasegawa 1989, Lam *et al.* 1994a), and inhibit chemically induced tumorigenesis on the skin (Lam *et al.* 1994a), various tissues in mice (Lam *et al.* 1989 and 1994a), and in the oral cavity of hamsters (Miller *et al.* 1989, 1994a and 1996). In addition, they also inhibit the proliferation, growth, and viability of hormone-independent human breast cancer cells in culture (Guthrie *et al.* 1997).

Recently, the biochemistry of citrus limonoids related to human nutrition has been reviewed (Hasegawa and Miyake 1996). This chapter discusses recent progress in citrus limonoid research including the anticarcinogenic activity of citrus limonoid aglycones and glucosides in mice and mammary tumor cells in culture, the bulk isolation of limonoid glucosides from citrus juice processing by-products, and the creation of transgenic citrus that will produce fruit that is free from limonoid bitterness.

Anticarcinogenic Activity

Limonoids are chemical compounds containing a furan moiety. Many plant-derived furan-containing natural products have been shown to induce the detoxifying enzyme system, glutathione S-transferase (GST), and inhibit or prevent chemically induced carcinogenesis (Lam *et al.* 1987). Lam *et al.* (1989) demonstrated for the first time that citrus limonoids, such as nomilin, obacunone, isoobacunoic acid, and ichangin, induce GST activity in liver and forestomach mucosa in mice. Since then, anticarcinogenic activities of limonoid aglycones and limonoid glucosides have been investigated. These anticarcinogenic activities are summarized in Table 7.1.

Anticarcinogenic activities of limonoid aglycones have been investigated more intensely than those of limonoid glucosides, particularly the two most abundant limonoids in *Citrus*, limonin (**1**) and nomilin (**4**) as illustrated in Figures 7.1 and 7.2 (Lam *et al.* 1994b, Miller *et al.* 1994b). Nomilin inhibitory potential appears to correlate with its ability to induce GST in mice. Nomilin induces GST in the liver, small intestinal mucosa, and forestomach in mice (Lam *et al.* 1989 and 1994a); it inhibits benz[a]pyrene (BP)-induced neoplasia in the forestomach of mice by >50% (Lam and Hasegawa 1989) and in the liver of rats (Lam *et al.* 1994) when given by gavage. Limonin inhibits the formation of 7,12-dimethylbenz-(a)anthracene (DMBA)-induced neoplasia in the oral cavity of hamsters (Miller *et al.* 1989). Limonin carboxymethoxime, a synthetic limonoid, also inhibits the formation of DMBA-induced tumorigenesis in the buccal pouch of hamsters (Miller *et al.* 1996). Citrus limonoids such as limonin, limonin glucoside, and limonoid glucoside mixtures, however, do not induce GST in hamsters (Miller *et al.* 1994a,

TABLE 7.1
Biological Functions of Citrus Limonoids

Function	Effective limonoid(s)	Animal	Organ	Reference
Induction of glutathione S-transferase	Nomilin, ichangin	Mice	Liver	Lam *et al.* (1989)
	Obacunone		Small intestinal mucosa	
	Isoobacunoic acid		Forestomach	
	Nomilin	Mice	Liver	Lam *et al.* (1994a)
			Small intestinal mucosa	
Anticarcinogenic activities	Nomilin	Mice	Forestomach	Lam and Hasegawa (1989)
	Limonin	Hamsters	Buccal pouch	Miller *et al.* (1989)
	Limonin glucoside	Hamsters	Buccal pouch	Miller *et al.* (1992)
	Nomilin	Mice	Forestomach	Lam *et al.* (1994a)
	Nomilin	Mice	Lung	Lam *et al.* (1994a)
	Nomilin	Mice	Skin, initial phase	Lam *et al.* (1994a)
	Limonin		Skin, promotion phase	
	Limonin carboxy-methoxime	Hamsters	Buccal pouch	Miller *et al.* (1994a)
	Limonoid glucosides	Hamsters	Buccal pouch	Miller *et al.* (1997)
Inhibition of DNA methylation	Limonin	Mice	Liver	Lam, unpublished data
	Nomilin			
Inhibition of proliferation and reduction of viability	Nomilin	Human	MDA-MB-435 estrogen receptor–negative breast cancer cells	Guthrie *et al.* (1997)
	Limonoid glucosides			
	Limonin			

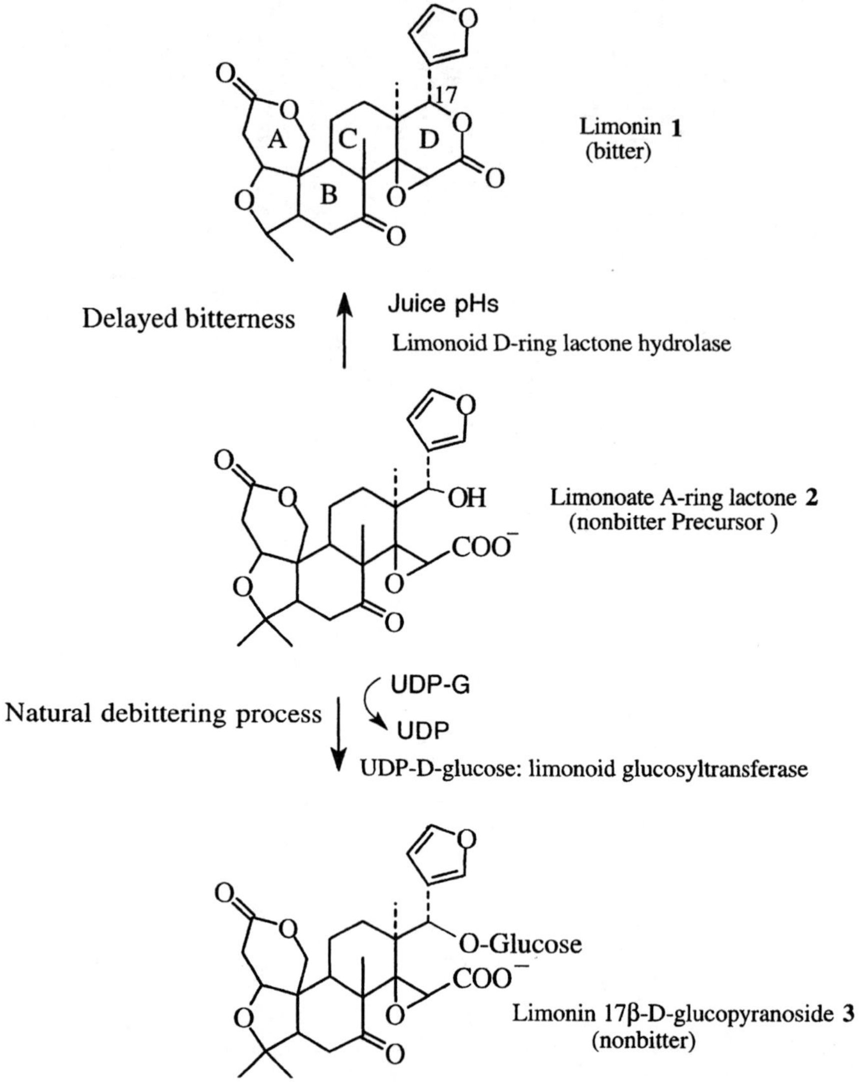

Fig. 7.1. Mechanism of the delayed bitterness and the natural limonoid debittering process. The delayed bitterness proceeds at the pH values of juice (acidic conditions) and is accelerated by the catalysis of limonoid D-ring lactone hydrolase. The natural debittering process occurs in mature fruit tissues and seeds, and is catalyzed by the action of UDP-D-glucose: limonoid glucosyltransferase.

1996 and 1997). The addition of limonin and nomilin to the diet inhibits BP-induced lung tumorigenesis in mice (Lam *et al.* 1994). The limonoids were also tested as chemopreventative agents on the two-stage skin carcinogenesis model (Lam *et al.* 1994). This model uses DMBA as the initiator and *O*-tetrade-

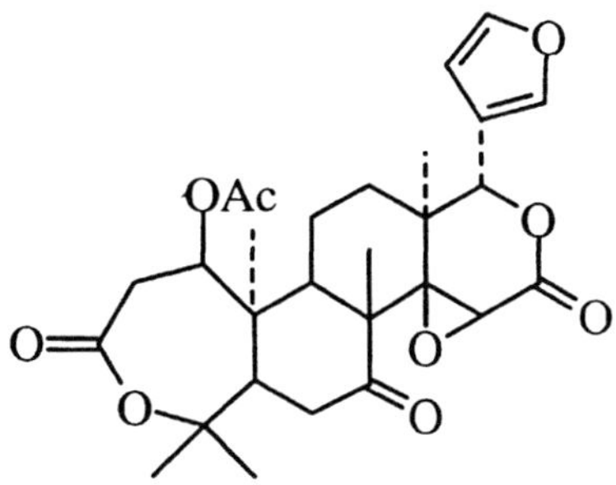

Nomilin **4**

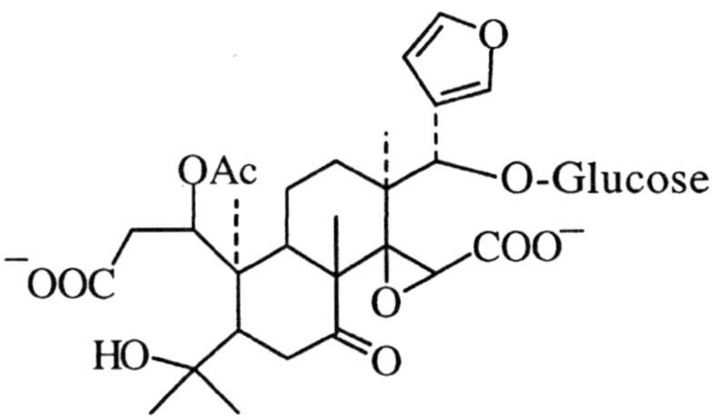

Nomilinic acid 17β-D-glucopyranoside **5**

Fig. 7.2. Structures of limonoids.

canoylphorbol 13-acetate as the promoter. Nomilin was found to be more effective than limonin as an inhibitor at the initiation stage of carcinogenesis, and limonin was found to be more effective than nomilin at the promotion stage. The latter finding appears to confirm the results of the hamster cheek pouch experiment in which limonin was more active than nomilin as a tumor inhibitor (Miller *et al.* 1989).

Limonin glucoside has been found to significantly inhibit the formation of DMBA-induced tumorigenesis in the oral cavity of hamsters when applied topically to the pouches (Miller *et al.* 1992). Limonoid glucosides also reduced the number of tumor formations in this oral carcinogenesis model when administered in the diet (Miller *et al.* 1997).

The inhibition of 4-(methylnitrosamino)-1-(3-pyridyl)-1-butanone (NNK)-induced methylation of DNA is correlated with tumor formation (Morse *et al.* 1989). Recently, citrus limonoid anticarcinogenic potential was examined by measuring its inhibitory effects on NNK-induced DNA methylation in mouse lung and liver (Lam 1997). Table 7.2 shows that all treatment groups, including the low-dose limonoids, demonstrated a reduction of O6-methyl guanine, indicating inhibition of carcinogen-DNA binding.

Recently, anticarcinogenic activity of citrus limonoids against human breast cancer cells in culture was examined. Guthrie *et al.* (1997) showed that limonoids effectively inhibited the proliferation of MDA-MB-435 estrogen receptor–negative human breast cancer cells in culture. The cells were incubated with or without limonoids for 48 h at 37°C. The amount of proliferation was determined by measuring the incorporation of [^{3}H]-thymidine into DNA. The concentration required to inhibit 50% of the proliferation (IC_{50}) was 0.4 µg/mL for nomilin and 12.5 µg/mL for limonin. Limonin glucoside was the least effective and its IC_{50} was 75 µg/mL. However, a mixture of limonoid glucosides isolated from orange molasses was extremely effective; its IC_{50} was 0.08 µg/mL. These results demonstrate the synergetic effects of limonoid glucosides as chemopreventative agents against MDA-MB-435 estrogen receptor–negative human breast cancer cells.

TABLE 7.2
Limonin and Nomilin Inhibition of NNK-Induced Methylation in the Liver of A/J Mice

Groups	Control	NNK	2.0 mmol Nomilin-NNK	1.0 mmol Nomilin-NNK	2.0 mmol Limonin-NNK	1.0 mmol Limonin-NNK
06-MeG[a]	0.00 ± 0.00	0.23 ± 0.03	0.12 ± 0.01[b]	0.12 ± 0.01[b]	0.09 ± 0.00[b]	0.12 ± 0.00[b]
Test/Control	—	—	0.52	0.53	0.40	0.49
7-MeG[c]	0.00 ± 0.00	3.51 ± 0.81	3.72 ± 0.64	3.23 ± 0.53	3.13 ± 0.64	3.70 ± 0.42
Test/Control	—	—	1.06	0.92	0.89	1.06

[a] Expressed as pmol 06-methylguanine/nmol guanine.
[b] Significantly different from 4-(methylnitrosamino)-1-(3-pyridyl)-1-butanone (NNK) control, $P < 0.05$.
[c] Expressed as pmol 7-methylguanine/nmol guanine. The chemicals were given by gavage in mmol/kg body weight per dose for a total of three doses.

Nomilin effectively reduced the viability of MDA-MB-435 estrogen receptor–negative human breast cancer cells (Guthrie *et al.* 1997). The cells were incubated with or without limonoids for 48 h. Viability was determined by measuring mitochondrial dehydrogenases that are active in living cells. Nomilin was very effective and reduced the viability of the cells to 50% at the concentration of 3 μg/mL and 37% at 6 μg/mL. The mixture of limonoid glucosides was less effective and reduced viability only to 65% at 6 μg/mL. The significance of this experiment is that it shows for the first time that certain citrus limonoids not only have chemopreventative effects against cancer, but also possess therapeutic effects.

Further research is required to understand how the limonoid glucosides are absorbed in the intestinal tract. Limonoids may be absorbed directly as glucosides by the cells in the gastrointestinal tract and/or after the limonoid moiety is liberated by the action of bacterial β-glucosidases. Several species of bacteria, which possess limonoid glucoside β-glucosidase activity, have been shown to be present in the human intestinal tract (Hasegawa, unpublished data). Because the liberated aglycones (the D-ring open) are soluble in aqueous solution at ≥pH 6.5, either the aglycone or glucoside form could be absorbed through the intestinal tract cells.

Toxicity Evaluation of Limonoid Glucosides

Limonoid glucosides appear to be safe to consume; they are consumed at relatively high levels by humans through digestion of citrus fruits, juices, and other processed citrus products. An average human may consume up to 50 g of limonoid glucosides annually. The exposure to high concentrations of limonoid glucosides has no negative effects on the general health of animals (Miller *et al.* 1997). When hamsters were exposed to the limonoid glucoside mixture with an amount equal to the amount an average adult would consume in 25 glasses of orange juice (1.6 g) per day for 45 d, there were no significant differences in food intake, weight-gain profile, blood count, or blood chemistry between animals consuming the control diet and those consuming limonoid glucoside diets.

The Ames bacterial tests, with the use of three revertant tester strains, TA98, TA100, and TA1535, and one forward mutation tester strain, TM677, showed that all four limonoids, limonin, nomilin, nomilinic acid, and limonin glucoside, did not show mutagenic activity at a 5 g/plate dose level (Lam, unpublished data).

Bulk Isolation of Limonoid Glucosides

There has been commercial interest in the utilization of limonoid glucosides in the food and agricultural industries. In contrast to limonoid aglycones, the nonbitter taste and water solubility properties of limonoid glucosides make them ideal for use as functional food additives. Due to the great potential biological activity, the demand for limonoid glucosides has increased considerably in recent years. The limonoid glucosides may be a new commercial by-product that the citrus industry can exploit.

Occurrence and Sources. Seventeen limonoid glucosides have been isolated from *Citrus* and hybrids (Hasegawa and Miyake 1996). Each has one D-glucose molecule attached to the 17-position of the limonoid molecule *via* a β-glucosidic linkage such as limonin 17β-D-glucopyranoside (**3**) (Fig. 7.1) or limonin glucoside. All limonoid glucosides isolated and characterized to date contain a single glucose sugar unit. Recently, however, we have found that a limonoid diglucoside form also occurs in the fruit tissues and seeds (Hasegawa, 1999). Limonoid glucosides are present only in mature fruit tissues and seeds (Fong *et al.* 1991, 1992a and 1992b, Hasegawa *et al.* 1989 and 1991, Herman *et al.* 1991). In mature seeds, however, both limonoid aglycones and glucosides are accumulated in high concentrations at a ratio of 2:l (Fong *et al.* 1992b, Ozaki *et al.* 1991).

The occurrence of limonoid glucosides in *Citrus* are summarized in Table 7.3. The citrus juice processing by-products, such as molasses, seeds, and peels, contain relatively high concentrations of limonoid glucosides. For example, orange molasses collected from seven different sources contained an average of 1980 ppm of total limonoid glucosides (Hasegawa *et al.* 1996a). One gallon of undiluted orange molasses contained up to 12.5 g of total limonoid glucosides. They are excellent sources for limonoid glucosides. Citrus fruits and juices, particularly sweet oranges, contain relatively high concentrations of limonoid glucosides.

Limonin glucoside is the predominant limonoid glucoside in fruit tissues, including juices and molasses, comprising >50% of total limonoid glucosides as summarized in Table 7.3 (Fong *et al.* 1989, Hasegawa *et al.* 1991, Herman *et al.* 1990). On the other hand, nomilin glucoside is the major limonoid glucoside in seeds, and limonin glucoside is a relatively minor constituent (Fong *et al.* 1992b, Ozaki *et al.* 1991).

TABLE 7.3
Occurrence and Sources of Limonoid Glucosides[a,b]

	Samples (n)	LG	NG	NAG	OG	DAG	DG	Total	Reference
Molasses	7	1015	710	175	80			1980	Hasegawa *et al.* (1997)
Seeds	8	725	1970	975	1105	530	890	5305	Ozaki *et al.* (1991)
Peels									
Navel	2							370	Hasegawa *et al.* (1991)
Valencia	2							760	Fong *et al.* (1992a)
Juices									
Orange	15	176	47	84	3	21		331	Herman *et al.* (1990)
Grapefruit	8	120						190	Fong *et al.* (1989)
Lemon	4	54						82	Fong *et al.* (1989)
Mandarin	11	129	50	42	0.5			223	Ozaki *et al.* (1992)

[a]Values are ppm.
[b]Abbreviations: G, glucoside; L, limonin; N, nomilin; NA, nomilinic acid; O, obacunone; D, deacetylnomilin; and DA, deacetylnomilinic acid.

Isolation. Limonoid glucosides are acidic, either monocarboxylic, such as limonin glucoside (**3**; see Fig. 7.1), or dicarboxylic, such as nomilinic acid glucoside (**5**; see Fig. 7.2). Their molecular weights are ~650 and they are soluble in aqueous solutions and in alcohols. On the basis of these chemical characteristics, limonoid glucosides can be isolated readily from citrus molasses or citrus juices by open column chromatography, high-performance liquid chromatography (HPLC), or a combination of both. At present, we are isolating limonoid glucosides from orange molasses by a sequence of open column chromatographic procedures using Dowex-50W, Amberlite XAD-2 and 4, Q-Sepharose resins (Hasegawa, 1999). All of the resins were obtained from Sigma Chemical, St. Louis, MO.

Orange molasses is first diluted to ~15° Brix and centrifuged at 10,000 × *g* for 20 min. The resulting solution is then filtered through celite to obtain a clear molasses solution. Passing the clarified molasses through a Dowex-50 column removes most of the dark pigments and positively charged compounds. Limonoid glucosides have practically no affinity to the column. Chromatography of the effluent through an absorbent column, XAD-4, removes sugars, acids, and many other constituents. After the column is washed thoroughly with water, limonoid glucosides are eluted from the column with 50% or higher concentrations of EtOH. The EtOH extract also contains flavonoid glucosides, which co-elute with limonoid glucosides during XAD-4 column chromatography. The EtOH extract is evaporated to near dryness and the residue is dissolved in H_2O. The solution is adjusted to pH 6.5 with 1 N NaOH and then loaded onto the weak anion exchange resin DEAE-Sepharose. Flavonoid glucosides are not retained on the column and can be washed off, whereas limonoid glucosides are retained and can then be eluted with a linear gradient of NaCl to 0.4 mol/L. Fractions containing limonoid glucosides are pooled. The pooled fractions are passed through a small Dowex 50 column to remove the sodium salt and then loaded onto a XAD-2 column. The column is eluted with a linear gradient of EtOH to 55%.

The use of nontoxic reagents such as EtOH and salt solutions can be employed in recognition of the food safety issue. The use of preparative HPLC would increase efficiency of purification and decrease the time of isolation. Scale-up procedures for isolation of limonoid glucoside mixtures from citrus juices or molasses to an industrial level have been developed (Ifuku *et al.* 1997). A Japanese company isolated 5 kg of a limonoid glucoside mixture from orange juices using this industrial procedure; the isolate was added to mandarin orange juice to raise the limonoid glucoside content from 220 to 1000 ppm. The product, LG 1000, was successfully test-marketed in Japan.

Limonoid glucosides can be isolated from citrus juices in ways similar to those outlined above. Citrus seeds are also excellent sources for both limonoid aglycones and glucosides (Ozaki *et al.* 1991). Both limonoid aglycones and glucosides can be extracted from the same batch of seed meals. The seed meal is first washed with hexane to remove the oils. The aglycones and glucosides are then extracted from the defatted meal with MeOH. The aglycones and glucosides can be separated by partitioning between H_2O and CH_2Cl_2. The H_2O fraction contains the glucosides, whereas the CH_2Cl_2 fraction contains the aglycones. The limonoid aglycones can then be isolated according to previously published procedures (Herman *et al.* 1992). The limonoid glucosides can be isolated by the procedures outlined above for molasses.

Increasing Limonoid Glucoside Concentrations in Citrus Fruit by Genetic Engineering

Limonoid glucosides are products of a naturally occurring limonoid debittering process and possess important pharmacologic properties. Promoting the accumulation of higher levels of limonoid glucosides in fruit tissues is particularly interesting for both the elimination of the bitterness problem and enhancement of the pharmacologic activity in the fruit. Transgenic citrus fruits that would enhance limonoid glucosyltransferase activity in the fruit tissues could be created with the use of genetic engineering techniques (Hasegawa *et al.* 1996b).

Biochemistry of Limonoids in Citrus. Significant progress has been made in understanding the biosynthesis of limonoids in *Citrus*. It is now established where, when, and how limonoids are biosynthesized and accumulated in various tissues of *Citrus* and its hybrids (Hasegawa and Herman, 1992; Hasegawa and Miyake, 1996).

Nomilin (**4**; see Fig. 7.2) is the precursor of all other limonoids found in *Citrus* and its hybrids. This precursor is biosynthesized in the phloem region of stems *via* terpenoid biosynthetic pathways from acetate, and mevalonate *via* farnesyl pyrophosphate (Hasegawa *et al.* 1984 and 1986, Ou *et al.* 1988). It is then translocated from stem to leaves, fruit tissues, and seeds, where it is further metabolized in each tissue to other limonoids through at least four different biosynthetic pathways, i.e., the limonin pathway (19 limonoids in *Citrus* and hybrids), the calamin pathway (12 limonoids in *Fortunella* and hybrids), the ichangensin pathway (one limonoid in *Papedocitrus* and hybrids) and the 7-acetate limonoid pathway (four limonoids in *Poncirus* and hybrids) (Hasegawa and Miyake 1996). At least five groups of enzymes are involved in the biosynthesis and biodegradatoin of limonoids.

The biosynthesis, glucosidation, and accumulation of limonoids occur independently in fruit tissues and seeds. There are significant differences in the composition of limonoid glucosides and the ratio of total glucosides and aglycones in the fruit tissues and the seeds. Limonin glucoside is the predominant limonoid glucoside in the fruit tissue (Fong *et al.* 1989, Herman *et al.* 1990), whereas nomilin glucoside is predominant in the seed (Ozaki *et al.* 1991). The ratio of total aglycones to total glucosides in fruit tissues averages 1:150, whereas in the seed, the ratio is 2:1. In addition, during late stages of fruit growth and maturation, the aglycone content in the fruit tissue continuously decreases as a result of its conversion to glucosides, but this decrease does not occur appreciably in seeds (Fong *et al.* 1991, 1992a and 1992b, Hasegawa *et al.* 1991, Herman *et al.* 1991, Ozaki *et al.* 1991).

Delayed Bitterness and Naturally Occurring Debittering Process. Intact fruit tissues do not normally contain the intensely bitter limonoid, limonin (**1**; see Fig 7.1), but instead contain the nonbitter limonoate A-ring lactone (**2**; see Fig. 7.1) (Maier and Beverly 1968). After juice is extracted, this precursor is gradually converted to limonin. This phenomenon is generally referred to as delayed bitterness. The reaction proceeds under acidic conditions and is accelerated by the enzyme limonoid D-ring lactone hydrolase (Maier *et al.* 1969). The delayed bitterness lowers the quality and value of some commercial juices.

Limonoid bitterness is a problem in juice from winter citrus fruits harvested in the early- to mid-season, but not in juices of fruits harvested in the late season. As the fruit ripens, the concentration of limonoid aglycones, such as limonoate A-ring lactone, decreases. This natural debittering process has been known for over a century, but the biochemical mechanism was not understood until the discovery of the limonoid glucosides. Hasegawa *et al.* (1989) determined that the natural debittering process is due to the conversion of limonoid aglycones to tasteless limonoid glucosides during late stages of fruit growth and maturation.

The glucosidation of the aglycones occurs in the fruit tissue and seeds during late stages of fruit growth and maturation. In both navel oranges and Valencia oranges grown in California, for example, the conversion of aglycones to their respective glucosides begins in September and continues until the fruits are harvested (Fong *et al.* 1992a, Hasegawa *et al.* 1991). The harvest season starts in November for navel oranges and in March for Valencia oranges in California. Juice from navel oranges harvested at early- to mid-season has the delayed bitterness problem because of the incomplete conversion of limonoid aglycones to the glucoside form. Juices from Valencia oranges, which are harvested in March and have a longer maturation time than navel oranges, do not have the delayed bitterness. During this extra time of maturation, the limonoid aglycones are almost completely converted to the tasteless glucosides. The enzyme responsible for the conversion of aglycones to glucosides has been identified, isolated, and characterized.

UDP-D-Glucose: Limonoid Glucosyltransferase. The glucosidation of aglycones is catalyzed by UDP-D-glucose: limonoid glucosyltransferase (Hasegawa *et al.* 1997). This enzyme activity is found in fruit tissues and seeds during the late stages of fruit growth and continues its activity until the fruit is harvested (Hasegawa *et al.* 1991). The limonoid glucosyltransferase enzyme has been chosen for genetic transfer to create transgenic citrus trees that produce fruit free from limonoid bitterness and simultaneously increase the overall content of limonoid glucosides in the fruit tissues (Hasegawa *et al.* 1996b).

Limonoid glucosyltransferase has been isolated from albedo tissues of two navel orange cultivars, Frost and Newhall, by a combination of $(NH_4)_2SO_4$ fractionation, UDP-glucuronic acid affinity chromatography and DEAE ion exchange HPLC (Hasegawa *et al.* 1997). SDS-polyacrylamide gel electrophoresis on the purified fraction revealed a single protein band having an apparent M_r between 56 and 58 kDa, which is very similar to flavanone glucosyltransferase from pummelo leaves (Bar-Peled *et al.* 1991) and grapefruit seeds (MacIntosh *et al.* 1990). This single enzyme appears to catalyze the glucosidation of all of the aglycones present in the system (Hasegawa *et al.* 1997).

The amino acid sequences of the N-terminal and three internal segments of this glucosyltransferase have been determined as follows. N-terminal: Gly-Thr-Glu-Ser-Leu-Val-His-Val-Leu. Three segments: (i) Phe-Pro-Leu-Ala-Gly-Val-Phe-Tyr-Gln-Gly-Phe-Ser-Ala-Thr-Pro-Ile-Ala-Thr-Phe-Pro-Thr-Leu-Thr-Thr-Pro; (ii) Ala-Glu-Glu-Ala-Val-Ala-Asp-Gly-Gly-Ser-Ser-Asp-Arg-Asn-Ile-Gln-Ala-Phe-Val; and (iii) Ala-Gly-Asn-Phe-Thr-Tyr-Glu-Pro-Thr-Pro-Val-Gly-Asp-Gly-Phe-Ile-(Arg)-Phe-Glu-Phe.

With the use of the amino acid sequence data, the gene for the limonoid glucosyltransferase has been isolated from DNA libraries created from the navel orange genome. The insertion of the gene with altered promoters into navel oranges is in progress (Omura, unpublished data). This should result in the creation of transgenic citrus varieties producing fruits with improved flavor and higher nutritional values.

Degradation of Limonoid Glucosides

Fruit Tissues and Seeds. Limonoid 17β-D-glucopyranoside β-glucosidase activity is found in both dormant and germinating citrus seeds (Ronneberg *et al.* 1995). This enzyme catalyzes the hydrolysis of limonoid glucosides and releases glucose and limonoid aglycones. The limonoid glucosides quickly disappear during the course of germination. The glucose moiety may be used by the plant as an energy source, whereas the aglycones may function as pesticides.

However, limonoid glucosides accumulated in fruit tissues are stable. Radioactive tracer work using [^{14}C]-nomilin glucoside showed the absence of this β-glucosidase activity in the fruit tissues. This explains why limonoid glucosides are present in high concentrations in the fruit tissue and why their concentration continues to increase as long as the fruit is attached to the tree.

Commercial enzymes, β-glucosidases from almonds and naringinases from molds, do not attack limonoid glucosides (Hasegawa, 1999). A species of bacterium isolated from soil (Hasegawa *et al.* 1989) and several species of bacteria isolated from human gastrointestinal tracts (Hasegawa, 1999) possess limonoid β-glucosidase activity.

Degradation During Processing and Storage. Limonoid glucosides in fruit and processed products are generally stable during processing and storage; there are no significant differences in the total and individual limonoid glucosides between freshly prepared orange juices, concentrated orange juices, and stored orange juices (Fong *et al.* 1989). A preliminary study has shown that most of the limonoid glucosides appear to be stable between pH 2 and 8. One known exception is nomilin glucoside, which is converted to nomilinic acid glucoside at pH values <3 and to obacunone glucoside at pH values >8 (Hasegawa, unpublished data).

Antifeedant Activity Against Insects

Citrus limonoid aglycones present in young leaves and fruit tissues may act as pest control agents. They have demonstrated antifeedant activity against a number of insects, including Colorado potato beetle, corn earworm, fall armyworm, spruce budworm, and tobacco budworm (Alford and Bentley 1986, Alford *et al.* 1987, Klocke and Kubo 1982). Young leaves and fruit tissues, which are vulnerable to insect attack, possess higher levels of limonoid aglycones, whereas older, mature tissues contain mainly limonoid glucosides, which have practically no antifeedant activity (Murray, unpublished data). This could be further exploited as a commercial product in the organic farming industry.

Conclusion

Research on the anticarcinogenic activity of citrus limonoids is still in the preliminary stages, and further research is required to provide comprehensive data, especially in clinical experiments related to possible roles in human health. Nonetheless, there has been significant commercial interest in limonoid glucosides. The potential health effects, non-bitter taste properties, and water solubility make limonoid glucosides ideal candidates for use as functional food additives. The possible role of limonoids as natural pesticides also has potential commercial applications. The demand for limonoid glucosides has increased considerably in recent years. The procedures for isolation of limonoid glucosides on an industrial scale have been established. By-products of the processing of citrus juice, i.e., molasses and seeds, contain high concentrations of limonoid glucosides, may represent new profitable commercial products for the citrus industry.

Basic research on the biochemistry of citrus limonoids has led to several opportunities for technology transfer research involving these compounds. One of our current interests is to improve the flavor quality of citrus juices and simultaneously increase nutritional values of citrus fruit by creating transgenic fruit with the use of genetic engineering techniques, fruit that is free from limonoid bitterness and possessing higher concentrations of limonoid glucosides.

References

Alford, A.R., and M.D. Bentley, Citrus Limonoids as Potential Antifeedants for the Spruce Budworm (*Lepidoptera tortricidae*), *J. Econ. Entomol. 79*: 35–38 (1986).

Alford, A.R., J.A. Cullen, R.H. Storch, and M.D. Bentley, Antifeedant Activity of Limonin Against the Colorado Potato Beetle, *J. Econ. Entomol. 80*: 575–578 (1987).

Bar-Peled, M., E. Lewinsohn, R. Fluhr, and J. Gressel, UDP-Rhamnose: Flavanone-7-O-glucoside-2Ó-O-rhamnosyltranferase, *J. Biol. Chem. 266*: 20953–20959 (1991).

Block, G., Vitamin C and Cancer Prevention: The Epidemiological Evidence, *Am. J. Clin. Nutr. 53*: 270–282 (1991).

Emerson, O.H., The Bitter Principle of Citrus Fruit. I. Isolation of Nomilin, a New Bitter Principle from the Seeds of Oranges and Lemons, *J. Am. Chem. Soc. 70*: 545–549 (1948).

Fong, C.H., S. Hasegawa, C.W. Coggins, Jr., D.R. Atkin, and M. Miyake, Contents of Limonoids and Limonin 17β-D-Glucopyranoside in Fruit Tissue of Valencia Orange During Fruit Growth and Maturation, *J. Agric. Food Chem. 40*: 1178–1181 (1992a).

Fong, C.H., S. Hasegawa, Z. Herman, and P. Ou, Limonoid Glucosides in Commercial Citrus Juices, *J. Food Sci. 54*: 1505–1506 (1989).

Fong, C. H., S. Hasegawa, Z. Herman, and P. Ou, Biosynthesis of Limonoid Glucosides in Lemon (*Citrus limon*), *J. Sci. Food Agric. 54*: 393–398 (1991).

Fong, C.H., S. Hasegawa, M. Miyake, Y. Ozaki, C.W. Coggins, Jr., and D.R. Atkin, Limonoids and Their Glucosides in Valencia Orange Seeds During Fruit Growth and Development, *J. Agric. Food Chem. 41*: 112–115 (1992b).

Guthrie, N., A.F. Chambers, and K.K. Carroll, Inhibition of MDA-MB-435 Estrogen Receptor- Negative Human Breast Cancer Cells by Citrus Limonoids, *Proc. Am. Assoc. Cancer Res. 38*: 113S (1997).

Hasegawa, S., R.D. Bennett, Z. Herman, C.H. Fong, and P. Ou, Limonoid Glucosides in *Citrus*, *Phytochemistry 28*: 1717–1720 (1989).

Hasegawa, S., R.D. Bennett, and V.P. Maier, Biosynthesis of Limonoids in Citrus Seedlings, *Phytochemistry 23*: 1601–1603 (1984).

Hasegawa, S., C.H. Fong, M. Miyake, and J.H. Keithly, Limonoid Glucosides in Orange Molasses, *J. Food Sci. 61*: 560–561 (1996a).

Hasegawa, S., and Z. Herman, Biosynthesis of Limonoids in *Citrus*, in *Secondary-Metabolite Biosynthesis and Metabolism*, edited by R.J. Petroski, and S.P. McCormick, Plenum Press, New York, 1992, pp. 305–317.

Hasegawa, S., Z. Herman, E.D. Orme, and P. Ou, Biosynthesis of Limonoids in *Citrus*: Sites and Translocation, *Phytochemistry 25*: 2783–2785 (1986).

Hasegawa, S., and M. Miyake, Biochemistry and Biological Functions of Citrus Limonoids, *Food Rev. Int. 12*: 413–435 (1996).

Hasegawa, S., P. Ou, C.H. Fong, Z. Herman, C.W. Coggins, Jr., and D.R. Atkin, Changes in the Limonoate A-Ring Lactone and Limonin 17β-D-Glucopyranoside Content of Navel Oranges During Fruit Growth and Maturation, *J. Agric. Food Chem. 39*: 262– 265 (1991).

Hasegawa, S., G.C. Suhayda, W. Hsu, and G.H. Robertson, Purification of Limonoid Glucosyltransferase from Navel Orange Albedo Tissues, *Phytochemistry 46*: 33–37 (1997).

Hasegawa, S., C. Suhayda, M. Omura, and M. Berhow, Creation of Transgenic *Citrus* Free from Limonin Bitterness, in *Biotechnology for Improved Foods and Flavors*, edited by G.R. Takeoka, R. Teranishi, P.J. Williams, and A. Kobayashi, ACS Symposium Series 637, American Chemical Society, Washington, DC, 1996b, pp. 79–86.

Hasegawa, S. Limonoid glucosides, in *Naturally Occurring Glycosides*, edited by R. Ikan, John Wiley & Sons, New York 1999, pp. 275–294.

Herman, Z., C.H. Fong, and S. Hasegawa, Biosynthesis of Limonoid Glucosides in Navel Orange, *Phytochemistry*, *30*: 1487–1488 (1991).

Herman, Z., C.H. Fong, and S. Hasegawa, Analysis of Limonoids in Citrus Seeds, in *Modern Methods of Plant Analysis. Seed Analysis*, edited by H.F. Linskens, and J.F. Jackson, Springer-Verlag, Berlin, New Series, Vol. 14, 1992, pp. 361–375.

Herman, Z., C.H. Fong, P. Ou, and S. Hasegawa, Limonoid Glucosides in Orange Juices by HPLC, *J. Agric. Food Chem. 38*: 1860–1861 (1990).

Ifuku, Y., H. Maeda, M. Miyake, N. Inaba, S. Ayano, Y. Ozaki, K. Maruyama, and S. Hasegawa, Methods for Manufacturing Limonoid Glucosides, U.S. Patent 5,734,046 (1997).

Klocke, J.A., and I. Kubo, Citrus Limonoid By-Products as Insect Control Agents, *Entomol. Exp. Appl. 32*: 299–301 (1982).

Lam, L.K.T., and S. Hasegawa, Inhibition of Benzo(a)pyrene-Induced Forestomach Neoplasia in Mice by Citrus Limonoids, *Nutr. Cancer 12*: 43–47 (1989).

Lam, L.K.T., Y. Li, and S. Hasegawa, Effects of Citrus Limonoids on Glutathione S-Transferase Activity in Mice, *J. Agric. Food Chem. 37*: 878–880 (1989).

Lam, L.K.T., V.L. Sparnings, and L.W. Watternberg, Effects of Kahweol and Cafestol on the Activity of Glutathione S-Transferase in Mice, *J. Med. Chem. 30*: 1399–1403 (1987).

Lam, L.K.T., J. Zhang, and S. Hasegawa, Citrus Limonoid Reduction of Chemically Induced Tumorigenesis, *Food Technol. 48*: 104–109 (1994a).

Lam, L.K.T., J. Zhang, S. Hasegawa, and H.A.J. Schut, Inhibition of Chemically Induced Carcinogenesis by Citrus Limonoids, in *Food Phytochemicals for Cancer Prevention I*, edited by M. Huang, T. Osawa, C. Ho, and R.T. Rosen, ACS Symposium Series 546, American Chemical Society, Washington, DC, 1994b, pp. 209–219.

MacIntosh, C.A., L. Latchinian, and R.L. Mansell, Flavanone-Specific 7-O-Glucosyltransferase Activity in *Citrus paradisi* Seedlings: Purification and Characterization, *Arch. Biochem. Biophys. 282*: 50–57 (1990).

Maier, V.P., and G.D. Beverly, Limonin Monolactone, the Nonbitter Precursor Responsible for Delayed Bitterness in Certain Citrus Juices, *J. Food Sci. 33*: 488–492 (1968).

Maier, V.P., S. Hasegawa, and E. Hera, Limonin D-Ring Lactone Hydrolase. A New Enzyme from Citrus Seeds, *Phytochemistry 8*: 405–407 (1969).

Miller, E.G., R. Fanous, F. Rivera-Hidalgo, W.H. Binnie, S. Hasegawa, and L.T.K. Lam, The Effect of Citrus Limonoids on Hamster Buccal Pouch Carcinogenesis, *Carcinogenesis. 10*: 1535–1537 (1989).

Miller, E.G., A.P. Gonzales-Sanders, A.M. Couvillon, W.H. Binnie, S. Hasegawa, and L.K.T. Lam, Citrus Limonoids as Inhibitors of Oral Carcinogenesis, *Food Technol. 48*: 110 (1994a).

Miller, E.G., A.P. Gonzales-Sanders, A.M. Couvillon, J.M. Wright, S. Hasegawa, and L.K.T. Lam, Inhibition of Hamster Buccal Pouch Carcinogenesis by Limonin 17-β-D-Glucopyranoside, *Nutr. Cancer 17*: 1–7 (1992).

Miller, E.G., A.P. Gonzales-Sanders, A.M. Couvillon, J.M. Wright, S. Hasegawa, L.K.T. Lam, and G.I. Sunahara, Inhibition of Oral Carcinogenesis by Green Coffee Beans and Limonoid Glucosides, in *Food Phytochemicals for Cancer Prevention I*, edited by M. Huang, T. Osawa, C. Ho, and R.T. Rosen, ACS Symposium Series 546, American Chemical Society, Washington, DC, 1994b, pp. 220–229.

Miller, E.G., J.L. Porter-Wright, W.H. Binnie, I.Y. Guo, and S. Hasegawa, The Importance of the B-Ring of the Limonoid Nucleus to the Cancer Chemopreventative Activity of Citrus Limonoids, in *Hypernutritious Foods*, edited by J.W. Finley, D.J. Armstrong, S. Nagy, and S.F. Robinson, Agscience Inc., Auburndale, FL, 1996, pp. 191–204.

Miller, E.G., M.T. Record, W.H. Binnie, and S. Hasegawa, Limonoid Glucosides: Systemic Effects on Oral Carcinogenesis, 5th Chemical Congress of North America, Cancun, Mexico, Nov. 11–15, 1997.

Morse, M.A., S.G. Amin, S.S. Hecht, and F.L. Chung, Effects of Aromatic Isothiocyanates on Tumorigenicity, O6-Methylguanine Formation, and Metabolism of the Tobacco-Specific Nitrosamine 4-(Methylnitrosamino)-1-(3-pyridryl)1-butanone in A/J Mouse Lung, *Cancer Res.* 2894–2897 (1989).

Ou, P., S. Hasegawa, Z. Herman, and H.C. Fong, Limonoid Biosynthesis in the Stem of *Citrus limon, Phytochemistry 27*: 115–118 (1988).

Ozaki, Y., S. Ayano, M. Miyake, H. Maeda, Y. Ifuku, and S. Hasegawa, Limonoid Glucosides in Juices of Satsuma Mandarin (*Citrus unshu* Marc.), *Biosci. Biotechnol. Biochem. 56*: 836–837 (1992).

Ozaki, Y., C.H. Fong, Z. Herman, H. Maeda, M. Miyake, Y. Ifuku, and S. Hasegawa, Limonoid Glucosides in Citrus Seeds, *Agric. Biol. Chem. 55*: 137–141 (1991).

Ronneberg, T.A., S. Hasegawa, C. Suhayda, and Y. Ozaki, Limonoid Glucoside β-Glucosidase Activity in Lemon Seeds, *Phytochemistry 39*: 1305–1307 (1995).

Chapter 8

Limonoid Glucosides: Systemic Effects on Oral Carcinogenesis

E.G. Miller[a], M.T. Record[a], W.H. Binnie[a], and S. Hasegawa[b]

[a]Department of Biomedical Sciences, Baylor College of Dentistry, Texas A&M University System, Dallas, TX 75246

[b]Western Regional Research Center, U.S. Department of Agriculture, Agricultural Research Service, Albany, CA 91106

Introduction

Data from over 20 separate epidemiological studies have shown that the consumption of citrus is protective against a wide range of cancers (Block 1991). Significant reductions in risk were found for cancers of the oral cavity, larynx, esophagus, stomach, pancreas, lung, colon, and rectum. The data also suggest that this activity is due to vitamin C plus one or more as yet unidentified cancer chemopreventive agents in citrus fruits and juices.

During the last decade, several laboratories have studied the anticancer activity of citrus limonoids. The initial studies concentrated on limonin and nomilin (Lam and Hasegawa 1989, Lam *et al.* 1994a and 1994b, Miller *et al.* 1989). The data from these experiments showed that both limonin and nomilin could inhibit the development of carcinogen-induced neoplasia in animal models for cancer of the forestomach, lung, skin, and oral cavity. Recent studies (Guthrie *et al.* 1997, So *et al.* 1996) have also suggested that limonin and nomilin may be able to inhibit the development of mammary tumors. In some of the tumor models. nomilin was the more effective agent, whereas in others, limonin proved to be more active. In the model for oral cancer, 12 limonoids were tested for antineoplastic activity (Miller *et al.* 1989, 1992b, 1994a and 1994b, Wright *et al.* 1996). Four of the limonoids (limonin, limonin 17β-D-glucopyranoside, limonin carboxymethoxime, and deoxylimonin) demonstrated significant activity, resulting in a significant reduction in either tumor number or tumor burden. Three structurally similar limonoids (nomilin, nomilin 17β-D-glucopyranoside, and obacunone) were classified as having partial activity, giving a consistent reduction of 20–30% in tumor number. Five of the limonoids (nomilinic acid 17β-D-glucopyranoside, ichangensin, limonol, deoxylimonic acid, and 17,19-didehydrolimonate) were inactive.

Of all of the chemicals tested for cancer chemopreventive activity, several properties associated with the limonoid glucosides suggest that these chemicals might be contributing to the anticancer activity of citrus and might be used as food additives for the prevention of cancer in humans (Fong *et al.* 1990, Hasegawa *et al.* 1989). Unlike most limonoids, the limonoid glucosides are soluble in water.

The aglycones, like limonin and nomilin, are often bitter, whereas the limonoid glucosides are tasteless. Compared with other limonoids, the concentration of limonoid glucosides in citrus fruits and products is extremely high. In commercial orange juice, the average concentration of the mixed limonoid glucosides is 320 ppm. The average concentration of limonin 17β-D-glucopyranoside and nomilin 17β-D-glucopyranoside, the two limonoid glucosides that have demonstrated anticancer activity in the model for oral carcinogenesis, is 176 ppm and 47 ppm, respectively. In addition, mixed limonoid glucosides can also be isolated in large quantities from two by-products of juice processing plants, the seeds and citrus molasses.

The positive features associated with the limonoid glucosides have led to the formulation of a test product by a Japanese company (Miller 1996). The scientists at this company isolated large quantities of a limonoid glucoside mixture from citrus molasses. This mixture of citrus chemicals was then added to one of the company's products, a mandarin orange juice. This process artificially raised the limonoid glucoside content (normally 200 ppm) to 1000 ppm. The test product was designated *LG1000*. The results from consumer focus groups suggested that this change in product composition had no effect on color, texture, appearance, or taste. It is not known at this time whether *LG1000* or a similar product will be marketed.

Additional research will be required before a product such as *LG1000* can be produced and marketed in other countries. More studies on the antineoplastic activity of these chemicals in animal models must be performed. Other studies are required on the uptake, turnover, distribution, and metabolism of these citrus chemicals in mammalian systems. Finally, clinical trials must establish whether these chemicals have anticancer activity in humans.

In most of the earlier studies on tumorigenesis, the limonoids and limonoid glucosides either were applied topically to the tissues or could have produced their effects through topical exposures. The primary purpose of this study was to see if similar effects might be produced systemically. In these experiments, a limonoid glucoside mixture was incorporated into the diets of the experimental animals. At the same time, we were also interested in determining what effect, if any, this long-term exposure to the limonoid glucoside mixture might have on the general health of the animals. Data were collected on food intake and weight gain. Blood samples were also taken from some of the animals and were used for complete blood counts and a series of blood chemistries. Some of the oral tissues from the hamsters were also used to examine the effect of the limonoid glucoside mixture on the glutathione S-transferase activity of the oral epithelial cells.

Experimental Procedures

Experimental Diets

For the experiments, female Syrian golden hamsters (Lak:LVG strain, n = 84) weighing ~100 g were purchased from the Charles River Breeding Laboratories

(Wilmington, MA). The hamsters were housed in stainless-steel cages in a temperature-controlled room (22°C) with a 12-h light:dark cycle. Water and food were furnished *ad libitum*.

The hamsters were given 1 wk to acclimate. During this time, the hamsters were fed a Purina Lab Chow (St. Louis, MO) specifically formulated for small rodents. At the end of this first week, the hamsters were weighed and separated into three equal groups (n = 28/group) and given one of three diets. The hamsters in group I were fed the semipurified AIN76A diet (ICN Biochemicals, Aurora, OH). The hamsters in groups II and III were fed the same diet supplemented with 0.05 and 0.5%, respectively, of the limonoid glucoside mixture. The limonoid glucoside mixture was isolated from citrus molasses, which was provided by a commercial juice processing plant. The composition of the limonoid glucoside mixture is given in Table 8.1.

Weight Gain, Food Intake, and Carcinogenesis

Once a week, the hamsters in the three groups were weighed. On six separate occasions, 24-h food intake data were collected. Sixty hamsters were used in the experiment on oral carcinogenesis. The left buccal pouches of these hamsters were treated 3 times a week. Three hamsters per group were used as controls. The pouches in these hamsters were treated with heavy mineral oil. The left buccal pouches of the remaining hamsters (n = 17/group) were treated with a 0.5% solution of the carcinogen, 7,12-dimethylbenz[a]anthracene (DMBA), dissolved in a heavy mineral oil. Each application with a #5 camel hair brush placed ~0.05 mL of the DMBA solution on the surface of the inverted pouch. After a total of 39 treatments, the hamsters were killed by inhalation of an overdose of CO_2.

The treated pouches were excised. The tumors were counted and measured (length, width, and height). The sum of these three measurements divided by six was used to calculate an average radius for each tumor. Using these numbers, an approximation of the volume for each tumor was determined. The formula for the volume of a sphere, $4/3\pi r^3$, was used for this calculation. The sum of the volumes for all of the tumors in one pouch was defined to be that animal's total tumor bur-

TABLE 8.1
Composition of the Limonoid Glucoside Mixture[a]

Limonoid glucoside	%
Limonin 17β-D-glucopyranoside	35.9
Nomilinic acid 17β-D-glucopyranoside	32.0
Nomilin 17β-D-glucopyranoside	12.7
Deacetylnomilinic acid 17β-D-glucopyranoside	8.7
Deacetylnomilin 17β-D-glucopyranoside	6.6
Obacunone 17β-D-glucopyranoside	4.1

[a]The limonoid glucoside mixture was isolated from citrus molasses.

den (Miller *et al.* 1991b, Niukian *et al.* 1987). After the gross tumor data had been collected, the pouches were fixed in 10% formalin and processed by routine histological techniques. The microscopic and macroscopic data were used to determine tumor incidence, number, burden, and type. Student's *t*-test and the Mann-Whitney U test were used to compare the data from the three groups. These procedures are similar to those used in earlier studies with coffee bean fractions and coffee chemicals (Miller *et al.* 1991b, 1992a, 1993, 1994b).

Blood Assays and Enzyme Assay

The remaining hamsters (n = 8/group) were used in this part of the study. None of the pouches in these hamsters were treated with either heavy mineral oil or the solution of the carcinogen. The 24 hamsters received their respective diets for 45 d and then were killed. Blood samples (1–2 mL) were collected by heart puncture and sent to an outside laboratory (Professional Animal Laboratories, Irvine, CA), where a complete blood count and a series of blood chemistries were performed. The technicians at the laboratory had no information on how the hamsters had been treated. After the blood samples had been collected, the left buccal pouches were removed, frozen, and stored at –80°C.

By combining two known procedures, one for the isolation of intact oral epithelial sheets (Harris *et al.* 1980) and the second for the isolation of the enzyme fraction (Lam *et al.* 1981), a technique has been developed to measure the glutathione S-transferase activity of oral epithelial cells (Miller *et al.* 1991a). To isolate the epithelial sheets, the frozen pouches were thawed, cut into strips measuring 3–5 mm, and washed two times with a solution of phosphate-buffered saline (PBS), pH 7.4, containing 4 mmol/L EDTA. The pouches were resuspended in 20 mL of the same solution and incubated at 37°C for 3 h. The extended treatment with EDTA loosens the epithelial sheets. After the incubation, the epithelial sheets were teased off of the underlying connective tissue and washed two times with 10 mL of the PBS solution to remove the EDTA. The epithelial sheets were suspended in 2 mL of a cold (4°C) hypotonic solution, 1:10 PBS, and homogenized in a small hand-held tissue grinder. The combination of hypotonic shock and homogenization lyzes most of the basal cells. Differential centrifugation was used to remove cellular debris and prepare the cytoplasmic fractions.

The incubation mixture (Habig *et al.* 1974) contained 100 mmol/L potassium phosphate buffer, pH 6.5, 5 mmol/L glutathione, 1 mmol/L 1-chloro-2,4-dinitrobenzene (CDNB), and 0.250 mL of the cytoplasmic fraction. CDNB is the color reagent for the enzyme assay. The complete assay mixture, minus the enzyme fraction, was used for the blank. The incubation volume was 2.0 mL and the reaction temperature was 30°C. The assays were run in duplicate and read at 340 nm in a Beckman DU-64 spectrophotometer (Fullerton, CA). The activity of the enzyme was expressed as μmol CDNB metabolized/(min · mg protein). The bicinchoninic acid protein assay (Redinbaugh and Turley 1986) was used to measure the protein concentration of each sample. Student's *t*-test was used to compare the data for the three groups.

Results

Weight Gain and Food Intake

The data showed that there was no significant difference in the weight-gain profiles for the three groups. At the end of the experiment, the average weights ± SEM for the hamsters in the three groups being treated with DMBA were 195 ± 6 g (group I), 190 ± 4 g (group II), and 182 ± 6 g (group III). The mean values for food intake (n = 6) were 11.5 g/d (group I), 11.0 g/d (group II), and 10.8 g/d (group III). The differences in food intake were not significant.

Tumor Data

Three of the hamsters being treated with DMBA died before the end of the experiment. These deaths occurred before the hamsters had a chance to develop tumors. Each of these hamsters was excluded from the study. As indicated in Table 8.2, the number of hamsters per group ranged from a low of 15 for group III to a high of 17 for group II. Gross tumors were found in all of the pouches from the hamsters in group I. Two of the pouches from the hamsters in group II and also group III were free of any apparent tumors. Most of the hamsters, 14 of 16 in group I, 10 of 17 in group II, and 10 of 15 in group III, had multiple tumors (2–7 tumors). The total number of tumors ranged from a low of 37 for groups II and III to a high of 56 for group I.

The data for average tumor number, average tumor radii, and average tumor burden are given in Table 8.3. The mean values for tumor number were 3.5 (group I), 2.2 (group II), and 2.5 (group III). By using Student's *t*-test, it was found that the differences between groups I and II were significant ($P < 0.025$). A similar comparison between groups I and III showed that these differences were not significant ($P < 0.10$). Further analysis of the data (Mann Whitney U test) showed that there was a significant reduction ($P < 0.03$) in tumor number for the hamsters consuming the diets that contained the limonoid glucosides (groups II and III vs. group I). The average radii for the tumors in the three groups ranged from a low of 1.80 mm (group I) to a high of 2.05 mm (group II). Because the tumors in this animal model are exophytic, well defined, and tend to be spherical in shape, the values for tumor radii are used to calculate values for tumor volume. Because the values for tumor radii are cubed, a small difference in average tumor radii can have a relatively large effect on tumor burden

TABLE 8.2
Tumor Incidence

Group	Animals (n)	Tumor-bearing animals (n)	Tumors (n)
I	16	16 (100%)	56
II	17	15 (88%)	37
III	15	13 (87%)	37

TABLE 8.3
Tumor Number, Radii, and Burden

Group	Average number of tumors[a]	Average tumor radii (mm)	Average tumor burden[a] (mm^3)
I	3.5 ± 0.4	1.80	82 ± 15
II	2.2 ± 0.4[b]	2.05	78 ± 24
III	2.5 ± 0.5[b]	1.85	66 ± 23

[a]Values are means ± SEM.
[b]Using the Mann-Whitney U test, $P < 0.03$ (Groups II and III vs. Group I).

(see Table 8.3). Even though the differences in average tumor radii were relatively small and proved to be insignificant, the slight increase in average tumor radii for groups II and III meant that the overall effects on average tumor burden for these two groups were considerably less than the effects on average tumor number. The mean values for tumor burden showed that diets containing the limonoid glucosides reduced average tumor burden by 5–20%. These differences were not significant.

Complete Blood Counts and Blood Chemistries

The complete blood counts for the three groups were normal. The mean values (n = 8) for the seven blood chemistries [serum glutamate oxalate transferase (SGOT), lactate dehydrogenase (LDH), total protein, glucose, creatinine, uric acid, and calcium] that were tested are given in Table 8.4. The data for the three groups were normal and no significant differences were found (Student's *t*-test). The greatest variability was seen in the results for the SGOT assay and the LDH assay. Most of the variability for the SGOT assay was due to one extremely large value of 486 IU/dL for a hamster in group I. The next highest value was 196 IU/dL for one of the hamsters in group III. If this one value is removed, the mean value for group I is reduced to 89 IU/dL. The variability in the LDH assay was found in the results for groups I and III. This can be seen in the SEM for the three groups: 133 IU/dL (group I), 39 IU/dL (group II), and 199 IU/dL (group III).

TABLE 8.4
Blood Chemistries[a,b]

Group	SGOT	LDH	Total protein	Glucose	Creatinine	Uric acid	Calcium
I	139	630	6.5	10.0	0.29	5.5	12.6
II	83	438	6.6	10.2	0.30	4.7	12.3
III	94	731	6.2	10.9	0.24	6.0	12.1

[a]The mean values for SGOT and LDH are in IU/dL. The mean values for total protein are in g/dL. The mean values for glucose, creatinine, uric acid, and calcium are in mg/dL.
[b]Abbreviations: SGOT, serum glutamate oxalate transferase; LDH, lactate dehydrogenase.

Glutathione S-Transferase Assay

The mean values (n = 8) ± SEM for glutathione S-transferase activity in μmol/(min · mg protein) were 0.295 ± 0.014 (group I), 0.277 ± 0.006 (group II), and 0.306 ± 0.016 (group III). The differences between the three groups were not significant (Student's *t*-test).

Discussion

Tumorigenesis Data

It is difficult to compare the results of this experiment on DMBA-induced oral tumors to the results of our earlier studies with the limonoid glucosides (Miller *et al.* 1992b, 1994b). The hamster cheek pouch model was used in each of the experiments; however, in one study, 3.5% solutions of the isolated limonoid glucosides were applied repeatedly (36 times) to the surface of the pouches, whereas in this experiment, the limonoid glucosides were added to the diets of the experimental animals. Another basic difference can be seen in Table 8.1. In the earlier set of experiments, each of the limonoid glucosides was tested separately. In this experiment, the hamsters were exposed to six chemicals rather than one. Each of these chemicals would be expected to have varying degrees of cancer chemopreventive activity. From our earlier research (Miller *et al.* 1989, 1992b, 1994a, 1994b and 1996, Wright *et al.* 1996), one might predict that limonin 17β-D-glucopyranoside would be active, nomilin 17β-D-glucopyranoside and obacunone 17β-D-glucopyranoside would be partially active, and nomilinic acid 17β-D-glucopyranoside would be inactive. The remaining chemicals, deacetylnomilinic acid 17β-D-glucopyranoside and deacetylnomilin 17β-D-glucopyranoside, have not been tested for antineoplastic activity.

In the earlier experiment with limonin 17β-D-glucopyranoside, the topical treatments with this chemical reduced tumor number by 15% and tumor burden by 55%. The reduction in tumor burden was significant. With nomilin 17β-D-glucopyranoside, the overall reduction in tumor number was slightly greater (20%), but the effect on tumor burden was smaller (15% reduction). Neither effect proved to be significant. In this experiment, the diets containing the limonoid glucoside mixture (groups II and III) reduced tumor number by 30–35% and tumor burden by 5–20%. The reduction in tumor number was significant. From the three sets of results, it is obvious that the greatest effect on tumor number occurred in this experiment with the limonoid glucoside mixture. From the results, it can also be seen that the pattern for the data for the limonoid glucoside mixture is similar to the pattern for the earlier data with nomilin 17β-D-glucopyranoside.

Taken together, the data (Tables 8.2 and 8.3) from the experiment on DMBA-induced oral tumors indicated that the limonoid glucoside mixture was active systemically. In addition, the data on tumor number showed that the reduction was essentially the same for groups II and III, suggesting that the smaller dose of limonoid glucosides was just as effective as the higher dose. This possibility suggests that an even lower dose of the limonoid glucoside mixture (e.g., 0.01 or 0.005%) might be equally effective in reducing tumor number. Because nothing is known about the uptake, distribution, turnover, and metabolism of limonoid glucosides in mammalian systems, it is impossible to predict how effective this same mixture might be on the development of carcinogen-induced tumors in other tissues. In addition, it is not known which of these limonoid glucosides may have contributed directly or indirectly to the anticancer activity of the mixture. It is also possible that the active agents were breakdown products that were formed during the metabolism of these compounds. For example, the glucose residues may have been removed and the aglycones may have been the active agents.

Comparison to Human Consumption

Looking at the data for weight gain and food intake, it can be seen that the pattern for the two sets of results was similar. By the end of the experiment, the average weight for the hamsters in group I was greater than the average weight for the hamsters in group II, which was greater than the average weight for the hamsters in group III. The pattern for the food intake data was the same (group I > group II > group III). Even though the differences in weight gain and food intake were not significant, it appears that the limonoid glucoside mixture may have lowered food consumption by 4–6% which, in turn, led to a slight reduction (3–7%) in the weight of the hamsters in groups II and III.

With the use of the food intake data, it was calculated that the average intake of mixed limonoid glucosides was 5.5 mg/d (group II) and 54 mg/d (group III). Combining these data with the data on weight gain, it was calculated that an adult weighing 60 kg would have to consume 1.6 g of mixed limonoid glucosides per day to equal the exposure for the hamsters in group II and 16 g of mixed limonoid glucosides per day to equal the exposure for the hamsters in group III. It is estimated that a person who drinks two glasses of orange juice per day will have consumed 0.16 g of mixed limonoid glucosides. One problem with this comparison is in the composition of the two limonoid glucoside mixtures. The predominant limonoid glucosides in orange juice and citrus molasses are the same (limonin 17β-D-glucopyranoside, nomilinic acid 17β-D-glucopyranoside, and nomilin 17β-D-glucopyranoside). As indicated in Table 8.1, the concentration of these limonoid glucosides in the mixture isolated from citrus molasses was 35.9% limonin 17β-D-glucopyranoside, 32.0% nomilinic acid 17β-D-glucopyranoside, and 12.7% nomilin 17β-D-glucopyranoside. In orange juice, the figures are quite different, ranging from 53.2% for limonin 17β-D-glucopyranoside, to 25.4% for nomilinic acid 17β-D-glucopyranoside, to 14.2% for nomilin 17β-D-glucopyranoside (Herman *et al.* 1990). On the basis of the difference in the concentration of limonin 17β-D-glucopyranoside in orange juice and citrus molasses, one might expect that the anticancer activity of the limonoid glucoside mixture from orange juice would be greater than the anticancer activity of the limonoid glucoside mixture from citrus molasses.

Blood Assays

The assays performed at Professional Animal Laboratories indicated that the blood profiles for the three groups were normal. Some variability was noted in the data for the blood chemistries (Table 8.4) but no significant differences or trends were seen. The data from the blood chemistries suggested that the basic function of tissues such as liver, bone, heart, pancreas, and skeletal muscle were normal. The data from the blood assays, the food intake studies, and even the data on weight gain indicated that the long-term exposure to the limonoid glucoside mixture had no apparent effect on the general health of the hamsters.

Glutathione S-Transferase Activity

Previous research has suggested that some limonoids (nomilin, obacunone, and possibly nomilin 17β-D-glucopyranoside) can increase the activity of glutathione S-transferases (Lam and Hasegawa 1989; Lam *et al.* 1989, 1994a and 1994b). The glutathione S-transferase enzymes catalyze the adduct formation of glutathione with electrophiles, including the reactive forms of carcinogens. This reaction makes the compound more water soluble and easier to excrete. Any increase in the activity of these enzymes should increase the rate at which carcinogens are safely removed from the cell and from the body (Wattenberg 1983). The data from this experiment showed that the limonoid glucoside mixture had no effect on the glutathione S-transferase activity of the oral epithelial cells, supporting an earlier conclusion (Lam *et al.* 1994b, Miller *et al.* 1994b and 1996) that some of these limonoids are producing their effects through another, as yet undetermined, mechanism of action.

Taken together, all of these data support the idea that citrus limonoids may be of benefit to humans. Considering the high concentration of limonoid glucosides in citrus fruits and juices, it is possible that these chemicals are already contributing to the protective effect of citrus consumption on a wide variety of cancers in humans.

Acknowledgments

The authors thank Josephine Taylor and Panagiota Laghios for their technical assistance. The research was funded through a grant from the Citrus Research Board (#5600-105b).

References

Block, G., Vitamin C and Cancer Prevention: The Epidemiologic Evidence, *Am. J. Clin. Nutr. 53*: 270S–282S (1991).

Fong, C.H., S. Hasegawa, Z. Herman, and P. Ou, Limonoid Glucosides in Commercial Citrus Juices, *J. Food Sci. 54*: 1505–1506 (1990).

Guthrie, N., A.F. Chambers, and K.K. Carroll, Inhibition of MDA-MB-435 Estrogen Receptor-Negative Human Breast Cancer Cells by Citrus Limonoids, *Proc. Am. Assoc. Cancer Res. 38*: 113S (1997).

Habig, W.H., M.J. Pabst, and W.B. Jakoby, Glutathione S-Transferase, *J. Biol. Chem. 249*: 7130–7139 (1974).

Harris, R.R., I.C. MacKenzie, and R.A.D. Williams, An Examination of the Relationship Between Experimentally Altered Rates of Epidermal Proliferations and Rates of Epidermal Metabolism Assayed *in vitro*, *J. Investig. Dermatol. 74*: 402–406 (1980).

Hasegawa, S., R.D. Bennett, Z. Herman, C.H. Fong, and P. Ou, Limonoid Glucosides in Citrus, *Phytochemistry 28*: 1717–1720 (1989).

Herman, Z., C.H. Fong, P. Ou, and S. Hasegawa, Limonoid Glucosides in Orange Juices by HPLC, *J. Agric. Food Chem. 38*: 1860–1861 (1990).

Lam, L.K.T., and S. Hasegawa, Inhibition of Benzo[a]pyrene-Induced Forestomach Neoplasia in Mice by Citrus Limonoids, *Nutr. Cancer 12*: 43–47 (1989).

Lam, L.K.T., Y. Li, and S. Hasegawa, Effects of Citrus Limonoids on Glutathione S-Transferase Activity in Mice, *J. Agric. Food Chem. 37*: 878–880 (1989).

Lam, L.K.T., V.L. Sparnins, J.D. Hochalter, and L.W. Wattenberg, Effects of 2- and 3-*tert*-Butyl- 4-hydroxyanisole on Glutathione S-Transferase and Epoxide Hydrolase Activities and Sulfhydryl Levels in Liver and Forestomaches of Mice, *Cancer Res. 41*: 3940–3943 (1981).

Lam, L.K.T., J. Zhang, and S. Hasegawa, Citrus Limonoid Reduction of Chemically Induced Tumorigenesis, *Food Technol. 48*: 104–108 (1994a).

Lam, L.K.T., J. Zhang, S. Hasegawa, and H.A.J. Schut, Inhibition of Chemically Induced Carcinogenesis by Citrus Limonoids, in *Food Phytochemicals for Cancer Prevention I. Fruits and Vegetables*, edited by M.T. Huang, T. Osawa, C.-T. Ho, and R.T. Rosen, American Chemical Society, Washington, DC, 1994b, pp. 209–219.

Miller, E.G., Nutrition—The Primary Risk Factor That Nobody Talks About, *Texas Dent. J. 113*: 35–38 (1996).

Miller, E.G., A.M. Couvillon, A.P. Gonzales-Sanders, W.H. Binnie, H.P. Wurzner, and G.I. Sunahara, Evidence for the Presence of Multiple Cancer Chemopreventive Agents in Green Coffee Beans, in *14th International Scientific Colloquium on Coffee,* Association Scientifique Internationale du Café, Paris, 1992a, pp. 46–51.

Miller, E.G., R. Fanous, F. Rivera-Hidalgo, W.H. Binnie, and S. Hasegawa, The Effect of Citrus Limonoids on Hamster Buccal Pouch Carcinogenesis, *Carcinogenesis 10*: 1535–1537 (1989).

Miller, E.G., A.P. Gonzales-Sanders, A.M. Couvillon, W.H. Binnie, S. Hasegawa, and L.K.T. Lam, Citrus Limonoids as Inhibitors of Oral Carcinogenesis, *Food Technol. 48*: 110–114 (1994a).

Miller, E.G., A.P. Gonzales-Sanders, A.M. Couvillon, W.H. Binnie, G.I. Sunahara, and R. Bertholet, Inhibition of Oral Carcinogenesis by Roasted Coffee Beans and Roasted Coffee Bean Fractions, in *15th International Scientific Colloquium on Coffee,* Association Scientifique Internationale du Café, Paris, 1993, pp. 420–425.

Miller, E.G., A.P. Gonzales-Sanders, A.M. Couvillon, J.M. Wright, S. Hasegawa, L.K.T. Lam, Inhibition of Hamster Buccal Pouch Carcinogenesis by Limonin 17β-D-Glucopyranoside, *Nutr. Cancer 17*: 1–7 (1992b).

Miller, E.G., A.P. Gonzales-Sanders, A.M. Couvillon, J.M. Wright, S. Hasegawa, L.K.T. Lam, and G.I. Sunahara, Inhibition of Oral Carcinogenesis by Green Coffee Beans and

Limonoid Glucosides, in *Food Phytochemicals for Cancer Prevention I. Fruits and Vegetables,* edited by M.T. Huang, T. Osawa, C.T. Ho, and R.T. Rosen, American Chemical Society, Washington, DC, 1994b, pp. 220–229.

Miller, E.G., S.J. Manek, J.M. Wright, A.P. Gonzales-Sanders, and A.M. Couvillon, Isolation of Glutathione S-Transferase from Buccal Pouch Epithelial Cells, *J. Dent. Res. 70*: 548S (1991a).

Miller, E.G., K. McWhorter, F. Rivera-Hidalgo, J.M. Wright, P. Hirsbrunner, and G.I. Sunahara, Kahweol and Cafestol: Inhibitors of Hamster Buccal Pouch Carcinogenesis, *Nutr. Cancer 15*: 41–46 (1991b).

Miller, E.G., J.L. Porter-Wright, W.H. Binnie, I.Y. Guo, and S. Hasegawa, The Importance of the B Ring of the Limonoid Nucleus to the Cancer Chemopreventive Activity of Citrus Limonoids, in *Hypernutritious Foods*, edited by J.W. Finley, D. J. Armstrong, S. Nagy, and S.F. Robinson, Agscience Incorporated, Auburndale, FL, 1996, pp. 191–204.

Niukian, K., J. Schwartz, and G. Shklar, Effects of Onion Extract on the Development of Hamster Buccal Pouch Carcinomas as Expressed in Tumor Burden, *Nutr. Cancer 9*: 171–176 (1987).

Redinbaugh, M.G., and R.D. Turley, Adaptation of the Bicinchoninic Acid Protein Assay for Use with Microtiter Plates and Sucrose Gradient Fractions, *Anal. Biochem. 153*: 267–271 (1986).

So, F.V., N. Guthrie, A.F. Chambers, M. Moussa, and K.K. Carroll, Inhibition of Human Breast Cancer Cell Proliferation and Delay of Mammary Tumorigenesis by Flavonoids and Citrus Juices. *Nutr. Cancer 26*: 167–181 (1996).

Wattenberg, L.W., Inhibition of Neoplasia by Minor Dietary Constituents, *Cancer Res. 43*: 2448S–2453S (1983).

Wright, J.L., E.G. Miller, W.H. Binnie, I.Y. Guo, and S. Hasegawa, Further Studies on the Anticancer Activity of Citrus Limonoids, *J. Dent. Res. 75*: 436S (1996).

Chapter 9

Bioactive Components of Flaxseed: Occurrence and Health Benefits

B. Dave Oomah and G. Mazza

Agriculture and Agri-Food Canada, Pacific Agri-Food Research Center, Summerland, British Columbia, Canada V0H 1Z0

Introduction: Flaxseed Oil

In addition to being one of the richest sources of α-linolenic acid oil, flaxseed is an essential source of high quality protein and soluble fiber, and has considerable potential as a source of phenolic compounds, especially flavonoids, lignans and phenolic acids. The demonstration of clinical activity associated with the consumption of flaxseed has led the National Cancer Institute (NCI) to target flaxseed as one of the six plant materials for study as cancer-preventive foods (Caragay 1992). Recently, the preventive and therapeutic potential of flaxseed has received increased scientific attention from evidence of beneficial biological effects in reducing the risk of degenerative diseases. The result has been a proliferation of literature on flaxseed with emphasis on human nutrition and disease prevention (Cunnane and Thompson 1995, Oomah and Mazza 1998).

Flaxseed generally contains 5% palmitic acid (16:0); 3% stearic acid (18:0); 17% oleic acid (18:1n-9); 15% linoleic acid (18:2n-6); and 59% α-linolenic acid (ALA; 18:3n-3) (DeClerq 1996). The successful modification of flaxseed oil to improve its stability and competitiveness as a salad or cooking oil has led to the introduction of solin-type varieties with low (<3%) α-linolenic acid, and concomitant increases in linoleic acid (~72%) (Anonymous 1997). Further alterations of the fatty acid composition of flaxseed, such as the development of high oleic, palmitic and palmitoleic acid cultivars are currently in progress (Miller 1993; Ntiamoah *et al.* 1995).

The effect of flaxseed oil on disease states is due to its high level of polyunsaturated fatty acids (PUFA) and α-linolenic acid in particular. The reported biological properties of α-linolenic acid (Table 9.1) demonstrate a broad spectrum of potential health benefits associated with the consumption of this fatty acid. Studies with deuterated α-linolenic acid suggest that humans are capable of desaturating and chain elongating it to eicosapentaenoic acid (EPA; 20:5n-3) and decosahexaenoic acid (DHA; 22:6n-3) (Emken *et al.* 1992). Elongation and desaturation of ALA to EPA and DHA occur in human leukocytes and livers (Simopoulos 1994). DHA can be synthesized from dietary ALA and can be found in membrane phospholipids of practically all cells of individuals consuming n-3 fatty acids. DHA is one of the most abundant components of the brain structural lipids, phosphatidylethanolamine (PE), phosphatidyl-

TABLE 9.1
Reported Biological Effects of α-Linolenic Acid

Biological effect	Reference
Modulation of synthesis and metabolism of eicosenoids derived from arachidonic acid	Hwang and Carroll (1980)
Reduction of production of "4 series" leukotrienes	Henry *et al.* (1990)
Contribution to the total pool of lipogenic precursors for brain lipid synthesis	Cook (1991), Holman *et al.* (1982)
Alleviation of clinical neurological symptoms, including numbness, visual blurring and paresthesil	de Lorgeril *et al.* (1994)
Protection against cardiovascular disease and myocardial infarction	Dyerberg (1986), Singer (1992)
Reduction of high blood pressure in hypertensives and lower plasma triglyceride and cholesterol levels	Burr *et al.* (1989), Cunnane *et al.* (1991)
Reduction of mortality due to cancer	Serraino *et al.* (1992)
Potential of antithrombogenic and antiarrhythmic effects	Kelly (1995)
Inhibition of lymphocyte proliferation with benefits to individuals with immune disorders	Johnston (1995)
Retardation of tumor growth	Serraino *et al.* (1992)
Essential for optimal neurological development in humans	Hwang and Carroll (1980)

choline (PC), and phosphatidylserine (PS). Hence, an increase in the consumption of ALA is important to provide adequate amounts of DHA to meet the needs of humans.

ALA may be an important physiologic modulator in the synthesis and metabolism of eicosanoids derived from arachidonic acid (AA) (Cunnane 1995) and may have inhibitory effects on basal and stimulated production of thromboxane (Tx) A_2, leukotrienes B_4 and C, and 12-hydroxyeicosatetraenoic acid (Henry *et al.* 1990, Hwang and Carroll 1980). α-Linolenic acid also alters the *in vitro* production of several prostanoids, including prostaglandin PGE_2, PGF, PGI_2, TxB_2, and leukotriene E_5 (Calder *et al.* 1992, Marshall and Johnston 1985). Animal studies suggest that ALA can be actively recycled, as a result of β-oxidation, into other lipids, particularly in the developing brain (Anderson and Connor 1994, Cook, 1991). ALA supplementation (~1.6 g/d) resolved the clinical neurological symptoms of α-linolenic deficiency, including numbness, visual blurring, and paresthesia over 3 mo (Holman *et al.* 1982). When ALA intake increased threefold in the Lyon "Mediterranean Diet" heart study (de Lorgeril *et al.* 1994), a 76% reduction in cardiovascular death, apparently due to the higher intake of ALA, was observed. Recently, an ALA-rich Mediterranean diet was shown to result in a 70% reduction in cardiovascular mortality over a 2-y observation period compared with a control diet (de Lorgeril *et al.* 1994). ALA has been shown to reduce blood pressure in hypertensives (Singer 1992) and lower triacylglycerols and cholesterol (Cunnane *et al.* 1991, Dyerberg 1986). The potential antithrombogenic effect of dietary ALA, possibly *via* EPA formation and/or damping the formation of AA-derived eicosanoids, was suggested by de Lorgeril *et al.* (1994), but reduced mortality could also have been mediated *via* an antiarrhythmic effect as suggested previously from the DART trial (Burr *et al.* 1989). Diets containing ALA inhibit lymphocyte proliferation in animals and immune response in healthy humans, which led Kelly (1995) to conclude that a high intake of ALA may be beneficial to individuals with autoimmune disorders. In addition to the evidence that dietary ALA may retard tumor growth, a number of reports indicate that it may also have a role in metastasis (Johnston 1995).

Although the physiologic benefits of flaxseed oil are attributed primarily to the high content of ALA, the functional properties of α-linolenic acid are not the same as those of flaxseed oil, because flaxseed oil contains other fatty acids ranging from 40 to 60%. Thus, flaxseed oil as a food ingredient has functional properties that go beyond the effects of α-linolenic acid. Even low dietary levels of flaxseed oil lead to deposition of ALA and its metabolite, docosapentaenoic acid (DPA; 22:5n-3), in mammary glands and tumors of carcinogen-treated rats and in the mammary glands of untreated animals (Serraino *et al.* 1992). Flaxseed oil feeding also reduces DMBA-induced mammary tumorigenesis in mice, and the effect is both fat and concentration dependent (Cameron *et al.* 1989).

Eicosapentaenoic acid produced *in vivo* from ALA from flaxseed (Johnston 1988) can compete for both the cyclooxygenase and lipoxygenase pathways, leading to the reduction of prostaglandins (PGE_2) and 4-series leukotrienes, which is reflected consequently in tumor inhibition. Recently, secondary products of peroxidation, especially from highly unsaturated fatty acids, have been implicated in inducing profound damage when constantly in contact with cancerous cells. This inhibitory effect of fatty acids such as ALA may suppress mammary tumor growth, in part as the result of the formation of toxic compounds by lipid peroxidation (Haumann 1997).

The biological effects of flaxseed oil may reflect the particular fatty acid profile of the oil itself, including the amounts of individual fatty acids and their ratio. Corn, canola, and soybean oils have n-6:n-3 fatty acid ratios of approximately 35:1, 2:1, and 7:1, respectively, whereas that of flaxseed oil is ~0.3:1 (Johnston 1995). It is believed that the optimal ratio of n-6:n-3 PUFA in the human diet should be 5:1; in modern Western diets, however, the ratio has shifted toward the n-6 fatty acids, leading to a ratio of ~40:1 (Eaton and Konner 1985). Hence flaxseed oil is one of the first choices as an essential fatty acid supplement because it provides both n-3 and n-6 fatty acids with a relatively larger amount of n-3 PUFA than is found in most foods or commercial oils. The low n-6:n-3 fatty acid ratio of flaxseed oil can lead to profound changes in biological processes that involve eicosanoids, including inflammation and blood clotting (Berdanier 1994); it may alter mammary gland development, puberty onset, and breast cancer risk (Hilakivi-Clarke *et al.* 1997). The high dietary n-3 fatty acids of flaxseed oil may be more important in infancy and early childhood than in adulthood when adequate bodily stores of these compounds exist (Nelson and Chamberlain 1995). The low n-6:n-3 fatty acid ratio of flaxseed oil permits significant conversion of ALA to EPA, which may result in beneficial reduction of platelet reactivity (Holub 1995).

Dietary Proteins

Flaxseed proteins, like other dietary proteins, play both obvious and subtle roles in the maintenance of good health in humans. Dietary proteins are believed to have an effect on coronary heart disease, cancer, kidney disease, and gallstone disease (Liepa *et al.* 1992). The composition, functionality, and uses of flaxseed proteins have been reviewed recently (Oomah and Mazza 1993 and 1995). The protein fraction contains a favorable ratio of amino acids (Table 9.2) with lysine, threonine, and tyrosine as the limiting amino acids. It is a good source of sulfur amino acids, methionine, and cystine. Flaxseed meal has an essential amino acid index of 69 compared with 79 and 75 for soybean and canola meals, respectively. Similarly, the protein score, based on the most limiting amino acid, relative to Food and Agriculture Organization (FAO) nutritional requirements, is 82 for flaxseed meal and 67 for soybean meal (Sosulski and Sarwar 1973). The lysine/arginine ratio, which is a determinant of the cholesterolemic and artherogenic effects of a protein (Czarnecki and Kritchevsky 1992), is 0.37 for flaxseed, suggesting that it is less lipidemic and artherogenic than either soybean or canola proteins, which have a lysine/arginine ratio of 0.88.

There are two ways in which flaxseed protein may influence blood glucose. The protein in flaxseed may stimulate insulin secretion, which could result in a reduced glycemic response (Nuttall *et al.* 1984). Furthermore, protein may be important in determining the glycemic responses of foods because of its interaction with polysaccharides. Lignans are also known to have strong protein-binding properties (Aldercreutz *et al.* 1993), which may suggest some partial chemopreventive effect of flaxseed protein in conjunction with lignans. The interaction between

TABLE 9.2
Essential Amino Acid Composition of Defatted Flaxseed, Soybean and Canola[a]

Amino acid	Flaxseed	Soybean (g/100 g)	Canola	Requirement
Arginine	11.5	7.5	7	5
Histidine	3	3.1	3.8	2.5
Isoleucine	5.1	4.7	3.7	4.6
Leucine	7	7.8	7.2	6.2
Lysine	4.2	6.6	6.1	7.5
Met + Cys	6.4	3	4.7	5
Phe + Tyr	8.3	8.3	6.8	6.7
Threonine	5	4.2	4.6	4.2
Tryptophan	1.8	1.1	1.1	1.2
Valine	5.6	4.9	5	5

[a]Values were calculated on an oil- and moisture-free basis.

flaxseed protein and soluble polysaccharides may play a significant role in reducing colon luminal ammonia, thereby protecting against the known tumor-promoting effects of ammonia (Clinton 1992).

A novel class of pathogenesis-related proteins, i.e., a 25-kDa protein isolated from flaxseed, has been reported to have antifungal properties against an agronomically important pathogen of tomato that causes early blight (*Alternaria solani*) (Borgmeyer *et al.* 1991). This protein also acts in synergism with nikkomycin Z to strongly inhibit the growth of the human pathogen *Candidas albicans*. Another class of native seed proteins, the oleosins, with both lipophilic and hydrophilic sites, is present in flaxseed (Huang 1992, Wilen *et al.* 1991). The oleosins may act as receptors for lipase binding in a manner analogous to apolipoprotein C-II binding of lipoprotein in mammalian transport lipid bodies. These proteins are being used in oleosin-partitioning technology for the production and purification of biologically active pharmaceuticals, such as the anticoagulant hirudin and human cytokine-interlukin 1 (Van Rooijen and Moloney 1995).

Polysaccharides

Flaxseed is well known for its high dietary fiber content (Bhatty and Cherdkiatgumchai 1990, Mazza and Oomah 1995). Flaxseed polysaccharide gum, also referred to as mucilage, comprises about 8% of the seed weight (Bhatty and Cherdkiatgumchai 1990). It may have nutritional value as a dietary fiber; as such, it appears to play a role in reducing diabetes and coronary heart disease risk, preventing colon and rectal cancer, and reducing the incidence of obesity (Jenkins *et al.* 1985, Mazza and Oomah 1995). The therapeutic use and biological activity of dietary fiber depend on the type and monomer composition of the polysaccharide. Flaxseed mucilage is predominantly of the rhamnogalacturonan- and arabinoxylan-type polysaccharides, whereas xyloglucans predominate in cereals (Table 9.3) (Cui *et al.* 1996). Hence, the response to oat

TABLE 9.3
Monosaccharide Composition of the Cell-Wall Polysaccharides (NSP) from Various Plant Seeds[a]

	Monosaccharides (% total NSP)						
Plant source (seed)	6-Deox hexose[b]	Ara	Xyl	Man	Gal	Glc	UA[c]
NorMan Flaxseed	26.2	13.5	37.4	0	20	2.1	23.2
Solin[d] (Linola)	15.8	18.1	42.5	0	18.4	3.7	15.7
Barley	0.5	14.7	25.1	2.7	1.9	51.8	3.3
Oats	0.4	7.1	33.7	1.6	2.4	50.5	4.2
Wheat	0	20.9	33.8	4.1	4.3	33	3.9

[a]*Source:* (Chesson 1995, Cui *et al.* 1996). Abbreviations: NSP, non-starch polysaccharides; Ara, arabinoxylan; Xyl, xyloglucan; Man, mannose; Gal, galactose; Glc, glucose; UA, uronic acids.
[b]Predominantly rhamnose with small amounts of fucose.
[c]Galacturonic and (4-*O*-methyl)glucuronic acids.
[d]Low linolenic flax

and barley (reductions of 3–26% total serum cholesterol) is more variable and generally less than that achieved with soluble polysaccharides such as those present in flaxseed (Chesson 1995).

Flaxseed soluble polysaccharides behave like typical viscous fibers with the ability to reduce postprandial blood glucose response (Wolever and Jenkins 1993). Soluble fiber is believed to reduce glycemic responses by increasing the viscosity of the small intestinal contents and delaying the digestion and absorption of carbohydrates (Edwards *et al.* 1987, Jenkins *et al.* 1978). The viscosity of flaxseed gum reduces serum cholesterol and flattens the blood glucose profile in a manner similar to that of guar gum, psyllium, oat gum, and other viscous fibers intensively studied for this purpose (Jenkins 1995). The addition of flaxseed gum to a glucose-containing solution reduces the area under the blood response curve by 27% compared with oral glucose alone (Cunnane *et al.* 1993). The beneficial effects of flaxseed gum on glucose metabolism are not completely understood. The hypothetical mechanism by which flaxseed reduces acute blood glucose response is by slowing carbohydrate digestion and absorption due to a combination of factors, including soluble fiber, and through the presumed effects of starch-protein interactions and antinutrients (Blackburn *et al.* 1984). Reducing blood glucose response contributes to improving overall blood glucose control and is likely to be beneficial for individuals with glucose intolerance.

The hypolipidemic, cholesterolemic, and atherogenic effects of flaxseed have been demonstrated in animal and human studies (Bierenbaum *et al.* 1993, Cunnane *et al.* 1995, Ranhotra *et al.* 1993). The effects of flaxseed gums as a dietary fiber on serum cholesterol and lipid risk factors for coronary heart disease (CHD) may be due in part to their physicochemical properties, such as viscosity, water-holding capacity, apparent solubility, binding ability, degradability, particle size, and alterations by processing (Anderson *et al.* 1990) and in part to the high production of short-chain fatty acids from fermentation in the colon (Berggren *et al.* 1993). The unusual combination of rheological and surface chemical properties make flaxseed polysaccharide extremely suitable for application as a saliva substitute for the prevention of xerostomia and offers practical utility as a bioadhesive for skin and mucous membranes of the body (Attström *et al.* 1993, O'Mullane and Hayter 1993). Recent experiments carried out in the authors' laboratory have shown that flaxseed gum possesses high antioxidant activity, which is significantly correlated with phenolic content (Table 9.4).

Phenolic Acids

Several bioactive functions, such as antioxidant, antimicrobial, and anticancer effects, have been attributed to phenolic acids. The levels of total and esterified phenolic acids are 81 and 73.9 mg/100 g, respectively, for dehulled defatted flaxseed meal (Dabrowski and Sosulski 1984). The major phenolic acids for dehulled defatted flaxseed are *trans*-ferulic (46%), *trans*-sinapic (36%), *p*-coumaric (7.5%), and *trans*-caffeic (6.5%), for both total and esterified phenolic acids. Flaxseed contains 8–10 g/kg of total phenolic acids, ~5 g/kg of esterified phenolic acids, and 3–5 g/kg of etherified phenolic acids.

TABLE 9.4
Antioxidant Activity and Total Phenolics of Methanolic Extracts of Flaxseed and Flaxseed Gum

Source	Antioxidant activity (%)	Total phenolic (g/kg)
Flaxseed	60.6	5.1
Solin	61.6	4.7
Flaxseed gum	78.6	13.5
Solin gum	82.1	14.2

The relative amounts of the three classes of phenolic acids present in flaxseed differ significantly among cultivars. The esterified phenolic acids represent 48–66% of the total phenolic acids and are not dependent on cultivar (Table 9.5). Variation in phenolic acids are due mainly to seasonal effect (Oomah *et al.* 1995). Ferulic acid (10.9 mg/g), chlorogenic acid (7.5 mg/g), gallic acid (2.8 mg/g), and traces of 4-hydroxybenzoic acid are the phenolics identified in defatted flaxseed powder (Harris and Haggerty 1993).

Flavonoids

Flavonoids are a group of polyphenolic compounds that contain 15 carbon atoms in their basic skeleton, arranged in a C_6-C_3-C_6 configuration. These compounds have been reported to exhibit a wide range of biological effects including antibacterial, antiviral, anti-inflammatory, antiallergenic, and vasodilatory actions. In addition, flavonoids inhibit lipid peroxidation, platelet aggregation, capillary permeability and fragility, and the activity of enzyme systems including cyclooxygenase and lipoxygenase (Cook and Samman 1996). The major flavonoids in flaxseed cotyledons are flavone C- and O-glycosides (Ibraham and Shaw 1970). The content of flavonoids in flaxseed ranges from 35 to 71 mg/100 g and varies with cultivar and environment (Oomah *et al.* 1996b). Genetic transformation conferring herbicide resistance to flaxseed does not appear to affect flavonoid content significantly (Oomah *et al.* 1996b).

Phytic Acid

Phytic acid is a natural plant inositol hexaphosphate commonly found in seeds; it represents the principal form of stored phosphate. Phytic acid has been reported to exhibit a variety of physiologic effects including hypocholesterolemic, antioxidative, anticarcinogenic, hypolipidemic, and hypoglycemic effects. The mechanisms by which these effects occur, however, are not known. It has been suggested that this phytochemical may be responsible in part for some of the health benefits attributed to high-fiber foods (Thompson 1994a). A reduction in mammary and colon cancer risk has also been attributed to the presence of phytic acid and phenolic acids in foods. The mechanism of the inhibitory effect of these phytochemicals on the progression of carcinogenesis seems to be multifaceted and has been summarized elsewhere (Thompson 1993 and 1994b). Phytin from defatted flaxseed flour contains 13% organic phosphorous and constitutes ~6% of the flour (Bolley and McCormack 1952). Defatted flaxseed contains 1.8–3.0% phytic acid, representing ~70% of the total phosphorous. (Bhatty and Cherdkiatgumchai 1990). Flaxseed contains 23–33 g phytic acid/kg meal, depending on the cultivar and environment (Table 9.5) (Oomah *et al.* 1996a).

Tocopherols

Tocopherols are important natural fat-soluble antioxidants. A commercial flaxseed variety grown in 1994 contained 0.88, 2.42, 9.2, 0.24 of α-, β-, γ-, δ- and 12.74 mg/100 g of seed (on a wet basis) of total tocopherols, respectively (Budin *et al.* 1995). Total tocopherol contents of flaxseed from 9 varieties grown at 13 locations worldwide ranged from 40 to 50 mg/100 g of oil; these differences were ascribed primarily to location (Marquard *et al.* 1977). Flaxseed contained an average of 9.3 mg/100 g of total tocopherol in the seed with γ-tocopherol representing >80% of the total. The level of tocopherol in flaxseed is cultivar specific and regulated by environmental conditions (Table 9.5) (Oomah *et al.* 1997a). Flaxseed hull contains ~26% of the total tocopherol found in the whole seed. The ratio of α-, γ- and δ-tocopherols of the hull (17:61:22) is different from that of the seed (1:94:5) and could be due to the preponderance of γ-tocopherol in embryo lipids and the presence of δ- and γ-tocopherol in seed coat lipids (Oomah *et al.* 1997a). Recently, high levels of plastochromanol-8 (18–26 mg/100 g flaxseed oil), with twice the antioxidative activity of α-tocopherol, have been reported in flaxseed (Olejnik *et al.* 1997). Tocopherols protect polyunsaturated fatty acids in cell membranes from oxidation. Furthermore, tocopherols keep selenium (which protects against oxidative tissue damage) in the reduced state, thus facilitating the antioxidant capacity of selenium. Additionally, vitamin E has been shown to decrease the formation of nitrosamines, which may be carcinogenic, particularly in the stomach (Bertram *et al.* 1987, Fiala *et al.* 1985).

Lipoxygenase

The physiologic function of lipoxygenase as a phytohormone in plant host-pathogen interaction is well known. A prostaglandin-like metabolite (phytotonic acid) produced on incubation of linolenic acid with an extract of flaxseed acetone powder (presumably containing lipoxygenase) has been characterized (Rabinovitch-Chable *et al.* 1978, Zimmerman and Vick 1970). Through purification, the activity of flaxseed lipoxygenase has been identified as that of an iron-containing protein of 130 kDa (Rabinovitch-Chable *et al.* 1992). Large variability exists in lipoxygenase content of flaxseed cultivars with a range from 1.6 to 6.0 g/kg (Table 9.5) (Oomah *et al.* 1997b). The production of phytohormones such as prostaglandin, a potent mediator of inflammation and autoimmune disease, *via* lipoxygenase enzyme has been suggested (Zimmerman

TABLE 9.5
Concentration of Selected Phytochemicals in Flaxseed Cultivars

Cultivar	Phenolic acids (g/kg)			Phytic acid (g/kg)	Tocopherol (mg/100g)				Lipoxygenase content (g/kg)
	Total	Esterified (1 mol/L NaOH)	Etherified (calculated)	Mean	α-	γ-	δ-	Total	
AC Emerson	9.92^{b}	4.79^{c}	5.13^{a}	28.33^{b}	0.05^{b}	8.80^{de}	0.22^{abc}	9.08^{cd}	3.62^{b}
AC Linora	10.34^{a}	5.38^{a}	4.97^{a}	22.75^{g}	0.07^{b}	9.72^{a}	0.22^{abc}	10.01^{a}	1.86^{f}
Flanders	8.46^{e}	5.12^{b}	3.34^{c}	24.23^{f}	0.10^{a}	8.96^{cd}	0.30^{a}	9.35^{bc}	2.36^{d}
Linola 947	7.86^{g}	5.19^{ab}	2.67^{d}	32.49^{a}	0.06^{b}	9.28^{b}	0.18^{bc}	9.52^{b}	5.98^{a}
McGregor	8.13^{f}	5.42^{a}	2.72^{d}	25.99^{d}	0.02^{c}	9.37^{b}	0.17^{c}	9.56^{b}	2.25^{e}
NorLin	9.48^{c}	5.21^{ab}	4.27^{b}	27.27^{c}	0.10^{a}	8.45^{f}	0.29^{a}	8.84^{d}	1.63^{g}
Somme	8.85^{d}	5.42^{a}	3.44^{c}	23.99^{f}	0.05^{b}	8.63^{ef}	0.22^{abc}	8.90^{d}	3.57^{b}
Vimy	9.52^{c}	5.30^{ab}	4.22^{b}	25.27^{e}	0.06^{b}	9.12^{bc}	0.26^{ab}	9.44^{b}	3.14^{c}

$^{a\text{-}g}$Means followed by the same superscript are not significantly different by Duncan's multiple range test at 5% level.

and Feng 1978). Because lipoxygenase is involved in cytokine production, it may have immunomodulatory effects on dietary ALA (Eaton and Konner 1985).

Cyanogenic Compounds

Cyanogenic compounds are nitrogenous secondary plant metabolites capable of liberating hydrogen cyanide upon hydrolysis. The composition, detection, and distribution of cyanogenic glycosides, linamarin, linustatin, and neolinustatin, present in flaxseed, have recently been reviewed (Mazza and Oomah 1995). Linustatin, 2-[(6-0-β-D-glucopyranosyl-β-D-glucopyranosyl)oxy]-2-methylpropane-nitrile, accounts for 54–76% of the total cyanogenic glycoside content in flaxseed. The concentration of neolinustatin, (2R)-[(6-0-β-D-glucopyranosyl-β-D-glucopyranosyl)oxy]-2-methylbutanenitrile, ranges from 91 to 203 mg/100 g, depending on flaxseed cultivars. The monoglycoside linamarin is present at low levels, 14–32 mg/100 g of flaxseed.

The biological activity of the diglycosides, linustatin and neolinustatin, extracted from flaxseed meal, in protecting against selenium toxicity in animals is well documented (Palmer 1995). The protective effect against selenosis appears to involve the release of CN^- from the glycoside, with a subsequent reaction of the CN^- and a selenium metabolite catalyzed by rhodanase (thiosulfate:cyanide sulfur transferase, EC 2.8.1.1) to form $SeCN^-$, which is readily excreted in the urine, lowering the effective level of selenium in the animal (Palmer 1995).

The anticancer effect of cyanogenic glycosides has been suggested in the past (Way 1981). The antitumor effect has been postulated to occur by the selective hydrolysis of cyanogenic glycoside by a β-glucosidase, liberating cyanide and benzaldehyde at the neoplastic site. The cyanide then selectively attacks the cancer cells, which are presumed to be low in rhodanase (Way 1981). Recently, epidemiologic studies have demonstrated the cancer-modulating effects of glucosinolates (Jongen 1996). Because cyanogenic glycosides and glucosides are derivatives of amino acids and are inter- and intramolecularly related (Butzenlechner *et al.* 1996), the cancer protective effects of cyanogenic glycosides is therefore more evident.

Lignans

Lignans are diphenolic compounds, $(C_6\text{-}C_3)_2$, formed by the tail-to-tail linkage of two cinnamic acid residues. Flaxseed is one of the richest sources of plant lignans (Caragy 1992), containing >52,000 μg/100 g of seed (Cunnane and Thompson 1995). Also, flaxseed is unique among foodstuffs in that its major lignan precursor is secoisolariciresinol, rather than matairesinol. Secoisolariciresinol is metabolized by intestinal bacteria to the mammalian lignan enterodiol (Kurzer *et al.* 1995). This lignan and the related compound, enterolactone, have been reported to have estrogen properties and to influence estrogen synthesis, availability, receptor binding, and action. Because of the strong link between a woman's hormonal profile and breast cancer, it has been suggested that flaxseed or its lignans may play a role in cancer prevention or treatment (Aldercreutz 1995).

Conclusions

The biological activities of flaxseed have been attributed to the high levels of α-linolenic acid, soluble polysaccharides, and lignans present in the seed. In the current rush for determining which phytochemical has the most biological activity, other major components and the synergistic effects of component interaction have received very little attention. Some phytochemicals may have beneficial, harmful, or no effects, depending on the conditions of the studies. Furthermore, isolating components from the seed may separate them from substances that enhance or reduce their true effects. In fact, most of the human studies to date that show beneficial effects have used whole flaxseed, flaxseed flour, or defatted flaxseed meal. The specific components responsible for the observed biological effects have yet to be confirmed definitively and the combined effects of several compounds that may be responsible for such effects to be elucidated. Because flaxseed with modified lipid composition has already been commercialized, it is likely that biotechnology and genetic engineering will be applied specifically toward the development of flaxseed and flaxseed products that contribute uniquely to disease prevention and health protection by selectively enhancing physiologic responses or deterring degenerative processes.

References

Aldercreutz, H., Phytoestrogens: Epidemiology and a Possible Role in Cancer Protection, *Environ. Health Persp. 103*: 103–112 (1995).

Adlercreutz, H., T. Fotsis, J. Lampe, K. Wahala, T. Makela, G. Brunnow, and T. Hase, Quantitative Determination of Lignans and Isoflavonoids in Plasma of Omnivorous and Vegetarian Women by Isotope Dilution Gas Chromatography-Mass Spectrometry, *Scand. J. Clin. Lab. Investig. 53*: 5–18 (1993).

Anderson, G.T., and W.E. Connor, Accretion of n-3 Fatty Acids in the Brain and Retina of Chicks Fed a Low-Linolenic Acid Diet Supplemented with Decosahexaenoic Acid, *Am. J. Clin. Nutr. 59*: 1338–1346 (1994).

Anderson, J.W., D.A. Deakins, T.L. Floore, B.M. Smith, and S.E. Whitis, Dietary Fiber and Coronary Heart Disease, *Crit. Rev. Food Sci. Nutr. 29*: 95–147 (1990).

Anonymous, Prairie Registration Recommending Committee for Grain, Oilseeds Subcommittee Report 1997, Report of Flax Co-operative Test 1996, 1997, pp. 1–26.

Attström, R., P.O. Glantz, H. Hakansson, and K. Larsson, Saliva Substitute, U.S. Patent 5,260,282 (1993).

Berdanier, C.D., ω-3 Fatty Acids: A Panacea? *Nutr. Today 29*: 28–32 (1994).

Berggren, A.M., I.M.E. Björck, E.M.G.L. Nyman, and B.O. Eggum, Short-Chain Fatty Acid Content and pH in Caecum of Rats Given Various Sources of Carbohydrates, *J. Sci. Food Agric. 63*: 397–406 (1993).

Bertram, J.S., L.N. Kolonel, and F.L. Meyskens, Rationale and Strategies for Chemoprevention of Cancer in Humans, *Cancer Res.* 47: 3012–3031 (1987).

Bhatty, R.S., and P. Cherdkiatgumchai, Compositional Analysis of Laboratory-Prepared and Commercial Samples of Linseed Meal and of Hull Isolated from Flax, *J. Am. Oil. Chem. Soc. 67*: 79–84 (1990).

Bierenbaum, M.L., R. Reichstein, and T.R. Watkins, Reducing Atherogenic Risk in Hyperlipemic Humans with Flax Seed Supplementation: A Preliminary Report, *J. Am. Coll. Nutr. 12*: 501–504 (1993).

Blackburn, N.A., J.S. Redfern, H. Jarjis, A.M. Holgate, I. Hanning, J.H.B. Scarpello, I.T. Johnson, and N.W. Read, The Mechanism of Action of Guar Gum in Improving Glucose Tolerance in Man, *Clin. Sci. 66*: 329–336 (1984).

Bolley, D.S., and R.H. McCormack, Separation of Phytin from Oil Seed Protein Flours, *J. Am. Oil Chem. Soc. 29*: 470–472 (1952).

Borgmeyer, J.R. C.E. Smith, and O.K. Huynh, Isolation and Characterization of a 25 kDa Antifungal Protein from Flax seeds, *Biochem. Biophys. Res. Commun. 187*: 480–487 (1992).

Budin, J.T., W.M. Breene, and D.H. Putnam, Some Compositional Properties of Camelina (*Camelina sativa* L. Crantz) Seeds and Oils, *J. Am. Oil Chem. Soc. 72*: 309–315 (1995).

Burr, M.L., A.M. Fehily, J.F. Gilbert, S. Rogers, R.M. Holliday, P.M. Sweetnana, P.C. Elwood, and N.M. Deadman, A.M. Fehily, S. Rogers, P.M. Sweetnam, and N.M. Deadman, Effects of Changes in Fat, Fish, and Fibre Intakes on Death, *Lancet 2*: 757–761 (1989).

Butzenlechner, M., S. Thimet, K. Kempe, H. Kexel, and H.-L. Schmidt, Inter- and Intramolecular Correlations in Some Cyanogenic Glycosides and Glucosinolates and Their Practical Importance, *Phytochem. 43*: 585–592 (1996).

Calder, P.C., S.J. Bevan, and E.A. Newsholme, The Inhibition of T-Lymphocyte Proliferation by Fatty Acids Is *via* an Eicosanoid Independent Mechanism, *Immunology 75*: 108–115 (1992).

Cameron, E., J. Bland, and R. Marcuson, Divergent Effects of Omega-6 and Omega-3 Fatty Acids on Mammary Tumor Development in C3H/Heston Mice Treated with DMBA, *Nutr. Res. 9*: 383–393 (1989).

Caragay, A.B., Cancer-Preventive Foods and Ingredients, *Food Technol. 46*: 65–68 (1992).

Chesson, A., *Dietary Fiber, Food Polysaccharides and Their Applications*, edited by A.M. Stephen, Marcel Dekker Inc., New York, 1995, pp. 547–576.

Clinton, S.K., Dietary Protein and the Origins of Human Cancer, in *Dietary Proteins: How They Alleviate Disease and Promote Better Health*, edited by G.U. Liepa, D.C. Bietz, A.C. Beynen, and M.A. Gorman, American Oil Chemists' Society, Champaign, IL, 1992, pp. 42–56.

Cook, H.W, Brain Metabolism of Alpha-Linolenic Acid During Development, *Nutrition 7*: 440–442 (1991).

Cook, N.C., and S. Samman, Flavonoids—Chemistry, Metabolism, Cardioprotective Effects and Dietary Sources, *J. Nutr. Biochem. 7*: 66–76 (1996).

Cui, W., E. Kenaschuk, and G. Mazza, Influence of Genotype on Chemical Composition and Rheological Properties of Flaxseed Gums, *Food Hydrocolloids 10*: 221–227 (1996).

Cunnane, S.C., Metabolism and Function of α-Linolenic Acid in Humans, in *Flaxseed in Human Nutrition*, edited by S.C. Cunnane, and L.U. Thompson, AOCS Press, Champaign, IL, 1995, pp. 99–127.

Cunnane, S.C., Z.Y. Chen, J. Yang, A.C. Liede, M. Hamadeh, and M.A. Crawford, Alpha-Linolenic Acid in Humans: Direct Functional Role or Dietary Precursor? *Nutrition 7*: 437–439 (1991).

Cunnane, S.C., S. Ganguli, C. Menard, A.C. Liede, M.J. Hamadeh, Z.Y. Chen, T.M. Wolever, and D.J. Jenkins, High Alpha-Linolenic Acid Flaxseed (*Linum usitatissimum*): Some Nutritional Properties in Humans, *Br. J. Nutr. 69*: 443–453 (1993).

Cunnane, S.C., M.J. Hamadeh, A.C. Liede, L.U. Thompson, T.M.S. Wolever, and D.J.A. Jenkins, Nutritional Attributes of Traditional Flaxseed in Healthy Young Adults, *Am. J. Clin. Nutr. 61*: 62–68 (1995).

Cunnane, S.C., and L.U. Thompson, *Flaxseed in Human Nutrition*, AOCS Press, Champaign, IL, 1995, p. 384.

Czarnecki, S.K., and D. Kritchevsky, Dietary Protein and Atherosclerosis, in *Dietary Proteins: How They Alleviate Disease and Promote Better Health*, edited by G.U. Liepa, D.C. Bietz, A.C. Beynen, and M.A. Gorman, American Oil Chemists' Society, Champaign, IL, 1992, pp. 42–56.

Dabrowski, K.J., and F.W. Sosulski, Composition of Free and Hydrolizable Phenolic Acids in Defatted Flours of Ten Oilseeds, *J. Agric. Food Chem. 32*: 128–130 (1984).

DeClercq, D.R., *1996 Preliminary Flaxseed Quality Estimate–Report No. 2*, Canadian Grain Commission, Winnipeg, MB, Canada, 1996, pp. 1–3.

de Lorgeril, M., S. Renaud, N. Mamelle, P. Salen, J.L. Martin, I. Monjaud, J. Guidollet, P. Touboul, and J. Delaye, Mediterranean Alpha-Linolenic Acid-Rich Diet in Secondary Prevention of Coronary Heart Disease, *Lancet 343*: 1454–1459 (1994).

Dyerberg, J., Linolenate-Derived Polyunsaturated Fatty Acids and Prevention of Atherosclerosis, *Nutr. Rev. 44*: 125–134 (1986).

Eaton, S.B., and M. Konner, Paleolithic Nutrition. A Consideration of Its Nature and Current Implications, *N. Engl. J. Med. 312*: 283–289 (1985).

Edwards, C.A., N.A. Blackburn, L. Craigen, P. Davison, J. Tomlin, K. Sugden, I.T. Johnson, and N.W. Read, Viscosity of Food Gums Determined *in vitro* Related to Their Hypoglycemic Actions, *Am. J. Clin. Nutr. 46*: 72–77 (1987).

Emken, E.A., R.O. Adolf, H. Rakoff, W.K. Rohwedder, and R.M. Gulley, Human Metabolic Studies with Deuterated Alpha-Linolenic Acid, *Nutrition 8*: 213–214 (1992).

Fiala, E.S., B.S. Reddy, and J.H. Weisburger, Naturally Occurring Anticarcinogenic Substances in Foodstuffs, *Annu. Rev. Nutr. 5*: 295–321 (1985).

Harris, R.K., and W.J. Haggerty, Assays for Potentially Anticarcinogenic Phytochemicals in Flaxseed, *Cereal Foods World 38*: 147–151 (1993).

Haumann, B.F., Nutritional Aspects of n-3 Fatty Acids, *INFORM 8*: 428–447 (1997).

Henry, M.M., J.N. Moore, E.B. Feldman, J.K. Fischer, and B. Russell, Effect of Dietary Alpha-Linolenic Acid on Equine Monocyte Procoagulant Activity and Eicosanoid Synthesis, *Circ. Shock 32*: 173–188 (1990).

Hilakivi-Clarke, L., R. Clarke, I. Onojafe, M. Raygada, E. Cho, and M. Lippman, A Maternal Diet High in n-6 Polyunsaturated Fats Alters Mammary Gland Development, Puberty Onset, and Breast Cancer Risk among Female Rat Offspring, *Proc. Natl. Acad. Sci. USA 94*: 9372–9377 (1997).

Holman, R.T., S.B. Johnson, and T.F. Hatch, A Case of Human Linolenic Acid Deficiency Involving Neurological Abnormalities, *Am. J. Clin. Nutr. 35*: 617–623 (1982).

Holub, B.J., Platelet Composition and Function, in *Flaxseed in Human Nutrition*, edited by S.C. Cunnane, and L.U. Thompson, AOCS Press, Champaign, IL, 1995, pp. 128–135.

Huang, A.H.C., Oil Bodies and Oleosins in Seeds, *Annu. Rev. Plant Physiol. Plant Mol. Biol. 43*: 177–200 (1992).

Hwang, D.H., and A.E. Carroll, Decreased Formation of Prostaglandins Derived from Arachidonic Acid by Dietary Linolenate in Rats, *Am. J. Clin. Nutr. 33*: 590–597 (1980).

Ibraham, R.K., and M. Shaw, Phenolic Constituents of the Oil Flax (*Linum usitatissimum*), *Phytochem. 9*: 1855–1858 (1970).

Jenkins, D.J.A., Incorporation of Flaxseed or Flaxseed Components into Cereal Foods, in *Flaxseed in Human Nutrition*, edited by S.C. Cunnane, and L.U. Thompson, AOCS Press, Champaign, IL, 1995, pp. 281–294.

Jenkins, D.J.A., T.M.S. Wolever, J. Kalmusky, S. Guidici, C. Giordani, G.S. Wong, J.N. Bird, R. Patten, M. Hail, G. Buckley, and J.A. Little, Low Glycemic Index Carbohydrate Foods in the Management of Hyperlipidemia, *Am. J. Clin. Nutr. 42*: 604–617 (1985).

Jenkins, D.J.A., T.M.S. Wolever, A.R. Leeds, M.A. Gassul, J.B. Dilawari, D.V. Goff, G.L. Metz, and K.G.M.M. Alberti, Dietary Fiber, Fiber Analogues, and Glucose Tolerance: Importance of Viscosity, *Br. Med. J. 1*: 1392–1394 (1978).

Johnston, P.V., Lipid Modulation of Immune Responses, in *Nutrition and Immunology*, edited by R.K. Chondra, Alan R. Liss, Inc., New York, NY, 1988, pp. 37–86.

Johnston, P.V., Flaxseed Oil and Cancer: α-Linolenic Acid and Carcinogenesis, in *Flaxseed in Human Nutrition*, edited by S.C. Cunnane, and L.U. Thompson, AOCS Press, Champaign, IL, 1995, pp. 207–218.

Jongen, W.M.F., Glucosinolates in *Brassica*: Occurrence and Significance as Cancer-Modulating Agents. *Proc. Nutr. Soc. 55*: 433–446 (1996).

Kelly, D.S., Immunomodulatory Effects of Flaxseed and Other Oils Rich in α-Linolenic Acid, in *Flaxseed in Human Nutrition*, edited by S.C. Cunnane, and L.U. Thompson, AOCS Press, Champaign, IL, 1995, pp. 145–156.

Kurzer, M., J.L. Slavin, and H. Adlercreutz, Flaxseed, Lignans, and Sex Hormones, in *Flaxseed in Human Nutrition*, edited by S.C. Cunnane, and L.U. Thompson, AOCS Press, Champaign, IL, 1995, pp. 136–144.

Liepa, G.U., D.C. Bietz, A.C. Beynan, and M.A. Gorman, *Dietary Proteins: How They Alleviate Disease and Promote Better Health*, American Oil Chemists' Society, Champaign, IL, 1992, p. 282.

Marquard, V.R., W. Schuster, and H. Iran-Nejad, Studies on Tocopherol and Thiamine Content of Linseed from Worldwide Cultivation and from Phytotron under Definite Climatic Conditions, *Fette Seifen Anstrichm. 79*: 265–270 (1977).

Marshall, L.A., and P.V. Johnston, The Influence of Dietary Essential Fatty Acids on Rat Immunocompetent Cell Prostaglandin Synthesis and Mitogen-Induced Blastogenesis, *J. Nutr. 115*: 1572–1580 (1985).

Mazza, G., and B.D. Oomah, Flaxseed, Dietary Fiber, and Cyanogens, in *Flaxseed in Human Nutrition*, edited by S.C. Cunnane, and L.U. Thompson, AOCS Press, Champaign, IL, 1995, pp. 56–81.

Miller, K.L., High-Stability Oils, *Cereal Foods World 38*: 478, 480–482 (1993).

Nelson, G.J., and J.G. Chamberlain, The Effect of Dietary α-Linolenic Acid on Blood Lipids and Lipoproteins in Humans, in *Flaxseed in Human Nutrition*, edited by S.C. Cunnane, and L.U. Thompson, AOCS Press, Champaign, IL, 1995, pp. 187–206.

Ntiamoah, C., G.G. Rowland, and D.C. Taylor, Inheritance of Elevated Palmitic Acid in Flax and Its Relationship to the Low Linolenic Acid, *Crop Sci. 35*: 148–152 (1995).

Nuttall, F.Q., A.D. Mooradian, M.C. Gannon, C. Billington, and P. Krezowski, Effect of Protein Ingestion on the Glucose and Insulin Response to a Standardized Oral Glucose Load, *Diabetes Care 7*: 465–470 (1984).

O'Mullane, J.E., and J.P. Hayter, Linseed Mucilage, International Patent PCT/GB93/00343 (1993).

Olejnik, K., M. Gogolewski, and M. Nogala-Katucka, Isolation and Some Properties of Plastchromanol-8, *Nahrung 4*: 101–104 (1997).

Oomah, B.D., E.O. Kenaschuk, and G. Mazza, Phenolic Acids in Flaxseed, *J. Agric. Food Chem. 43*: 2016–2019 (1995).

Oomah, B.D., E.O. Kenaschuk, and G. Mazza, Phytic Acid Content of Flaxseed as Influenced by Cultivar, Growing Season, and Location, *J. Agric. Food Chem. 44*: 2663–2666 (1996a).

Oomah, B.D., E.O. Kenaschuk, and G. Mazza, Tocopherols in Flaxseed, *J. Agric. Food Chem. 45*: 2076–2080 (1997a).

Oomah, B.D., E.O. Kenaschuk, and G. Mazza, Lipoxygenase Enzyme in Flaxseed, *J. Agric. Food Chem. 45*: 2426–2430 (1997b).

Oomah, B.D., and G. Mazza, Flaxseed Proteins—A Review, *Food Chem. 48*: 109–114 (1993).

Oomah, B.D., and G. Mazza, Functional Properties, Uses of Flaxseed Protein, *INFORM 6*: 1246–1252 (1995).

Oomah, B.D., and G. Mazza, Flaxseed Products for Disease Prevention, in *Functional Foods: Biochemical and Processing Aspects*, edited by G. Mazza, Technomic Publishing Co. Inc., Lancaster, PA, 1998, pp. 91–138.

Oomah, B.D., G. Mazza, and E.O. Kenaschuk, Flavonoid Content of Flaxseed. Influence of Cultivar and Environment, *Euphytica 90*: 163–167 (1996b).

Palmer, J.S., Effects of Flaxseed on Selenium Toxicity, in *Flaxseed in Human Nutrition*, edited by S.C. Cunnane, and L.U. Thompson, AOCS Press, Champaign, IL, 1995, pp. 187–206.

Rabinovitch-Chable, M., J. Cook-Moreau, J.C. Breton, and M. Rigaud, Purification of Lipoxygenase and Hydroperoxide Dehydrase in Flaxseeds: Interaction Between These Enzymatic Activities, *Biochem. Biophys. Res. Commun. 188*: 858–864 (1992).

Ranhotra, G.S., J.A. Gelroth, B.K. Glaser, and P.S. Potnis, Lipidemic Responses in Rats Fed Flaxseed Oil and Meal, *Cereal Chem. 70*: 364–368 (1993).

Serraino, M.R., L.U. Thompson, and S.C. Cunnane, Effect of Low Level Flaxseed Supplementation on the Fatty Acid Composition of Mammary Glands and Tumors in Rats, *Nutr. Res. 12*: 767–772 (1992).

Simopoulos, A.P., Fatty Acids, *Functional Foods: Designer Foods, Pharmafoods, Nutraceuticals*, edited by J. Goldberg, Chapman and Hall, Inc., New York, 1994, pp. 355–392.

Singer, P., Alpha-Linolenic Acid vs. Long-Chain n-3 Fatty Acids in Hypertension and Hyperlipidemia, *Nutrition 8*: 133–135 (1992).

Sosulski, F.W., and G. Sarwar, Amino Acid Composition of Oilseed Meals and Protein Isolates, *Can. Inst. Food Technol. J. 6*: 1–5 (1973).

Thompson, L.U., Potential Health Benefits and Problems Associated with Antinutrients in Foods, *Food Res. Int. 26*: 131–149 (1993).

Thompson, L.U., Phytic Acid and Other Nutrients: Are They Partly Responsible for Health Benefits of High Fiber Foods? in *Dietary Fiber*, edited by D. Kritchevsky, and C. Bonefield, Plenum, New York, 1994a, pp. 305–317.

Thompson, L.U., Antioxidants and Hormone-Mediated Health Benefits of Whole Grains, *Crit. Rev. Food Sci. Nutr. 34*: 473–497 (1994b).

van Rooijen, G.J.H., and M.M. Moloney, Plant Seed Oil-Bodies as Carriers for Foreign Proteins, *Bio/Technology 13*: 72–77 (1995).

Way, I.L., Pharmacological Aspects of Cyanide and Its Antagonism, in *Cyanide in Biology*, edited by B. Vennesland, E.E. Conn, C.J. Knowles, J. Westley, and F. Wissing, Academic Press, London, 1981, pp. 28–49.

Wilen, R.W., G.J.H. van Rooijen, D.W. Pierce, R.P. Pharis, L.A. Holbrook, and M.M. Moloney, Effects of Jasmonic Acid on Embryo-Specific Processes in *Brassica* and *Linum* Oilseeds, *Plant Physiol. 95*: 399–405 (1991).

Wolever, T.M.S., and D.J.A. Jenkins, Effect of Dietary Fiber and Foods on Carbohydrate Metabolism, in *CRC Handbook of Dietary Fiber in Human Nutrition*, edited by G.A. Spiller, CRC Press, Boca Raton, FL, 1993, pp. 111–152.

Zimmerman, D.C., and P. Feng, Characterization of a Prostaglandin-Like Metabolite of Linolenic Acid Produced by a Flaxseed Extract, *Lipids 13*: 313–316 (1978).

Zimmerman, D.C., and B.A. Vick, Specificity of Flaxseed Lipoxidase, *Lipids 5*: 392–397 (1970).

Chapter 10

Medicinal Lignans from Flaxseed: Isolation and Purification

Neil D. Westcott, and Alister D. Muir

Crop Utilization Section, Saskatoon Research Center, Agriculture and Agri-Food Canada, Saskatoon, SK, Canada, S7N 0X2

Introduction

Lignans are formed by dimerization of two cinnamic acids or their biogenetic equivalents. There are hundreds of different lignans each differing in degree of oxidation, cyclization, or derivatization (Ayers and Loike 1990). They have been isolated from roots, stems, seeds, fruits, and leaves of many plants. The so-called mammalian lignans, enterolactone and enterodiol, are formed by the action of gut bacteria on plant lignans such as secoisolariciresinol diglucoside (SDG, Fig. 10.1) (Setchell *et al.* 1980, 1981 and 1982). SDG was first reported as a phytochemical in flaxseed in 1956 (Bakke and Klosterman 1956). The presence of SDG in flaxseed has been confirmed by several independent studies (Axelson *et al.* 1982, Bambagiotti-Alberti *et al.* 1994, Harris *et al.* 1992 and 1994, Harris and Haggerty 1993, Jenkins *et al.* 1993, Obermeyer *et al.* 1993 and 1995, Westcott and Muir 1996). Apart from the original report by Bakke and Klosterman, the isolation of significant quantities of purified SDG has previously been reported partially by only one other group (Thompson *et al.* 1994). Recently, Westcott and Muir patented a process for the production of SDG from flaxseed (Westcott and Muir 1998).

Lignans or plants containing lignans have a long history of use as traditional folk medicines (Ayers and Loike 1990, MacRae and Towers 1984). Flaxseed or SDG has attracted interest for intervention in some cancers (Thompson 1995, Thompson *et al.* 1995, 1996a, 1996b and 1996c), kidney disorders (Clark *et al.* 1994 and 1995, Clark and Parbtani 1996), atherosclerosis (Prasad 1997, Prasad *et al.* 1998), and as an antioxidant (Prasad 1996).

Although global production of flaxseed is cyclical, Canada has always been a major producer. In 1994, of the 2,500,000 tons produced worldwide, Canada produced 964,000 tons (Anonymous 1995). Within Canada, flaxseed production takes place mainly in the provinces of Saskatchewan and Manitoba. With the ready availability of flaxseed and increasing interest in its lignan, SDG, we embarked on a project to improve the isolation process for SDG.

Optimization of Extraction

The original investigators used alcohol/dioxane to prepare an extract from defatted flaxseed (Klosterman and Smith 1954). From this extract, an amorphous, gum-like

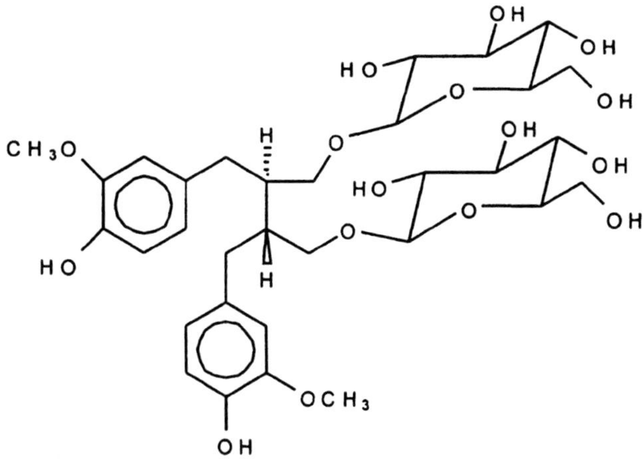

Fig. 10.1. The structure of secoisolariciresinol diglucoside (SDG).

substance that contained bound lignan and other substances was precipitated. Alkaline methanolysis of the precipitate resulted in the release of cinnamic acid derivatives (Klosterman *et al.* 1955, Klosterman and Muggli 1959), methyl hydroxymethylglutarate (Klosterman and Smith 1954), and SDG (Bakke and Klosterman 1956).

The solubility properties of the amorphous gum that contains SDG suggested that aqueous alcohols might be the most suitable extraction solvents. In preliminary experiments (not reported), pure and aqueous mixtures of several simple aliphatic alcohols were evaluated for their extraction capabilities. It was determined that methanol and ethanol were the best alcohols, and that addition of water improved the extraction efficiency. The high-performance liquid chromatography (HPLC) chromatogram of an aqueous alcohol extract is shown in Figure 10.2. With the use of an evaporative light scattering detector and a photodiode array detector (280 nm) the presence of two main peaks was detected. The first peak was due to a very polar material as indicated by its nonretention on reverse-phase resins. Its relatively weak ultraviolet (UV) absorption suggested that this material most likely comprises free sugars, polar amino acids, and salts. The second peak (retention time of ~21 min) had a much more intense absorption at 280 nm and was due to the complex that contained bound SDG. After our study was completed, Jenkins (1995) reported on the use of aqueous ethanol for extraction of supercritical carbon dioxide defatted meal.

Considering these preliminary findings, a study was undertaken to evaluate extraction temperature, liquid to solid ratio, and solvent composition on the extraction of lignan-containing material from defatted flaxseed meal. The experiment was carried out according to a central composite design with three factors and five levels (Haaland 1989, Oomah *et al.* 1994). The temperature range was selected from room temperature (25°C) to near the boiling point of the alcohol (65°C for

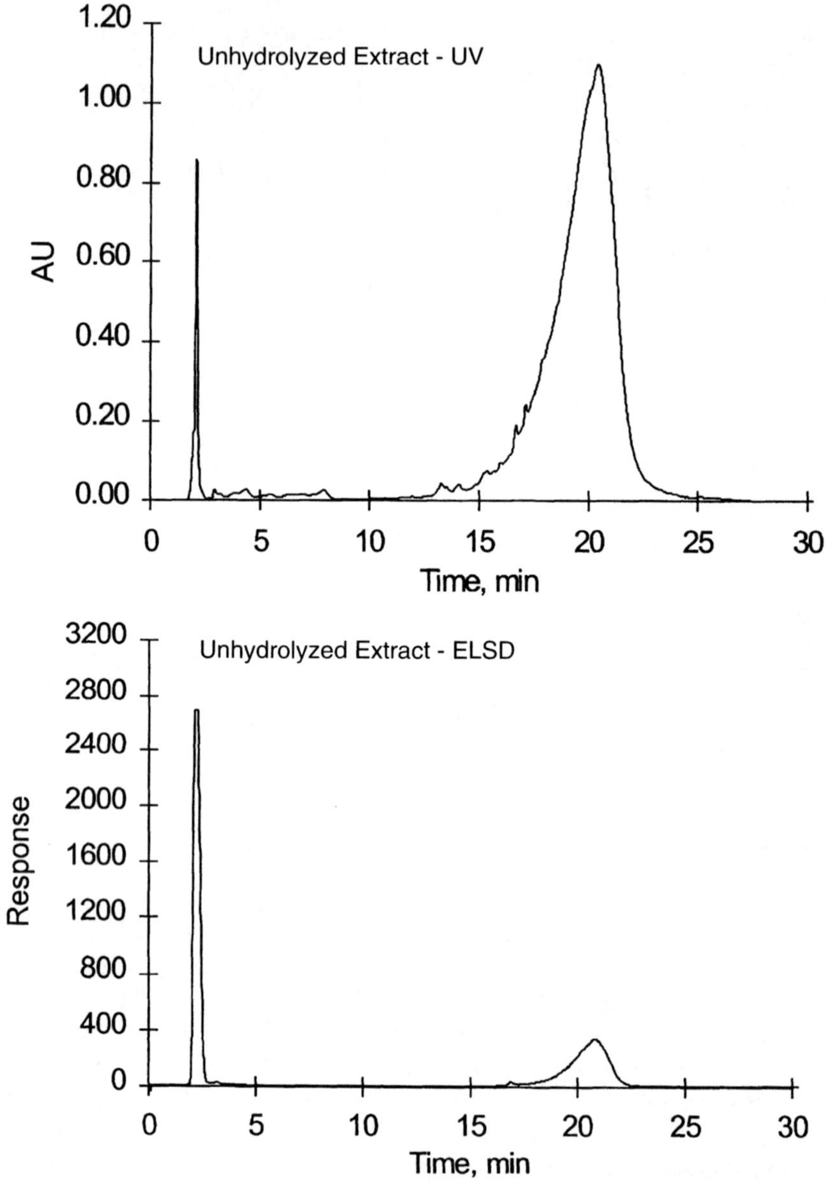

Fig. 10.2. Chromatogram of the aqueous alcohol extract of defatted flaxseed meal. Upper panel: Ultraviolet (UV) detection at 280 nm. Lower panel: Evaporative light scattering detector (ELSD). [Conditions: Symmetry C-18 (5 μm), 150 × 3 mm; solvent A: 0.05% (vol/vol) trifluoroacetic acid in water; solvent B: Acetonitrile; flow rate: 0.4 mL/min; gradient, 0–5 min: 90% A + 10% B; linear gradient to 60% A + 40% B at 20 min; linear gradient to 90% A + 10% B at 30 min].

methanol, 79°C for ethanol). The liquid/solids ratio was selected from a minimum volume needed to adequately slurry the meal (6:1, vol/wt) to a maximum ratio of 30:1. The solvent composition was selected on the basis of the preliminary experiments and the desire to minimize the extraction of flaxseed gums and other oligosaccharides by using at least 50% alcohol (vol/vol) in any solvent mixture.

The extracts were hydrolyzed with base, neutralized, and analyzed by liquid chromatography. Data were analyzed by multiple regression to fit a second-order polynomial model using a RSREG (Response Surface REGression) program (SAS Institute Inc. 1990). The models developed for ethanol and methanol had R^2 values of 0.99 and 0.92, respectively, and significant F-values (probability of $F < 0.005$). Thus the model developed was very satisfactory.

The calculated effects of liquid/solid ratio and temperature on the yield of lignan with 75% methanol extractant are presented in Figure 10.3. Increasing extraction temperature had a major influence on increasing yield, whereas increasing liquid/solid ratio was of lesser importance, particularly beyond the midpoint of the ratios studied. Using a fixed value of liquid/solids of 18:1, the calculated effects of temperature and percentage aqueous modifier in methanol on lignan yield are presented in Figure 10.4. As before, temperature had a positive influence in extraction efficiency, but the more noticeable effect was due to solvent composition. Lowest yields of lignan were obtained with pure alcohol and with increasing yields when 30–40% water was added as modifier. Addition of more water resulted in a slight decrease in yield of the lignan. Similar results were obtained with ethanol as extractant.

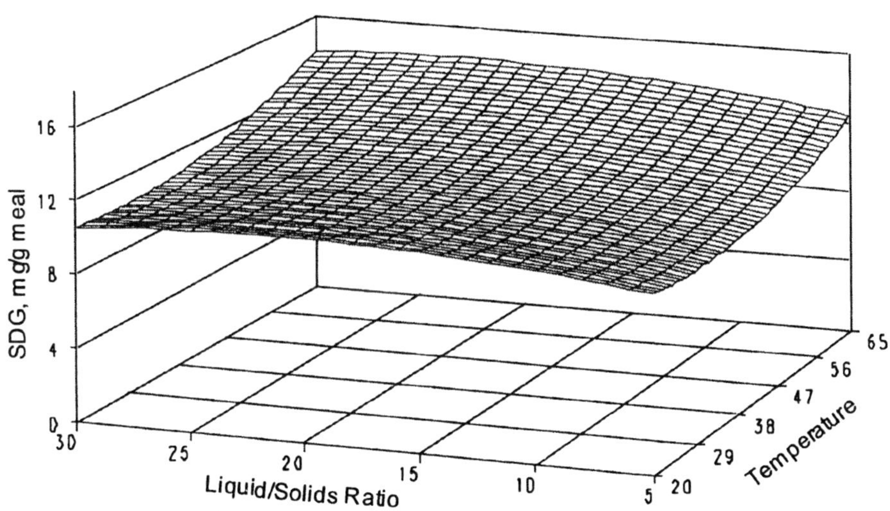

Fig. 10.3. Calculated response surface graph on secoisolariciresinol diglucoside (SDG) extracted with 75% aqueous methanol from defatted flaxseed meal vs. liquid/solids ratio and temperature.

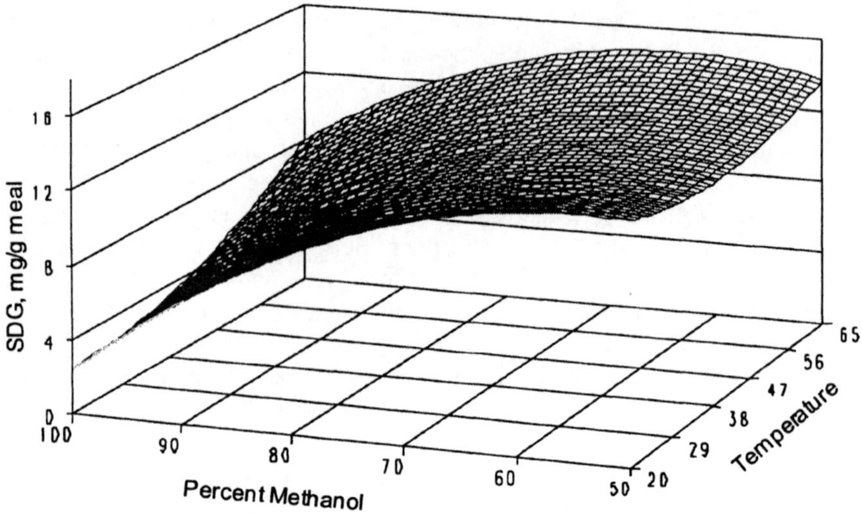

Fig. 10.4. Calculated response surface graph on secoisolariciresinol diglucoside (SDG) extracted from defatted flaxseed meal at an 18:1 liquid/solids ratio vs. percentage methanol and temperature.

Anhydrous Methanolysis of Flaxseed Meal Extract

The use of the aqueous alcohol extraction procedure allowed the precipitation of the amorphous, gum-like substance according to the published procedure of Klosterman and Smith (1954). Analysis of the mother liquor remaining after precipitation indicated that more SDG-containing material was still present. Thus, to maximize the amount of free lignan obtainable, we eliminated the precipitation step and used the total extract for methanolysis.

The compound desired is the first peak of the doublet with retention time of 33 min and is just resolved under the listed conditions (Fig. 10.5). It was found that the methanolyzed extract could be partitioned between ethyl acetate and water. This separated the methyl cinnamate derivatives from the desired SDG, which was retained in the aqueous phase (Fig. 10.6). The aqueous phase containing the SDG also contained both UV- and non-UV–absorbing substances and required further purification as described below.

Aqueous Hydrolysis of Flaxseed Meal Extract

For routine operations, it was not practicable to have to evaporate extracts to dryness and keep the methanolysis reaction free of water. In addition, sodium methoxide absorbs both water and carbon dioxide, which could limit its effectiveness. The use of aqueous hydrolysis has not been reported previously. Sodium hydroxide in an aqueous solution was proven to be equally effective in hydrolyzing the alcoholic extract obtained from the defatted flaxseed (Fig. 10.7). This modification

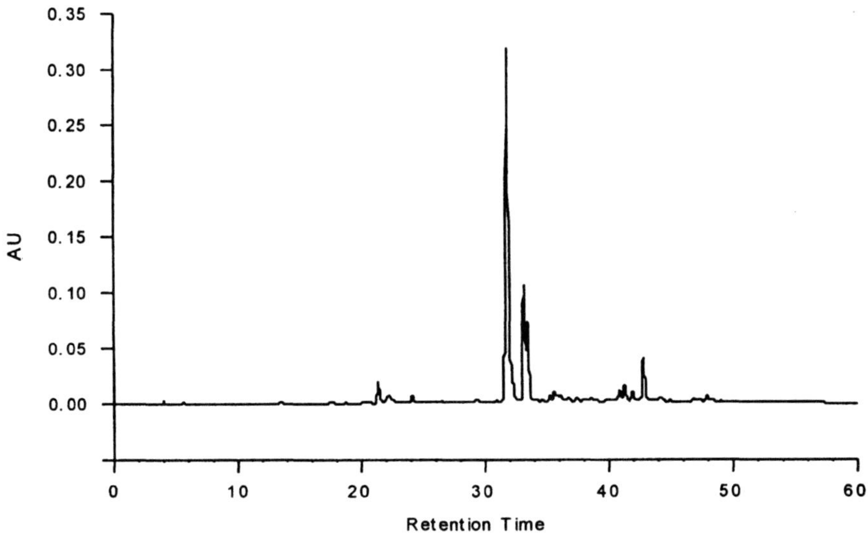

Fig. 10.5. Chromatogram of the anhydrous methanolysis of the total defatted flaxseed meal extract. [Conditions: LiChrospher 100 RP-18 (5 μm), 250 × 4 mm; Solvent A: 1% (vol/vol) acetic acid in water; Solvent B: Methanol; flow rate: 1 mL/min; gradient, 0 min: 100% A; linear gradient to 40% A + 60% B at 44 min; hold 40% A and 60% B until 48 min; linear gradient to 100% A at 55 min].

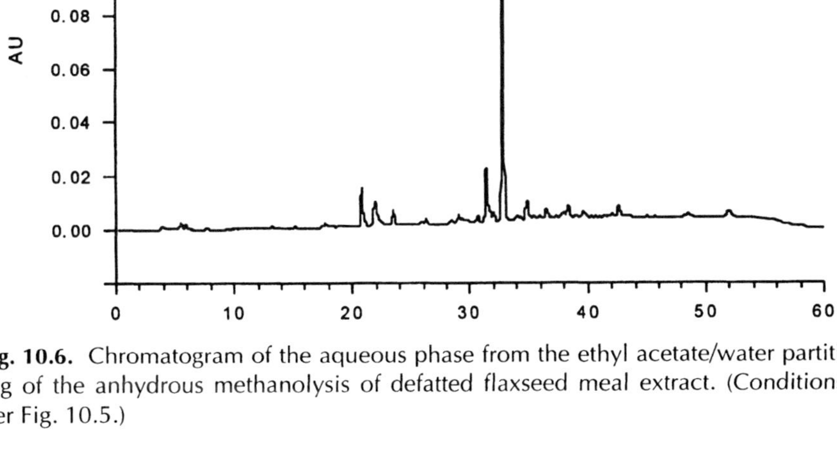

Fig. 10.6. Chromatogram of the aqueous phase from the ethyl acetate/water partitioning of the anhydrous methanolysis of defatted flaxseed meal extract. (Conditions as per Fig. 10.5.)

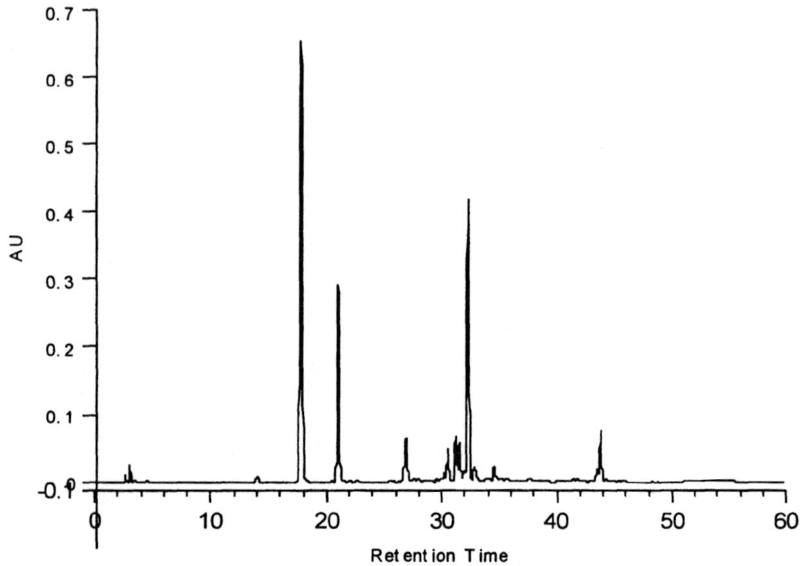

Fig. 10.7. Chromatogram of the sodium hydroxide hydrolysis of the total extract of defatted flaxseed meal extract. (Conditions as per Fig. 10.5.)

does not affect the yield of SDG but does result in the formation of cinnamic acid derivatives, as opposed to the methyl esters that were formed in the alkaline methanolysis reaction. Because the cinnamic acid derivatives produced were no longer soluble in ethyl acetate, alternative purification strategies were developed.

Anion-Exchange Chromatography

Because the aqueous alkali degradation produced carboxylic acid derivatives, the use of ion exchange chromatography for component separation was possible. Use of any of the several commercially available anion-exchange resins was found to be suitable. Elution of the column with water removed the SDG plus other neutral substances. The carboxylic acid derivatives were eluted with acidic aqueous alcohol. The purity of the lignan fraction was similar to that of the aqueous phase from the ethyl acetate purification step above and was purified in a similar manner as described below.

Reverse-Phase Resin Chromatography

The SDG-containing fractions prepared from either the anhydrous methanolysis or the aqueous hydrolysis could be conveniently purified using C-18 reverse-phase resins. SDG was retained on the resin, whereas the more polar co-extractives were eluted with aqueous acetic acid containing low concentrations of methanol. The SDG could then be eluted from the column with solvents containing more methanol. By using a column with a larger diameter (36 cm) and relatively short

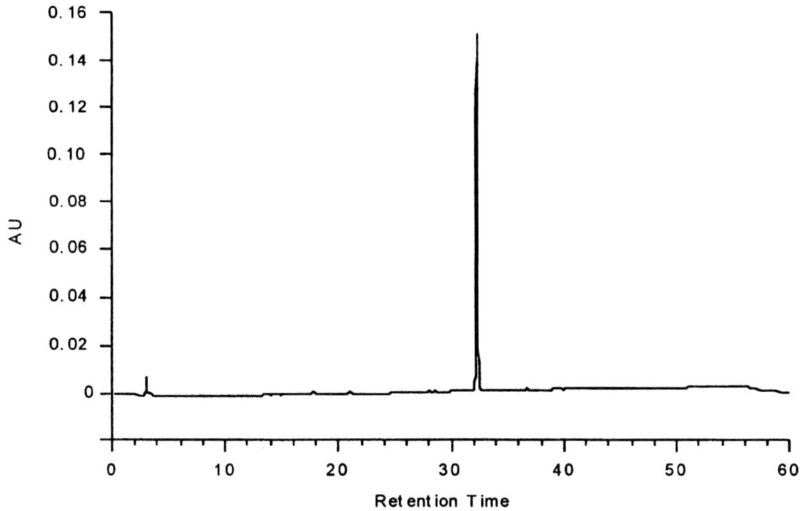

Fig. 10.8. Chromatogram of the purified secoisolariciresinol diglucoside (SDG). (Conditions as per Fig. 10.5.)

height (15 cm) packed with reverse-phase resin, it is possible to purify the extract from up to 20 kg of defatted meal. Depending on the elution gradient, 200 g of SDG with a purity of ≥90% can be obtained. Pure SDG can be obtained by semi-preparative HPLC on a 40 × 300 mm column (Fig. 10.8).

Conclusions

Aqueous methanol or ethanol effectively extracts the complex containing SDG from defatted flaxseed meal. It was determined that it is possible to use the total extract rather than just the material that precipitates for production of SDG. Anhydrous bases such as sodium methoxide may be used to free the SDG if the water is completely removed. Partial purification of anhydrous base released material can be accomplished by liquid-liquid partitioning between water and ethyl acetate. The aqueous layer that contains free SDG may be further purified by the use of reverse-phase chromatography. A more practical procedure for producing SDG is to use aqueous sodium hydroxide on the original aqueous alcohol extract, thus avoiding the costly step of removing all of the water from the extract. The aqueous hydrolyzed material can then be purified by a combination of ion exchange and reverse-phase chromatography. The methods described can be used to produce hundreds of grams of purified SDG in a laboratory environment.

Acknowledgments

We wish to acknowledge the technical assistance provided by Sandra Northrup and Andy Aubin. This paper is Contribution 1273 of the Saskatoon Research Center.

References

Anonymous, *FAO Yearbook: Production*, FAO, Rome, 1995.

Axelson, M., J. Sjövall, B.E. Gustaisson, and K.D.R. Setchell, Origin of Lignans in Mammals and Identification of a Precursor from Plants, *Nature 298*: 659–670 (1982).

Ayers, D.C., and J.D. Loike, *Lignans: Chemical, Biological and Clinical Properties,* Cambridge University Press, Cambridge, 1990.

Bakke, J.E., and H.J. Klosterman, A New Diglucoside from Flaxseed, *Proc. N. Dakota Acad. Sci. 10*: 18–22 (1956).

Bambagiotti-Alberti, M., S.A. Coran, C. Ghiara, V. Giannellini, and A. Raffaelli, Revealing the Mammalian Lignan Precursor Secoisolariciresinol Diglucoside in Flax Seed by Ionspray Mass Spectrometry, *Rapid Commun. Mass Spectrom. 8*: 595–598 (1994).

Clark, W.F., and A. Parbtani, Flaxseed in Experimental and Clinical Lupus Nephritis, in *Proc. 56th Flax Institute of the United States*, Fargo, ND, 1996, pp. 17–24.

Clark, W.F., A. Parbtani, H. de Salis, L. Rudzitis, M. Keeney, I. Chin-Yee, D.J. Philbrick, and B.J. Holub, Flaxseed-Derived Lignan in MRL/lpr Lupus Mice, 27th Ann. Meeting Am. Soc. Nephrol., Orlando, FL, Abst. 743, 1994.

Clark, W.F., A. Parbtani, M.W. Huff, E. Spanner, D.J. Philbrick, and B.J. Holub, Flaxseed: A Nutraceutical for Lupus Nephritis? Ann. Meeting Canadian Soc. Clin. Investig. R. Coll. Phys. Surg. of Canada, Montreal, Quebec, Abst. B103, 1995.

Haaland, P.D., *Experimental Design in Biotechnology,* Marcel Dekker, New York, 1989.

Harris, R.K., J. Greaves, D. Alexander, and W.J. Haggerty, Preliminary Research on the Development of Stability-Indicating Analytical Methods for Flaxseed Lignans and Their Precursors, in *Proc. 54th Flax Institute of the United States*, Fargo, ND, 1992, pp. 24–29.

Harris, R.K., J. Greaves, D. Alexander, T. Wilson, and W.J. Haggerty, Development of Stability-Indicating Analytical Methods for Flaxseed Lignans and Their Precursors, in *Food Phytochemicals for Cancer Prevention II. Teas, Spices, and Herbs,* edited by C.-T. Ho, T. Osawa, M.-T. Huang, and R.T. Rosen, ACS Symposium Series 547, American Chemical Society, Washington, DC, 1994, pp. 295–305.

Harris, R.K., and W.J. Haggerty, Assays for Potentially Anticarcinogenic Phytochemicals in Flaxseed, *Cereal Foods World 38*: 147–151 (1993).

Jenkins, D.J.A., Incorporation of Flaxseed or Flaxseed Components into Cereal Foods, in *Flaxseed in Human Nutrition*, edited by S.C. Cunnane, and L.U. Thompson, AOCS Press, Champaign, IL, 1995, pp. 281–294.

Jenkins, D.J.A., T.M.S. Wolever, A.V. Rao, R.A. Hegele, S.J. Mitchell, T.P.P. Ransom, D.L. Boctor, P.J. Spadafora, Effect on Blood Lipids of Very High Intakes of Fiber in Diets Low in Saturated Fat and Cholesterol, *N. Engl. J. Med. 329*: 21–26 (1993).

Klosterman, H.J., and R.Z. Muggli, The Glucosides of Flaxseed II: Linocaffein, *J. Am. Chem. Soc. 81*: 2188–2191 (1959).

Klosterman, H.J., and F. Smith, The Isolation of β-Hydroxy-β-methylglutaric Acid from the Seed of Flax (*Linum usitatissimum*), *J. Am. Chem. Soc. 76*: 1229–1230 (1954).

Klosterman, H.J., F. Smith, and C.O. Clagett, The Constitution of Linocinnamarin, *J. Am. Chem. Soc. 77*: 420–421 (1955).

MacRae, W.D., and G.H.N. Towers, Biological Activities of Lignans, *Phytochemistry 23*: 1207–1220 (1984).

Obermeyer, W.R., S.M. Musser, J.M. Betz, R.E. Casey, A.E. Pohland, and S.W. Page, Chemical Studies of Phytoestrogens and Related Compounds in Dietary Supplements: Flax and Chapparral, *Proc. Soc. Exp. Biol. Med. 208*: 6–12 (1995).

Obermeyer, W.R., C. Warner, R.E. Casey, and S.M. Musser, Flaxseed Lignans. Isolation, Metabolism and Biological Effects, *FASEB J. 7*: A863 (1993).

Oomah, B.D., G. Mazza, and W. Cui, Optimization of Protein Extraction from Flaxseed Meal, *Food Res. Int. 27*: 355–361 (1994).

Prasad, K., Hydroxy Radical-Scavenging Property of Secoisolariciresinol Diglucoside (SDG) Isolated from Flax-Seed, *Mol. Cell. Biochem. 168*: 117–123 (1997).

Prasad, K., Dietary Flaxseed in the Prevention of Hypercholesterolemic Atherosclerosis, *Atherosclerosis 132*: 69–75 (1997).

Prasad, K., S.V. Mantha, J. Kalra, A.D. Muir, and N.D. Westcott, Reduction of Hypercholesterolemic Atherosclerosis by CDC-Flaxseed with Very Low Alpha-Linolenic Acid, *Atherosclerosis 136*: 367–375 (1998).

SAS Institute Inc., SAS/STAT User's Guide, Version 6, 4th ed., SAS Institute, Cary, NC, 1990.

Setchell, K.D.R., A.M. Lawson, S.P. Borriello, H. Adlercreutz, and M. Axelson, Formation of Lignans by Intestinal Microflora, in *Falk Symposium 31, Colonic Carcinogenesis*, edited by R.A. Malt, and R.C.N. Williamson, MTP Press Ltd., Lancaster, 1982, pp. 93–97.

Setchell, K.D.R., A.M. Lawson, E. Conway, N.F. Taylor, D.N. Kirk, G. Cooley, R.D. Farrant, and S. Wynn, The Definite Identification of the Lignans *trans*-2,3-bis(3-Hydroxybenzyl-γ-butyrolactone and 2,3-*bis*(3-Hydroxybenzyl)butane-1,4-diol in Human and Animal Urine, *Biochem. J. 197*: 447–458 (1981).

Setchell, K.D.R., A.M. Lawson, F.L. Mitchell, H. Adlercreutz, D.N. Kirk, and M. Axelson, Lignans in Man and Animal Species, *Nature 287*: 740–742 (1980).

Thompson, L.U., Flaxseed, Lignans and Cancer, in *Flaxseed in Human Nutrition*, edited by S.C. Cunnane, and L.U. Thompson, AOCS Press, Champaign, IL, 1995, pp. 219–236.

Thompson, L.U., L. Orcheson, S. Rickard, M. Jenab, M. Serraino, M. Seidl, and F. Cheung, Anticancer Effects of Flaxseed Lignans, in *Natural Antioxidants and Food Quality in Atherosclerosis and Cancer Prevention*, edited by J.T. Kumpulainen, and J.T. Salonen, The Royal Society of Chemistry, London, 1996a, pp. 356–364.

Thompson, L.U., L. Orcheson, S. Rickard, and M. Seidl, Anticarcinogenic Effect of a Mammalian Lignan Precursor from Flaxseed, in *Proc. 55th Flax Institute of the United States*, Fargo, ND, 1994, pp. 46–50.

Thompson, L.U., S.E. Rickard, L.J. Orcheson, and M.M. Seidl, The Role of Lignans and Oil in Flaxseed on Mammary Tumorigenesis, in *Proc. 86th Ann. Meeting Am. Assoc. Cancer Res.*, Toronto, ON, Abst. 114, 1995.

Thompson, L.U., S.E. Rickard, L.J. Orcheson, and M.M. Seidl, Flaxseed and Its Lignan and Oil Components Reduce Mammary Tumor Growth at a Late Stage of Carcinogenesis, *Carcinogenesis 17*: 1373–1376 (1996b).

Thompson, L.U., M.M. Seidl, S.E. Rickard, L.J. Orcheson, and H.H.S. Fong, Antitumorigenic Effect of a Mammalian Lignan Precursor from Flaxseed, *Nutr. Cancer 26*: 159–165 (1996c).

Westcott, N.D., and A.D. Muir, Variation in the Concentration of the Flax Seed Lignan Concentration with Variety, Location and Year, in *Proc. 56th Flax Institute of the United States*, Fargo, ND, 1996, pp. 77–80.

Westcott, N.D., and A.D. Muir, Process for Extracting Lignans from Flaxseed, U.S. Patent 5,705,618, (1998).

Chapter 11

Reaction of α-Tocopherol During the Peroxidation of Phospholipids in Liposomes

Ryo Yamauchi

Department of Food Science, Faculty of Agriculture, Gifu University, Gifu City, Gifu 501-1193, Japan

Introduction

Alpha--Tocopherol (α-TH, Fig.11.1) is generally considered to protect biological membranes from lipid peroxidation (Burton and Ingold 1981 and 1986, Niki *et al.* 1984). The antioxidative activity of α-TH is related to the scavenging of the peroxyl radicals of unsaturated lipids in the membranes. α-TH efficiently transfers a hydrogen atom to a peroxyl radical, giving a hydroperoxide and an α-tocopheroxyl radical (α-T$^{\bullet}$) (Burton and Ingold 1986). The α-T$^{\bullet}$, once formed, reacts with a second peroxyl radical to form a nonradical product. Thus, each molecule of α-TH that is consumed traps two peroxyl radicals. The reaction products of α-TH with peroxyl radicals have been studied extensively to explain the mechanism for autoxidation inhibition by α-TH (Ham and Liebler 1995 and 1997, Liebler *et al.* 1990 and 1991, Liebler and Burr 1992, Matsuo *et al.* 1989, Winterle *et al.* 1984, Yamauchi *et al.* 1989a, 1990, 1993, 1994 and 1996). The reaction of peroxyl radicals with the α-T$^{\bullet}$ forms 8a-substituted α-tocopherones, and often involves the coupling of alkylperoxyl radicals with the tocopheroxyl radical to form 8a-alkyldioxy-α-tocopherones [**1**] (Liebler *et al.* 1990, 1991 and 1992, Matsuo *et al.* 1989, Winterle *et al.* 1984, Yamauchi *et al.* 1989a, 1990, 1993, 1994 and 1996).

Phospholipids are important structural components in biological membranes; they are also targets of lipid peroxidation. Phosphatidylcholine (PC) liposomal systems have been employed to model the peroxyl radical-trapping reactions of α-TH in biological membranes (Fujii *et al.* 1991, Liebler *et al.* 1991, Liebler and Burr 1992, Yamauchi *et al.* 1994 and 1996). This paper reports the preparation and characterization of 8a-(PC-dioxy)-α-tocopherones [**1a**] from the reaction between α-TH and peroxyl radicals of 1,2-dilinoleoyl-3-*sn*-phosphatidylcholine (DLPC), 1-palmitoyl-2-linoleoyl-3-*sn*-phosphatidylcholine (PLPC), 1-palmitoyl-2-arachidonoyl-3-*sn*-phosphatidylcholine (PLPC), and 1-palmitoyl-2-arachidonoyl-3-*sn*-phosphatidylcholine (PAPC). The antioxidant efficiency of α-TH against the peroxidation of PC was assessed by the formation of α-TH oxidation products. Peroxidation of the PC liposomes was started in the lipid phase by lipid-soluble 2,2′-azo*bis*(2,4-dimethylvaleronitrile) (AMVN) for the multilamellar liposomal system, and started in the aqueous phase by water-soluble 2,2′-azo*bis*(2-amidinopropane) dihydrochloride (AAPH)

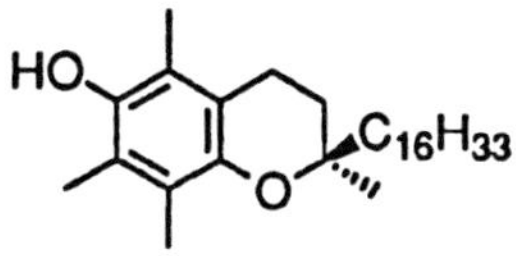

α-TH

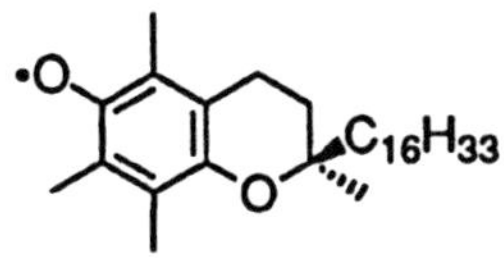

α-T•

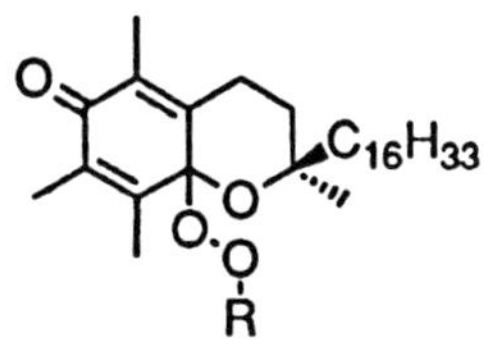

[1a] R = phosphatidylcholine

[1b] R = $C(CN)(CH_3)CH_2CH(CH_3)_2$

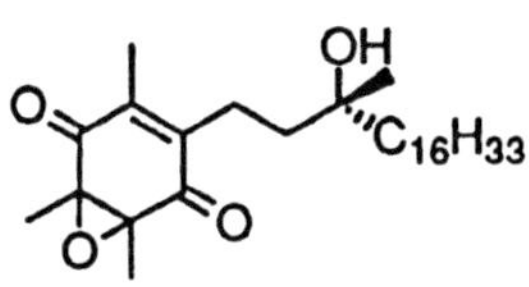

[2]

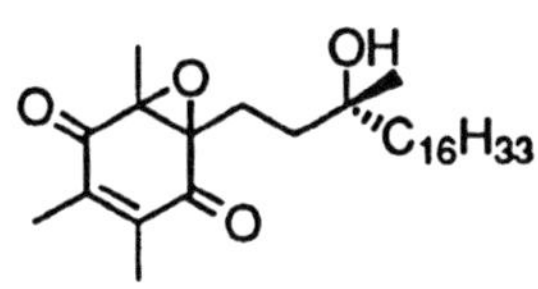

[3]

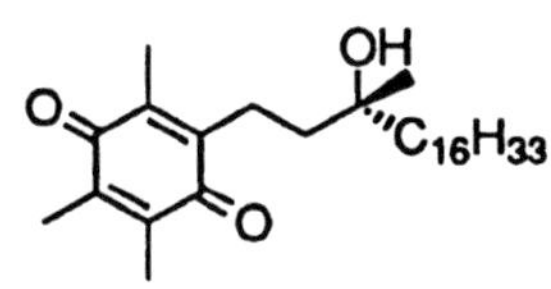

[4]

Fig. 11.1. Structure of compounds referred to in the text.

or a mixture of ferric monosodium ethylenediaminetetraacetate [Fe(III)-EDTA] and ascorbate for the unilamellar liposomal system. Concentrations of α-TH in biological membranes are approximately one part per 1000 lipid molecules (Kelley *et al.* 1995); thus, we have chosen the concentration of α-TH in PC liposomes to be 0.1 mol% relative to the PC concentration.

Materials and Methods

Materials

RRR-α-TH (type V) was purchased from Sigma Chemical (St. Louis, MO) and purified by reversed-phase high-performance liquid chromatography (HPLC) (Yamauchi *et al.* 1994). *RRR*-α-TH, which was used as an internal standard, was prepared from mixed isomers of tocopherol (Hohnen Oil, Tokyo, Japan). The AMVN-derived 8a-alkyldioxy-α-tocopherones [**1b**] were prepared by the reaction of α-TH with AMVN in benzene (Yamauchi *et al.* 1989a). 5,6-Epoxy-tocopherylquinone (5,6-epoxy-α-TQ, [**2**]) and 2,3-epoxy-α-tocopherylquinone (2,3-epoxy-α-TQ,[**3**]) were obtained as the oxidation products of α-TH during the peroxidation of methyl linoleate in an aqueous dispersion (Yamauchi *et al.* 1989b). α-Tocopherylquinone (α-TQ, [**4**]) was the oxidation product of α-TH with $FeCl_3$ in ethanol. The spirodiene dimer of α-TH was synthesized by following the procedure of Nelan and Robeson (1962). DLPC, PLPC, and PAPC were synthesized according to the method of Gupta *et al.* (1977). The synthesized PC was purified by reversed-phase HPLC using an Inertsil Prep-ODS column (20 × 250 mm, GL Sciences, Tokyo, Japan) with methanol to remove the contaminating hydroperoxides. Authentic PC hydroperoxides (PC-OOH) were prepared by the oxidation of PC liposomes in the presence of 1 mol% α-TH, initiated by AMVN (Weenen and Porter 1982), and purified by HPLC under the same condition as those already described. Free-radical initiators, AAPH and AMVN, were obtained from Wako Pure Chemical Industries (Osaka, Japan). Fe(III)-EDTA was from Dojindo Laboratories (Kumamoto, Japan).

Apparatus

HPLC was carried out with a Waters Model 600E system (Waters, Milford, MA). A Waters Model 486 UV/VIS detector or a Jasco Model 820-FP spectrofluorometer (Japan Spectroscopic, Tokyo, Japan) was used as the detector. Ultraviolet (UV) spectrum was measured with a Jasco LThest-30 UV/VIS spectrophotometer. Fast atom bombardment mass spectrometry (FAB-MS) was done using a Shimadzu 9020-DF instrument (Shimadzu, Kyoto, Japan) with a xenon fast atom bombardment gun. Samples were placed in the 5.5-kV xenon atom beam on a probe with triethanolamine as the matrix. The proton nuclear magnetic resonance (^{1}H NMR) spectrum was recorded at 270.17 MHz with a JEOL GX-270 instrument (JEOL, Tokyo, Japan) with $CDCl_3$ as the solvent and tetramethylsilane as the internal standard.

Preparation of 8a-(PC-Dioxy)-α-Tocopherones

The 8a-(PC-dioxy)-α-tocopherones [**1a**] were prepared by the reaction of α-TH with PC, initiated by AMVN (Yamauchi *et al.* 1994). Multilamellar liposomes of DLPC, PLPC, or PAPC (1 mmol) containing α-TH (5 μmol) and AMVN (0.1 mmol) in 100 mL of 50 mmol/L sodium phosphate buffer (pH 7.4, containing 50 mmol/L NaCl and 0.1 mmol/L Na_2EDTA) were incubated at 37°C for 15 (PAPC) and 20 h (DLPC and PLPC) under air with mechanical shaking. The reaction mixture was extracted with 200 mL of chloroform/methanol (1:1, vol/vol). The chloroform layer was evaporated *in vacuo* and dissolved in 1 mL of ethanol. The ethanol solution was injected onto reversed-phase HPLC. HPLC was done using an Inertsil C8 column (4.6 × 150 mm) or 7.6 × 250 mm, GL Sciences) in a solvent system of methanol/water (99:1, vol/vol) at a flow rate of 1.0 or 4.0 mL/min at 240 nm. The fractions of 8a-(PC-dioxy)-α-tocopherones were collected and the solvent was evaporated *in vacuo*. The isolated compounds were dissolved in ethanol and stored at –20°C until analysis.

Peroxidation of Liposomes

Multilamellar liposomes of DLPC (10 mmol/L), containing α-TH (0.1 mol%, based on DLPC) and AMVN (20 mol%, based on DLPC), were prepared by coevaporating from ethanol and then vigorously mixing in a 50 mmol/L sodium phosphate buffer (pH 7.4, containing 50 mmol/L NaCl), and incubated at 37°C under air. In this case, the liposomes were prepared quickly at 4°C to prevent the start of AMVN decomposition. Large unilamellar liposomes containing α-TH were prepared by the extrusion method (MacDonald *et al.* 1991). Mixtures of PC and α-TH (0.1 mol%, based on PC) were suspended by mixing vigorously for 2 min in a 50 mmol/L sodium phosphate buffer (pH 7.4, containing 50 mmol/L NaCl) and then by sonicating for 2 min. The milky suspension was transferred into a LiposoFast apparatus (Avestin, Ottawa, Canada), extruded 19 times back and forth through a polycarbonate membrane (100-nm pore size), and diluted with the same buffer to give a final PC concentration of 10 mmol/L. After preincubation for 5 min at 37°C in a shaking water bath, either AAPH (a final concentration of 2 mmol/L) or a mixture of Fe(III)-EDTA and ascorbic acid (final concentrations of 0.1 and 2 mmol/L, respectively) was added to the liposomes. The peroxidation was carried out at 37°C in air.

Quantification

At a given time, 500-μL aliquots of the reaction mixtures were withdrawn; 25 μL of 0.2 mol/L Na_2EDTA was added and extracted twice with two volumes of hexane/2-propanol (3:2, vol/vol). The extracts were evaporated to dryness *in vacuo;* the lipid residues were dissolved in ethanol and applied to HPLC. The amount of PC-OOH was determined by HPLC using a TSK-gel Octyl-80TS column (4.6 × 150 mm, Tosoh, Tokyo, Japan) with methanol/water (95:5, vol/vol) at 1.0 mL/min, with α-TH quantified by HPLC with fluorescence detection (Yamauchi *et al.* 1994). The amount of 8a-(PC-dioxy)-α-tocopherones [**1a**] was determined with the use of an Inertsil C-8 column (4.6 × 150 mm) under the same conditions as described previously.

To analyze the other oxidation products of α-TH [**1b**], [**2**]–[**4**], PC and its oxidation products, which coeluted with the oxidation products of α-TH, were removed by solid-phase extraction before the HPLC analysis. Each lipid extract was dissolved in 0.5 mL of hexane/ethyl acetate (1:1, vol/vol) and applied to a Sep-Pak silica cartridge (Waters) that had been prewetted with the same solvent. The oxidation products of α-TH were eluted with 4 mL of the same solvent. The eluted fractions were pooled and evaporated under nitrogen. The residue was dissolved in ethanol and analyzed by reversed-phase HPLC with a Puresil 5-μg C-18 120Å column (4.6 × 150 mm. Nihon Waters, Tokyo, Japan) using methanol/acetonitrile/water (87:10:3, by vol) at 1.0 mL/min. Absorbance was monitored at 285 nm, and γ-TH was used as an internal standard.

Results

Isolation and Characterization of 8a-(PC-Dioxyl)-α-Tocopherones

AMVN decomposes at 37°C to form alkyl radicals which, in turn, react with molecular oxygen to form alkylperoxyl radicals (Niki *et al.* 1984). The reaction mixture contains a larger amount of PC than α-TH. Therefore, most of the alkylperoxyl radicals attack PC to generate the pentadienyl radical of PC, which reacts rapidly with oxygen to give the PC-peroxyl radical (Yamauchi *et al.* 1994). The products of PC-peroxyl radicals with α-TH were separated by reversed-phase HPLC and obtained as compound **A** from DLPC, **B** from PLPC, and **C** from PAPC, respectively. The structures of compounds **A**–**C** were identified as described below (Fig. 11.2).

The structure of **A** was identified as being a mixture of 1-[9-(8a-dioxy-α-tocopherone)-10,12-octadecadienoyl]-2-linoleoyl]-2-linoleoyl-, 1-[13-(8a-dioxy-α-tocopherone)-9,11-octadecadienoyl]-2-linoleoyl-, 1-linoleoyl-2-[9-(8a-dioxy-α-tocopherone)-10,12-octadecadienoyl]-, and 1-linoleoyl-2-[13-(8a-dioxy-α-tocopherone)-9,11-octadecadienoyl]-3-*sn*-phosphatidylcholines. UV (ethanol)λ_{max} 235 nm (ε35400); FAB-MS *m/z* 1243 (MH^+), 814 ($[MH - 429]^+$), 781 ($[MH - 462]^+$), 520 ($[MH - 723]^+$), 430 ([α-tocopherone]$^+$), 224 ([CH=CH-CH_2-O-phosphocholine]$^+$), 184 ([phosphocholine]$^+$), and 104 ([choline]$^+$); ^{1}H NMR ($CDCl_3$)δ 0.80–0.95 (m, 18H), 1.07 and 1.25 (s, 3H), 1.15 (m, 3H), 1.30 (m, 48H), 1.53–1.65 (m, 4H), 1.68 (m, 2H), 1.86 (s, 6H), 1.93 and 1.95 (s, 3H), 2.01–2.08 (two q, J = 7 Hz, 2H), 2.17 (broad q, 2H), 2.27 (t, J, = 7 Hz), 2.30 (t, J = 7 Hz, 2H), 2.54–2.61 (m, 2H), 2.77 (t, J = 6 Hz, 2H), 3.37 (s, 9H), 3.81 (m, 2H), 3.94 (m, 2H), 4.10 (m, 1H), 4.15 (m, 1H), 4.37 (m, 2H), 4.41 (m, 1H), 5.20 (m, 1H), 5.27–5.36 (m, 4H), 5.39–5.52 (m, 2H), 5.91–6.03 (m, 1H), and 6.29–6.43 (m, 1H). The isomeric compositions of compound **A** were determined by normal-phase HPLC after $NaBH_4$ reduction and transmethylation (Chan and Levett 1977). Compound **A** consisted of the following isomers: the 9-(10*E*,12*Z*) (49%), the 9-(10*E*,12*E*) (9%), the 13-(9*Z*,11*E*) (34%), and the 13-(9*E*,11*E*) (8%).

The structure of compound **B** was identified as a mixture of 1-palmitoyl-2-[9-(8a-dioxy-α-tocopherone)-10,12-octadecadienoyl]- and 1-palmitoyl-2-[13-(8a-dioxy-α-tocopherone)-9,11-octadecadienoyl]-3-*sn*-phosphatidylcholines: UV (ethanol)λ_{max} 233 nm (ε35400); FAB-MS *m/z* 1219 (MH^+), 790 ($[MH - 429]^+$), 757 ($[MH - 462]^+$), 430 ([α-tocopherone]$^+$), 224 ([CH=CH-CH_2-O-phosphocholine]$^+$), and 104 ([choline]$^+$), ^{1}H NMR ($CDCl_3$)δ 0.80–0.90 (m, 18H), 1.07 and 1.25 (s, 3H), 1.15 (m, 3H), 1.20–1.40 (m, 56H), 1.40–1.80 (m, 6H), 1.86 (s, 6H), 1.95 (s, 3H), 2.0–2.2 (m, 4H), 2.28 and 2.30 (t, J = 7 Hz, 4H), 2.60 (m, 2H), 3.35 (s, 9H), 3.77 (m, 2H), 3.97 (m, 2H), 4.10 (m, 1H), 4.16 (m, 2H), 4.36 (m, 2H), 4.42 (m, 1H), 5.20 (m, 1H), 5.3–5.6 (m, 2H), 5.9–6.0 (m, 1H), and 6.3–6.4 (m, 1H). Compound **B** consisted of the following isomers: the 9-(10*E*,12*Z*) (44%), the 9-(10*E*,12*E*) (12%), the 13-(9*Z*,11*E*) (32%), and the 13-(9*E*,11*E*) (11%).

The structure of compound **C** was identified as a mixture of 1-palmitoyl-2-[5-(8a-dioxy-α-tocopherone)-6,8,11,14-eicosatetraenoyl]-, 1-palmitoyl-2-[8-(8a-dioxy-α-tocopherone)-5,9,11,14-eicosatetraenoyl]-, 1-palmitoyl-2-[11-(8a-dioxy-

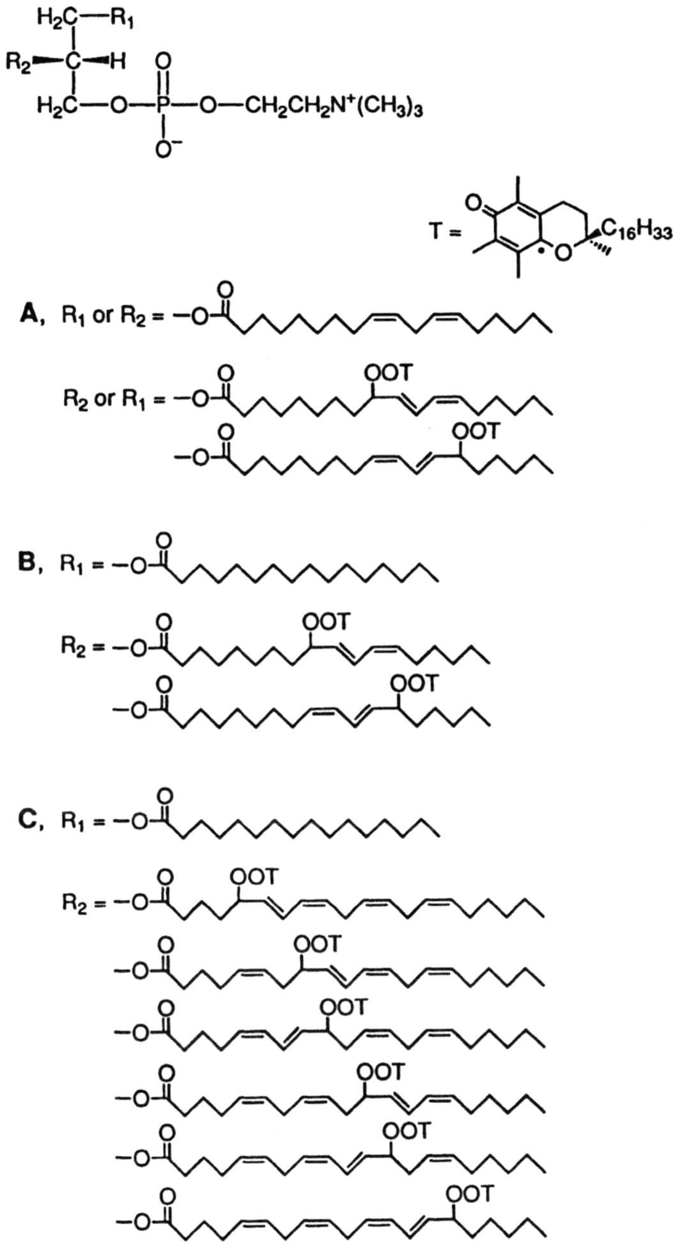

Fig. 11.2. Structures of the addition products of α-tocopherol (α-TH) with peroxyl radicals of (A) 1,2-dilinoleoyl-3-*sn*-phosphatidylcholine (DLPC), (B) 1-palmitoyl-2-linoleoyl-3-*sn*-phosphatidylcholine (PLPC), and (C) 1-palmitoyl-2-arachidonoyl-3-*sn*-phosphatidylcholine (PAPC).

α-tocopherone)-5,8,12,14–eicosatetraenoyl]-, 1-palmitoyl-2-[12-(8a-dioxy-α-tocopherone)5,8,10,14–eicosatetraenoyl]-, and 1-palmitoyl-2-[15-(8a-dioxy-α-tocopherone)-5,8,11,13-eicosatetraenoyl]-3-*sn*-phosphatidylcholines: UV (ethanolλ_{max} 235 nm (ε 32200); FAB-MS *m/z* 1243 (MH^+), 814 ($[MH - 429]^+$), 781 ($[MH - 462]^+$), 430 ($[\alpha\text{-tocopherone}]^+$), 224 ($[CH{=}CH\text{-}CH_2\text{-O-phosphocholine}]^+$), 165 ($[\text{phosphocholine}]^+$), and 104 ($[\text{choline}]^+$); ^1H NMR ($CDCl_3$)δ0.80–0.91 (m, 18H), 1.07 and 1.25 (s, 3H), 1.15 (m, 3H), 1.2–1.4 (m, 48H), 1.4–1.8 (m, 6H), 1.86 (s, 6H), 1.95 (s, 3H), 2.05 (m, 4H), 2.28 and 2.30 (t, $J = 7$ Hz, 4H), 2.60 (m, 2H), 2.81 (m, 4H), 3.36 (s, 9H), 3.79 (m, 2H), 3.98 (m, 2H), 4.15 (m, 1H), 4.15 (m, 1H), 4.37 (m, 2H), 4.43 (m, 1H), 5.15 (m, 1H), 5.2–5.6 (m, 6H), 5.9–6.1 (m, 1H), and 6.3–6.6 (m, 1H). The isomeric compositions of compound **C** were analyzed by HPLC after reduction and transmethylation (Yamagata *et al.*, 1984). Compound **C** consisted of the following isomers: the 5-(6*E*,8*Z*,11*Z*,14*Z*) (5%), the 8-(5*Z*,9*E*,11*Z*,14*Z*) (15%), the 9-(5*Z*,7*E*,11*Z*,14*Z*) (16%), the 11-(5*Z*,8*Z*,12*E*,14*Z*) (16%), the 12-(5*Z*,8*Z*,10*E*,14*Z*) (17%), and the 15-(5*Z*,8*Z*,11*Z*,13*E*) (31%).

Reaction of α-TH During the Peroxidation of DLPC in Liposomes

DLPC liposomes containing 0.1 mol% α-TH were reacted with peroxyl radicals generated in the lipid phase by thermal decomposition of AMVN, and the oxidation products of α-TH were analyzed by reversed-phase HPLC (Fig. 11.3). The sample extracts in ethanol were directly injected onto a C-8 column, and the peaks corresponding to the isomeric 8a-(DLPC-dioxyl)-α-tocopherones [**1a**] appeared in the 3-h sample (Fig. 11.3A). After treatment and the solid-phase extraction, the samples were injected onto a C-18 column (Fig. 11.3B). Along with unreacted α-TH and the internal standard (γ-TH), peaks corresponding to 5,6-epoxy-α-TQ [**2**], 2,3-epoxy-α-TQ [**3**], α-TQ [**4**], and a mixture of four diastereomers of 8a-alkyldioxy-α-tocopherones [**1b**], which were the addition products of the α-T• with AMVN-derived peroxyl radicals (Yamauchi *et al.* 1989a), were detected as the oxidation products. Because the concentration of α-TH added to the liposomes was relatively low, only a trace amount of the spirodiene dimer of α-TH, which was not analyzed further, appeared in the reaction mixture.

Figure 11.4A shows the results of the formation of PC-OOH, α-TH decay, and production of the oxidation products of α-TH [**1**]–[**4**] during AMVN-initiated peroxidation of DLPC liposomes. α-TH suppressed the formation of PC-OOH until the former had disappeared. The main products first formed were DLPC- and AMVN-derived 8a-alkyldioxy-α-tocopherones [**1a**] and [**1b**], after which epoxy-α-TQ [**2**] and [**3**] and α-TQ [**4**] were accumulated in the liposomes. The unilamellar liposomes of DLPC containing 0.1 mol% α-TH were reacted with peroxyl radicals that had been generated in the aqueous phase by thermal decomposition of AAPH (Fig. 11.4B). α-TH suppressed the formation of PC-OOH until all of the α-TH had been depleted. The products first formed were 8a-(DLPC-dioxy)-α-tocopherones [**1a**]; then epoxy-α-TA [**2**] and [**3**] and α-TQ [**4**] were accumulated in the liposomes.

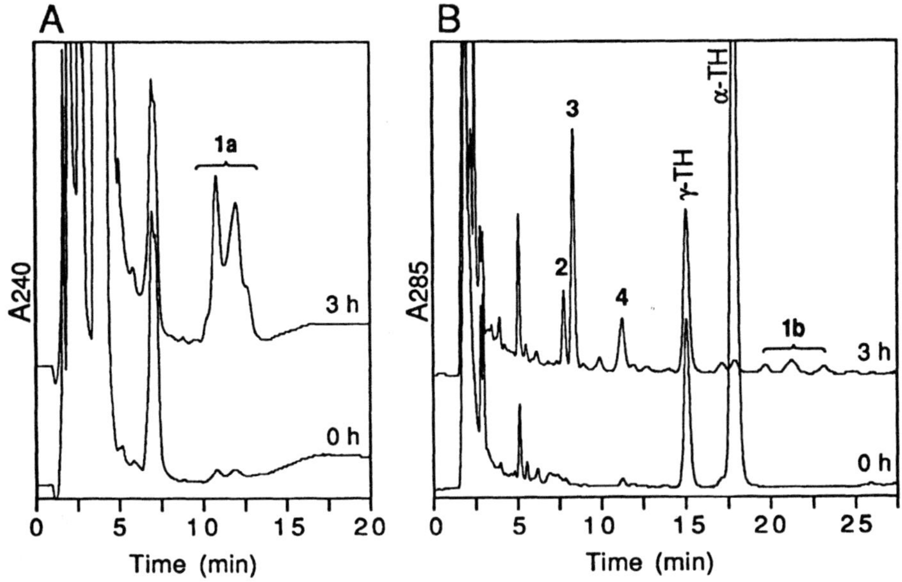

Fig. 11.3. Reversed-phase high-performance liquid chromatography (HPLC) of the reaction products of α-tocopherol (α-TH) in 1,2-dilinoleoyl-3-*sn*-phosphatidylcholine (DLPC) liposomes initiated by 2,2′-azo*bis*(2,4-dimethylvaleronitrile) (AMVN). Multilamellar liposomes of DLPC (10 mmol/L) containing 0.1 mol% α-TH and 20 mol% AMVN were incubated at 37°C for 0 and 3 h. (A) Product extracts were analyzed by HPLC without further treatment. HPLC was conducted with an Inertsil C-8 column (4.6 × 150 mm) developed with methanol/water (99:1, vol/vol). (B) Product extracts were treated by solid-phase extraction before the HPLC analysis as described in the Materials and Methods section. HPLC was conducted with a Puresil C-18 column (4.6 × 150 mm) developed with methanol/acetonitrile/water (87:10:3, by vol).

Reaction products between α-TH and AAPH-derived peroxyl radicals could also be expected, although we could not detect such compounds under the present HPLC conditions.

Reaction of α-TH During Peroxidation of PLPC and PAPC

DLPC is relatively rare in biological membranes; animal tissues contain phospholipids of molecular species defined by a saturated fatty acid esterified at the *sn*-1 position of glycerol and an unsaturated fatty acid at the *sn*-2 position (Takamura *et al.* 1986). Therefore, the reaction products of α-TH in unilamellar liposomes of PLPC and PAPC were examined. Table 11.1 compares the relative product yields of α-TH during peroxidation of DLPC, PLPC, and PAPC. In all oxidation systems, 8a-(PC-dioxy)-α-tocopherones [**1a**], epoxy-α-TQ [**2**] and [**3**], and α-TQ [**4**] accumulated with the depletion of α-TH. The oxidation products detected in AMVN-

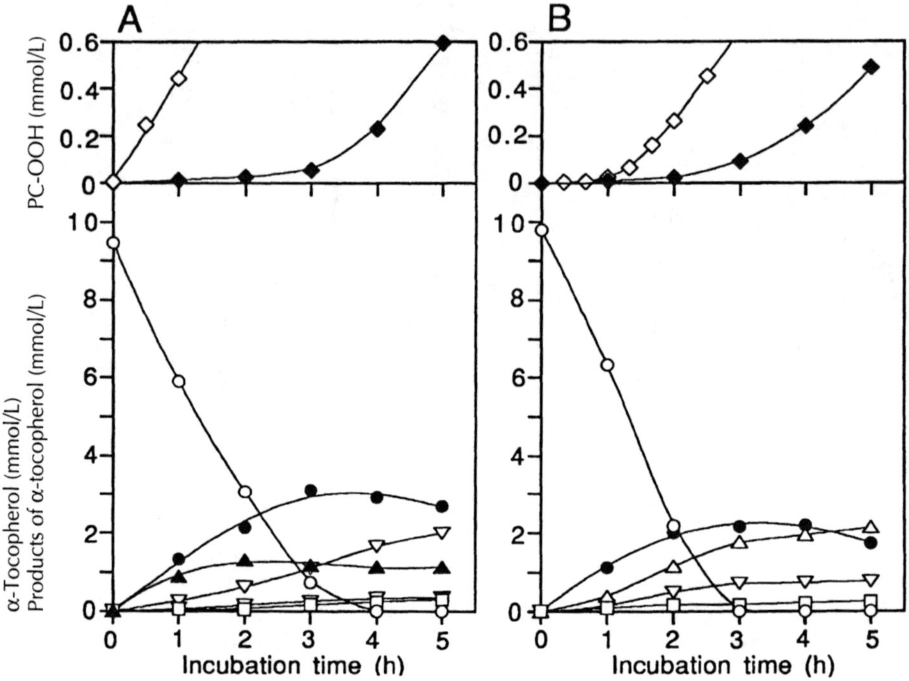

Fig. 11.4. Reaction of α-tocopherol (α-TH) during the peroxidation of 1,2-dilinoleoyl-3-*sn*-phosphatidylcholine (DLPC). (A) Multilamellar liposomes of DLPC (10 mmol/L) containing 0.1 mol% and α-TH were peroxidized by 2,2′-azo*bis*(2,4-dimethylvaleronitrile) (AMVN) (20 mol%) at 37°C. (B) Unilamellar liposomes of DLPC (10 mmol/L) containing 0.1 mol% of α-TH were peroxidized by 2,2′-azo*bis*(2-amidinopropane) dihydrochloride (AAPH) (2 mmol/L) at 37°C. PC-OOH, without (◇) or with (◆) α-TH; (○) α-TH; and the reaction products of α-TH, (●) [**1a**], (▼) [**1b**], (▽) [**2**], (△) [**3**], (□) [**4**].

initiated peroxidation accounted for ~65% of the α-TH consumed, whereas the oxidation products in AAPH-initiated peroxidation processes accounted for 45–50% of the α-TH consumed. Incubation of PC liposomes containing α-TH with Fe(III)-EDTA/ascorbate also caused the consumption of α-TH. The 8a-(PC-dioxy)-α-tocopherones in Fe(III)-EDTA/ascorbate-initiated peroxidation were relatively low yields compared with those in the AAPH-initiated peroxidation system. The products accounted for only 30% of the α-TH consumed.

Discussion

α-Tocopherol acts as the major lipid-soluble chain-breaking antioxidant in biological systems. The effectiveness of α-TH depends not only on its ability to trap peroxyl radicals, but also on the ability of the resulting α-T$^{\bullet}$ to trap additional peroxyl radi-

TABLE 11.1
Relative Yields of α-TH and Its Oxidation Products During the Peroxidation of DLPC, PLPC, and PAPC in Liposomes at 37°C[a]

		Yield[b] (%)					
Incubation time (h)		α-TH	**[1a]**	[**1b**]	[**2**]	[**3**]	[**4**]
AMVN-initiated peroxidation							
in DLPC	24	32.4	22.6	13.8	1.6	6.5	0.5
		ND	30.8	12.1	3.5	17.6	2.4
AAPH-initiated peroxidation							
in DLPC	24	22.4	20.5		5.0	11.8	1.7
		ND	22.6		7.6	20.0	2.2
in PLPC	24	14.6	8.6		9.4	20.7	2.8
		ND	7.7		11.0	27.7	3.9
in PAPC	24	31.9	3.8		5.6	12.4	4.5
		1.9	3.9		8.1	23.7	9.3
Fe(III)-EDTA/ascorbate-initiated peroxidation							
in PLPC	24	65.8	2.0		1.0	1.9	5.5
		ND	4.3		2.6	12.0	11.7
in PAPC	24	7.6	4.1		2.4	7.1	3.2
		ND	0.8		4.0	14.4	7.7

[a]Abbreviations: α-TH, α-tocopherol; DLPC, 1,2-dilinoleoyl-3-*sn*-phosphatidylcholine; PLPC, 1-palmitoyl-2-linoleoyl-3-*sn*-phosphatidylcholine ; PAPC, 1-palmitoyl-2-arachidonoyl-3-*sn*-phosphatidylcholine; AMVN, 2,2′-azo*bis*(2,4-dimethylvaleronitrile); AAPH, 2,2′-azo*bis*(2-amidinopropane) dihydrochloride; ND, not detectable.

[b]Mol% for each theoretical yield is based on the starting material.

cals (Burton and Ingold 1986). Therefore, products of the latter trapping reaction may be useful markers for the antioxidative activity of α-TH in a biological system (Ham and Liebler 1995). Much interest has been shown in the products of its reactions with lipid peroxyl radicals (Ham and Liebler 1995 and 1997, Liebler *et al.* 1990 and 1991, Liebler and Burr 1992, Matsuo *et al.* 1989, Winterle *et al.* 1984, Yamauchi *et al.* 1989a, 1990, 1993, 1994 and 1996). In this study, we isolated and characterized the 8a-(PC-dioxy)-α-tocopherones as the reaction products of α-T$^{\bullet}$ with peroxyl radicals of DLPC, PLPC, and PAPC. The PC-peroxyl radicals derived from each molecular species of PC could attack the α-T$^{\bullet}$ at the 8a-position to form the corresponding 8a-(PC-dioxy)-α-tocopherones (Fig. 11.2).

The autoxidation of unsaturated lipid (LH) in the presence of initiator (In) and inhibition by α-TH in liposomes proceeds according to the following reactions (Porter and Wagner 1986, Porter *et al.* 1995).

Initiation:

$$LH + In \rightarrow L^{\bullet} + InH \tag{1}$$

Propagation:

$$L^{\bullet} + O_2 \rightarrow LOO^{\bullet} \tag{2}$$

$$LOO^{\bullet} + LH \rightarrow LOOH + L^{\bullet} \tag{3}$$

Termination:

$$LOO^{\bullet} + \alpha\text{-TH} \rightarrow LOOH + \alpha\text{-T}^{\bullet} \quad (4)$$

$$LOO^{\bullet} + \alpha\text{-T}^{\bullet} \rightarrow \text{nonradical products} \quad (5)$$

If LH is PC, peroxyl radicals (LOO$^{\bullet}$) can be produced by oxygen addition of the pentadienyl radicals (L$^{\bullet}$) of PC (Porter *et al.* 1995), i.e., the 9- and 13-peroxyl radicals of the linoleoyl moiety from DLPC and PLPC, and the 5-, 8-, 9-, 11-, 12-, and 15-peroxyl radicals of the arachidonoyl moiety from PAPC, respectively. α-TH can terminate the propagation reaction by reactions (4) and (5). Reaction (5) leads to the formation of addition products of α-T$^{\bullet}$ with PC-peroxyl radicals. This study indicates that peroxyl radicals of DLPC, PLPC, and PAPC can attack the α-T$^{\bullet}$ at the 8a-position to form all possible 8a-(PC-dioxy)-α-tocopherones as the non-radical products. Therefore, this mechanism is consistent with the behavior of α-TH as a chain-breaking antioxidant in PC liposomes.

AMVN decomposes at 37°C to form alkyl radicals that react with molecular oxygen to form alkylperoxyl radicals in the hydrophobic inner part of PC liposomes. Because PC exists in a larger amount than α-TH, most of the AMVN-derived peroxyl radicals would attack PC, and the resulting PC-peroxyl radicals would react with α-TH. However, both of the peroxyl radicals produced from AMVN and DLPC reacted with α-TH to form AMVN- and DLPC-derived 8a-alkyldioxy-α-tocopherones [**1a**] and [**1b**], respectively (Figs. 11.3 and 11.4A). It is now accepted that α-TH is retained in membranes in such a way that the hydroxyl group is placed at or near the surface and the phytyl side chain, which is embedded into the membranes, so that the long phytyl side chain must suppress the mobility of α-TH (Fukuzawa *et al.* 1992). This indicates that α-TH and DLPC, which have long side chains, cannot move as readily as AMVN-derived peroxyl radicals, which do not have a long side chain. Because of the high concentration of DLPC in the liposomes, the first-formed peroxy radicals from AMVN reacted primarily with DLPC to form the corresponding DLPC-peroxyl radicals. If the mobility of the DLPC-peroxyl radicals within the liposomes is restricted by their fatty acyl chains, the reaction between DLPC-peroxyl radicals and α-TH must be suppressed. Consequently, AMVN-derived peroxyl radicals, as well as DLPC-peroxyl radicals, could react with α-TH to form the corresponding 8a-alkyldioxy-α-tocopherones. The AAPH-derived peroxyl radicals produced in the aqueous phase may also have been scavenged by α-TH before reacting with PC to form AAPH-derived 8a-alkyldioxy-α-tocopherones. Although AAPH-derived tocopherones could not be detected, the low yield of the oxidation products from AAPH-initiated oxidation, compared with that from the AMVN-initiated oxidation, might indicate the formation of such products. Furthermore, the amount of 8a-(PC-dioxy)-α-tocopherones formed in AAPH-initiated peroxidation of PLPC was about half that seen in the DLPC liposomes (Table 11.1). The liposomal concentration of linoleate is two times higher for DLPC than for PLPC, and the formation of 8a-(PC-dioxy)-α-tocopherones might thus be favored for DLPC.

The 8a-(PC-dioxy)-α-tocopherones were not the only reaction products during peroxidation of PC in liposomes. In this study, epoxy-α-TQ [**2**] and [**3**], and α-TQ [**4**] were also detected during peroxidation of PC liposomes (Fig. 11.3 and Table 11.1). The peroxyl radical-trapping reaction of the α-T$^{\bullet}$ forms 8a-alkyldioxy-α-tocopherones [**1**], which are then hydrolyzed to α-TQ [**4**] (Liebler *et al.* 1989). The formation of these products, therefore, is consistent with the behavior of α-TH as a chain-breaking antioxidant in liposomes. On the contrary, significant amounts of epoxy-α-TQ [**2**] and [**3**] were observed during peroxidation of PC liposomes (Table 11.1). This suggests that the formation of epoxy-α-TQ occurred *via* a pathway other than that from the decomposition of 8a-alkyldioxy-α-tocopherones. The formation of epoxy-α-TQ and their precursors, epoxyhydroperoxy-α-tocopherones, is thought to involve peroxyl-radical–dependent epoxidation of the α-T$^{\bullet}$, oxygen addition to the 8a-position, and subsequent hydrogen addition (Liebler *et al.* 1991, Matsuo *et al.* 1989). Although the mechanism by which these epoxides were formed remains unknown, oxidation of α-TH to epoxides resulted in the net consumption of two peroxyl radicals (Liebler and Burr 1995).

Ham and Liebler (1997) investigated the antioxidant reactions of α-TH in a perfused rat liver model. The principal products formed were 8a-substituted tocopherones and epoxy-tocopherones, rather than their hydrolysis products. However, their finding was evident only from the release of additional α-TQ or epoxy-α-TQ in HCl-treated samples. Further studies on the identification of 8a-substituted tocopherones in the intact organ system are required to elucidate the antioxidative activity of α-TH.

Acknowledgment

This work was supported in part by a Grant-in-Aid for Scientific Research (No. 08660151) from the Ministry of Education, Science, Sports and Culture of Japan.

References

Burton, G.W., and K.U. Ingold, Autoxidation of Biological Molecules, I. The Antioxidant Activity of Vitamin E and Related Chain-Breaking Phenolic Antioxidants *in vitro*, *J. Am. Chem. Soc. 103*: 6472–6477 (1981).

Burton, G.W., and K.U. Ingold, Vitamin E: Application of the Principles of Physical Organic Chemistry to the Exploration of Its Structure and Function, *Acc. Chem. Res. 19*: 194–201 (1986).

Chan, H.W.-S., and G. Levett, Autoxidation of Methyl Linoleate: Separation and Analysis of Isomeric Mixtures of Methyl Linoleate Hydroperoxides and Methyl Hydroxylinoleates, *Lipids 12*: 99–104 (1977).

Fujii, T., Y. Hiramoto, J. Terao, and K. Fukuzawa, Site-Specific Mechanisms of Initiation by Chelated Iron and Inhibition by α-Tocopherol of Lipid-Peroxide-Dependent Lipid Peroxidation in Charged Micelles, *Arch. Biochem. Biophys. 284*: 120–126 (1991).

Fukuzawa, K., W. Ikebata, A. Shibata, I. Kumadaki, T. Sakanaka, and S. Urano, Location and Dynamics of α-Tocopherol in Model Phospholipid Membranes with Different Charges, *Chem. Phys. Lipids 63*: 69–75 (1992).

Gupta, C.M., R. Radhakrishnan, and H.G. Khorana, Glycerophospholipid Synthesis: Improved General Method and New Analog Containing Photoactivable Groups, *Proc. Natl. Acad. Sci. USA 74:* 4315–4319 (1977).

Ham, A.J.L., and D.C. Liebler, Vitamin E Oxidation in Rat Liver Mitochondria, *Biochemistry 34*: 5754–5761 (1995).

Ham, A.J.L., and D.C. Liebler, Antioxidant Reactions of Vitamin E in the Perfused Rat Liver: Product Distribution and Effect of Dietary Vitamin E Supplementation, *Arch. Biochem. Biophys. 339*: 157–164 (1997).

Kelley, E.E., G.R. Buettner, and C.P. Burns, Relative α-Tocopherol Deficiency in Cultured Cells: Free Radical-Mediated Lipid Peroxidation, Lipid Oxidizability, and Cellular Polyunsaturated Fatty Acid Content, *Arch. Biochem. Biophys. 319*: 102–109 (1995).

Liebler, D.C., P.F. Baker, and K.L. Kaysen, Oxidation of Vitamin E: Evidence for Competing Autoxidation and Peroxyl Radical Trapping Reactions of the Tocopheroxyl Radical, *J. Am. Chem. Soc. 112*: 6995–7000 (1990).

Liebler, D.C., and J.A. Burr, Oxidation of Vitamin E During Iron-Catalyzed Lipid Peroxidation: Evidence for Electron-Transfer Reactions of the Tocopheroxyl Radical, *Biochemistry 31*: 8278– 8284 (1992).

Liebler, D.C., and J.A. Burr, Antioxidant Stoichiometry and the Oxidative Fate of Vitamin E in Peroxyl Radical Scavenging Reactions, *Lipids 30*: 789–793 (1995).

Liebler, D.C., K.L. Kaysen, and J.A. Burr, Peroxyl Radical Trapping and Autoxidation Reactions of α-Tocopherol in Lipid Bilayers, *Chem. Res. Toxicol. 4*: 89–93 (1991).

Liebler, D.C., K.L. Kaysen, and T.A. Kennedy, Redox Cycles of Vitamin E: Hydrolysis and Ascorbic Acid Dependent Reduction of 8a-(Alkyldioxy)tocopherones, *Biochemistry 28:* 9772–9777 (1989).

MacDonald, R.C., R.I. MacDonald, B. Ph.M. Menco, K. Takeshita, N.K. Subbarao, and L. Hu, Small-Volume Extrusion Apparatus for Preparation of Large, Unilamellar Vesicles, *Biochim. Biophys. Acta 1061*: 279–303 (1991).

Matsuo, M., S. Matsumoto, Y. Iitaka, and E. Niki, Radical-Scavenging Reactions of Vitamin E and Its Model Compound, 2,2,5,7,8-Pentamethylchroman-6-ol, in a *tert*-Butylperoxyl Radical-Generating System, *J. Am. Chem. Soc. 111*: 7179–7185 (1989).

Nelan, D.R., and C.D. Robeson, The Oxidation Product from α-Tocopherol and Potassium Ferricyanide and Its Reaction with Ascorbic and Hydrochloric Acids, *J. Am. Chem. Soc. 84*: 2963–2965 (1962).

Niki, E., T. Saito, A. Kawakami, and Y. Kamiya, Inhibition of Oxidation of Methyl Linoleate in Solution by Vitamin E and Vitamin C, *J. Biol. Chem. 259*: 4177–4182 (1984).

Porter, N.A., S.E. Caldwell, and K.A. Mills, Mechanisms of Free Radical Oxidation of Unsaturated Lipids, *Lipids 30*: 277–290 (1995).

Porter, N.A., and C.R. Wagner, Phospholipid Autoxidation, *Adv. Free Radic. Biol. Med. 2*: 283–323 (1986).

Takamura, H., Narita, H., R. Urade, and M. Kito, Quantitative Analysis of Polyenoic Phospholipid Molecular Species by High-Performance Liquid Chromatography, *Lipids 21*: 356–361 (1986).

Weenen, H., and N.A. Porter, Autoxidation of Model Membrane Systems: Cooxidation of Polyunsaturated Lecithins with Steroids, Fatty Acids, and α-Tocopherol, *J. Am. Chem. Soc. 104*: 5216–5221 (1982).

Winterle, J., D. Dulin, and T. Mill, Products and Stoichiometry of Reaction of Vitamin E with Alkylperoxyl Radicals, *J. Org. Chem. 49*: 491–495 (1984).

Yamagata, S., H. Murakami, J. Terao, and S. Matsushita, Decomposition Products of Methyl Arachidonate Monohydroperoxides, *Agric. Biol. Chem. 48*: 101–109 (1984).

Yamauchi, R., T. Matsui, K. Kato, and Y. Ueno, Reaction of α-Tocopherol with 2,2′-Azo*bis*(2,4-dimethylvaleronitrile) in Benzene, *Agric. Biol. Chem. 53*: 3257–3262 (1989a).

Yamauchi, R., T. Matsui, K. Kato, and Y. Ueno, Reaction Products of α-Tocopherol with Methyl Linoleate-Peroxyl Radicals, *Lipids 25*: 152–158 (1990).

Yamauchi, R., N. Miyake, K. Kato, and Y. Ueno, Reaction of α-Tocopherol with Alkyl and Alkylperoxyl Radicals of Methyl Linoleate, *Lipids 28*: 201–206 (1993).

Yamauchi, R., Y. Yagi, and K. Kato, Isolation and Characterization of Addition Products of α-Tocopherol with Peroxyl Radicals of Dilinoleoylphosphatidylcholine in Liposomes, *Biochim. Biophys. Acta 1212*: 43–49 (1994).

Yamauchi, R., Y. Yagi, and K. Kato, Oxidation of α-Tocopherol During the Peroxidation of Dilinoleoylphosphatidylcholine in Liposomes, *Biosci. Biotechnol. Biochem. 60*: 616–620 (1996).

Yamauchi, R., Y. Yasue, K. Kato, and Y. Ueno, Reaction Products of α-Tocopherol During the Peroxidation of Methyl Linoleate in Aqueous Dispersions, in *Proc. ISF-JOCS World Congress 1988*, The Japan Oil Chemists' Society, Tokyo, 1989b, pp. 575–581.

Chapter 12

Serum Antioxidant Capacity Assessed by an Iron-Dependent Assay in Carotid Stenotics Given Tocotrienols and Tocopherols

T.R. Watkins, D.K. Kooyenga, and M.L. Bierenbaum

Jordan Heart Research Foundation, Montclair, NJ 07042

Introduction

Oxygen's toxicity to living organisms is well known. The growth rate of aerobic organisms such as *Escherichia coli*, for example, is slowed when cultures are exposed to an atmosphere of 80% oxygen and 506 kPa atm total pressure (Brown and Yein 1978). In the plant kingdom, the soy plant exposed to elevated oxygen levels produces less tissue dry weight, with the seed less sensitive than the rest of the plant (Quebedeaux *et al.* 1975). Premature human infants kept in an incubator in an oxygen-enriched atmosphere develop retrolental fibroplasia (Johnson *et al.* 1974). Apparently, the elevated oxygen inhibits growth of retinal blood vessels. When the oxygen level decreases to normal, more blood vessel growth occurs; often this excessive growth results in retinal detachment. Concomitant bleeding from the weakened vessels often leads to blindness. The condition has been ameliorated, largely prevented with vitamin E supplementation and lower oxygen tension in the incubator.

To explain these kinds of toxic effects of oxygen, proportionately more severe than expected, Gershman *et al.* (1954) proposed the "superoxide theory of oxygen toxicity." A free radical form of oxygen, specifically superoxide, $O_2^{-\bullet}$, once formed, was more toxic than oxygen itself. Other free radicals, such as nitric oxide, $NO^{\bullet}$, although important in regulation of blood pressure, became toxic at levels above normal.

The body itself can produce superoxide, e.g., during phagocyte (eosinophil, macrophage, monocyte, neutrophil) activation. The phagocyte generates superoxide as a defense against bacterial infection to kill the invader (Babior and Woodman 1990). In some diseases, the mechanism becomes exacerbated and out of control, such as in rheumatoid arthritis and inflammatory bowel disease. In addition, some superoxide is formed as an incidental by-product of metabolism by coenzyme Q reductase (Fridovich 1986). The nitric oxide free radical is commonly produced by the endothelium to regulate vascular tone; hence it may regulate blood pressure. Superoxide reacts rapidly with $NO^{\bullet}$; excess superoxide may thus contribute to vasospasm and elevated blood pressure.

Although the body has natural defenses against superoxide, such as superoxide dismutase, catalase, and generic antioxidants, able to donate an electron or a hydrogen atom locally to stop free radical chain reactions, in the environment to which the body is exposed, it is presumed that such antioxidant reserves may be inadequate.

The body has a natural defense against oxygen and other free radicals, distinct from enzymatic means. Various antioxidants, such as ascorbic acid, tocopherols, tocotrienols, lipoic acid, uric acid, carotenoids, and flavonoids, can reduce the free radical, thus interrupting the chain reaction. The chemistry and roles of ascorbate and tocopherol have received considerably more attention than the others until quite recently. Once a free radical has formed (e.g., $L^\bullet$) and reacted with oxygen ($LOO^\bullet$), it may be reduced by an antioxidant such as tocopherol. (See model reactions in Scheme 12.1.) The antioxidant (AOH) interrupts the potential chain, and a less reactive radical forms, $AO^\bullet$. These "spent antioxidants" may be recycled by another reducer such as ascorbate or lipoate (Packer *et al.* 1995).

Should excess free radicals accumulate, such as superoxide or fatty acid peroxy radicals, one could expect to be able to measure the accumulation, associated tissue damage, and ultimately, degenerative disease. This approach has been used and documented; persons with diagnosed disease have been assessed in terms of limited or suboptimal reserve of a single antioxidant. Gey (1995) observed that the serum α-tocopherol level better predicts the risk of cardiovascular disease than any of the classical risk factors, such as serum cholesterol, smoking, or hypertension. Would a similar relation prevail in a clinical setting?

Antioxidant Reserves in the Stroke Patient

On the basis of these epidemiological data, would intervention with tocopherol supplements in hyperlipidemics increase their tissue antioxidant reserves? Would increased tissue reserves attenuate oxidation by-products, such as the 2-thiobarbituric acid reactive substances (TBARS)? And, would a generic test (here with iron) be useful in evaluating their antioxidant reserves?

$$LH \rightarrow L^\bullet + H^\bullet$$

$$L^\bullet + O_2 \rightarrow LOO^\bullet$$

$$LOO^\bullet + LH \rightarrow LOOH + L^\bullet$$

$$LOO^\bullet + AOH \rightarrow LOOH + AO^\bullet$$

$$AO^\bullet + AAH \rightarrow AOH + AA^\bullet$$

$$2\ AA^\bullet + GSH \rightarrow GSSG + 2\ AAH$$

$$GSSG + 2XH \rightarrow 2\ GSH + 2X$$

SCHEME 12.1. Suggested scheme of a free radical chain reaction and interruption by an antioxidant. Key: L, fatty acid; AOH, antioxidant; AAH, ascorbic acid; GSH, glutathione.

TABLE 12.1
Relation of Serum Levels of α-Tocopherol and Peroxidation Products (TBARS) and Thrombotic Risk in Hyperlipidemic Subjects Supplemented with α-Tocopherol[a,b]

Analyte	Base	Final
α-Tocopherol (μmol/L)	37.1 ± 4.0	59.4 ± 5.8[c]
TBARS (μmol/L)	4.10 ± 0.71	1.96 ± 0.23[c]
Platelet aggregation (Ω)	34 ± 1.3	26 ± 2.6[c]

[a]Source: Bierenbaum *et al.* 1992.
[b]The level of supplementation was 800 mg/d for 6 wk; n = 15 subjects. Agonist: Collagen, equine, 2 μg/mL. TBARS, thiobarbituric acid reactive substances.
[c]$P < 0.01$.

These questions were evaluated with hyperlipidemic subjects, using lipid, vitamin, and TBARS assay procedures published elsewhere (Bierenbaum *et al.* 1992, Kooyenga *et al.* 1997).

To test this hypothesis, we recruited hyperlipidemic subjects. Each was used as his or her own control in a supplementation trial in which each ingested 800 IU of α-tocopherol daily for 8 wk (Bierenbaum *et al.* 1992). The serum vitamin E level varied inversely with the peroxide level, i.e., as subjects' vitamin E levels rose, their serum peroxides decreased (Table 12.1). Raising antioxidant concentrations modulated serum peroxide, a well-known risk factor, measured as TBARS. Platelet aggregation tendency, an indicator of thrombotic risk, also decreased as serum antioxidant levels rose.

Many antioxidants other than the tocols (here α-tocopherol) also afford protection to the body. These include ascorbic acid (vitamin C), carotenoids, flavonoids, available sulfhydryl groups as in cysteine and lipoic acid, and uric acid. Other than measuring each of these compounds separately, how might one estimate total antioxidant reserves? Several approaches come to mind. Total oxygen uptake of the sample could be measured under standard conditions, gravimetrically, or with a Clark electrode. The sample could also be exposed to another oxidizer that would challenge and eventually consume the antioxidants present. Iron or copper would be apt candidates for this role. Benzie and Strain (1996) tested iron with a reporter dye for monitoring and reporting sample antioxidant reserves in a group of healthy subjects. Such a test for antioxidant reserves would be of great value as a diagnostic screening tool and also as a monitoring probe for patients at elevated risk of cardiovascular disease (CVD). Certain features might extend its utility (Table 12.2).

TABLE 12.2
Select Features of the Ferric Reducing Test

- Screening groups to identify individuals at risk of oxygen stress–related disease
- Titrating the need for dietary intervention and supplementation in individuals
- Monitoring the putative need for more effective thrombolytic therapy
- Monitoring the status of patients in *on-going* therapy, economically

In the case of CVD, heart disease and stroke presently account for one of every two deaths in the United States, with annual health care costs in the billions of dollars. Assessing risk of CVD would include measurement of antioxidant reserves in an accessible tissue, such as serum. A few desired features of such a test of total antioxidant reserves are shown in Table 12.2.

Stroke disease may be broadly characterized in terms of impaired cerebral blood flow in the carotid arteries, hemorrhage, enhanced blood viscosity attributed to hyperaggregable platelets, or a combination of these. Would antioxidant supplementation of the stroke patient enhance blood dynamics in terms of these known risk factors? Earlier work showed the value of supplementation on serum antioxidant levels in the hypercholesterolemic. Oxygen toxicity had been associated with increased atheroma formation leading to eventual atherosclerosis *via* peroxidation of low density lipoprotein (LDL) components, such as fatty acids and sterols (Steinberg *et al.* 1989). On the basis of NO's effect in relaxing the arterial wall, thereby attenuating blood pressure, and its effect in countering superoxide (Laurindo *et al.* 1991, Nakazono *et al.* 1991), leading to hyperaggregable platelets, we tested antioxidant supplementation (mixed tocotrienols and tocopherols) in the stroke patient.

Antioxidant Supplementation of the Stroke Patient

The test cohort, a group of 50 subjects who had suffered a transient ischemic attack (TIA), monocular blindness of <24 h duration, or temporary aphasia, were randomly assigned to receive a placebo (triolein), or mixed tocotrienols and tocopherols, 336 mg/d (Kooyenga *et al.* 1997). Subjects were matched for several characteristics, including age, gender, cigarette use, height, weight, blood pressure, use of antihypertensive medication, anticoagulants and antiplatelet medication, and diabetes. The average age was 66 y. The percentage with >49% carotid stenosis was 11 in the placebo and 15 in the test group. Subjects were randomly assigned to receive tocotrienols and tocopherols, or a placebo. The dosages were increased to 240 and 96 mg, respectively, by the end of y 1. No antilipidemic drugs (statin-type drugs) were taken, although all subjects continued their antiplatelet medication. Within 6 mo, the tocotrienol subjects showed improved blood flow in the carotid arteries. This improvement, evaluated at 6-mo intervals, has continued on into y 3.

Supplementation with tocotrienols and tocopherols did not modulate serum lipid levels. In the placebo group, serum total cholesterol (6.05 mmol/L at the start) held steady at 5.95 after 2 y. Similarly, the initial and final values in the tocotrienol group were 5.90 and 5.77 mmol/L, respectively. LDL, high density lipoprotein (HDL) and triacylglycerol concentrations did not change throughout this period. Hence, supplementation with tocotrienols at these levels in subjects eating self-selected diets did not modulate serum lipid levels.

Thrombotic risk plays a major role in stroke disease. Platelet aggregation has been widely used to assess thrombotic risk. The platelet response to the agonist collagen, a natural trigger protein, increased significantly in both groups, rising from 17.7 to 28.8 Ω in the placebo group and 16.9 to 25.5 Ω in the tocotrienol group ($P < 0.05$). The platelet response to thrombin did not change significantly.

A risk factor more recently associated with the risk of cardiovascular disease, in particular stroke, which indicates oxygen toxicity manifested in various molecules, has been measured as an adduct of thiobarbituric acid. When oxygen damage has occurred, this TBARS measure, in general, has decreased significantly with antioxidant supplementation in this clinic. In the reference group of stroke patients reported by Kooyenga *et al.* (1997), TBARS in subjects receiving placebos remained unchanged at a level of 1 μmol/L. In contrast, in subjects given tocotrienols, TBARS decreased from 1.08 to 0.80 μmol/L ($P < 0.05$).

Antioxidant Reserves

Daily supplementation with antioxidants would be expected to result in measurable elevation in serum antioxidant levels. This would be expected of tocotrienol (an E vitamer) antioxidant supplementation. Serum vitamin E values in the placebo group (receiving a soft gel capsule containing oil) decreased somewhat, from 38.1 to 27.6 μmol/L ($P < 0.05$). However, in the test group, the serum level increased from 37.1 to 74.3 μmol/L ($P < 0.01$) during this 2-y period of addition of tocotrienols and tocopherols to the daily diet.

With antioxidant supplementation, serum evidence of peroxidative damage declined. Abundant levels of peroxidative damage, signs of oxygen toxicity, accumulated in the placebo subjects. As serum antioxidant capacity rose in the test subjects given a daily allotment of tocotrienol and tocopherol antioxidant supplements for 2 y, serum oxygen-damaged pollutants decreased significantly. Levels of TBARS decreased by about one fourth, falling from 1.08 to 0.80 μmol/L ($P < 0.05$) as serum antioxidant levels rose.

What was the functional effect of antioxidant supplementation? Boosting serum levels of antioxidants conferred a distinct benefit upon carotid artery dynamics in these subjects over a 2-y period. The carotid artery stenosis in these subjects was reversed with antioxidant addenda as tocols. In the group given the antioxidant boost, carotid blood flow to the brain improved in 32% of the subjects, decreasing in only 8% of the subjects (Table 12.3). In the placebo group, none improved, and 44% showed marked progression of the disease (further impaired blood flow to the brain).

TABLE 12.3
Influence of Tocotrienol with Tocopherol Supplement on Carotid Stenosis after 2 Years[a]. (Data is percent of patients)

Group	Deterioration	No change (%)	Improvement
Placebo	44	56	0
Tocotrienol + tocopherol*	8	60	32

[a]Source: Kooyenga *et al.* (1997). *$p < 0.01$, test vs. placebo, Mann-Whitney.

How was such a disease condition improved? A boost in antioxidant reserves led to amelioration of the disease. The improvement was mediated by the tocols, chiefly tocotrienols. As reported above, the serum α-tocopherol concentration doubled in the test group; no change occurred in the placebo group. The implication of oxygen damage has been documented here primarily with TBARS. Hence, we attributed the improvement of the carotid artery and blood flow to the tocols' reversal of oxidant injury to the arterial intima and lesion mobilization.

We present some data about the use of ferric ion as a model probe to secure this kind of information from the stroke patient. We have evaluated serum of these patients with ferric ion and tripyridyltriazine. The principle of the test rests on the ability of ferric ion in serum (or plasma) to oxidize numerous electron donors. It proceeds as follows:

$$XH + Fe^{+3}/TPTZ \text{ (colorless)} \rightarrow Fe^{+2}/TPTZ \text{ (blue)} + X^{+}$$

Any substance present with a half-reaction redox potential less positive than Fe^{+3}/Fe^{+2}-TPTZ will drive the reduction of Fe^{+3} to Fe^{+2}, and lead to the development of an intense blue color (absorption maximum at 593 nm). Hence, under conditions with an excess of ferric ion and in the presence of TPTZ dye, color development will be proportional to the antioxidant present in the sample. Selected half-reaction redox potential values appear in Table 12.4, including values for several antioxidants.

Would the reaction prove to yield a linear response in a range usable for the cardiovascular patient? In a 4-min reaction time, the reaction produced a linear response from 0 to 2000 μmol/L, with a slope of 4.3×10^{-4} OD units/(μmol · L) of Fe^{+2}.

Does this assay generate data consistent with the health status of populations? The reaction performance was satisfactory for the evaluation of the stroke patient,

TABLE 12.4
Standard Redox Potentials of Selected Half-Reactions (E_o')[a]

Reduction half-reaction		Product	E_o', Voltage*
1/2 Oxygen + $2H^+ + 2e^-$	→	Water	0.816
Ferric + e^-	→	Ferrous	0.771
$CoQ_{10} + H^+ + e^-$	→	$CoQ_{10}H_2$	0.190
(FAD) + $2H^+ + 2e^-$	→	$(FAD)H_2$	0.003
Ascorbic acid + $2H^+ + 2e^-$	→	Dehydro ascorbic acid	–0.127
FMN + $2H^+ + 2e^-$	→	$FMNH_2$	–0.219
GSSG + 2H+ + $2e^-$	→	2 GSH	–0.230
α-Tocopherol + $2H^+ + 2e^-$	→	α-Tocopherol quinone	–0.273
Lipoic acid + $2H^+ + 2e^-$	→	Dihydro lipoic acid	–0.290
$NAD(P)^+ + 2H^+ + 2e^-$	→	$NAD(P)H + H^+$	–0.320
Uric acid + H_2O_2	→	Xanthine + $O_2 + H_2O$	–0.360

[a] *Source*: Mahler and Cordes (1966).
[b] Parentheses indicate protein-bound.
*Half-reaction potentials in volts.

whether supplemented at these levels, unsupplemented, or a healthy (non-stroke) reference subject, as seen in Table 12.5.

The mean value of reserves measured in the American subjects was 760 µmol/L (range, 550–1050 µmol/L), ~25% lower than the value of 1017 µmol/L (range, 612–1634 µmol/L) reported for healthy Chinese in Kowloon, Hong Kong (Benzie and Strain 1996). The lowest value obtained in our subjects was measured in one who claimed to "avoid eating vegetables." The difference undoubtedly reflects the paucity of vegetables reported eaten daily in the typical American diet, in contrast to the Chinese diet, noted for its abundance of vegetables. It may account, in part, for the lower incidence of atherosclerotic cardiovascular disease seen among the Chinese in comparison with Americans.

Conclusion

Antioxidants added to the diet in capsules will curb peroxidation reactions. Antioxidants as mixed tocols, chiefly tocotrienols, with some tocopherols, and placebo were fed at random to 50 stroke patients. The ultrasonography revealed that eight of the tocotrienol patients showed regression, two progression, whereas 11 of the placebo group showed progression in 3 y ($P < 0.01$). None in the placebo group showed regression. In this trial, intervention with palm-derived antioxidants did not result in altered serum lipid profiles. Collagen-induced platelet aggregation increased in both groups ($P < 0.05$). Oxygen damage to serum components decreased significantly when measured as TBARS ($P < 0.05$).

Further, an alternate approach for evaluating serum oxygen toxicity was evaluated as serum ferric ion reduction capacity. According to the preliminary data using this alternate index, placebo subjects showed the lowest reserves (559 µmol/L), subjects ingesting tocols more abundant reserves (828 µmol/L), and non-stroke, healthy subjects intermediate levels (760 µmol/L). Antioxidant supplements such as mixed tocotrienols and tocopherols may offer potential relief for the stroke patient, in part, by boosting serum antioxidant reserves. Also, on the basis of these preliminary serum ferric reduction data, assay of generic antioxidant reserves holds

TABLE 12.5
Serum TBARS, Vitamin E and Antioxidant Reserve of Stroke Subjects After 2 Years of Supplementation (Oil Vehicle or Tocopherols + Tocotrienols) or Healthy Subjects (Reference)[a], µmol/L.

Analyte	n	Placebo	Test	Reference
TBARS	25	0.83 ± 0.2	0.44 ± 0.1[b]	0.9 ± 0.3
Vitamin E (as α-tocopherol)	25	27.9 ± 2.3	74.8 ± 7.0[b]	16.2 ± 7.0[b]
Antioxidant reserve	15	559 ± 185	828 ± 434[c]	760 ± 193[c]

[a]*Source*: Kooyenga *et al.* (1997). TBARS, thiobarbituric acid reactive substances.
[b]$P < 0.01$; test vs. placebo, healthy reference (non-stroke) vs. placebo.
[c]$P < 0.05$; test vs. placebo, healthy reference (non-stroke) vs. placebo.

promise for evaluating antioxidant reserve in carotid stenosis. This colorimetric test with ferric ion and tripyridyltriazine as indicator may offer a rapid, spectrophotometric means to assess serum antioxidant reserves in other populations. Its features include its nonspecificity, speed, ease of detection, adaptability to automation, and economy.

References

Babior, B.M., and R.C. Woodman, Chronic Granulomatous Disease, *Semin. Hematol. 27*: 247–259 (1990).

Benzie, I.F.F., and J.J. Strain, The Ferric Reducing Ability of Plasma (FRAP) as a Measure of Antioxidant Power: The FRAP Assay, *Anal. Biochem. 239*: 70–76 (1996).

Bierenbaum, M.L., H. Bhagavan, and T.R. Watkins, Relationship Between Serum Lipid Peroxidation Products in Hypercholesterolemic Subjects and Vitamin E Status, *Int. J. Biochem. 10*: 57–66 (1992).

Brown, O.R., and F. Yein, Diydroxyacid Dehydratase: The Site of Hyperbaric Oxygen Poisoning in Branch-Chain Amino Acid Biosynthesis, *Biochem. Biophys. Res. Commun. 85*: 1219–1221 (1978).

Fridovich, I., Superoxide Dismutases, *Methods Enzymol. 58*: 61–97 (1986).

Gershman, K., D.L. Gilbert, S.W. Nye, P. Dwyer, and W.O. Fenn, Oxygen Poisoning and X-Irradiation: A Mechanism in Common, *Science 119*: 623–626 (1954).

Gey, K.F., Ten-Year Retrospective on the Antioxidant Hypothesis of Atherosclerosis, *J. Nutr. Biochem. 6*: 206–236 (1995).

Johnson, L., D. Schaeffer, and T.R. Boggs, Jr., The Premature Infant, Vitamin E Deficiency and Retrolental Fibroplasia, *Am. J. Clin. Nutr. 27*: 1158–1173 (1974).

Kooyenga, D.K., M. Geller, T.R. Watkins, A. Gapor, E. Diakoumakis, and M.L. Bierenbaum, Palm Oil Antioxidant Effects in Patients with Hyper-Lipidaemia and Carotid Stenosis—2 Year Experience, *Asian J. Clin. Nutr. 6*: 72–75 (1997).

Laurindo, F.R.M., P.L. Da Luz, L. Uint, T.F. Rocha, R.F. Jaeger, and E.A. Lopes, Evidence for Superoxide Radical-Dependent Coronary Vasospasm After Angioplasty in Intact Dogs, *Circulation 83*: 1705–1715 (1991).

Mahler, H.R., and E.H. Cordes, *Biological Chemistry,* New York, Harper and Row, 1966, p. 207.

Nakazono, K., N. Watanabe, K. Matsuno, J. Sasaki, T. Sato, and M. Inoue, Does Superoxide Underlay the Pathogenesis of Hypertension? *Proc. Natl. Acad. Sci. USA 88*: 10045–10048 (1991).

Packer, L., E.H. Witt, and H.J. Tritschler, Alpha-Lipoic Acid as a Biological Antioxidant, *Free Radic. Biol. Med. 19*: 227–250 (1995).

Quebedeaux, B., U.D. Havelka, K.L. Livak, and R.W.F. Hardy, Effect of Altered pO_2 in the Aerial Part of Soybean on Symbiotic N_2 Fixation, *Plant Physiol. 56*: 761–764 (1975).

Steinberg, D., S. Parthasarathy, T.E. Carew, J.C. Khoo, and J.L. Witzum, Beyond Cholesterol: Modifications of Low-Density Lipoprotein That Increase Its Atherogenicity, *N. Engl. J. Med. 320*: 915–924 (1989).

Chapter 13

Role of Flavonols and Anthocyanins from Fruits and Vegetables in Cancer Prevention

Barbara Raab, Gisela Jacobasch, Lutz Heinevetter, and Stefanie Maurer

German Institute of Human Nutrition Potsdam-Rehbrücke, D-14558 Bergholz-Rehbrücke, Germany

Introduction

Flavonoids are widely distributed in fruits and vegetables and thus serve as important bioactive phytochemicals in the human diet. They exert many effects on mammalian cell functions *in vitro* and *in vivo* (Formica and Regelson 1995, Middleton and Kandaswami 1994). In particular, their antioxidant properties (Bors *et al.* 1994 and 1997, Jovanovic *et al.* 1994, Kandaswami and Middleton 1994, Rice-Evans *et al.* 1995 and 1996) and their ability to modulate the activity of numerous enzymes are well documented (Kandaswami and Middleton 1994). Because of their influence on enzymes such as cyclooxygenase (Kalkbrenner *et al.* 1992), protein kinase C (Ferriola *et al.* 1989), protein tyrosine kinase (Chang and Geahlen 1992), cyclin-dependent kinases (Lepley and Pelling 1997), and topoisomerase II (Miyahara *et al.* 1997), flavonoids act on various signal transduction pathways and parameters of the cell cycle. These properties should contribute to the expected role of flavonoids in the prevention of degenerative diseases.

Our interest concentrated on the possibility of cancer prevention by flavonols and anthocyanins occurring in plant food. Therefore, we investigated the antioxidative activities of pure compounds and their influence on the antioxidative capacity of blood plasma (AOCP) *in vivo*, their protective effect against oxidative damage of DNA, their influence on proliferation and apoptosis, as well as the effect of rutin (R), on the development of intestinal polyps in Min mice to clarify whether these flavonoids are able to reduce DNA damage caused by oxidative stress, increase the AOCP *in vivo*, and exert anticarcinogenic effects.

Materials and Methods

Quercetin (Q), kaempferol (K), cyanidin (C), as well as their 3-*O*-rutinosides (Q-3-*O*-rut or R, K-3-*O*-rut, C-3-*O*-rut), Q-3-*O*-glucoside (Q-3-*O*-gluc), peonidin (P or 3′-methoxy-C) were purchased from Extrasynthèse (Genay, France), the 3-*O*-glucuronides from Q and K (Q-3-*O*-glucur and K-3-*O*-glucur) were isolated from French beans (*Phaseolus vulgaris* L.) by Hempel (Hempel and Böhm 1996); Q-4′-*O*-gluc and Q-3,4'-gluc were isolated from onions (*Allium cepa* L.) by Hempel (Hempel *et al.* 1997).

Enhanced chemiluminescence (ECL) immunoassay signal reagent and horseradish peroxidase (HRP) conjugate were obtained from Amersham Buchler (Braunschweig, Germany), the kits for integral antioxidant capacity of water- and lipid-soluble substances from FAT (Berlin, Germany), the annexin V-FITC kit from Boehringer Ingelheim Bioproducts (Heidelberg, Germany), and the *in situ* cell death detection kit, POD (TUNEL) from Boehringer Mannheim (Mannheim, Germany).

The chemiluminescence methods used allow the determination of antioxidant properties of pure compounds as well as biological fluids. Ascorbic acid (asc) or trolox are suitable as calibration standards. The ECL technique according to Whitehead *et al.* (1992) was conducted with the help of an LS-50B Luminescence Spectrophotometer (Perkin-Elmer, Ueberlingen, Germany). The photoinduced chemiluminescence (PCL) method was performed as described by Popov and Lewin (1994 and 1996) for water- and lipid-soluble substances, respectively, on a Photochem apparatus (FAT).

The murine fibrosarcoma cell line L929 (DSM ACC 2, Braunschweig, Germany) was cultured according to Leist *et al.* (1996) and used for the [1-(4,5-dimethylthiazol-2-yl)-2,5-diphenyltetrazolium bromide] (MTT)-test as described or for the comet assay according to Raab *et al.* (1997).

Culture flasks containing 4×10^4 cells/cm^2 in each case of the human colon adenocarcinoma cell line HT-29 (ATCC HTB 38, Rockville, MD, USA) were grown at 37°C in a 5% carbon dioxide atmosphere in Dulbecco's modified Eagle's medium (DMEM)/Ham's F-12 medium with stable glutamine, supplemented with 10% calf serum (Biochrom, Berlin, Germany) over 3 d and then treated with flavonoids as indicated in the figures.

Bivariate flow cytometric analysis of the phosphatidylserine (PS) translocation in early apoptotic cells was similar to the method of Vermes *et al.* (1995). After treatment with Q annexin V-FITC/propidium iodide (PI), stained floating and trypsinized adherent HT-29 cells were analyzed with a FACScan (Becton Dickinson, Heidelberg, Germany) with the use of the software CellQuest version 3.1.

C57BL/6J-Apc-min/+ and C57BL/6J-Apc+/+ mice supplied by Bomholtgard Breeding and Research Center (Bomgard, Denmark) were provided with a semi-synthetic diet or the same diet supplemented with rutin (R) and water *ad libitum* for ≤60 d.

Results and Discussion

As shown in Table 13.1, the structural determinants for the antioxidative activities of the polyphenolic flavonoids investigated are of different influence in the various chemiluminescence systems used. The 3-*O*-glycosides with hydroxyl groups in the 3′ and 4′ positions of ring B exert the strongest antioxidative properties in the aqueous PCL system, whereas C and K rank first in the methanolic system, and Q and its 3-*O*-glucuronide and 3-*O*-rutinoside are most active under the conditions of

TABLE 13.1
Comparison of the Relative Antioxidative Activities of Flavonols and Anthocyanins (Trolox Equivalents) Determined by Different Chemiluminescence Methods[a,b]

	PCL 0.4 μmol/L solutions		ECL 0.5 μmol/L solutions
Flavonoid	Aqueous, pH 10.7	Methanolic, pH 10.7	Aqueous, pH 7.2
C-3-*O*-rut	4.4	ND	ND
Q-3-*O*-rut	4.0	1.4	1.6
Q-3-*O*-gluc	3.8	ND	1.8
C	2.8	1.8	ND
P	1.9	ND	ND
Q	1.9	1.4	2.4
K	1.2	1.6	0.6
Q-4′-*O*-gluc	0.3	ND	ND
K-3-*O*-rut	0	0	0
Q-3,4′-*O*-bis-gluc	0	ND	ND
Q-3-*O*-glucur	ND	1.3	0.1
K-3-*O*-glucur	ND	0.1	0

[a]Interserial SD < 5%.
[b]Trolox, 6-hydroxy-2,5,7,8-tetramethylchromane-2-carbonic acid; PCL, photoinduced chemiluminescence; ECL, enhanced chemiluminescence; C-3-*O*-rut, cyanidin-3-*O*-rutinoside; Q-3-*O*-rut, quercetin-3-*O*-rutinoside; Q-3-*O*-gluc, quercetin-3-*O*-glucoside; C, cyanidin; P, Peonidin or 3′-methoxy-cyanidin; Q, quercetin; K, kaempferol; K-3-*O*-rut, kaempferol-3-*O*-rutinoside; Q-4′-*O*-gluc, quercetin-4′-*O*-glucoside; Q-3,4′-bis-gluc, quercetin-3,4′-bis-glucoside; Q-3-*O*-glucur, quercetin-3-*O*-glucuronide; K-3-*O*-glucur, kaempferol-3-*O*-glucuronide; ND, not determined.

the ECL technique. The different antioxidant activities of flavonols and anthocyanins, depending on the method of determination, may be due to the polarities, pH, as well as the kinds of radicals present in the respective mixtures. A variety of radicals should be present in the ECL measuring solution, whereas the PCL systems very specifically generate superoxide anion radicals (Popov and Lewin 1994).

Unrepaired damage to DNA caused by reactive oxygen species (ROS) might be a first step in the initiation of cancer. We used the comet assay, a very sensitive method to determine single strand breaks in the DNA, to evaluate the preventive effect of the flavonols and anthocyanins of interest against oxidative stress. Figure 13.1 demonstrates a significant reduction of DNA damage due to the genotoxic action of the ROS hydrogen peroxide (H_2O_2), linoleic acid hydroperoxide (LAOOH), and *tert*-butyl hydroperoxide (*t*-BuOOH) by C, an aglycon with antioxidative activity approximately two times higher than that of trolox. The compound C itself does not show genotoxicity in the test system used. The same proves true for the aglycon Q (Table 13.2), but despite the antigenotoxic activities of these flavonoids, a residual level of single strand breaks remains with both concentrations of the aglycons. That might be caused by radical intermediates produced by autoxidation (Sahu and Washington 1991a, b), adduct formation with DNA, or free radicals resulting from metabolism of the compounds. Although K and most of the

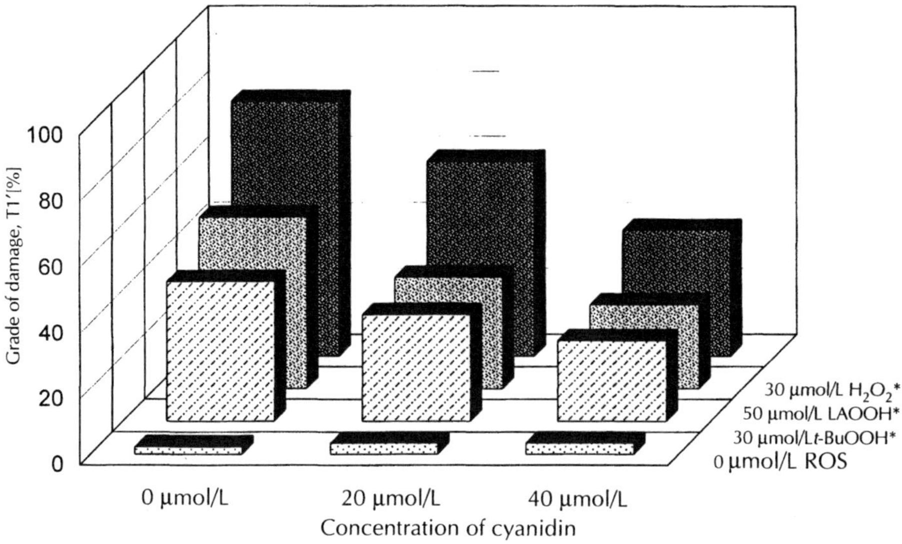

Fig. 13.1. Reduction of DNA damage by Cyanidin (Comet assay); $*P < 0.01$. Abbreviations: H_2O_2, hydrogen peroxide; LAOOH, linoleic acid hydroperoxide; *t*-BuOOH, *tert*-butyl hydroperoxide; ROS, reactive oxygen species; TI, tail intensity.

glycosides possess antioxidative properties, they actually increased the effects of ROS or had no influence (Table 13.2). Kaempferol-induced nuclear DNA damage and lipid peroxidation were also reported by Sahu and Gray (1994). These results point to a dual role of flavonols and anthocyanins, presumably depending on their concentration, the redox-state of the test system, or the complex biological environment and could explain the contradictory findings with respect to (anti)mutagenic or (anti)carcinogenic properties of flavonoids.

Proliferation and apoptosis are key events in the normal turnover of epithelial cells. Imbalance between these processes plays an important role in colorectal carcinogenesis. To characterize relevant flavonoids, we investigated their influence on proliferation and apoptosis of cultured tumor cells.

Figure 13.2 shows the influence of Q, K, and C on the proliferation of the murine fibrosarcoma cell line L929. The decrease in growth rate was measured by the reduced capacity of cells to form formazan, a tetrazolium bromide derivative, from MTT. The flavonol aglycons Q and K drastically reduced the proliferation of L929 cells in 5–50 µmol/L concentrations. The effects were more pronounced after 50 h compared with 24 h of treatment. In the case of C, the growth rate was diminished only to ~59% of the control with 50 µmol/L solutions after 50 h of treatment. All glycosides investigated had no significant effect on the proliferation (results not shown).

With the human HT-29 colon adenocarcinoma cell line, Q also inhibited proliferation as determined by cell counting after staining with trypan blue, but treatment of almost confluent cells requires much higher concentrations for inhibition.

TABLE 13.2
Influence of Flavonols and Anthocyanins on DNA Damage by Reactive Oxygen Species[a,b]

Flavonoid	H_2O_2	*t*-BuOOH	LAOOH
Q	–	–	(–)
K	+	+	+
Q-3-*O*-rut	+	=	=
K-3-*O*-rut	+	=	=
Q-3-*O*-glucur	+	=	=
K-3-*O*-glucur	=	=	=
Q-3-*O*-gluc	=	=	=
Q-4′-*O*-gluc	=	(–)	=
Q-3,4′-*O*-bis-gluc	+	=	(+)
P	+	+	+
C-3-*O*-rut	+	+	ND

[a]The symbol + indicates significant ($P < 0.05$) increase; – indicates significant decrease of DNA damage; (+) indicates a nonsignificant increase; (–) indicates a nonsignificant decrease; = indicates no influence.
[b]Abbreviations: H_2O_2, hydrogen peroxide; *t*-BuOOH, *tert*-butyl hydroperoxide; LAOOH, linoleic acid hydroperoxide; for abbreviations of flavonoids see Table 1; ND, not determined.

After 3 d of treatment with up to 400 µmol/L Q in the culture medium, cell numbers were reduced to ~50% (results not shown).

We could demonstrate that floating cells and, to a lesser degree, adherent cells, show morphological changes of nuclei, such as chromatin condensation, membrane blebbing, and apoptotic bodies, after DNA staining, as well as DNA laddering after electrophoresis (results not shown). These qualitative features of apoptosis were quantified by flow cytometric analysis.

Changes in the plasma membrane of cells are one of the earliest features of cells undergoing apoptosis (Vermes et *al.* 1995). In that process, the membrane phospholipid phosphatidylserine (PS) is translocated from the inner leaflet of the plasma membrane to the outer one without damage to the membrane integrity. Coupling of this PS with annexin-V, conjugated with FITC, and additional staining of the cells with propidium iodide (PI) enables bivariate analysis by flow cytometry to distinguish between cells undergoing apoptosis (PI negative) and those that are necrotic or in late apoptosis (annexinV-FITC positive, PI positive).

In adherent HT-29 cells (Fig. 13.3), treatment of ~95% confluent cells with 200 µmol/L Q over 3 d or an increase of the Q concentration to 400 mol/L for the last 24 h increases the level of apoptosis from 5.5 in control cells to 6.8 and 13.1%, respectively. Approximately 10–15% of late apoptotic cells are also present. After the same treatment, floating cells (Fig. 13.3.) contain 15.4–20.4% of early apoptotic cells with intact cell membranes. The content of late apoptotic cells is increased to ~60 %.

The results concerning the influence of Q on proliferation and apoptosis in the cultured tumor cells suggest a positive effect on the maintenance of the balance

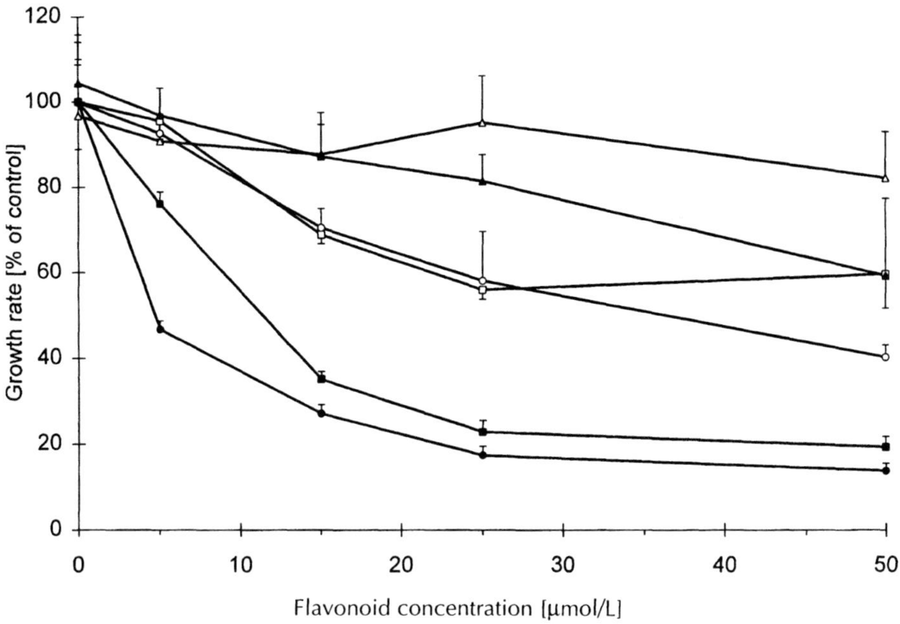

Fig. 13.2. Influence of flavonoids on the proliferation of L929 cells as determined by the [1-(4,5-dimethylthiazol-2-yl)-2,5-diphenyltetrazolium bromide] (MTT)-test: (○), quercetin (Q), 24 h; (●), Q, 50 h; (□), kaempferol (K), 24 h; (■), K, 50 h; (△), cyanidin (C), 24 h; (▲), C, 50 h.

between both processes and thus exertion of anticarcinogenic effects *in vivo*. We used the Min mouse, a multiple intestinal neoplasia mouse model (Moser *et al.* 1990, Su *et al.* 1992), to test this hypothesis.

Min mice are characterized by a mutation in the apc gene and display a neoplastic phenotype similar to that observed in familial adenomatous polyposis coli (FAP) in humans. Individuals with mutation of the apc gene develop numerous intestinal polyps at an early age; additional mutations lead to transformed states of tumor formation. Min/+ heterozygous mice die of carcinomas, tumor-related anemia, or bowel obstruction by 120 d of age.

Supplementation of the standard diet of Min mice with 94 μgR/g standard diet added as dry, pulverized buckwheat leaves influenced the AOCP and the development of intestinal polyps as shown in Table 13.3. The AOCP parallels the severity of cancerogenesis in Min mice. It decreases significantly from 166 μmol asc/L in controls to 89, and 79 μmol/L after 34–40 and 57–61 days, respectively. In contrast to the untreated Min mice, those with R supplementations maintain high levels of AOCP. Concomitant with reduced oxidative stress in the test animals after administration of R, the number of intestinal polyps was significantly lower (34 vs. 69 in the controls).

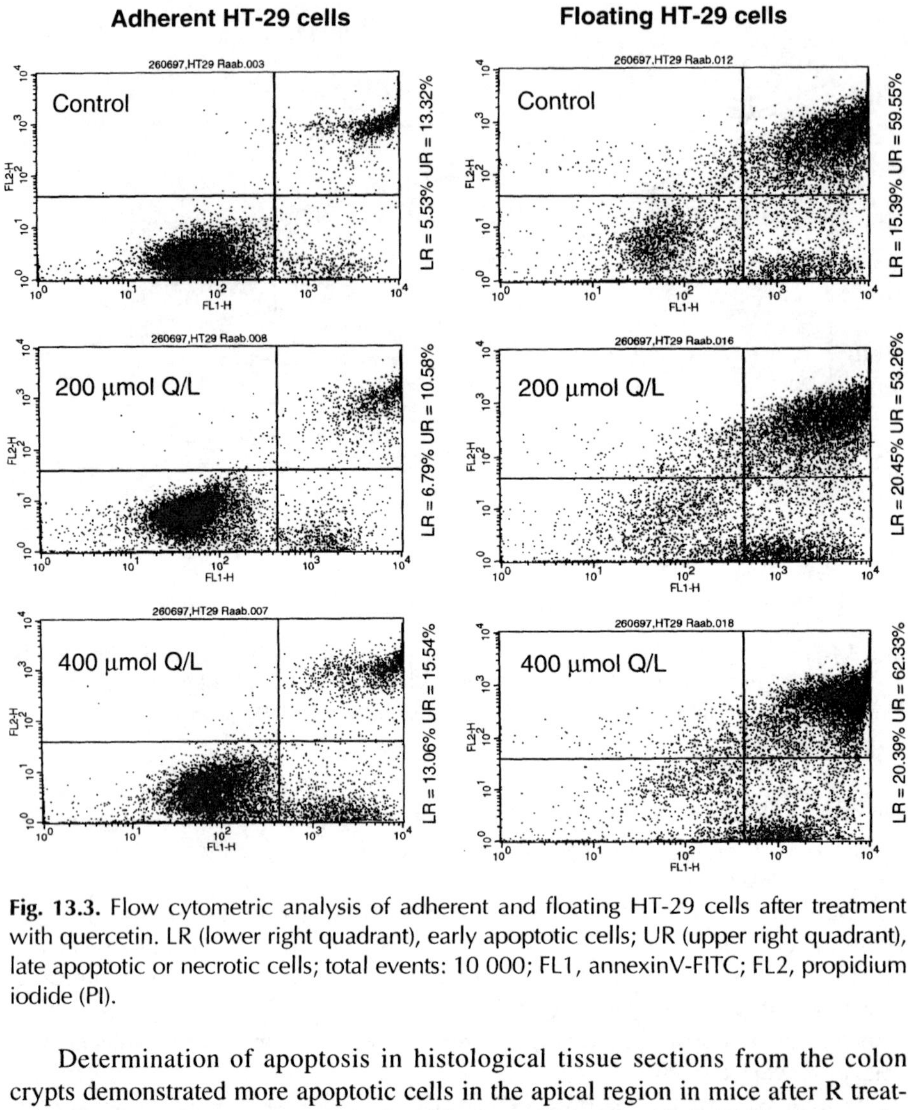

Fig. 13.3. Flow cytometric analysis of adherent and floating HT-29 cells after treatment with quercetin. LR (lower right quadrant), early apoptotic cells; UR (upper right quadrant), late apoptotic or necrotic cells; total events: 10 000; FL1, annexinV-FITC; FL2, propidium iodide (PI).

Determination of apoptosis in histological tissue sections from the colon crypts demonstrated more apoptotic cells in the apical region in mice after R treatment in comparison to untreated mice (black nuclei in Fig. 13.4). This points to the influence of the glycoside R or its aglycone Q, released by the gut microflora, on cell cycle parameters; such an effect warrants further investigation.

Conclusions

The expected correlation between antioxidative activities of the flavonols and anthocyanins investigated and their protective effects against oxidative stress *in vitro* could not be confirmed. Growth inhibition and increased apoptosis of cul-

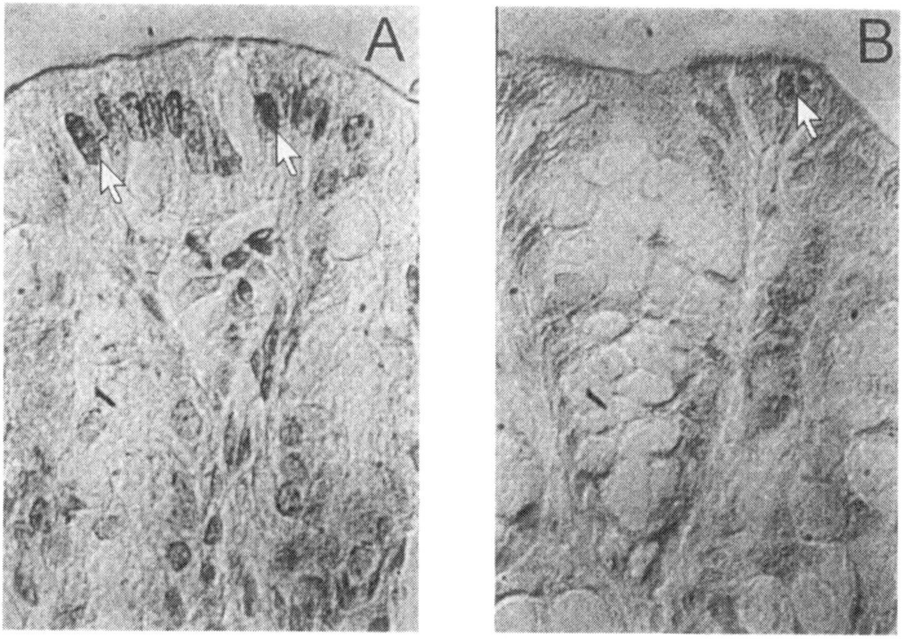

Fig. 13.4. Detection of apoptosis in murine colon crypts by the TUNEL assay. (A) Min mouse after R administration; (B) control mouse; arrows indicate apoptotic cells.

tured tumor cells suggest the influence of aglycons on parameters of the cell cycle. Administration of R in the standard diet of Min mice, a genetically defined model for inherited and sporadic forms of human colorectal tumorigenesis, significantly increased, *in vivo*, the AOCP and decreased the intestinal polyposis. These findings suggest an anticarcinogenic effect of R delivered by food.

TABLE 13.3
Influence of Rutin Supplementation on the Antioxidative Capacity of Blood Plasma and the Development of Intestinal Polyps[a,b]

Mice (μg R/g standard diet)	Days of treatment	AOCP (mol asc/L)	Mice (n)	Intestinal polyps	Mice (n)
Control	60	166 ± 49[c)]	28		
Control (94)	60	143 ± 22	4		
Min	27	142 ± 2	3		
Min	34–40	89 ± 40	7		
Min	57–61	79 ± 15[c,d)]	7	69 ± 37[e]	7
Min (94)	60	193 ± 104[d]	6	34 ± 13[e]	6

[a]Values marked by *c, d,* or *e* differ significantly from *c), d),* or *e)* with $P < 0.05$.
[b]Abbreviations: R, rutin or quercetin-3-*O*-rutinoside; R was applied as dry, pulverized buckwheat leaves with 6.5% R mixed with standard diet; AOCP, antioxidative capacity of blood plasma; asc, ascorbic acid; Control, C57BL/6J-Apc+/+ mice; Min, C57BL/6J-Apc min/+-mice.

Acknowledgments

The technical assistance of B. Kunkel, E. Chudoba, E. Wendt, and B. Streich is gratefully acknowledged.

References

Bors, W., C. Michel, and M. Saran, Flavonoid Antioxidants: Rate Constants for Reactions with Oxygen Radicals, *Methods Enzymol. 234*: 420–429 (1994).

Bors, W., C. Michel, K. Stettmaier, and W. Heller, EPR Studies of Plant Polyphenols, in *Natural Antioxidants: Chemistry, Health Effects and Applications*, edited by F. Shahidi, AOCS Press, Champaign, IL, 1997, pp. 346–357.

Chang, C.-J., and R.L. Geahlen, Protein-Tyrosine Kinase Inhibition: Mechanism-Based Discovery of Antitumor Agents, *J. Nat. Prod. 55*: 1529–1560 (1992).

Ferriola, P.C., V. Cody, and E. Middleton, Jr., Protein Kinase C Inhibition by Plant Flavonoids—Kinetic Mechanisms and Structure-Activity Relationships, *Biochem. Pharmacol. 38*: 1617–1624 (1989).

Formica, J.V., and W. Regelson, Review of the Biology of Quercetin and Related Bioflavanoids, *Food Chem. Toxicol. 33*: 1061–1080 (1995).

Hempel, J., and H. Böhm, Quality and Quantity of Prevailing Flavonoid Glycosides of Yellow and Green French Beans (*Phaseolus vulgaris* L.), *J. Agric. Food Chem. 44*: 2114–2116 (1996).

Hempel. J., B. Raab, and H. Böhm, Flavonoid Status of Bean (*Phaseolus vulgaris* L.) and Onion (*Allium cepa* L.) Varieties and Antioxidative Properties of the Main Components, in *Flavonoids and Bioflavonoids,* edited by S. Antus, M. Gábor, and K. Vetschera, Akadémiai Kiadó, Budapest, 1996, pp. 287–291.

Jovanovic, S.V., S. Steenken, M. Tosic, B. Marjanovic, and M.G. Simic, Flavonoids as Antioxidants, *J. Am. Chem. Soc. 116:* 4846–4851 (1994).

Kalkbrenner, F., G. Wurm, F. von Bruchhausen, *In Vitro* Inhibition and Stimulation of Purified Prostaglandin Endoperoxide Synthase by Flavonoids, *Pharmacology 44:* 1–12 (1992).

Kandaswami, C., and E. Middleton, Jr., Free Radical Scavenging and Antioxidant Activity of Plant Flavonoids, *Adv. Exp. Med. Biol. 366*: 351–376 (1994).

Leist, M., B. Raab, S. Maurer, U. Rösick, and R. Brigelius-Flohé, Conventional Cell Culture Media Do Not Adequately Supply Cells with Antioxidants and Thus Facilitate Peroxide-Induced Genotoxicity, *Free Radic. Biol. Med. 21*: 297–306 (1996).

Lepley, D.M., and J.C. Pelling, Induction of p21/WAF1 and G_1 Cell-Cycle Arrest by the Chemopreventive Agent Apigenin, *Mol. Carcinog. 19:* 74–82 (1997).

Middleton, E., Jr., and C. Kandaswami, The Impact of Plant Flavonoids on Mammalian Biology, in *The Flavonoids: Advances in Research Since 1986*, edited by J.B. Harborne, Chapman and Hall, London, 1994, pp. 619–652.

Miyahara, M., M. Kawasaki, H. Akiyama, T. Narui, M. Toyoda, T. Okuyama, and Y. Saito, Some Phytochemicals and Related Compounds in Vegetables as Potent Inhibitors of Human DNA Topoisomerase II, in *Food Factors for Cancer Prevention*, edited by H. Ohigashi, T. Osawa, J. Terao, S. Watanabe, and T. Yoshikawa, Springer-Verlag, Tokyo, 1997, pp. 182–187.

Moser, A.R., H.C. Pitot, and W.F. Dove, A Dominant Mutation That Predisposes to Multiple Intestinal Neoplasia in the Mouse, *Science 247*: 322–324 (1990).

Popov, I.N., and G. Lewin, Photochemiluminescent Detection of Antiradical Activity: II. Testing of Nonenzymic Water-Soluble Antioxidants, *Free Radic. Biol. Med. 17*: 267–271 (1994).

Popov, I.N., and G. Lewin, Photochemiluminescent Detection of Antiradical Activity: IV. Testing of Lipid-Soluble Antioxidants, *Biochem. Biophys. Methods 31*: 1–8 (1996).

Raab, B., J. Hempel, and H. Böhm, Antioxidative and Antigenotoxic Properties of Flavonoids from Beans (*Phaseolus vulgaris* L.), in *Food Factors for Cancer Prevention*, edited by H. Ohigashi, T. Osawa, J. Terao, S. Watanabe, and T. Yoshikawa, Springer-Verlag, Tokyo, 1997, pp. 309–312.

Rice-Evans, C.A., N.J. Miller, P.G. Bolwell, P.M. Bramley, and J.B. Pridham, The Relative Antioxidant Activities of Plant-Derived Polyphenolic Flavonoids, *Free Radic. Res. 22:* 375–383 (1995).

Rice-Evans, C.A., N.J. Miller, and G. Paganga, Structure-Antioxidant Activity Relationships of Flavonoids and Phenolic Acids, *Free Radic. Biol. Med. 20*: 933–956 (1996).

Sahu, S.C., and G.C. Gray, Kaempferol-Induced Nuclear DNA Damage and Lipid Peroxidation., *Cancer Lett. 85*: 159–164 (1994).

Sahu, S.C., and M.C. Washington, Quercetin-Induced Lipid Peroxidation and DNA Damage in Isolated Rat Liver Nuclei, *Cancer Lett. 58*: 75–79 (1991a).

Sahu, S.C., and M.C. Washington, Effects of Antioxidants on Quercetin-Induced Nuclear DNA Damage and Lipid Peroxidation, *Cancer Lett. 60*: 259–264 (1991b).

Su, L.-K., K.W. Kinzler, B. Vogelstein, A.C. Preisinger, A.R. Moser, C. Luongo, K.A. Gould, and W.F. Dove, Multiple Intestinal Neoplasia Caused by a Mutation in the Murine Homolog of the Apc Gene, *Science 256*: 668–670 (1992).

Vermes, I., C. Haanen, H. Steffens-Nakken, and C. Reutlingsberger, A Novel Assay for Apoptosis: Flow Cytometric Detection of Phosphatidylserin Expression on Early Apoptotic Cells Using Fluorescein Labelled Annexin V, *J. Immunol. Methods 184*: 39–51 (1995).

Whitehead, T.P., G.H.G. Thorpe, and S.R.J. Maxwell, Enhanced Chemiluminescence Assay for Antioxidant Capacity in Biological Fluids, *Anal. Chim. Acta 266*: 265–277 (1992).

Chapter 14

Dietary Intake of Flavonoids and Isoflavonoids by Japanese and Their Pharmacokinetics and Bioactivities

Shaw Watanabe[a], Yusuke Arai[a], Ryota Haba[a], Mariko Uehara[a,b], Herman Adlercreutz[b], Kayako Shimoi[c], and Naohide Kinae[c]

[a]Department of Nutritional Science and Epidemiology, Tokyo University of Agriculture, Tokyo, Japan

[b]Folkhälnen Research Center, Institute for Preventive Medicine, Nutrition and Cancer, Helsinki, Finland

[c]Department of Food Science, Shizuoka Municipal University, Shizuoka, Japan

Introduction

Nonnutrient food chemicals (factors) have various pharmacologic activities in the human body. The antioxidant properties of flavonoids, such as quercetin and kaempferol, are believed to contribute to the low risk of coronary heart diseases (Frankel *et al.* 1993, Frankel 1997, Hertog *et al.* 1993). The Japanese consume many soybean products containing phytoestrogens, such as daidzein and genistein, which may lead to the low incidence of estrogen-related cancers (Adlercreutz *et al.* 1995, Watanabe and Koessel 1993). More than 600 food factors in vegetables are thought to influence the various metabolic stages in the body. An accurate estimate of these chemical intakes is becoming more important, yet a proper Japanese food consumption database has not been developed. A preliminary database was developed in this study to estimate the amount of flavonoid and isoflavonoid intake through a field study conducted by volunteers.

To identify the biological effects of these phytochemicals, pharmacokinetic studies are necessary. Because the Japanese consume a large amount of isoflavonoids, the pharmacokinetics of the absorption and excretion of isoflavonoids were evaluated after a single oral ingestion of soybean powder (kinako) in humans (Watanabe *et al.* 1998).

The antioxidant activity of isoflavonoid-rich soy-hypocotyl tea was then tested by volunteers. Urinary excretion of ^{8}OHdG, which is considered to be a biological marker for DNA damage by various free radicals, was also measured to clarify the effect of isoflavone in humans.

Methods

Nutritional Study

Intake of food factors was calculated from a field study in I-city of Iwate prefecture. In early September, 1995, housewives (volunteers) from 50 families were asked to record all food and beverage items consumed and cooking methods used for 3 consecutive days (Kimira *et al.* 1998). The dietary records were checked through interviews by trained dietitians on d 4 of recording. Blood samples (10 mL) were also collected for analyses.

Intake of flavonoids and isoflavonoids was estimated by calculation from Shimoi's database. Optimization of a quantitative high-performance liquid chromatography (HPLC) determination was employed for measuring flavonoids (Hertog *et al.* 1992a). Quercetin, kaempferol, myricetin, fisetin, eriodictyol, and luteolin were measured in freeze-dried vegetables and fruits after acid hydrolysis of the parent glycosides. In some foods, such as bean sprouts, grapefruit, moroheiya, onion, tomato, and green tea, methanol extraction yielded more flavonoids, such as rutin and myricitrin; however, we used the value of acid hydrolyzates because of their similarity in cooking and absorption in the body. The database of these concentrations (Table 14.1) was used in the calculation of intakes.

Each food item was recorded, and consumption dose and frequency were summarized. Food nutrients were calculated using the Standard Tables of Food Composition in Japan. In addition, values for each food item were multiplied by the flavonoids/isoflavonoids concentrations table (Table 14.1).

Pharmacokinetic Study

Seven healthy men participated in the study on pharmacokinetics after the consumption of kinako. The study design was approved by the ethics committee of the National Institute of Health and Nutrition (Watanabe *et al.* 1998). The participants stayed in the dormitory of the institute from d 4 to 6; specially prepared meals were served.

The participants were asked to avoid tofu, natto (fermented soybean), miso (fermented soybean paste), and shoyu (soy sauce) during the study period. All foods and beverages were reported in dietary records and checked by trained dietitians. On the morning of d 4, 60 g of kinako (baked soybean powder) dissolved in 200 g of 3% agar water was added to the subjects' breakfast. From d 4 to 6, meals adjusted to the caloric requirement of the individual subjects were provided by the dietitian.

On d 4, a blood sample was collected every 2 h after kinako ingestion, and an additional fasting blood sample was collected in the morning on d 5. Blood samples were collected every other day at 11:00 A.M. A 24-h urine sample (from the second urination of the day to the first urination on the next day) was collected by the urimate (R). On d 4, each urine sample was separately stored and labeled; the time measured and volume were recorded. Feces were collected for 3 d, from d 5 to 7. Indigocarmine (1 g) was administered orally when the subject ingested the kinako so that defecation of meals containing kinako could be confirmed by the red color of the stools. All subjects had normal daily bowel movements.

Plasma, urine, and fecal samples were analyzed in the Department of Clinical Chemistry, University of Helsinki (Adlercreutz *et al.* 1994). Deuterium-labeled phytoestrogens were used as an internal standard. Eight phytoestrogens, i.e., daidzein, genistein, O-DMA, equol, enterolactone, enterodiol, matairesinol, and secoisolacriresinol,

TABLE 14.1
Concentration of Flavonoids and Isoflavonoids in Legumes, Vegetables, and Fruits[a] (μg/wet g)

Food name	Da	Ge	Mytrin	Ru	Mytin	Fi	Eri	Qu	Lurin	Ka	Iso total	Flav total
Kidney bean	—	—	ND	4.67	11.10	ND	ND	12.60	10.10	18.60	—	57.81
Green soybean	—	—	ND	ND	ND	ND	ND	0.03	ND	12.30	—	12.33
Soybean seed	846.20	1106.80	—	—	—	—	—	—	—	—	1953.00	—
Soyflour	654.70	1122.60	—	—	—	—	—	—	—	—	1777.30	—
Bean sprout	—	—	1.29	2.78	ND	ND	43.70	1.47	ND	3.26	—	60.93
Tofu (momen)	169.00	253.00	ND	ND	ND	ND	ND	ND	ND	11.90	422.00	11.90
Silken tofu (kinu)	151.00	225.00	—	—	—	—	—	—	—	—	376.00	—
Packed tofu	220.00	294.00	—	—	—	—	—	—	—	—	514.00	—
Baked tofu	175.00	256.00	—	—	—	—	—	—	—	—	431.00	—
Fried tofu	335.00	401.00	—	—	—	—	—	—	—	—	736.00	—
Grilled tofu	118.00	317.00	—	—	—	—	—	—	—	—	435.00	—
Ganmodoki (mixed fried tofu)	336.00	412.00	—	—	—	—	—	—	—	—	748.00	—
Natto	277.20	327.00	—	—	—	—	—	—	—	—	604.20	—
Soybean paste (white)	116.00	172.00	—	—	—	—	—	—	—	—	288.00	—
Soybean paste (red)	179.00	278.00	—	—	—	—	—	—	—	—	457.00	—
Potato	—	—	ND	ND	ND	ND	ND	ND	ND	23.40	—	23.40
Tomato	—	—	ND	3.38	ND	0.12	ND	1.58	ND	7.50	—	12.58
Green pepper	—	—	ND	ND	ND	ND	ND	14.10	14.70	3.17	—	31.97

Continued

TABLE 14.1
Continued

Food name	Da	Ge	Mytrin	Ru	Mytin	Fi	Eri	Qu	Lurin	Ka	Iso total	Flav total
Eggplant	—	—	ND	ND	ND	ND	ND	1.58	ND	ND	—	1.58
Carrot	—	—	ND	0.02	ND	ND	ND	ND	ND	7.20	—	15.32
Parsley	—	—	ND	ND	216.00	ND	ND	7.03	3.09	45.10	—	271.22
Radish	—	—	ND	0.55	ND	ND	ND	ND	ND	3.36	—	3.91
Cabbage	—	—	ND	ND	ND	ND	ND	ND	ND	7.20	—	7.20
Broccoli	—	—	ND	0.56	7.05	ND	ND	9.77	ND	16.10	—	33.48
Moroheiya	—	—	ND	99.40	19.30	ND	ND	154.00	ND	118.00	—	390.70
Spinach	—	—	ND	0.24	37.50	ND	ND	ND	7.13	ND	—	44.87
Lettuce	—	—	1.23	1.24	ND	ND	ND	4.78	5.23	ND	—	12.48
Onion	—	—	ND	2.37	2.95	4.78	ND	33.70	1.91	14.09	—	59.80
Lotus root	—	—	ND	ND	5.89	5.80	ND	4.41	3.61	7.57	—	27.28
Cucumber	—	—	ND	0.94	ND	0.14	ND	ND	ND	7.61	—	8.69
Kiwi fruit	—	—	ND	ND	ND	2.03	ND	2.07	ND	30.60	—	34.70
Watermelon	—	—	ND	ND	ND	ND	ND	ND	ND	18.10	—	18.10
Orange	—	—	ND	ND	14.40	ND	ND	17.50	0.96	31.50	—	64.36
Peach	—	—	ND	0.54	ND	0.58	ND	1.08	ND	6.54	—	8.74
Apple	—	—	ND	ND	ND	26.90	ND	5.27	ND	26.70	—	58.87
Persimmon	—	—	ND	ND	10.60	10.50	ND	ND	1.43	ND	—	22.53
Grape	—	—	ND	ND	ND	3.93	ND	ND	ND	16.80	—	20.73
Strawberry	—	—	ND	27.10	ND	160.00	ND	6.91	ND	19.40	—	213.41
Green Tea (Infusion)	—	—	ND	3.12	ND	ND	ND	1.10	ND	0.60	—	4.82

[a]Abbreviations: Da, daidzein; Ge, genistein; Mytrin, myricitrin; Ru, rutin; Mytin, myricetin; Fi, ficetin; Eri, eriodic; Qu, quercetin; Lu, luteorin; Ka, kaempferol; ND, not detected.

were measured. Hydrolysis was conducted with β-glucuronidase before quantitative analysis for serum, urine, and feces.

Bioactivity Study

The preliminary experiment to measure the biological effects of soy-hypocotyl tea was conducted with 35 volunteers. Fifteen female students consumed >1 L (daidzin × 12 mg/L, daidzein 1.1 mg/L, genistin × 1.7 mg/L, genistein × 0.2 mg/L) per day of the tea for 4 wk. Blood and 24-h urine samples were taken after 0, 2, and 4 wk. Phosphatidylcholine peroxides (PCOOH) and phosphatidylethanolamine peroxides (PEOOH) were measured by chemoluminescence (HPLC-CL) after extraction from red blood cells (Miyazawa *et al.* 1990). Urinary excretion of ^{8}OHdG was measured by enzyme-linked immunosorbent assay (ELISA) (Lee *et al.* 1993).

Statistical Analysis

Statistical analysis was performed with the use of SPSS version 6.1 (SPSS, Chicago, IL).

Results

Nutritional Study in the Field

Food intake of I-city people met the recommended dietary allowances (RDA) for Japanese. Average nutrient intakes were as follows: 1886 kcal for energy; 80.2 g for protein; 56.2 g for fat; and 256.7 g for carbohydrate. Food composition over 3 d consisted of 55 vegetables in 15 species, 6 tuber crops in 5 species, 5 seeds in 5 species, and 4 crops. On average, the study population consumed 365.7 g vegetables, 155.5 g fruits, 85.5 g pulses, 217.7 g cereal, and 58.0 g potato. Per capita per day consumption was as follows: 203.8 g for rice, noodles and bread; 84.5 g for pulse crops; 58.9 g for tuber crops; 2.3 g for nuts and seeds; and 15.2 g for oil, most of which was plant oil.

Kaempferol, quercetin, myricetin, and rutin were common flavonoids obtained from vegetables; the average intakes were 16.2 mg/d (3.7–48.3 mg). Average intakes of flavonoid per day per capita were as follows: 4.91 mg for kaempferol; 8.82 mg for quercetin; 1.50 mg for rutin; 0.57 mg for myricetin; 0.28 mg for luteolin; 0.08 mg for myricitrin; 0.39 mg for fisetin; and 0.27 mg for eriodictyol. Attributable rates of various foods for quercetin were 78.6% from onion, 14.0% from moroheiya, 1.4% from green tea, and 0.36 % from orange. Those for kaempferol were 4.1% from apples, 5.5% from onion, and 0.2% from carrot.

Tofu, miso (fermented soy bean), and natto were the main soybean food products; average intakes of genistein and daidzein were 23.27 (4.62–52.12) and 16.2 (3.18–35.61) mg/d, respectively. Attributable rates of foods for genistein were as follows: 41.7% for tofu; 20.3% for soybean paste; 6% for natto; and 2.1% for cooked soybean. Those for daidzein were very similar. Intake of flavonoids was significantly associated with intake of green/yellow vegetables, whereas intake of isoflavonoids was associated with tocopherol intake. Intake amounts of flavonoids, isoflavonoids, vitamins, and carotenoids are summarized in Table 14.2.

Pharmacokinetic Study

Kinako powder is a soybean product rich in daidzin and genistin. Average concentration of isoflavone in 60 g kinako after hydrolysis was 103 μmol for daidzein and 112 μmol for genistein.

The plasma level of genistein increased after 2 h and reached its highest value of 2.44 ± 0.65 μmol/L 6 h later. The plasma level of daidzein reached 1.56 ± 0.34 μmol/L at the same time. O-DMA and/or equol peaks appeared after the peak for daidzein in four and two subjects, respectively. In contrast to plasma, daidzein was the main component in urinary excretion. Its concentration in urine started to increase shortly after the rise in its plasma level and reached 7.38 ± 3.75 μmol per urination at 6 h after ingestion of kinako. Almost all subjects showed two or three peaks in urinary excretion over 3 d, probably due to enterohepatic recirculation. Genistein levels in urine paralleled those of daidzein, but the concentrations were about half (3.69 ± 2.72 μmol). Fecal isoflavonoids were excreted mainly on d 2 and 3 after intake of kinako. Total recovery of isoflavonoids from urine and feces was 54.7% for daidzein and 20.1% for genistein. Genistein seems to be most efficiently absorbed from the intestinal tract and

TABLE 14.2
Intake of Vitamins, Isoflavonoids, and Flavonoids[a]

Variable	Unit	Mean	SD	Median	Minimum	Maximum
Retinol	μg	370	635	183	23	3589
Carotenoid	μg	3351	1481	2968	836	7222
Vitamin B_1	mg	1.25	0.4	1.18	0.8	3.39
Vitamin B_2	mg	1.51	0.39	1.45	0.93	2.86
Niacin	mg	18.8	5.5	17.3	7.9	31.8
Vitamin C	mg	117	47	105	47	280
Tocopherol	mg	21.15	5.48	21.15	9.4	31
Daidzein	mg	16.20	7.65	13.72	3.18	35.61
Genistein	mg	23.27	11.25	19.35	4.62	52.12
Myricitrin	mg	0.01	0.01	0.00	0.00	0.06
Rutin	mg	1.50	1.68	0.91	0.19	7.90
Myricetin	mg	0.57	0.55	0.40	0.00	2.17
Fisentin	mg	0.39	0.54	0.17	0.01	2.76
Eriodictyol	mg	0.27	0.48	0.00	0.00	1.94
Quercetin	mg	8.28	6.63	7.46	0.46	26.90
Luteolin	mg	0.28	0.20	0.30	0.00	0.76
Kaempferol	mg	4.91	2.84	4.48	0.84	12.82

[a]n = 50.

stays longer in blood. The half-life of isoflavonoids in blood from the highest level was 8 h for genistein and 5.99 h for daidzein (Fig. 14.1).

Bioactivity Study

PCOOH and PEOOH from red blood cells were lower among soy-hypocotyl tea drinkers than nondrinkers, but the difference was not significant. Urinary ^{8}OHdG, on the other hand, was significantly different between tea drinkers and nondrinkers ($P < 0.05$) (Table 14.3).

Discussion

Average intake of flavonoids was inversely associated with mortality from coronary heart disease and explained about 25% of the variance in coronary heart disease rates in the 16 cohorts from 7 countries (Hertog *et al.* 1995). In multivariate analysis, intake of saturated fat (73%, $P = 0.001$) and flavonoid (8%, $P = 0.01$), and percentage of smokers per cohort (9%, $P = 0.03$) explained 90% of the variance in coronary heart disease rates. These were independent of intake of alcohol and antioxidant vitamins. Average flavonoid intake may contribute in part to coronary heart disease mortality across populations, but not to cancer mortality (Hertog *et al.* 1995).

These prevention properties of flavonoids are considered to be due to their strong antioxidant action (Ho 1997). Hertog *et al.* (1993) measured quercetin, kaempferol,

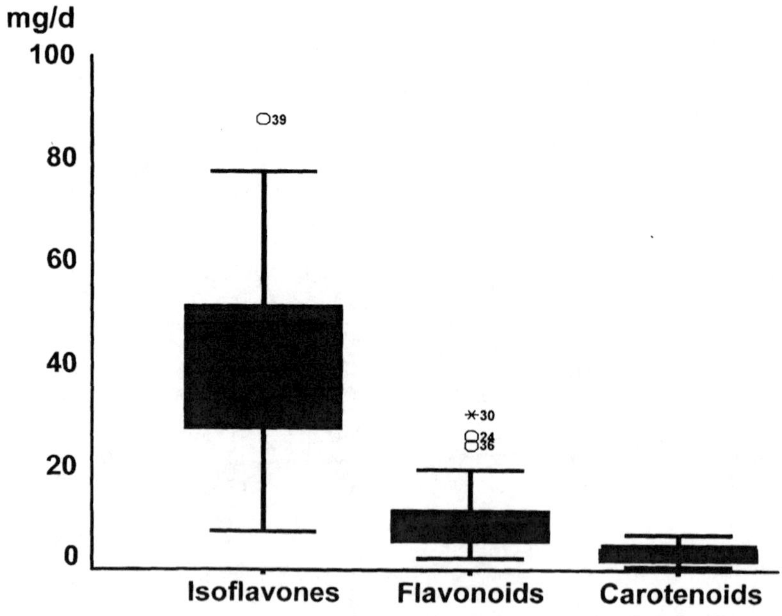

Fig. 14.1. Intake of isoflavones, flavonoids, and carotenoids by Japanese.

TABLE 14.3
PCOOH and PEOOH of Red Blood Cells and Urinary ^{8}OHdG Excretion in Nondrinkers and Drinkers of Soy-Hypocotyl Tea[a]

	Nondrinker	Drinker	*t*-test
PCOOH	101.0 ± 29.5	87.8 ± 22.7	0.197
PEOOH	92.4 ± 43.8	88.5 ± 39.6	0.854
^{8}OHdG	597.9 ± 347.8	363.7 ± 166.0	0.016[b]

[a]Abbreviations: PCOOH, phosphatidylcholine peroxide; PEOOH, phosphatidylethanolamine peroxide; ^{8}OHdG, 8-hydroxydeoxyguanin.
[b]$p < .05$.

myricetin, apigenin, and luteolin in the Zutphen Elderly Study. The relative risk ratio of coronary heart disease mortality in the highest over the lowest tertial flavonoid intake was 0.42 [95% confidence interval (CI); 0.2–0.88]. After controlling various factors, the risk was still significant [0.32 (0.15–0.71)]. Intake of tea, onions, and apples as major sources of flavonoids was also inversely related to the coronary heart disease mortality, but these associations were weaker.

Sample bias should also be considered. In this study, the participants were volunteers with high interest in nutrition; thus it is not surprising for them to meet the RDA for Japanese, except for calcium intake. The standardized mortality rate (SMR) of cancer in I-city in 1990 was 0.93, but that of cerebrovascular diseases was 1.40 (Vital Statistics 1990–1995). Knowing that flavonoids are associated with those diseases, we tried to estimate the amount of intake of flavonoids in this area to create reference data for the Japanese. Isoflavonoid intake was also estimated because high consumption of isoflavonoids by Japanese in other locations was reported (Adlercreutz *et al.* 1991). The database for the concentrations of flavonoids and isoflavonoids among Japanese foods, however, has been quite insufficient for nutrient calculation. Hertog *et al.* (1992b) determined the content of quercetin, kaempferol, myricetin, apigenin, and luteolin among 28 vegetables and 9 fruits by reversed-phase (RP)-HPLC with ultraviolet (UV) detection. Quercetin levels in the edible parts of most vegetables were generally <10 mg/kg except for onions (284–486 mg/kg), kale (110 mg/kg), and broccoli (30 mg/kg).

Although our database is still preliminary, in this study 40 food items have been chosen to represent 77% of flavonoids and isoflavonoids consumption from vegetables, 99.5% from legumes, 58.4% from potatoes, and 43.5% from fruits. In the Zutphen Elderly Study, mean baseline flavonoids intake was 25.9 mg/d (Hertog *et al.* 1993). The major sources of intake were tea (61%), onion (13%), and apples (10%). Rimm *et al.* (1996) reported the average intake of flavonoids of Americans was 20 mg/d, composed of black tea 25%, onion 25%, apple 10%, and broccoli 7%.

Our subjects consumed 16.22 mg flavonoids; this is less than the amount consumed by other ethnic groups such as Caucasians in the United States. We did not calculate catechin intake from tea, but average tea consumption was 300 mL; the amount of flavonoids in tea thus may be <3 mg. Isoflavonoid intake, however, was much higher than that of flavonoids. Although few studies have reported isoflavonoid intake of Caucasians in the United States (Adlercreutz *et al.* 1993), Japanese consumption was higher. The health effects of these chemicals are still under observation, and the mechanism of the biological effects of different metabolites must also be clarified.

We reported the blood level of isoflavonoids among Japanese, and found that it was higher than that of the endogenous 17-β estradiol level, indicating the presence of a complicated interaction in the body of food-derived chemicals (factors) and body constituent biomolecules.

The blood level of isoflavones increased several hours after kinako intake, and genistein levels were always higher than those of daidzein (Fig. 14.2). Although some investigators have reported higher daidzein levels after soy milk consumption, it may be due to the different composition during food processing (Xu *et al.* 1994, 1995).

Xu *et al.* (1994) reported that the bioavailability of isoflavones in soy milk varies from 13 to 35% depending upon the species of gut microflora of the individual. Our study revealed the importance of enterohepatic circulation; fecal inactive aglycone would be derived from deconjugated aglycones excreted in bile. Both endogenous and bacterial enzymes in intestinal flora should influence the absorption, excretion, and metabolism of isoflavonoids. Genistein seemed to be the most efficiently absorbed from the intestinal tract and remained longer in the body.

Bioactivity resulting from isoflavone administration in humans is seldom reported (Xu *et al.* 1995). Isoflavone-rich soy-hypocotyl tea drinking reduced the

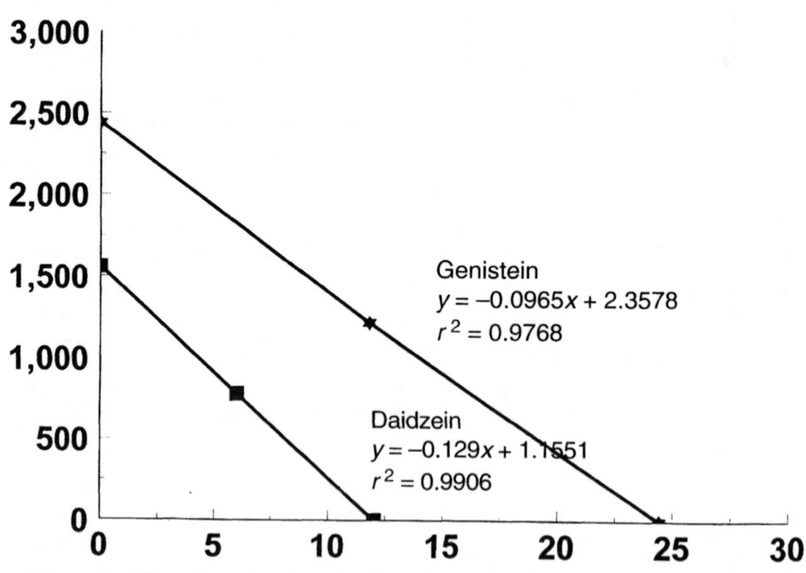

Fig. 14.2. The regression line for the plasma peak value of isoflavonoids.

level of PCOOH and PEOOH in red blood cells and urinary excretion of ^{8}OHdG. This suggests the possibility that habitual drinking of the tea may decrease the level of free radicals that affect the lipid membrane and DNA molecules.

References

Adlercreutz, H., B.R. Foldin, S.L. Gorbach, K.A.V. Hockerstedt, S. Watanabe, E.K. Hamalainen, M.H. Markkanen, T.H. Makela, T.A. Hase, and T. Fotsis, Soybean Phytoestrogen Intake and Cancer Risk, *J. Nutr. 125*: 757S–770S (1995).

Adlercreutz, H., T. Fotsis, S. Watanabe, J. Lampe, K. Wahala, T. Makela, and T.A. Hase, Determination of Lignans and Isoflavonoids in Plasma by Isotope Dilution Gas Chromatography-Mass Spectrometry, *Cancer Detect. Prev. 18*: 259–271 (1994).

Adlercreutz, H., H. Honjo, A. Higashi, T. Fotsis, E. Hamalainen, T. Hasegawa, and H. Okada, Urinary Excretion of Lignans and Isoflavonoid Phytoestrogens in Japanese Men and Women Consuming a Traditional Japanese Diet, *Am. J. Clin. Nutr. 54*: 1093–1100 (1991).

Adlercreutz, H., H. Markkanen, S. Watanabe, Plasma Concentrations of Phytoestrogens in Japanese Men, *Lancet 342*: 1209–1210 (1993).

Frankel, E.N., Nutritional Benefits of Flavonoids, in *Food Factors for Cancer Prevention,* edited by H. Ohigashi, T. Osawa, J. Terao, S. Watanabe, and T. Yoshikawa, Springer-Verlag, Tokyo, 1997, pp. 613–616.

Frankel, E.N., J. Kanner, J.B. German, E. Parks, and J.E. Kinsella, Inhibition of Oxidation of Human Low-Density Lipoprotein by Phenolic Substances in Red Wine, *Lancet 341:* 454–457 (1993).

Hertog, M.G.L., E.J.M. Feskens, P.C.H. Hollman, M.B. Katan, and D. Kromhout, Dietary Antioxidant Flavonoids and Risk of Coronary Heart Disease: The Zutphen Elderly Study, *Lancet 342:* 1007–1011 (1993).

Hertog, M.G.L., P.C.H. Hollman, and M.B. Katan, Content of Potentially Anticarcinogenic Flavonoids of 28 Vegetables and 9 Fruits Commonly Consumed in The Netherlands, *J. Agric. Food Chem. 40:* 2379–2383 (1992a).

Hertog, M.G.L., P.C.H. Hollman, and D.P. Venema, Optimization of a Quantitative HPLC Determination of Potentially Anticarcinogenic Flavonoids in Vegetables and Fruits, *J. Agric. Food Chem. 40*: 1591–1598 (1992b).

Hertog, M.G.L., D. Kromhout, C. Aravanis, H. Blackburn, R. Buzina, F. Fidanza, S. Giampaoli, A. Jansen, A. Menotti, S. Nedeljkovic, M. Pekkarinen, B.S. Simic, H. Toshima, E.J.M. Feskens, P.C.H. Hollman, and M.B. Katan, Flavonoid Intake and Long-Term Risk of Coronary Heart Disease and Cancer in the Seven Countries Study, *Arch. Intern. Med. 155:* 381–386 (1995).

Ho, C.-T., Antioxidant Properties of Plant Flavonoids, in *Food Factors for Cancer Prevention,* edited by H. Ohigashi, T. Osawa, J. Terao, S. Watanabe and T. Yoshikawa, Springer-Verlag, Tokyo, 1997, pp. 593–597.

Kimira, M., Y. Arai, Y. Iwase, and S. Watanabe, Intake of Flavonoids and Isoflavonoids from Foods by Japanese, *J. Epidemiol. 8:* 168–175 (1998).

Lee, Y.S., H.S. Lee, M.K. Park, E.S. Hwang, E.M. Park, H. Kasai, and M.H. Chung, Identification of 8-Hydroxyguanine Glycosylase Activity in Mammalian Tissues Using 8-Hydroxyguanin Specific Monoclonal Antibody, *Biochem. Biophys. Res. Commun. 196:* 1545–1551 (1993).

Miyazawa, T., T. Suzuki, K. Fujimoto, and T. Kaneda, Phospholipid Hydroperoxide Accumulation in Liver of Rats Intoxicated with Carbon Tetrachloride and Its Inhibition by Dietary Alpha-Tocopherol, *J. Biochem. 107:* 689–693 (1990).

Rim, E.B., M.B. Katan, A. Archerio, M.J. Stampler, and W.C. Wilett, Relationship Between Intake of Flavonoids and Risk of Coronary Heart Disease in Male Health Professionals, *Ann. Intern. Med. 125:* 384–389 (1996).

Vital Statistics, Ministry of Health and Welfare, Bureau of Vital Statistics, Tokyo, 1990–1995.

Watanabe, S., and S. Koessel, Colon Cancer; Approaches from Molecular Epidemiology, *J. Epidemiol. 3:* 47–61 (1993).

Watanabe, S., M. Yamaguchi, T. Sobue, T. Takahashi, T. Miura, Y. Arai, W. Mazur, K. Wahala, and H. Adlercreutz, Pharmacokinetics of Soybean Isoflavones in Plasma, Urine and Feces of Men After Ingestion of 60 g Baked Soybean Powder (Kinako), *J. Nutr. 128*: 1710–1715 (1998).

Xu, X., H.-J. Wang, P.A. Murphy, L. Cook, S. Hendrich, Daidzein Is a More Bioavailable Soymilk Isoflavone Than Is Genistein in Adult Women, *J. Nutr. 124:* 825–832 (1994).

Xu, X., K.S. Harris, H.J. Wang, P.A. Murphy, and S. Hendrich, Bioavailability of Soybean Isoflavones Depends upon Gut Microflora in Women, *J. Nutr. 125*: 2307–2315 (1995).

Chapter 15

Membrane Localizations of Carotenoids and α-Tocopherol and Their Antioxidant Properties for Singlet Oxygen

Kenji Fukuzawa[a], Yasuki Inokami[a], and Junji Terao[b]

[a]Faculty of Pharmaceutical Sciences and [b]Department of Nutrition, School of Medicine, Tokushima University, Tokushima 770, Japan

Introduction

Singlet oxygen ($^1O_2$) oxidizes many biological substances and causes cell damage that results in a number of diseases. α-Tocopherol (α-Toc; vitamin E) and carotenoids are known to act as biological antioxidants. They function not only as radical scavengers but also as effective quenchers of $^1O_2$. There have been many studies on quenching of $^1O_2$ by α-Toc and carotenoids, but in most of these studies, pure organic solvents were used, even though these substances are located in membranes in biological systems. Thus, we investigated their antioxidant properties in membranes (Fukuzawa *et al.* 1997, 1998a and 1998b).

In this study, we compared the rate constants for scavenging $^1O_2$ and activities for inhibiting lipid peroxidation of α-Toc and carotenoids [β-carotene (β-Car), astaxanthin, and canthaxanthin] in liposomes and EtOH solution. Factors influencing their antioxidant properties in membranes are discussed.

Experimental Methods

Preparation of Liposomes and Photosensitized Oxidation

Phosphatidylcholine (PC) liposomes were used as a model of membranes, and liposomes were prepared as described previously (Fukuzawa *et al.* 1997, 1998a and 1998b). $^1O_2$ was generated by a photoirradiation system using photosensitizers, water-soluble methylene blue (MB) or rose bengal (RB) and lipid-soluble pyrenedodecanoic acid (PDA). The liposome suspension was placed at a distance of 4 cm from an 85-W halogen-tungsten lamp (Toshiba JD 100W 110V 85 WN-E) and photoirradiated at 1546 lm at 37°C.

Quantitations of α-Toc and Peroxides of Egg PC

α-Toc was separated by high-performance liquid chromatography (HPLC) and detected with a Hitachi F-1000 fluorescence spectrophotometer with excitation and emission at 298 and 325 nm, respectively (Fukuzawa *et al.* 1997). Formation of egg PC hydroperoxides (PC-OOH) was followed by HPLC with an ultraviolet (UV)-detector at 234 nm (Fukuzawa *et al.* 1995).

Results and Discussion

Figure 15.1 shows the effect of membrane charge on oxidation of α-Toc in dimyristoylphosphatidylcholine (DMPC) liposomes with MB or PDA. In the water-soluble MB system, the oxidation rate of α-Toc was fast in negatively charged dicetylphosphate (DCP)-containing DMPC liposomes and slow in neutrally charged DMPC liposomes and positively charged stearyl amine (SA)-containing DMPC liposomes. On the contrary, in the lipid-soluble PDA system, there was no difference in the oxidation rates of α-Toc in the three types of liposomes. In the water-soluble RB system, the oxidation rate of α-Toc was fast in positively charged SA-DMPC liposomes and slow in negatively charged DCP-DMPC liposomes (data not shown).

In the MB system, positively charged MB is located at the negatively charged membrane surface of DCP-containing liposomes; $^1O_2$ is also generated at the membrane surface (Fig. 15.2), resulting in rapid oxidation of α-Toc. On the other hand, in positively charged SA-containing liposomes or neutrally charged DMPC liposomes, MB is located in the bulk water phase; $^1O_2$ is also generated in the

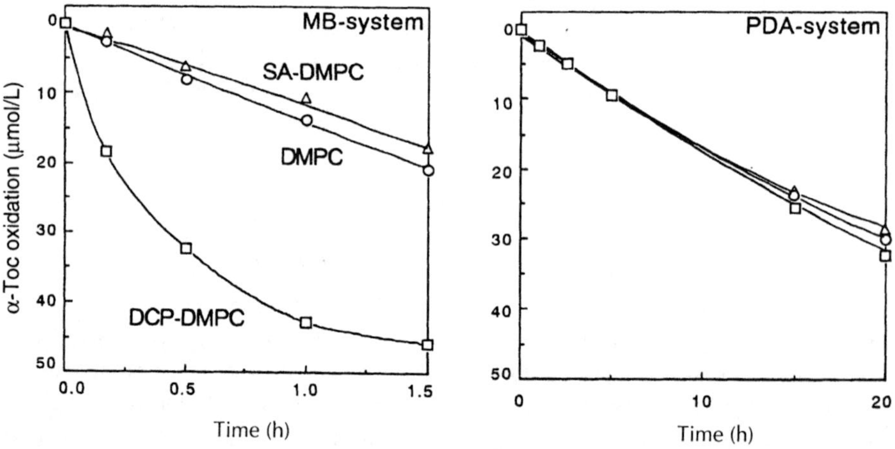

Fig. 15.1. Effect of membrane charges on oxidation of α-tocopherol (α-Toc) by photoirradiation with methylene blue (MB) or pyrenedodecanoic acid (PDA) in liposomes: (○) neutrally charged dimyristoylphosphatidylcholine (DMPC) liposomes; (□) negatively charged dicetylphosphate (DCP)-DMPC liposomes; (△) positively charged stearyl amine (SA)-DMPC liposomes. The concentrations of reactants in liposome solution were 50 μmol/L α-Toc, 5 mmol/L DMPC, 0.5 mmol/L DCP or SA, 0.5 mmol/L DTPA, 10 mmol/L HEPES buffer (pH 7.0), and 50 μmol/L MB or 250 μmol/L PDA. Incubation was carried out at 37°C.

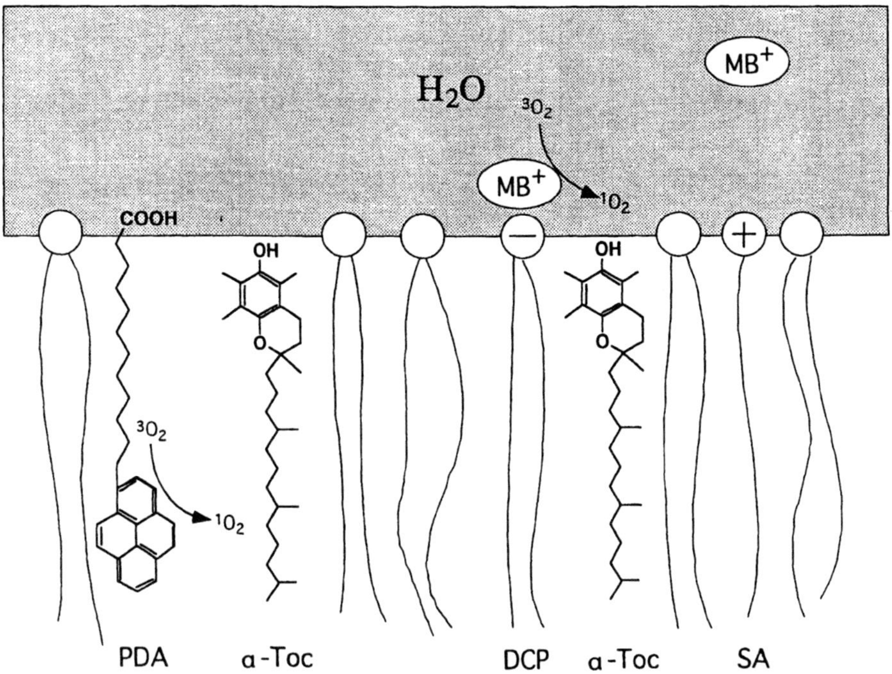

Fig. 15.2. Possible locations of α-tocopherol (α-Toc), methylene blue (MB), and pyrenedodecanoic acid (PDA) in liposomal membranes with different charges. The distributions of α-Toc in membranes are independent of the membrane charge. $^1O_2$ is generated in a site-specific manner by MB and PDA, whose distributions depend on the membrane charge. In the MB-system, $^1O_2$ is generated at the membrane surface of dicetylphosphate (DCP)-dimyristoylphosphatidylcholine (DMPC) liposomes and in the bulk water phase in stearyl amine (SA)-DMPC and DMPC liposome systems, but in the PDA system, $^1O_2$ is generated in the inner region of liposome membranes, irrespective of the membrane charge.

water phase, resulting in slow oxidation of α-Toc. In the RB system, negatively charged RB is located at the positively charged membrane surface of SA-containing liposomes; $^1O_2$ also generates at the membrane surface, resulting in rapid oxidation of α-Toc. The interaction of MB with the negatively charged membrane surface of DCP-liposomes was confirmed by the finding that MB increased the ς-potential of DCP-liposomes, but did not affect the ς-potential of neutrally charged or positively charged SA-liposomes (Fukuzawa *et al.* 1997). RB also decreased the ς-potential of positively charged SA-liposomes. However, in the PDA system, $^1O_2$ is generated in the inner hydrophobic region of membranes, irrespective of the membrane charge, resulting in no difference in the oxidation rates of α-Toc in the three types of liposomes.

Figure 15.3A shows the oxidations of α-Toc by $^1O_2$ generated with RB in liposomes and EtOH solution (Fukuzawa *et al.* 1998a). We used positively charged

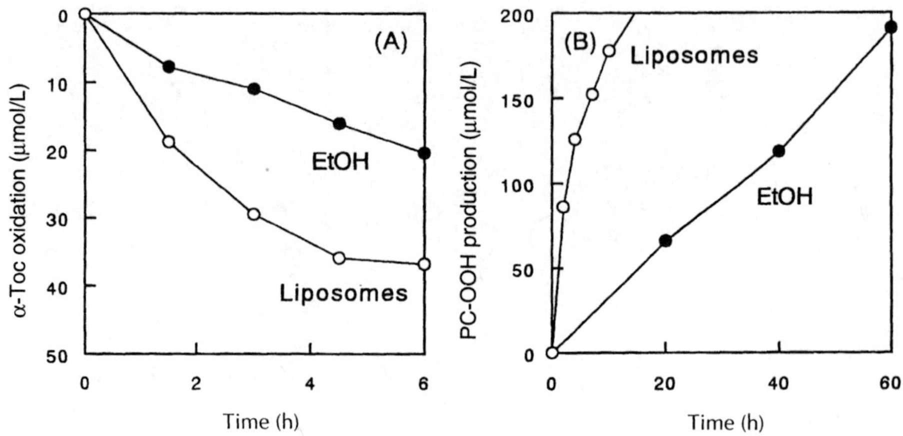

Fig. 15.3. Oxidation of α-tocopherol (α-Toc) and phosphatidylcholine hydroperoxide (PC-OOH) formation of photoirradiation with rose bengal (RB) in EtOH solution and liposomes. Open symbols are for liposomes and closed symbols for EtOH solutions. (A) The concentrations of reagents were 50 μmol/L α-Toc and 5 μmol/L RB in EtOH solution, and were 50 μmol/L α-Toc, 5 mmol/L dimyristoylphosphatidylcholine (DMPC), 0.5 mmol/L stearyl amine (SA), 5 μmol/L RB, 10 mmol/L HEPES buffer (pH 7.0) with 0.5 mmol/L DTPA in liposome solution. (B) The concentrations of reagents were 5 mmol/L egg PC and 5 μmol/L RB in EtOH solution, and 5 mmol/L egg PC, 0.5 mmol/L SA, 5 μmol/L RB, 10 mmol/L HEPES buffer (pH 7.0) with 0.5 mmol/L DTPA in liposome solution. Incubations were carried out at 37°C.

SA-liposomes because negatively charged RB should be located at their membrane surface. α-Toc was oxidized faster in liposomes than in EtOH solution because the local concentrations of α-Toc and RB were higher in liposomes than in EtOH solution. Similarly, as shown in Figure 15.3B, the production of PC-OOH was higher in liposomes than in EtOH solution (Fukuzawa *et al.* 1998a).

The effect of membrane fluidity on the oxidation rates of α-Toc was investigated with the use of two types of liposomes with different transition temperatures, i.e., DMPC liposomes and dipalmitoylphosphatidylcholine (DPPC) liposomes (Fukuzawa *et al.* 1997). At 30°C, DPPC liposomes are in a rigid gel state because their transition temperature is 41°C, whereas DMPC liposomes are in a liquid crystalline state because their transition temperature is 24°C. Figure 15.4 shows that α-TOC was oxidized by $^1O_2$ at 30°C more rapidly in DMPC liposomes than in DPPC liposomes. These results indicate that α-Toc is oxidized more quickly in the liquid crystalline phase than in the gel phase of membranes.

Next, we determined the $^1O_2$-scavenging rate constant (k_Q) of α-Toc from the competitive reaction between α-Toc and diphenylbenzofuran (DPBF). The k_Q value was calculated by the following equation (Young *et al.* 1971):

$$S_0/S_{Toc} = 1 + (k_Q/k_d)\ [\alpha\text{-Toc}]$$

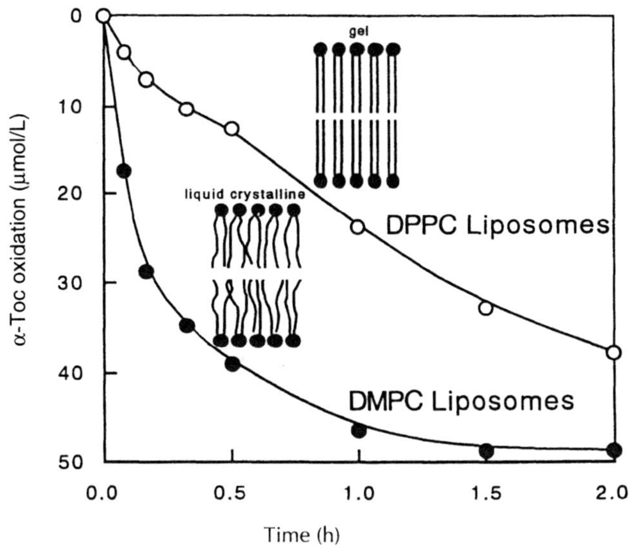

Fig. 15.4. Effect of membrane fluidity on oxidation of α-tocopherol (α-Toc) by photoirradiation with methylene blue (MB): (●) dimyristoylphosphatidylcholine (DMPC) liposomes; (○) dipalmitoylphosphatidylcholine (DPPC) liposomes. The concentrations of reactants were 50 μmol/L α-Toc, 5 mmol/L DMPC or DPPC, 0.5 mmol/L DTPA, 10 mmol/L HEPES buffer (pH 7.0), and 50 μmol/L MB or 250 mol/L pyrenedodecanoic acid (PDA). Incubations were carried out at 30°C.

where S_0 and S_{Toc} are the first-order plots of the decrease of DPBF in the absence and presence of α-Toc. The k_d value is the rate constant of natural decay of $^1O_2$ to ground state $^3O_2$. The k_d value of α-Toc in liposomes in the RB- or MB-system was estimated to be the same as that in EtOH (8.3×10^4 s^{-1}) (Merkel and Kearns 1972) because the micropolarity (dielectric constant) of the environment around the OH-group of α-Toc in liposomes is equal to that in EtOH (Bellemare and Fragata 1980). For calculation of the k_Q value of β-Car in liposomes, the k_d value in *tert*- BuOH (3.0×10^4 s^{-1}) (Young *et al.* 1973) was used because the active site of β-Car, the center of the conjugated polyene-chain, is probably localized in the hydrophobic region (Gabrielska and Gruszecki 1996).

Table 15.1 shows the k_Q values of α-Toc and carotenoids in EtOH solution and liposomes in the RB- and PDA-systems. The k_Q value of α-Toc in EtOH solution in the RB-system was 3.6×10^8 $[(mol \cdot s)/L]^{-1}$. Without considering the local concentration of α-Toc in liposomes, its k_Q value was 3.4×10^9 $[(mol \cdot s)/L]^{-1}$ (data not shown), which was higher than that in EtOH solution. However, assuming that α-Toc and RB are both concentrated in the liposomal membrane, the value was 1.4×10^7 $[(mol \cdot s)/L]^{-1}$, which was lower than that in EtOH solution. The lower k_Q value in liposomes than in EtOH solution may be due to the lower diffusion rates of reactants in membranes than in EtOH solution.

TABLE 15.1
$^1O_2$ Scavenging Rate Constants of α-Tocopherol and Carotenoids in EtOH Solution and Liposomes in the Rose Bengal (RB)-System[a] and Pyrenedodecanoic Acid (PDA)-System

	k_Q {[(mol · s)/L]$^{-1}$}			
	RB-system		*PDA-system*	
	EtOH	Liposomes[b]	EtOH	Liposomes
α-Tocopherol	3.6×10^8	1.4×10^7	2.1×10^8	0.25×10^7
β-Carotene	127×10^8	2.1×10^7	104×10^8	1.7×10^7
Astaxanthin	240×10^8	1.9×10^7	209×10^8	1.9×10^7
Canthaxanthin	128×10^8	1.8×10^7	136×10^8	1.7×10^7

[a]The concentrations of α-tocopherol and carotenoids were 25–100 and 2.5–15 μmol/L, respectively; those of diphenylbenzofuran (DPBF) and RB were 250 and 5 μmol/L, respectively. The concentrations of other reactants in the liposome systems were 5 mmol/L dimyristoylphosphatidylcholine (DMPC), 0.5 mmol/L stearyl amine (SA), and 10 mmol/L HEPES buffer (pH 7.0) containing 0.5 mmol/L DTPA. EtOH solutions and liposomes were photoirradiated for 20 and 30 s, respectively, at 37°C.

[b]Values in stearyl amine (SA)-DMPC liposomes were revised because the local concentrations of reactants were all located in membranes. Their concentrations in liposomes were 242 times that in the bulk phase, calculated assuming that the volume of one DMPC molecule was 1253 Å^3 (Huang and Mason 1978).

In EtOH solution in the RB-system, the k_Q values of carotenoids were ~40–80 times higher than that of α-Toc, but the k_Q values of liposomes were only twice that of α-Toc. The mobility of carotenoids would be suppressed more than that of α-Toc in membranes, resulting in a greater decrease in their k_Q values than that of α-Toc in liposomes compared with EtOH solution.

Table 15.2 shows the inhibitions by α-Toc and carotenoids of $^1O_2$-dependent lipid peroxidation as reciprocals of their IC_{50} values, that is, the concentrations required for 50% inhibition of lipid peroxidation. The antioxidant activities were lower in liposomes than in EtOH solution, possibly because of slower mobilities of the antioxidant and faster lipid peroxidation in membranes. Their lipid peroxidation–inhibiting activities are well correlated with their $^1O_2$- scavenging rate constants. In EtOH solution, β-Car showed ~7 times higher activity, and canthaxanthin and astaxanthin ~20 times higher activities than that of α-Toc. However, in the liposome system, the antioxidant activities of carotenoids were only about twice that of α-Toc (Table 15.3). The membrane structure may inhibit the interaction of $^1O_2$ with carotenoids more strongly than with α-Toc, resulting in a greater decrease in the $[IC_{50}]^{-1}$ values of carotenoids than of α-Toc in the membrane system.

The ratios of $[IC_{50}]^{-1}$ in liposomes to that in EtOH solution ($[IC_{50}]^{-1}_{Liposomes}$/$[IC50]^{-1}_{EtOH}$) are shown in Table 15.3. RB is known as a photosensitizer with two actions, generating mainly $^1O_2$ (Type II action) and some free radicals (Type I action) (Lee and Rodgers 1987, Saniabadi 1994). The order of the ratios was BHT (radical scavenger) > α-Toc (radical scavenger and $^1O_2$ quencher) > carotenoids ($^1O_2$ quencher), suggesting that some radical-type active oxygens generated by the minor Type I action of RB-photoirradiation participate in the progress of lipid peroxidation

TABLE 15.2
Inhibitions by α-Tocopherol, Carotenoids and BHT of Rose Bengal (RB)- and Pyrenedodecanoic Acid (PDA)-Sensitized Peroxidations of Egg Phosphatidylcholine (PC) in EtOH Solution and Liposomes[a]

	$[IC_{50}]^{-1}$ (10^{-3} (mol/L)$^{-1}$		
	RB-system		PDA-system
	EtOH	Liposomes[b]	Liposomes
α-Tocopherol	13.5	0.71	0.76
β-Carotene	83.3	1.38	5.6
Astaxanthin	238	1.97	11.1
Canthaxanthin	236	1.38	11.0
BHT	0.063	0.019	0.026

[a]The concentrations of α-tocopherol, carotenoids and butyl hydroxytoluene (BHT) were 2.5–200 μmol/L, 0.5–50 μmol/L, and 0.025–50 μmol/L, respectively. The concentration of RB was 5 μmol/L. The concentrations of other reactants in the liposome systems were 5 mmol/L egg PC, 0.5 mmol/L SA, and 10 mmol/L HEPES buffer (pH 7.0) containing 0.5 mmol/L DTPA. The EtOH solution and liposomes were photoirradiated for 60 and 6 min, respectively, at 37°C.
[b]Revised values in SA-egg PC liposomes, calculated assuming that the concentration of reactants in the membrane phase was 242 times that in the bulk phase.

more in liposomes than in EtOH solution because PC molecules are more suitably ordered and packed in membranes for progress of the radical chain reaction.

Next, we compared the effect of the generation site of $^1O_2$ in membranes on the k_Q and $[IC_{50}]^{-1}$ values of α-Toc and those of carotenoids using two different photosensitizers, RB and PDA. As shown in Table 15.4, the k_Q value of α-Toc in liposomes was ~5 times higher in the RB-system than in the PDA-system. However, carotenoids had similar k_Q values in the two systems. We suppose that the higher value of α-Toc in the RB-system was due to the localization of the OH-groups of α-Toc in the membrane. As shown in Figure 15.2, the OH-groups of α-Toc would be distributed at a rel-

TABLE 15.3
Ratios of the $[IC_{50}]^{-1}$ Values Carotenoids to Those of α-Tocopherol and Those of Liposomes to Those in EtOH Solution[a]

	Ratio of $[IC_{50}]^{-1}$ value[b]		
		Carotenoid/α-Tocopherol	
	Liposomes/ EtOH	EtOH	Liposomes
α-Tocopherol	1/19	1.0	1.0
β-Carotene	1/60	6.2	1.9
Astaxanthin	1/121	17.6	2.8
Canthaxanthin	1/171	17.1	1.9
BHT	1/3.3	—	—

[a]For abbreviations see Table 15.2.
[b]The values in RB-system.

TABLE 15.4
Ratios of the K_Q Values and the $[IC_{50}]^{-1}$ Values in the Rose Bengal (RB)-System to Those in the Pyrenedodecanoic Acid (PDA)-System in Liposomes[a]

	RB-system/PDA-system	
	K_Q value	[IC50]–1 value[b]
α-Tocopherol	5.6	5.4
β-Carotene	1.2	1.4
Canthaxanthin	1.1	0.70
Astaxanthin	1.0	1.0

[a]For abbreviations see Tables 15.1 and 15.2.
[b]Activity relative to that of astaxanthin: astaxanthin (1.0/1.0 = 1.0), α-tocopherol (0.36/0.066 = 5.4), β-Carotene (0.69/0.50 = 1.4), canthaxanthin (0.69/0.99 = 0.70).

atively high concentration in a region close to the membrane surface (Fragata and Bellemare 1980, Fukuzawa *et al.* 1992), where $^1O_2$ generated by RB is scavenged effectively, resulting in a higher k_Q value $\{1.4 \times 10^7$ $[(\text{mol} \cdot \text{s})/\text{L}]^{-1}\}$. But PDA generates $^1O_2$ in the inner hydrophobic region, where the OH-groups of α-Toc are present at a lower concentration (Fukuzawa *et al.* 1992), resulting in a lower k_Q value $\{2.5 \times 10^6$ $[(\text{mol} \cdot \text{s})/\text{L}]^{-1}\}$. The k_Q values of α-Toc in membrane described above were calculated on the assumption that the OH-groups of α-Toc are uniformly distributed in the membrane even though they are localized. We recalculated the k_Q values of α-Toc in DMPC liposomes on the assumption that 0, 60, and 40% of the OH-groups of α-Toc are localized at the polar zone (10 Å), hydrogen belt (4 Å), and hydrophobic core (16 Å), respectively (Fragata and Bellemare 1980, Fukuzawa *et al.* 1992, Takahashi *et al.* 1989); the values were 3.1×10^6 and 3.3×10^6 $[(\text{mol} \cdot \text{s})/\text{L}]^{-1}$ in the RB- and PDA-systems, respectively. On the contrary, the quenching site of carotenoids, probably the center of the conjugated polyene-chain, may be localized in the hydrophobic region (Fig. 15.5) (Gabrielska and Gruszecki 1996) where the $^1O_2$ concentration is also high, irrespective of its generation site, because $^1O_2$ is more soluble in a hydrophobic region than in a polar region (Battino 1981), resulting in similar k_Q values for carotenoids in the two photosensitizer systems.

The ratio of the $[IC_{50}]^{-1}$ value of carotenoids to that of α-Toc in liposomes was ~2 in the RB-system but 10 in the PDA-system (Table 15.2). Thus, α-Toc had higher antioxidant activity in the RB-system than in the PDA system. These ratios of carotenoids to α-Toc in the RB- and PDA-systems were well correlated with their ratios of k_Q values in these systems (Table 15.4), supporting our proposal described above that the $^1O_2$-inactivating activities of α-Toc and carotenoid depend on their membrane localization and $^1O_2$ generation sites.

Conclusion

The following five points are considered to be important factors for the k_Q and $[IC_{50}]^{-1}$ values of α-Toc and carotenoids in membranes: (i) The antioxidant concentration in

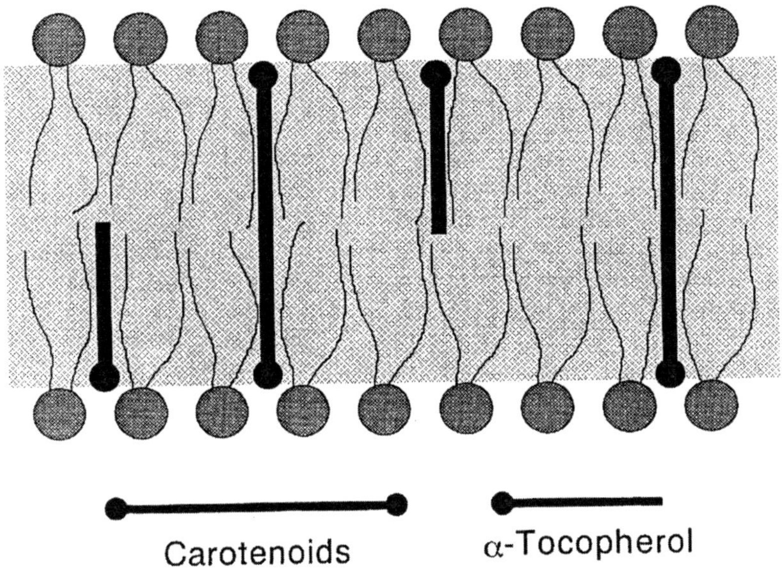

Fig. 15.5. Possible locations of β-Carotene and α-Tocopherol in liposome membranes.

membranes; (ii) the mobility of antioxidants in membranes, which is lower than that in EtOH solution; (iii) the membrane fluidity, i.e., fast in the liquid crystalline state and slow in the gel state; (iv) the membrane localization of active groups of antioxidants and their local concentration in membranes; and (v) the generation site of $^1O_2$ depending on the localization of photosensitizers in liposome systems, that is, (a) bulk water or membranes and (b) membrane surface or membrane inner region.

References

Battino, R., ed., *Oxygen and Ozone, Solubility Data Series,* Vol. 7, Pergamon Press, Oxford, 1981.

Bellemare, F., and M.J. Fragata, Polarity Studies on the Head Group of Single-Layered Phosphatidylcholinea-α-Tocopherol Vesicles, *Colloid Interface Sci.77*: 243–252 (1980).

Fragata, M., and F. Bellemare, Model of Singlet Oxygen Scavenging by α-Tocopherol in Biomembranes, *Chem. Phys. Lipids 27*: 93–99 (1980).

Fukuzawa, K., M. Iemura, and A. Tokumara, Lipid Peroxidation in Egg Phosphatidylcholine Liposomes: Comparative Studies on the Induction Systems Fe^{2+}/Ascorbate and Fe^{3+}-Chelates/Xanthine-Xanthine Oxidase, *Biol. Pharm. Bull. 19*: 665–671 (1995).

Fukuzawa, K., W. Ikebata, A. Shibata, I. Kumadaki, T. Sakanaka, and S. Urano, Location and Dynamics of α-Tocopherol in Model Phospholipid Membranes with Different Charges, *Chem. Phys. Lipids 63*: 69–75 (1992).

Fukuzawa, K., Y. Inokami, A. Tokumura, J. Terao, and A. Suzuki, Singlet Oxygen Scavenging by α-Tocopherol and β-Carotene: Kinetic Studies in Phospholipid Membranes and Ethanol Solution, *BioFactors 7*: 31–40 (1998a).

Fukuzawa, K., Y. Inokami, A. Tokumura, J. Terao, and A. Suzuki, Rate Constants for Quenching Singlet Oxygen and Activities for Inhibiting Lipid Peroxidation of Carotenoids and α-Tocopherol in Liposomes, *Lipids 33*: 751–756 (1998b).

Fukuzawa, K., K. Matsuura, A. Tokumura, A. Suzuki, and J. Terao, Kinetics and Dynamics of Singlet Oxygen Scavenging by α-Tocopherol in Phospholipid Model Membranes, *Free Radic. Biol. Med. 22*: 923–930 (1997).

Gabrielska, J., and W.I. Gruszecki, Zeaxanthin (Dihydroxy-β-Carotene) but Not β-Carotene Rigidifies Lipid Membranes: A ^{1}H-NMR Study of Carotenoid-Egg Phosphatidylcholine Liposomes, *Biochim. Biophys. Acta 1285*: 167–174 (1996).

Huang, C., and J.T. Mason, Geometric Packing Constraints in Egg Phosphatidylcholine Vesicles, *Proc. Natl. Acad. Sci. USA 75*: 308–310 (1978).

Lee, P.C.C., and M.A.J. Rodgers, Laser Flash Photokinetic Studies of Rose Bengal Sensitized Photodynamic Interactions of Nucleotides and DNA, *Photochem. Photobiol. 45*: 79–86.

Merkel, P.B., and D.R. Kearns, Radiationless Decay of Singlet Molecular Oxygen in Solution. An Experimental and Theoretical Study of Electronic-to-Vibrational Energy Transfer, *J. Am. Chem. Soc. 94*: 7244–7253 (1972).

Saniabadi, A.R., Photosensitizers and Photochemical Reactions, in *A Novel Photochemical Model for Thrombosis Research and Evaluation of Antithrombotic and Thrombolytic Agents*, edited by M. Nakashima, Churchill Livingstone, Tokyo, 1994.

Takahashi, M., J. Tsuchiya, and E. Niki, Scavenging of Radicals by Vitamin E in the Membranes as Studied by Spin Labeling, *J. Am. Chem. Soc. 111*: 6350–6353 (1989).

Young, R.H., D. Brewer, and R.A. Keller, The Determination of Rate Constants of Reaction and Lifetimes of Singlet Oxygen in Solution by a Flash Photolysis Technique, *J. Am. Chem. Soc. 95*: 375–379 (1973).

Young, R.H., K. Wehrly, and R.L. Martin, Solvent Effects in Dye Sensitized Photooxidation Reactions, *J. Am. Chem. Soc. 93*: 5774–5779 (1971).

Chapter 16

Carotenoid Regeneration of Phenolic Compounds from Phenoxyl Radicals in Relation to Oxidative Defense

Alan Mortensen[a], Steven A. Everett[b], and Leif H. Skibsted[a]

[a]Food Chemistry, Department of Dairy and Food Science, Royal Veterinary and Agricultural University, DK-1958 Frederiksberg C, Denmark

[b]Gray Laboratory Cancer Research Trust, Mount Vernon Hospital, Northwood, Middlesex HA6 2JR, UK

Introduction

Antioxidants exert their effects by a number of different mechanisms that include scavenging of radicals, chelation of transition metals, and photoprotection. The primary role of some antioxidants may be to regenerate other antioxidants by reducing the one-electron oxidized antioxidant. It was proposed, therefore, (Tappel 1968) that one of the antioxidant functions of vitamin C (ascorbate) is the regeneration of vitamin E (α-tocopherol) from one-electron oxidized α-tocopherol (the α-tocopheroxyl radical). This kind of regeneration may be the basis of antioxidant synergism, i.e., the fact that certain mixtures of antioxidants provide better protection against oxidation than that calculated by addition of the effects of the individual antioxidants.

It has been demonstrated recently that α-tocopherol and β-carotene acted synergistically in a membrane model system (Palozza and Krinsky 1992). An increased consumption of α-tocopherol in the presence of β-carotene was observed, an observation that was suggested to imply that α-tocopherol protected β-carotene from being oxidized. Synergism between carotenoids and tocopherol was also observed in the oxidation of triacylglycerols. Lutein or β-carotene in combination with either γ-tocopherol or α-tocopherol (Fig. 16.1) was thus found to inhibit oxidation more efficiently than γ-tocopherol or α-tocopherol alone (Haila and Heinonen 1994, Haila *et al.* 1996, Heinonen *et al.* 1997).

However, it has been suggested that a number of carotenoids, with the exception of astaxanthin, are able to reduce the α-tocopheroxyl radical generated by pulse radiolysis in a solution of α-tocopherol and β-carotene in hexane (Böhm *et al.* 1997). As a result, the carotenoid radical cation given below was generated.

$$\text{Car} + \text{TO}^{\bullet} + \text{H}^{+} \rightarrow \text{Car}^{\bullet +} + \text{TOH}$$

Furthermore, ascorbate was found to reduce the carotenoid radical cation, and it was thus suggested that carotenoids play a role in the regeneration of vitamin E by

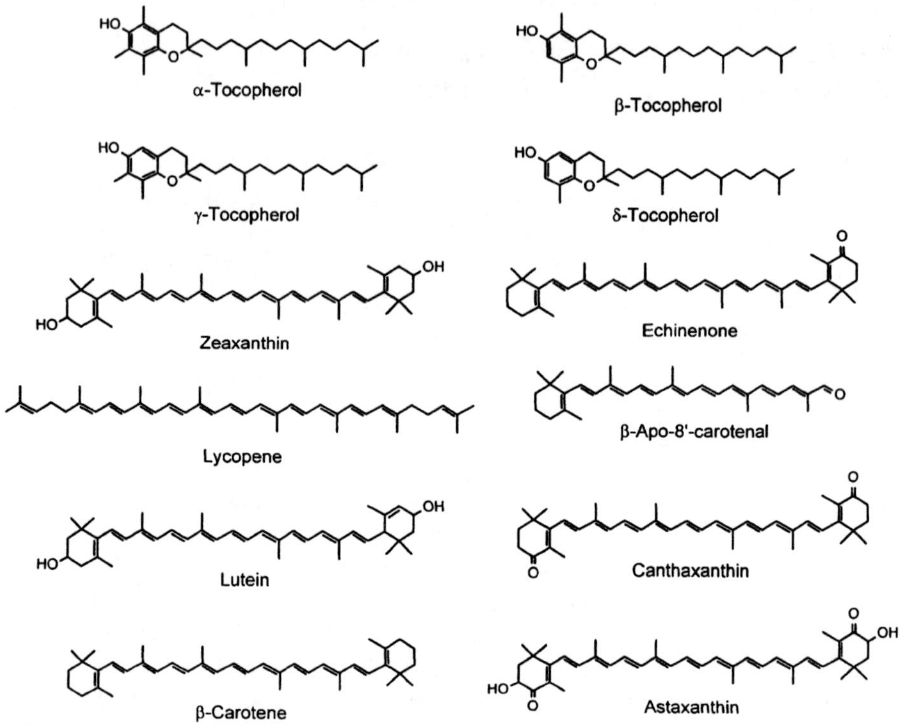

Fig. 16.1. Chemical structures of tocopherols and carotenoids.

vitamin C (Truscott 1996). To determine whether carotenoids are able to reduce the α-tocopheroxyl radical or whether α-tocopherol may reduce the carotenoid radical cations, we performed a series of laser flash photolysis, pulse radiolysis, and steady-state radiolysis experiments.

Carotenoid-Tocopherol Interaction

A mixture of α-tocopherol (1 mmol/L), β-carotene (10 μmol/L), and phenol (1.75 μmol/L) in di-*tert*-butylperoxide/benzene was subjected to laser flash photolysis. This generates the phenoxyl radical

$$(CH_3)_3COO(CH_3)_3 + h\upsilon \rightarrow 2\,(CH_3)_3CO^{\bullet}$$

$$(CH_3)_3CO^{\bullet} + C_6H_5OH \rightarrow (CH_3)_3COH + C_6H_5O^{\bullet}$$

which goes on to react with α-tocopherol and β-carotene to generate the α-tocopheroxyl radical and two types of β-carotene radicals (Mortensen and Skibsted 1996a) believed to be the β-carotene radical cation and an adduct between β-carotene and the phenoxyl radical, as shown below.

$$C_6H_5O^\bullet + TOH \rightarrow C_6H_5OH + TO^\bullet$$

$$C_6H_5O^\bullet + Car \rightarrow Car^{\bullet+} + C_6H_5O$$

$$C_6H_5O^\bullet + Car \rightarrow [CarOC_6H_5O]^\bullet$$

The lifetime of the β-carotene radicals in this system is usually rather long, i.e., they can be observed on a subsecond time scale. However, in the presence of α-tocopherol, the lifetime is rather short, i.e., the β-carotene radicals disappear in <50 μs (Mortensen *et al.* 1998). Furthermore, permanent bleaching of β-carotene is observed in the absence of α-tocopherol, but not in its presence, i.e., α-tocopherol prevents degradation of β-carotene. These two observations strongly suggest that α-tocopherol is able to reduce carotenoid radical cations. The rate constant of this reaction was determined to be 1.7×10^7 μmol/(L · s) (Mortensen *at al.* 1998). It is possible to generate the α-tocopheroxyl radical directly in hexane by laser flash photolysis yielding a hydrogen atom.

$$TOH + h\upsilon \rightarrow TO^\bullet + H^\bullet$$

The hydrogen atom reacts with hexane to generate hexyl radicals which, in the presence of β-carotene, react with β-carotene to generate the β-carotene radical cation, as shown below.

$$H^\bullet + C_6H_{12} \rightarrow H_2 + C_6H_{11}^\bullet$$

$$C_6H_{11}^\bullet + Car \rightarrow C_6H_{11}^- + Car^{\bullet+}$$

In this system, the bleaching of β-carotene is reversible, again indicating that α-tocopherol can reduce the carotenoid radical cation. Using pulse radiolysis, the rate constants of reaction between hexyl radicals and α-tocopherol and between hexyl radicals and β-carotene were determined (Mortensen *et al.* 1998); the latter reaction was 400 times faster than the former. Despite the fact that β-carotene reacts faster with hexyl radicals than does α-tocopherol, α-tocopherol was found to be degraded before β-carotene in a solution of these two compounds in hexane exposed to (steady-state) γ-irradiation, a fact that strongly implies that α-tocopherol protects β-carotene from being degraded by reducing the carotenoid radical cation.

The reaction between α-tocopheroxyl radicals and β-carotene has also been studied by monitoring the decay of the α-tocopheroxyl radical by electron spin resonance (ESR) spectroscopy (Valgimigli *et al.* 1997). α-Tocopheroxyl radicals were generated either photolytically or by reaction with the 1,1-diphenyl-2-picrylhydrazyl radical. In either case the presence of β-carotene did not influence the decay of the α-tocopheroxyl radical, demonstrating that β-carotene was not able to react with the α-tocopheroxyl radical. Furthermore, the phenoxyl radical of 2,2,5,7,8-pentamethyl-6-hydroxychromane, a tocopherol homolog, did not react with β-carotene (Tyurin *et al.* 1997). In sum, the majority of experiments have shown that α-tocopherol can reduce carotenoid radical cations in a number of different systems and that the reverse reaction, i.e., reduction of the α-tocopheroxyl radical by carotenoids, does not take place.

Pulse radiolysis of α-tocopherol in hexane leads to generation of both tocopheroxyl radicals ($TO^{\bullet}$) and tocopheroxyl radical cations ($TOH^{\bullet+}$); quite recently, it has been suggested that tocopheroxyl radical cations, but not tocopheroxyl radicals, may be reduced by a number of carotenoids with the exception of astaxanthin (Edge *et al.* 1998).

$$TOH^{\bullet+} + Car \rightarrow TOH + Car^{\bullet+}$$

However, it was not taken into account that carotenoids may also react with strong acids (such as the tocopheroxyl radical cation) and that the absorption spectrum of $CarH^{+}$ is very similar to that of $Car^{\bullet+}$ (Mortensen and Skibsted, unpublished results). The exact mechanisms by which carotenoids and tocopherols interact are thus not fully elucidated.

$$TOH^{\bullet+} + Car \rightarrow TO^{\bullet} + CarH^{+}$$

Hierarchy of Tocopherols and Carotenoids

Previous results (Mortensen and Skibsted 1996a and 1997a) had shown that phenoxyl radicals may be reduced by a number of carotenoids, thereby generating the carotenoid radical cation. Carotenoids with carbonyl groups, however, did not react efficiently with phenoxyl radicals, whereas lycopene was found to be the most efficient quencher of phenoxyl radicals. It appears, therefore, that at least some carotenoids may be able to reduce certain one-electron oxidized phenolic antioxidants in living systems even though they cannot reduce the α-tocopheroxyl radical. Accordingly, we decided to study the reactions among four tocopherol homologs and eight different carotenoids (Fig. 16.1).

Carotenoid radical cations can be generated by laser flash photolysis of chloroform solutions yielding chloroform-derived radicals by electron transfer from excited state carotenoid to chloroform (Mortensen and Skibsted 1996b and 1997b).

$$Car + CHCl_3 \rightarrow Car^{\bullet+} + CHCl_2^{\bullet} + Cl^{-}$$

These chloroform-derived radicals react with the excess carotenoid yielding more carotenoid radicals.

$$CHCl_2^{\bullet} + Car \rightarrow CHCl_2^{-} + Car^{\bullet+}$$

In the presence of tocopherols, the chloroform-derived radicals also react with the tocopherols to yield tocopheroxyl radicals.

$$CHCl_2^{\bullet} + TOH \rightarrow CH_2Cl_2 + TO^{\bullet}$$

With this system, it is thus possible to study simultaneously both the reaction between carotenoid radical cations and tocopherols and between carotenoids and tocopheroxyl radicals, and thereby determine the direction of electron transfer, i.e., from tocopherol to carotenoid radical cation or from carotenoid to tocopheroxyl radical.

The reactivity of the lycopene radical cation was found to be comparable to that of the four tocopheroxyl radicals. α-Tocopherol could thus reduce the lycopene radical cation, whereas lycopene reduced the α-tocopheroxyl radical, generating the lycopene radical cation (Mortensen and Skibsted 1997c).

$$Lyc^{\bullet+} + \alpha\text{-TOH} \rightarrow Lyc + \alpha\text{-TO}^{\bullet} + H^+$$

$$Lyc + \delta\text{-TO}^{\bullet} + H^+ \rightarrow Lyc^{\bullet+} + \delta\text{-TOH}$$

The lycopene radical cation and the β- and γ-tocopheroxyl radicals were in equilibrium (see below).

$$Lyc + \beta/\gamma\text{-TO}^{\bullet} + H^+ \rightleftarrows Lyc^{\bullet+} + \beta/\gamma\text{-TOH}$$

Subjecting carotenoids in chloroform to laser flash photolysis in the presence of one of the four tocopherols allowed a hierarchy of relative reactivity of the carotenoid radical cations and tocopheroxyl radicals to be established (Mortensen and Skibsted 1997d). This hierarchy (Table 16.1) shows that α-tocopherol may recycle all carotenoids by reduction of the radical cations and may also be expected to be able to reduce β-, γ-, and δ-tocopheroxyl radicals, and that lycopene is the strongest reductant of the carotenoids. The order of reactivity of the carotenoids is the same as that observed toward the phenoxyl radical (Mortensen and Skibsted 1997a). Bleaching of the carotenoids is not fully reversible in chloroform, which may be due to the fact that tocopherols do not seem to react with a transient carotenoid species formed before formation of the carotenoid radical cation upon laser flash photolysis of carotenoids in chloroform (Mortensen and Skibsted 1997d). Alternatively, this may be due to formation of an adduct between carotenoid radical cation and tocopherol, a type of adduct that was observed in the reaction of phenoxyl radicals with carotenoids (Mortensen and Skibsted 1996a and 1997a).

In low-density lipoproteins (LDL) subjected to oxidative stress, the order of consumption of a number of tocopherols and carotenoids is in close agreement with the antioxidant hierarchy established above, i.e., α-tocopherol is consumed

TABLE 16.1
Relative Reduction Potentials of Carotenoids and Tocopherols[a]

α-Tocopherol
Lycopene, β-tocopherol, γ-tocopherol
β-Carotene
Zeaxanthin, δ-tocopherol
Lutein
Echinenone
Canthaxanthin
β-Apo-8′-carotenal, canthaxanthin
Astaxanthin

[a]A compound is able to reduce the radical of a compound below it in the table.

before any of the other antioxidants (Esterbauer *et al.* 1993). This does not constitute proof that recycling of antioxidants does take place *in vivo*, but our results show that this certainly may be the case.

Acknowledgments

Roche A/S is kindly thanked for providing the carotenoids. This work is funded by the FØTEK program through the LMC-Center for Advanced Food Studies. S.A.E. is supported by the Cancer Research Campaign (CRC).

References

Böhm, F., R. Edge, E.J. Land, D.J. McGarvey, and T.G. Truscott, Carotenoids Enhance Vitamin E Antioxidant Efficiency, *J. Am. Chem. Soc. 119*: 621–622 (1997).

Edge, R., E.J. Land, D. McGarvey, L. Mulroy, T.G. Truscott, Relative One-Electron Reduction Potentials of Carotenoid Radical Cations and the Interactions of Carotenoids with the Vitamin E Radical Cation, *J. Am. Chem. Soc. 120*: 4087–4090 (1998).

Esterbauer, H., H. Puhl, G. Waeg, A. Krebs, and M. Dieber-Rotheneder, The Role of Vitamin E in Lipoprotein Oxidation, in *Vitamin E in Health and Disease*, edited by L. Packer, and J. Fuchs, Marcel Dekker, New York, 1993, pp. 649–671.

Haila, K., and M. Heinonen, Action of β-Carotene on Purified Rapeseed Oil During Light Storage, *Lebensm.-Wiss. Technol. 27*: 573–577 (1994).

Haila, K.M., S.M. Lievonen, and M.I. Heinonen, Effects of Lutein, Lycopene, Annatto, and γ-Tocopherol on Autoxidation of Triglycerides, *J. Agric. Food Chem. 44*: 2096–2100 (1996).

Heinonen, M., K. Haila, A.-M. Lampi, and V. Piironen, Inhibition of Oxidation in 10% Oil-in-Water Emulsions by β-Carotene with α- and γ-Tocopherols, *J. Am. Oil Chem. Soc. 74*: 1047–1052 (1997).

Mortensen, A., and L.H. Skibsted, Kinetics of Parallel Electron Transfer from β-Carotene to Phenoxyl Radical and Adduct Formation Between Phenoxyl Radical and β-Carotene, *Free Radic. Res. 25*: 515–523 (1996a).

Mortensen, A., and L.H. Skibsted, Kinetics of Photobleaching of β-Carotene in Chloroform and Formation of Transient Carotenoid Species Absorbing in the Near Infrared, *Free Radic. Res. 25*: 355–368 (1996b).

Mortensen, A., and L.H. Skibsted, Importance of Carotenoid Structure in Radical-Scavenging Reactions, *J. Agric. Food Chem. 45*: 2970–2977 (1997a).

Mortensen, A., and L.H. Skibsted, Free Radical Transients in Photobleaching of Xanthophylls and Carotenes, *Free Radic. Res. 26*: 549–563 (1997b).

Mortensen, A., and L.H. Skibsted, Real Time Detection of Reactions Between Radicals of Lycopene and Tocopherol Homologues, *Free Radic. Res. 27*: 229–234 (1997c).

Mortensen, A., and L.H. Skibsted, Relative Stability of Carotenoid Radical Cations and Homologue Tocopheroxyl Radicals. A Real Time Kinetic Study of Antioxidant Hierarchy, *FEBS Lett. 417*: 261–266 (1997d).

Mortensen, A., L.H. Skibsted, A. Willnow, and S.A. Everett, Reappraisal of the Tocopheroxyl Radical Reaction with β-Carotene: Evidence for Oxidation of Vitamin E by the β-Carotene Radical Cation, *Free Radic. Res. 28*: 69–80 (1998).

Palozza, P., and N.I. Krinsky, β-Carotene and α-Tocopherol Are Synergistic Antioxidants, *Arch. Biochem. Biophys. 1*: 184–187 (1992).

Tappel, A.L., Will Antioxidant Nutrients Slow Aging Processes? *Geriatrics 23*: 97–105 (1968).

Truscott, T.G., β-Carotene and Disease: A Suggested Pro-Oxidant and Anti-Oxidant Mechanism and Speculations Concerning Its Role in Cigarette Smoking, *J. Photochem. Photobiol. B Biol. 35*: 233–235 (1996).

Tyurin, V.A., G. Carta, Y.Y. Tyurina, S. Banni, B.W. Day, F.P. Corongiu, and V.E. Kagan, Peroxidase-Catalyzed Oxidation of β-Carotene in HL-60 Cells and in Model Systems: Involvement of Phenoxyl Radicals, *Lipids 32*: 131–142 (1997).

Valgimigli, L., M. Lucarini, G.F. Pedulli, and K.U. Ingold, Does β-Carotene Really Protect Vitamin E from Oxidation? *J. Am. Chem. Soc. 119*: 8095–8096 (1997).

Chapter 17

Inhibition of Carcinogenesis by Tea Constituents and the Mechanisms Involved

Chung S. Yang, Janelle M. Landau, Guang-yu Yang, Sungbin Kim, and Jie Liao

Laboratory for Cancer Research, College of Pharmacy, Rutgers University, Piscataway, NJ 08854

Introduction

In recent years, the general public has heard a great deal about tea as a drink that may reduce the risk of a number of chronic conditions, including cancer and cardiovascular diseases. Tea (*Camellia sinensis*) has been considered a healthful drink for 4000 years by Eastern cultures. Most historically claimed benefits of drinking tea, however, were anecdotal. Scientific investigations are now beginning to ask questions concerning the effect of tea on health by considering data from epidemiology, clinical studies, laboratory research utilizing animal models, as well as mechanistic studies *in vitro*.

Epidemiologic data have been inconclusive regarding the relationship between tea consumption and human cancer incidence. Studies in animal models in many laboratories, however, have demonstrated an inhibitory effect of tea against carcinogenesis. The cancer preventive activity of tea has been demonstrated in many animal models including those with skin, lung, esophagus, stomach, liver, duodenum and small intestine, pancreas, and mammary gland cancers (Katiyar and Mukhtar 1996, Yang and Wang 1993, Yang *et al.* 1996). Tea is one of the few chemopreventive agents known to have protective effects at different stages of the carcinogenic process.

The beneficial health effects of tea are thought to be due to its polyphenolic compounds. Green tea contains (-) epigallocatechin-3-gallate (EGCG), (-) epigallocatechin (EGC), (-) epicatechin-3-gallate (ECG), and (-)epicatechin (EC). These compounds are usually referred to as tea catechins and account for ~10% of the dry weight of the tea leaves and 30% of the dry weight of the solids in brewed tea or tea water extracts (Balentine 1992). These catechins are colorless, water-soluble compounds that contribute astringency to green tea. The fermentation process of black tea produces other groups of polyphenols, theaflavins and thearubigens, which are oxidation and condensation products of the catechins. The oxidative condensation of the catechins reduces the astringent taste of the tea. Theaflavin, theaflavin gallate, and theaflavin digallate account for 1–2% of the dry weight of the water-extractable material. Thearubigins, which are poorly characterized chemically, may account for 60% of the dry weight of black tea leaves and 20% of the solid weight of water extract of black tea (Balentine 1992). Both green and black teas also contain caffeine and flavonoids such as quercetin, rutin, kaempferol, and myricitin.

In this paper, we discuss the inhibitory activities of green and black tea preparations as well as purified tea polyphenols against chemically induced lung carcinogenesis in A/J mice. The lung tumorigenesis model with A/J mice has been very useful in characterizing the inhibitory action of different tea preparations against tumorigenesis. This model, with discernible stages of initiation, promotion, and progression, is suitable for mechanistic studies. Experiments were also conducted with cell culture systems to obtain additional information on the growth inhibitory activities of tea and tea polyphenol preparation.

Inhibition of N*-Nitrosodiethylamine (NDEA)-Induced Lung Tumorigenesis in A/J Mice by Green Tea*

In this study, *N*-nitrosodiethylamine (NDEA) was used to induce lung tumors, and green tea infusions were administered to the mice as the sole source of drinking fluid both during or after the carcinogen treatment (Wang *et al.* 1992). The green tea leaves (Gunpowder brand from China, 12.5 g) were steeped twice for 15 min each time in 500 mL of boiled deionized water; the water extracts were cooled on ice and combined (referred to as 1.25% green tea). The amount of tea solids in this preparation was 4.69 mg/mL, which contained 15.1, 6.9, 3.0, 1.8, and 8.1% of EGCG, EGC, ECG, EC, and caffeine, respectively.

Female A/J mice were administered NDEA intragastrically at a dose of 10 mg/kg once weekly for 8 wk and were killed 16 wk after the last dose (Wang *et al.* 1992). More than 90% of the mice developed adenomas in the lung; on average, there were 2.5 lung tumors per mouse. When 0.63 or 1.25% green tea infusion was used as the sole source of drinking fluid starting 2 wk before the first dose of NDEA and continuing until 1 wk after the last dose of NDEA (the carcinogen treatment period), the tumor incidence was decreased by 13 or 39%, and the tumor multiplicity was decreased by 36 or 56%, respectively, compared with positive controls. When 0.63 or 1.25% green tea infusion was used starting 1 wk after the last dose of NDEA until the end of the experiment (postinitiation phase), the tumor incidence was reduced by 50 and 36%, and the tumor multiplicity was reduced by 52 and 44%, respectively (see Table 17.1 for summary). Apparently, tea administration to the mice inhibits the tumorigenesis process in both the carcinogen treatment and postcarcinogen treatment periods. It was noted that the body weights of the mice in the 1.25% tea treatment group were lower (up to 12% lower) than those of control mice, but such a difference was not observed in the 0.63% tea group.

In this carcinogenesis model, forestomach tumors (papilloma, carcinoma *in situ*, and squamous cell carcinoma) were also induced by the treatment with NDEA (Wang *et al.* 1992). The tumor incidence was 98%, and the tumor multiplicity was 8.3 forestomach tumors per mouse. When 0.63 or 1.25% green tea infusion was given to the mice during the carcinogen treatment period (as described above), the tumor multiplicity was decreased by 31 and 58%, respectively, and the tumor incidence was decreased slightly (~15%) by 1.25% green tea infusion. When given to

TABLE 17.1
Summary of Studies on the Anticarcinogenic Effects of Tea in A/J Mouse Lung[a]

Carcinogen	Tea treatment	Inhibition of lung tumor, % Incidence	Multiplicity
	Initiation phase		
NDEA	0.63% GT	13	36
10 mg/kg (i.g.)	1.25% GT	39	56
	Postinitiation phase		
NDEA	0.63% GT	50	52
	1.25% GT		
	Initiation phase		
NNK	0.6% DGT	15	65
103 mg/kg (i.p.)	0.6% DBT	15	65
	Initiation phase		
NNK	0.6% DGT	30	85
103 mg/kg (i.p.)	0.6% DBT	—	63
	60 wk		
Spontaneous	1.0% BT	23	28
(60 wk)	2.0% BT	48	54
	1.0% GT	32	43

[a]Abbreviations: NDEA, *N*-nitrosodiethylamine; GT, green tea; i.g., intragastric; NNK, 4-(methylnitrosamino)-1-(3-pyridyl)-1-butanone; DGT, decaffeinated green tea; i.p., intraperitoneal; DBT, decaffeinated black tea; BT, black tea.

mice starting 1 wk after the NDEA treatment period, 0.63 or 1.25% green tea infusion decreased tumor multiplicity by 35 and 47%, respectively, but did not significantly affect the forestomach tumor incidence.

Inhibition of Lung Tumorigenesis Induced by 4-(Methylnitrosamino)-1-(3-pyridyl) -1- butanone (NNK) in Mice by Tea Preparations

In these studies, 4-(methylnitrosamino)-1-(3-pyridyl)-1-butanone (NNK) was used to induce lung tumors, and decaffeinated green tea (DGT) or decaffeinated black tea (DBT) was administered to the mice as the sole source of drinking fluid during different time periods. The tea solutions were reconstituted from DGT and DBT solids as 0.6 or 0.3% solutions. The tea solids (from Thomas J. Lipton, Englewood Cliffs, NJ) were prepared from tea water extracts of decaffeinated tea leaves by spray drying. The decaffeination was accomplished by extraction with supercritical CO_2. In the DGT solids, EGCG, EGC, ECG, EC, and caffeine account for 11.1, 11.0, 2.6, 3.1, and 0.3% of the dry weight, respectively, as measured by high-performance liquid chromatography (HPLC). In the DBT solids, the corresponding values were 2.2, 0.9, 1.3, 0.9, and 0.4%. In addition, the DBT solids also contain theaflavins (1.7%) and thearubigins (12–18%).

Treatment of female A/J mice with a single dose (103 mg/kg) of NNK resulted in the formation of pulmonary adenomas in almost all of the mice, with an average of

9.3 tumors per mouse after 16 wk; tumorigenesis was inhibited by the administration of DGT or DBT (Wang *et al.* 1992). When administered to mice starting 2 wk before the NNK dose for 3 wk, 0.6% DGT and DBT decreased tumor multiplicity similarly by ~65% and decreased tumor incidence slightly (~15%, but not statistically significant). When administered to mice starting 1 wk after the NNK dose until the termination of the experiment, 0.6% DGT decreased tumor multiplicity by 85% and tumor incidence by 30%, whereas 0.6% DBT decreased tumor multiplicity by 63% but did not influence tumor incidence. In this latter protocol, 0.3% DGT and DBT appeared to be less effective than the 0.6% solutions, but a clear dose-response relationship was not observed. To our knowledge, this is the first demonstration that decaffeinated tea preparations are effective in inhibiting lung tumorigenesis and that black tea has an inhibitory activity comparable to that of green tea.

In a second study, 0.6% DGT was administered to mice for different time periods in the same NNK-induced lung tumorigenesis model (Shi *et al.* 1994). When DGT was administered to mice for 2 d before the NNK dose or for 7 d after the NNK dose, reduction of tumor multiplicity (by 31 or 20%, respectively) was observed, but the treatment was not as effective as the 3-wk treatment starting 2 wk before the NNK dose (which decreased tumor multiplicity by 56% and tumor incidence by 20%). Treatment of mice with 0.6% DGT for 12 wk starting 5 wk after the NNK dose also decreased tumor multiplicity by 54%; this was slightly less effective than the 16-wk treatment (starting 1 wk after the NNK dose), which decreased tumor multiplicity by 65% and tumor incidence by 20% (Shi *et al.* 1994).

These results suggest that tea preparations can inhibit lung tumorigenesis at the initiation and promotion stages. The inhibition of the metabolic activation of NNK and other carcinogens has been demonstrated *in vitro* (Shi *et al.* 1994). However, inhibition of NNK-induced DNA methylation in the lungs by green and black tea was not demonstrated *in vivo* under conditions in which inhibition of tumorigenesis has been demonstrated (Shi *et al.* 1994). Xu *et al.* (1992) showed that administration of 2% green tea or 560 ppm of EGCG in drinking water to mice inhibited the NNK-induced formation of 8-hydroxydeoxyguanosine, but not the formation of O^6-methylguanine in the lungs. In this animal model, either tea or EGCG was given from wk 2 to 13, and NNK was administered by gavage three times weekly for 10 wk from wk 3 to 12. The tumor multiplicity, measured at wk 19, was decreased 46 and 28% by green tea and EGCG, respectively, in comparison to the control group (with 22.5 lung adenomas per mouse).

Antiproliferative Effects and Inhibition of Tumorigenesis by Green and Black Tea Constituents

Perhaps one of the most important factors in the anticarcinogenic actions of tea is the antiproliferative effect of tea and tea constituents (Yang *et al.* 1997a and 1997b). Not only is hyperproliferation a necessary factor for tumor development, but cell division must occur also for DNA damage to be expressed as mutations. The antiproliferative effect of tea may inhibit tumor initiation after the treatment with NNK or NDEA and may also be important for the inhibition of tumor promotion and progression in which increased cell proliferation is an important event.

To provide a more mechanistic understanding of the inhibitory actions of tea constituents against tumorigenesis, their effect on NNK-induced early hyperproliferation was examined (Yang *et al.* 1997a and 1997b). The NNK-induced early hyperproliferation was defined as the increase of proliferation index without changes in morphology. The black tea preparation used contained theaflavins, EGCG, EGC, ECG, and EC at concentrations of 13, 17.5, 1.6, 10.6, and 5.6%, respectively. The preparation of theaflavins contained theaflavin, theaflavin-3-gallate, theaflavin-3'-gallate, and theaflavin-3,3'-digallate at concentrations of 21, 30, 15, and 28%, respectively. These preparations and those of pure EGCG and EGC were supplied by Thomas J. Lipton. Treatment of female A/J mice with a dose of NNK (103 mg/kg) on d 0 induced hyperproliferation of the bronchiolar epithelial cells, but not the alveolar cells, as measured by 5-Bromo-2′-deoxyuride-labeling. The hyperproliferation was pronounced between d 2 and 14, with the highest proliferation index observed on d 5. On d 35, the cell proliferation index returned to the normal level; however, upon thorough examination of all of the slides from the lung, proliferative foci were found in the bronchiolar epithelium, particularly in the bronchiolalveolar region. These foci, aggregates of more than five BrdU-labeled cells, were detected in the lungs of NNK-treated mice (average, 5 foci per lung) but not in those from the control group. Further development of these proliferative foci may lead eventually to tumors, especially the papillary type of adenocarcinoma. Alternatively, hyperproliferative cells in the foci may grow along the alveolar wall, differentiating into the bronchiolar cell type and entrapping resident alveolar type II cells in tumorigenesis.

Administration of a 0.3% black tea polyphenol preparation (as the sole source of drinking fluid) starting 24 h after the NNK dose significantly decreased the hyperproliferation (by 65%) (Yang *et al.* 1997a). Theaflavins and green tea polyphenols EGCG and EGC, when administered in drinking fluid as 0.3 or 0.6% solution starting 24 h after the NNK dose, also effectively inhibited bronchiolar epithelial hyperproliferation (by 30–35%) in this model (Yang *et al.* 1997a).

When a 0.1% solution of theaflavins was administered as drinking fluid to mice starting 2 d after the NNK dose and continuing for 16 wk, inhibition of lung adenoma formation was observed (Yang *et al.* 1997a). The tumor multiplicity and volume were reduced by 23 and 34%, respectively. The proliferation index of lung adenomas was also significantly inhibited by theaflavins. Similar results on the inhibition of adenoma formation and cell proliferation by a black tea polyphenol fraction were also observed (Yang *et al.* 1997a).

Inhibition of the Progression of Pulmonary Adenomas to Adenocarcinomas by Black Tea

In the NNK-induced lung tumorigenesis model, adenomas were observed in almost all of the mice 16 wk after the NNK administration, and 80% of the mice developed malignant tumors with an average of 2.4 adenocarcinomas per mouse 52 wk after the NNK dose. Treatment of the mice with a 0.6% solution of black tea solids, starting 16 wk after the NNK dose until the termination of the experiment, decreased the adenocarcinoma incidence by 27% and multiplicity by 51% (Yang *et al.* 1997b). The treatment also decreased the tumor volume by 56%. The progression of lung tumorigenesis is associated with an increase in proliferation with adenoma, dysplasia within adenoma, and adenocarcinoma (defined when nuclear crowding, pleomorphism, and atypia were seen), showing proliferation indices of 25, 42, and 53%, respectively. In the black tea–treated group, the proliferation index was significantly lower than the control group in adenoma, but not in the two higher-grade lesion categories (Yang *et al.* 1997b).

Inhibition of Spontaneous Lung Tumors and Rhabdomyosarcomas by Tea

Both black and green tea infusions inhibited the spontaneous formation of lung tumors and rhabdomyosarcomas in A/J mice (Landau *et al.* 1998). Female A/J mice, 6 wk of age, were given 0.5, 1, or 2% black tea or 1% green tea as the sole source of drinking fluid. After 60 wk, the lung tumor incidence, multiplicity, and volume were significantly lower in the 2% black tea group compared with the controls (27 vs. 52%, 0.33 vs. 0.72 tumors/mouse, and 4.27 vs. 38.3 mm^3, respectively). The 1% green tea group had significantly lower lung tumor multiplicity (0.41/mouse), whereas the 1% black tea group had significantly decreased tumor volume (7.17 mm^3) (see Table 16.1 for summary). Rhabdomyosarcomas were found in 34% of the mice in the control group, and both the 1 and 2% black tea groups had significantly lower incidences (13 and 14%, respectively). The mice in the 2% black tea group weighed 16% less than those in the control group, although they consumed more food than the control group. The other tea-consuming groups also weighed less than the control group (7.8–11%) while consuming more food and fluid. Although lower body weight could be a factor, studies have shown that the inhibition of tumorigenesis using decaffeinated tea or tea polyphenolic fractions does not lower the body weight of the animals. In a separate experiment, similar carcinogenesis inhibition was also observed in female A/J mice that were given 0.6% and then 0.3% instant black tea for 52 wk. These results demonstrate the inhibitory activity of black tea against the spontaneous formation of lung tumors and rhabdomyosarcomas in mice.

Growth Inhibition and Mechanistic Studies with Cell Lines

The growth inhibition of three human lung cancer cell lines has been studied recently (Yang *et al.* 1998) EGCG and EGC displayed strong growth inhibition against lung H661 and H1299 cells with estimated IC_{50} values (concentration required for 50% inhibition) of 22 μmol/L, but were less effective against H441 cells with an IC_{50} value twofold higher. ECG had lower inhibition activity, and EC was even less effective. Decaffeinated green tea extract had similar activities as regular green tea extract in the inhibition of the growth of H661 and H1299 cells, but was less effective than regular green tea extract with H441. Preparations of green tea polyphenols and theaflavins had inhibitory activities similar to those of green tea extracts. The results suggest that, in addition to EGCG, other green tea catechins, theaflavins, and other black tea polyphenols also possess growth inhibitory activities. Additional studies indicated that, at 30 and 100 μmol/L, EGCG, EGC, and theaflavins induced apoptosis in H661 cells when measured 24 h after addition of the tea compound. The addition of catalase (50 units/mL) to the culture medium eliminated the apoptosis completely and eliminated most of the growth inhibitory effects of tea polyphenols. The results suggest that hydrogen peroxide is involved in inducing apoptosis and that tea polyphenols serve as prooxidants. It remains to be determined whether such a mechanism of apoptosis-induction occurs *in vivo*.

Inhibition of the Activator protein-1 activation by tea polyphenols was demonstrated in our recent studies with JB6 mouse epidermal cell line (Dong *et al.* 1997). EGCG and theaflavins, at the concentration range of 5–20 μmol/L, inhibited epidermal growth factor–induced or 12-*O*-tetradecanoylphorbol-13-acetate–induced cell transformation by 20–80%. At this concentration range, EGCG and theaflavins also inhibited AP-1–dependent transcriptional DNA binding activity. The inhibition of AP-1 activation was due to the inhibitory action against a c-Jun, N-terminal kinase–dependent pathway.

Green tea extracts, black tea polyphenols, and EGCG have been shown to inhibit the growth of rat hepatoma, mouse erythroleukemia, and several human cancer cell lines (Lea *et al.* 1993, Okabe *et al.* 1997, Valcic *et al.* 1996). Inhibition of tumor necrosis factor α (TNF-α)–dependent and other signal transduction pathways has also been proposed as a possible mechanism for the anticancer activities of tea polyphenols (Lin and Lee 1995, Suganuma *et al.* 1996), but most of the activities were observed at 20–50 μmol/L concentrations.

Bioavailability and Relevance to Humans

Our previous studies indicated that, under conditions in which inhibition could be demonstrated in mice and rats, the average EGCG and EGC concentrations in the blood were in the range of 0.1–0.3 μmol/L, and the levels of these catechins in the lung tissues were even lower (Yang *et al.* 1996). After ingestion of 2 or 3 cups of tea by human subjects, the average peak plasma values of EGCG and EGC were also in this range (Lee *et al.* 1995). These levels appear to be lower than the concentrations required for exerting effective growth inhibition *in vitro*, even though the exact cellular concentrations in mouse lung cells and in cultured cancer cells have yet to be determined. It is not clear whether the mechanisms of growth inhibition against cancer cells in culture are different from the cancer-preventive effects of tea and tea polyphenols demonstrated in animals. The metabolism of the tea polyphenols may occur at different rates in the animal compared with cell culture, which may account for the higher concentrations required *in vitro*. Further investigation is warranted in this area to determine the effect of metabolism on tea polyphenols *in vitro* as well as the mode of action exerted by the polyphenols *in vivo*, particularly concerning the signal transduction pathways.

For humans to benefit from the effects of tea, the active compounds must be absorbed and available to the appropriate tissues. The bioavailability and metabolic fate of tea polyphenols in humans are not completely understood. One obstacle is the ability to detect the polyphenolic compounds in human fluids and tissues. Although recent advances have been made with the major green tea polyphenols (Chen *et al.* 1997, Hollman *et al.* 1997, Lee *et al.* 1995), theaflavins and thearubigens are more difficult to extract from biological fluids and to analyze. Although these compounds seem to be absorbed, it is not known if theaflavins and thearubigens are absorbed intact or if they are degraded before absorption.

Conclusion

These studies demonstrate the antiproliferative and anticarcinogenic effects of different tea and tea polyphenol preparations in the mouse lung. EGCG, EGC, and theaflavins are shown to be the main active components for these activities, although other tea constituents may also possess such activities. EGCG is considered to be the most abundant catechin in green tea, but EGC levels could be high, i.e., comparable to those of EGCG, in many green tea preparations. Judging from the fact that EGC is more bioavailable in rats and humans, EGC could be an important component for the inhibition of lung tumorigenesis. EGCG, EGC, and ECG are present in substantial quantities in black tea. These catechins and theaflavins contribute to the inhibitory action of black tea. The possible antitumorigenic activities of thearubigens remain to be demonstrated.

The mechanism(s) by which tea inhibits tumorigenesis remains to be further elucidated. The antioxidant properties likely play a role, but several other potentially important characteristics can be attributed to tea. Tea inhibits the activity of several enzymes related to tumor promotion and cell proliferation, including ornithine decarboxylase, protein kinase C, cyclooxygenase, and lipoxygenase. This antiproliferative property of tea deserves further investigation. Other mechanisms such as the quenching of activated carcinogens, inhibition of signal transduction, and enhancement of immune function have been suggested, but their relevance to human carcinogenesis is not clear.

Tea is one of the few agents that has been demonstrated to inhibit lung tumorigenesis at the postinitiation stage. This property provides the potential for tea and tea polyphenol preparations to be chemopreventive agents against lung carcinogenesis in humans, for example, among ex-smokers. Whether tea and tea polyphenols have antitumorigenic activities in humans depends on whether the mechanisms of inhibiting animal carcinogenesis are also applicable to humans and on the concentrations of tea polyphenols that can be delivered to the target tissues. Both issues require further investigation.

References

Balentine, D.A., Manufacturing and Chemistry of Tea, in *Phenolic Compounds in Food and Their Effects on Health I: Analysis, Occurrence, and Chemistry*, edited by C.-T. Ho, M.-T. Huang, and C.Y. Lee, American Chemical Society, Washington, DC, 1992, pp. 102–117.

Chen, L, M.-J. Lee, H. Li, and C.S. Yang, Absorption, Distribution, and Elimination of Tea Polyphenols in Rats, *Drug Metab. Dispos. 25:* 1045–1050 (1997).

Dong, Z., W.-Y. Ma, C. Huang, and C.S. Yang, Inhibition of Tumor Promoter-Induced AP-1 Activation and Cell Transformation by Tea Polyphenols, (-)-Epigallocatechin Gallate and Theaflavins, *Cancer Res. 57:* 4414–4419 (1997).

Hollman, P.C.H., L.B.M. Tijburg, and C.S. Yang, Bioavailability of Flavonoids from Tea, *Crit. Rev. Food Sci. Nutr. 37:* 719–738.

Katiyar, S.K., and H. Mukhtar, Tea in Chemoprevention of Cancer: Epidemiological and Experimental Studies (Review), *Int. J. Oncol. 8:* 221–238 (1996).

Landau, J.M., Z.-Y. Wang, G.-Y. Yang, W. Ding, and C.S. Yang, Inhibition of Spontaneous Formation of Lung Tumors and Rhabdomyosarcomas in A/J Mice by Black and Green Tea, *Carcinogenesis 19:* 501–507 (1998).

Lea, M.A., Q. Xiao, A.K. Sadhukhan, S. Cottle, Z.Y. Wang, and C.S. Yang, Inhibitory Effects of Tea Extract and (-)-Epigallocatechin Gallate on DNA Synthesis and Proliferation of Hepatoma and Erythroleukemia Cells, *Cancer Lett. 68:* 231–236 (1993).

Lee, M.-J., Z.-Y. Wang, H. Li, L. Chen, Y. Sun, S. Gobbo, D.A. Balentine, and C.S. Yang, Analysis of Plasma and Urinary Tea Polyphenols in Human Subjects, *Cancer Epidemiol. Biomark. Prev. 4:* 393–399 (1995).

Lin, J.K., and S.F. Lee, Inhibition of Tumor Promotion Through Blocking Signal Transduction, *Zool. Stud. 34:* 67–81 (1995).

Okabe, S., M. Suganuma, M. Hayashi, E. Sueoka, A. Komori, and H. Fujiki, Mechanisms of Growth Inhibition of Human Lung Cancer Cell Line, PC-9, by Tea Polyphenols, *Jpn. J. Cancer Res. 88:* 639–643 (1997).

Shi, S.T., Z.-Y. Wang, T.J. Smith, J.-Y. Hong, W.-F. Chen, C.-T. Ho, and C.S. Yang, Effects of Green Tea and Black Tea on 4-(Methylnitrosamino)-1-(3-pyridyl)-1-butanone Bioactivation, DNA Methylation, and Lung Tumorigenesis in A/J Mice, *Cancer Res. 54:* 4641–4647 (1994).

Suganuma, M., S. Okabe, E. Sueoka, N. Iida, A. Komori, S.-J. Kim, and H. Fujiki, A New Process of Cancer Prevention Mediated Through Inhibition of Tumor Necrosis Factor Expression, *Cancer Res. 56:* 3711–3715 (1996).

Valcic, S., B.N. Timmermann, D.S. Alberts, G.A. Wachter, M. Krutzsch, J. Wymer, and J.M. Guillen, Inhibitory Effect of Six Green Tea Catechins and Caffeine on the Growth of Four Selected Human Tumor Cell Lines, *Anticancer Drugs 7:* 461–468 (1996).

Wang, Z.Y., J.Y. Hong, M.T. Huang, K.R. Reuhl, A.H. Conney, and C.S. Yang, Inhibition of *N*-Nitrosodiethylamine and 4-(Methylnitrosamino)-1-(3-pyridyl)-1-butanone-Induced Tumorigenesis in A/J Mice by Green Tea and Black Tea, *Cancer Res. 52:* 1162–1170 (1992).

Xu Y, C.-T. Ho, S.G. Amin, C. Han, and F.-L. Chung, Inhibition of Tobacco-Specific Nitrosamine-Induced Lung Tumorigenesis in A/J Mice by Green Tea and Its Major Polyphenol as Antioxidants, *Cancer Res. 52:* 3875–3879 (1992).

Yang, C.S., L. Chen, M.-J. Lee, and J.M. Landau, Effects of Tea on Carcinogenesis in Animal Models and Humans, in *Dietary Phytochemicals in Cancer Prevention and Treatment*, edited under the auspices of the American Institute for Cancer Research, Plenum Press, New York, 1996, pp. 51–61.

Yang, C.S., and Z.-Y. Wang, Tea and Cancer: A Review, *J. Natl. Cancer Inst. 58:* 1038–1049 (1993).

Yang, G.-Y., J. Liao, K. Kim, E.J. Yurkow, and C.S. Yang, Inhibition of Growth and Induction of Apoptosis in Human Cancer Cell Lines by Tea Polyphenols, *Carcinogenesis 19:* 611–616 (1998).

Yang, G.-Y., Z. Liu, D.N. Seril, J. Liao, W. Ding, S. Kim, F. Bondoc, and C.S. Yang, Black Tea Constituents, Theaflavins, Inhibit 4-(Methylnitrosmaino)-1-(3-pyridyl)-1-butanone (NNK)-Induced Lung Tumorigenesis in A/J Mice, *Carcinogenesis 19:* 611–616 (1998).

Yang, G-Y, Z.-Y. Wang, S. Kim, J. Liao, D. Seril, X. Chen, T.J. Smith, and C.S. Yang, Characterization of Early Pulmonary Hyperproliferation, Tumor Progression and Their Inhibition by Black Tea in a 4-(Methylnitrosamino)-1-(3-pyridyl)-1-butanone (NNK)-Induced Lung Tumorigenesis Model with A/J Mice, *Cancer Res. 57:* 1889–1894 (1997b).

Chapter 18

Antioxidative Activity of (-)-Epicatechin in Rat Plasma: Intragastric Administration and Its Metabolites

Mariusz K. Piskula[a], Edson L. da Silva[b], and Junji Terao[c]

[a]Institute of Animal Reproduction and Food Research, 10-718 Olsztyn, Poland

[b]Universidade Federal de Santa Catarina, Florianopolis-Santa Catarina 88040-900, Brazil

[c]Department of Nutrition, School of Medicine, The University of Tokushima, 770 Tokushima, Japan

Introduction

Oxidative damage in biological systems is known as one of the triggers of the aging process and degenerative diseases such as atherosclerosis and cancer. Therefore, consumption of foods containing antioxidative compounds is supposed to provide protection against uncontrolled oxidation. A substantial amount of dietary antioxidants are of plant origin. They are primarily polyphenols, of which flavonoids have attracted considerable attention. Recently, epidemiologic evidence has emerged showing an inverse association between the intake of dietary flavonoids and coronary heart disease (Knekt *et al.* 1996, Hertog *et al.* 1995). The so-called French paradox, a lack of positive correlation between high intake of saturated fat and coronary heart disease, is related at least in part to the consumption of red wine, which is rich in phenolic compounds (Renaud and deLorgeril 1992). Although the *in vitro* experiments demonstrated the antioxidative activity of red wine phenolics in inhibiting oxidation of human low-density lipoproteins (Frankel *et al.* 1993), the *ex vivo* trials gave contradictory results (deRijke *et al.* 1996, Fuhrman *et al.* 1995). In addition, *in vitro* studies showed that flavonoids act as antioxidants by lipid peroxidation inhibition, oxygen radical scavenging, or metal ion chelating (Afanas'ev *et al.* 1989, Cotelle *et al.* 1992, Terao *et al.* 1994, Terao and Piskula1998). This activity is essentially related to the number and arrangement of phenolic hydrogens in the flavonoid molecule, whose loss was found to decrease flavonoid antioxidative potential (Rice-Evans *et al.* 1995, Van Acker *et al.* 1996).

A number of new studies on absorption of flavonoids have recently been carried out in humans (Fuhr *et al.* 1995, Hollman *et al.* 1996) and in rats (Unno *et al.* 1996, Piskula and Terao 1998, Hacket and Griffiths 1982, Manach *et al.* 1997, Okushio *et al.* 1996). However, the question remains whether the flavonoids retain any of their antioxidative properties after absorption and metabolic alteration. This study aims to clarify this question. We used (-)-epicatechin (EC), a flavonoid commonly present in food of plant origin and consumed mainly in green and black tea, wine, apples, and spices. In the experiment, EC was administered orally to rats;

levels of its metabolites in plasma were measured as well as their effect on plasma susceptibility to *ex vivo* oxidation.

Experimental Procedures

Chemicals

(-)-Epicatechin, β-glucuronidase type VII-A, sulfatase type VIII, sulfatase type H-5 and D-saccharic acid 1,4-lactone were obtained from Sigma Chemical (St. Louis, MO); 2,2′-azobis(2-amidinopropane) dihydrochloride (AAPH) was purchased from Wako Chemicals (Tokyo, Japan). Other reagents and solvents were of analytical or HPLC grade.

Animals and (-)-Epicatechin Administration

Twenty-seven male Wistar rats, weighing 195–210 g, were obtained from Charles River Japan (Atsugi, Japan). The animals were divided randomly into one group of three rats and three groups of eight rats and kept in a temperature- (23 ± 1°C) and light- (0700-1900 h) controlled room for 3–4 d before the commencement of the experiments; rats had free access to tap water and standard MF diet (Oriental Bioservice, Kanto, Tsukuba, Japan). The day before the experiment, at 20:00 h, food was withheld; the next day, at 09:00 h, rats were administered EC orally by direct stomach intubation. To obtain plasma for determination of the EC absorption profile, a group of three rats was administered EC (50 mg/kg of body weight) in 2 mL water/propylene glycol (3:1, vol/vol). Before (control) and after administration, ~400 µL blood samples were collected from the tail vein at certain intervals into heparinized tubes; plasma was prepared by centrifugation (800 × *g* for 20 min at 4 °C). To obtain blood plasma for oxidation experiments, two groups of eight animals were administered EC, 50 or 250 mg/kg body weight, in 2 mL of water. The third group of eight rats (control) was administered 2 mL water. Four rats out of each group of eight animals were anesthetized with diethyl ether at 1 and 6 h after administration, and blood was collected from the abdominal aorta into heparinized tubes. Blood plasma was prepared by centrifugation (800 × *g* for 20 min at 4°C), then pooled and used immediately for experiments or stored at –0°C for no longer than 1 wk. During the experiments, rats received humane care consistent with institutional guidelines.

Determination of (-)-Epicatechin and Its Metabolites in Plasma

During absorption and metabolism, (-)-epicatechin is extensively conjugated with sulfate and glucuronide moieties and is also methylated (methyl-EC) (Piskula and Terao 1998). For the determination of EC, methyl-EC, and their conjugates in rat plasma, a method based on enzymatic hydrolysis of conjugates followed by high-performance liquid chromatography (HPLC) with electrochemical detector (HPLC-ECD) analysis was used (Piskula and Terao 1998). Briefly, glucuronide, sulfate, or total glucuronide and/or sulfate conjugates were hydrolyzed into free EC and methyl-EC by β-glucuronidase (type VII-A), sulfatase (type VIII with 4 mg/mL of D-saccharic acid 1,4-lactone as inhibitor of contaminant β-glucuronidase), or a mixture of β-glucuronidase and sulfatase (commercial sulfatase type H-5). The difference in EC or methyl-EC contents after and before hydrolysis with a particular enzyme was assumed to be the amount of EC and methyl-EC conjugates in the sample.

Plasma Oxidation Induced by Copper Ion or Radical Generator

Pooled plasma was diluted fourfold with phosphate-buffered saline (PBS) buffer (pH 7.4); 0.9 mL of this solution was oxidized by adding 0.1 mL of 1 mmol/L $CuSO_4$ or 0.1 mL of 50 mmol/L AAPH and incubating at 37°C in a shaking water bath for 6 h. The oxidation process was controlled by the measurement of cholesteryl ester hydroperoxides (CE-OOH) produced.

Determination of Cholesteryl Ester Hydroperoxides

Aliquots (100 μL) were withdrawn from the incubation systems at certain time intervals and mixed with 3 mL methanol containing 2.5 mmol/L 3,5-di-*tert*-butyl-4-hydroxytoluene (BHT). The mixture was sonicated (Heat Systems Sonifier, Model W-380, Farmingdale, NY) for 1 min; neutral lipids were extracted twice with 3 mL *n*-hexane by vortexing vigorously for 1 min. The hexane extracts were combined and evaporated using a rotary evaporator at room temperature. The remaining lipids were dissolved in 100 μL methanol/chloroform (95:5, vol/vol), and CE-OOH content was determined by reversed-phase (RP)-HPLC with a TSK-gel Octyl-80Ts column (Tosoh, Tokyo, Japan) and ultraviolet (UV) detector set at 235 nm (Shimadzu SPD-10AV, Shimadzu, Kyoto, Japan). The flow rate of the mobile phase composed of methanol/water (95:5, vol/vol) was 1.0 mL/min. The concentration of CE-OOH was calculated from the standard curve of cholesteryl linoleate hydroperoxide (Arai *et al.* 1996).

Plasma Autoxidation

Plasma pooled from control rats or those administered EC (250 mg/kg, 1 h after administration) was autoxidized without dilution, by incubation under air at 37°C in a shaking water bath for 4 d without addition of oxidation initiators. During oxidation, contents of EC and methyl-EC metabolites as well as CE-OOH were measured at certain intervals as described above.

Results

(-)-Epicatechin Absorption Profile and EC Metabolites in Pooled Plasma

After administration of EC to rats [50 mg/kg in 2 mL of water/propylene glycol (3:1, vol/vol)], EC metabolite content was determined as a function of time. Plasma EC metabolites were grouped as methylated and nonmethylated, conjugated and nonconjugated with glucuronide and/or sulfate (Fig. 18.1). All groups of metabolites reached their maximum concentrations within 2 h after administration. After this time, plasma levels started to decline. Only free EC was cleared from the plasma before the end of the observation. Nonconjugated methyl-EC was not detected. On the basis of the absorption profile obtained, we decided to collect the blood plasma for oxidation experiments at 1 and 6 h after administration of the EC water solution (50 or 250 mg/kg body weight) or water. Because there was a high individual variation in absorption, plasma was pooled and contents of glucuronides and/or sulfates conjugates of EC and methyl-EC were determined (Table 18.1). One hour after EC administration (50 mg/kg), rat plasma contained 42.3 µmol/L of total EC metabolites; 31% of that was methylated, 77% was glucuronized, and only 7% was in nonconjugated form. After 6 h, the total content of EC plasma metabolites, compared with the 1-h level, decreased by 84%. A substantial decrease occurred particularly in the level of nonmethylated EC metabolites (97%), whereas the level of methyl-EC metabolites decreased by only 55%.

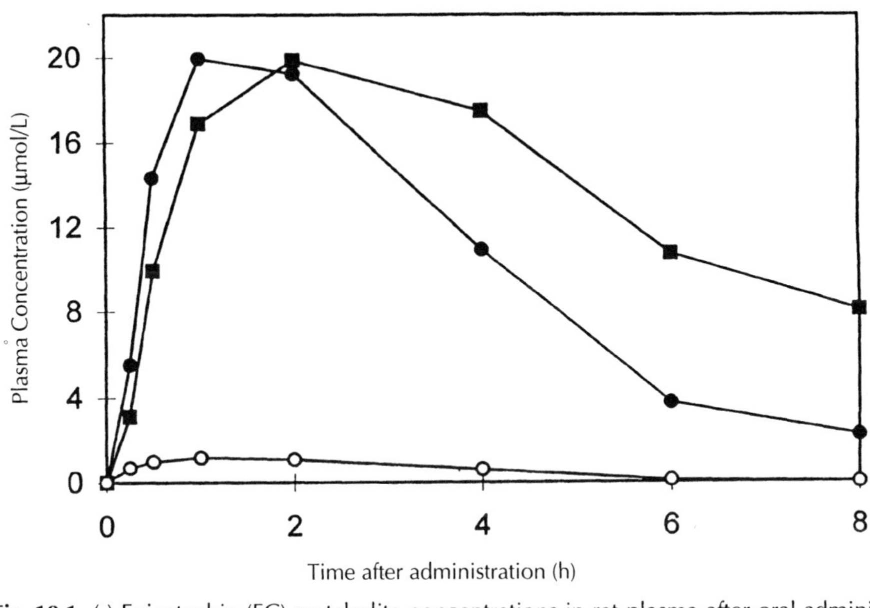

Fig. 18.1 (-)-Epicatechin (EC) metabolite concentrations in rat plasma after oral administration of 50 mg EC/kg rat body weight. Nonconjugated EC is marked as (○), conjugates possessing basic structure of EC are marked as (●), and those with basic structure of 3′-methylated (-)-epicatechin (methyl-EC) as (■). Metabolites were extracted from plasma obtained from rat tail blood taken before and for 8 h after EC administration. Before and after enzymatic hydrolysis of conjugates, EC and methyl-EC were quantified by high-performance liquid chromatography (HPLC). For determination details see section on Experimental Procedures.

TABLE 18.1
Concentration of (-)-Epicatechin Metabolites in Rat Plasma Following Intragastric Administration of Two Doses of (-)-Epicatechin to Rats[a]

	Time after administration			
	1 h		6 h	
Conjugates	EC[b] (μmol/L)	Methyl-EC[c] (μmol/L)	EC[b] (μmol/L)	Methyl-EC[c] (μmol/L)
Dose, 50 mg/kg[d]				
Free	3.0	0.0	0.0	0.0
Sulfate	1.0	1.3	0.0	0.9
Glucuronide	25.2	7.4	0.9	2.8
Sulfate/glucoronide	0.0	4.4	0.0	2.2
Total	29.2	13.1	0.9	5.9
Dose, 250 mg/kg[d]				
Free	13.3	0.9	2.6	0.3
Sulfate	0.0	7.6	3.6	9.3
Glucuronide	52.1	5.9	11.9	11.5
Sulfate/glucuronide	0.0	23.2	5.4	28.7
Total	65.4	37.6	23.5	49.8

[a]*Source*: da Silva *et al.* (1998).
[b]EC, nonmethylated (-)-epicatechin.
[c]Methyl-EC, methylated (-)-epicatechin.
[d]mg/kg, mg of (-)-epicatechin per kg of rat body weight.

(-)-Epicatechin administration to rats at a dose of 250 mg/kg body weight resulted in a total plasma EC metabolite concentration of 103.0 and 73.3 μmol/L at 1 and 6 h after treatment, respectively. After administration (1 and 6 h) of this high dose, ~14 and 4%, respectively, of the total EC metabolites were found to be nonconjugated. Nevertheless, compared with the EC dose of 50 mg/kg, a similar pattern of molar ratio of nonmethylated to methylated EC metabolites was observed 1 and 6 h after administration.

Oxidation of Plasma Lipids by Copper Ion

Pooled plasma was obtained after oral administration of EC in water or water oxidized (see above) in the presence of 100 μmol/L $CuSO_4$. During oxidation, the formation of CE-OOH was measured (Fig. 18.2). The effect of EC administration on the rat plasma resistance to copper ion oxidation is clear; both plasmas from rats administered 50 mg/kg EC (1 and 6 h after administration) were remarkably more resistant to CE-OOH formation compared with plasmas of their respective controls. The lag phase in CE-OOH accumulation in both control plasmas was ~2 h; however, the final level of CE-OOH in oxidized control plasma obtained 6 h after water administration (Fig. 18.2B) was higher than that in the 1-h control plasma (Fig. 18.2A). This suggests a prooxidative effect of a prolonged fasting condition. Indeed, it was reported that during starvation, circulating levels of free fatty acids

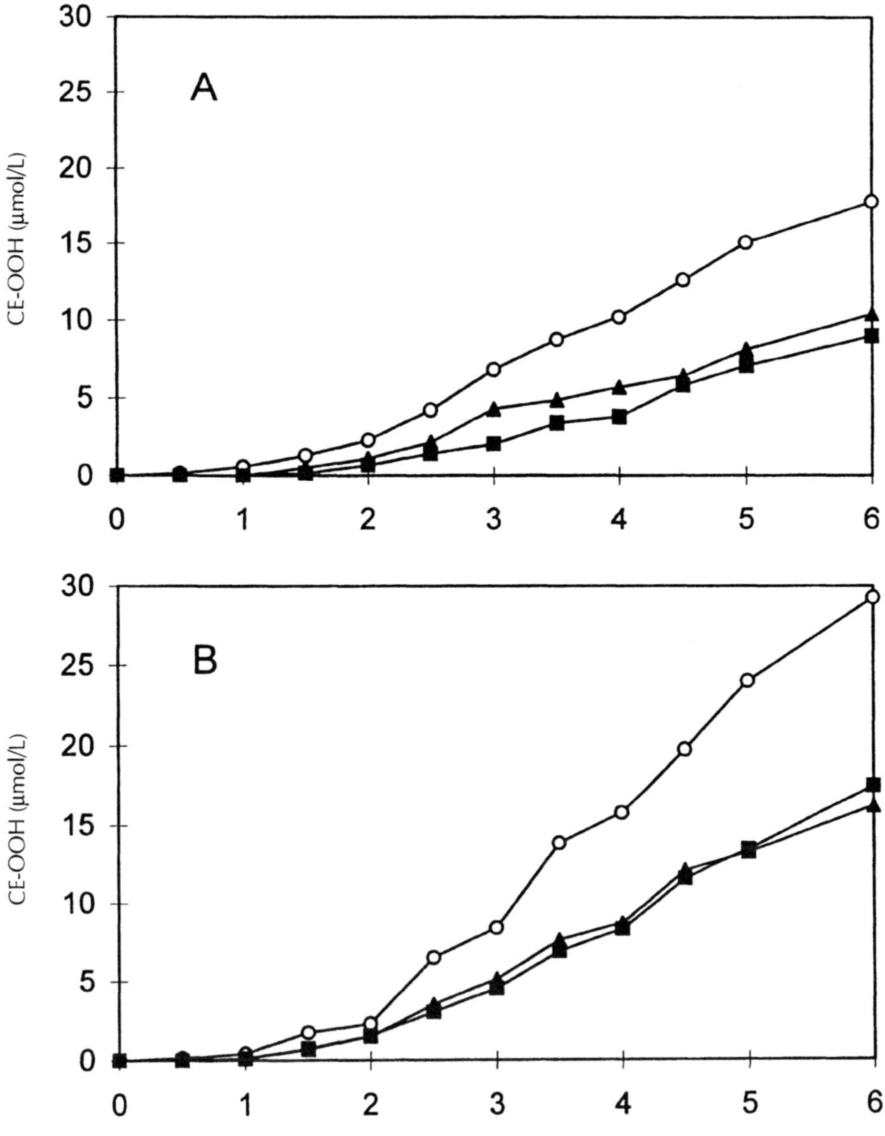

Fig. 18.2 Accumulation of cholesteryl esters hydroperoxides (CE-OOH) in $CuSO_4$-induced rat plasma oxidation. (A) Plasma taken 1 h after administration of water (control) or (-)-epicatechin (EC) to rats. (B) Plasma taken 6 h after administration of water or EC to rats. Pooled plasma from four rat groups was diluted fourfold with phosphate-buffered saline (PBS) and incubated with 100 μmol/L $CuSO_4$ at 37°C for 6 h. (○) Control group; (▲) group administered 50 mg EC/kg body weight; (■) group administered 250 EC mg/kg body weight. *Source*: da Silva *et al.* (1998).

rise as a result of the action of adipose tissue triacylglycerol lipase in response to an elevated glucagon/insulin ratio (Eaton *et al.* 1996). Nevertheless, the influence of other components cannot be excluded.

The increase to 250 mg/kg of the EC dose administered neither elevated the plasma resistance to copper ion oxidation nor lowered the α-tocopherol consumption, indicating that there was a saturation effect at the dose of 50 mg/kg. In a separate experiment, standard EC was added to a control plasma sample (taken 1 and 6 h after water administration) and oxidized with copper ion for 6 h. To achieve the same CE-OOH levels as in the case of oxidation of plasmas taken from EC-treated rats, i.e., ~10 and 17 μmol/L (Fig. 18.2), it was necessary to add a standard EC at levels as high as >80 and 8–40 μmol/L, respectively. As shown in Table 18.1, the content of free EC in each of the rat plasmas administered EC was much lower than that of the standard EC added *in vitro* to the control plasmas as described above. Moreover, no free EC was detected in the plasma obtained 6 h after 50 mg/kg EC administration. Therefore, it seems that elevation of rat plasma antioxidative resistance by EC administration is derived not only from free EC but also from EC metabolites or other phenomena.

Oxidation of Plasma Lipids by Water-Soluble Radical Generators

Pooled plasma obtained after oral administration of EC in water solution or water as described above was oxidized in the presence of 5 mmol/L AAPH. During oxidation, the formation of CE-OOH was measured (Fig. 18.3). For both control plasmas, taken at 1 and 6 h after water administration, a lag phase of 2 h was observed. (-)-Epicatechin administration to rats at doses of 50 and 250 mg/kg resulted in an increase of the lag phase of up to 3 and 5 h, respectively, in the oxidation of plasmas taken at 1 h after administration. In contrast, this lag phase increase was not observed in the oxidation of plasmas taken 6 h after EC administration of either 50 or 250 mg/kg. As in copper ion oxidation and in AAPH-induced oxidation, to achieve an inhibition of CE-OOH formation in control plasmas similar to that in plasmas from EC-treated rats, it was necessary to add standard EC at levels from 8 to 80 μmol/L. In each case, the EC amount added to control plasmas significantly exceeded the amount of free EC found in the respective plasmas of EC-treated rats. Again, this indicates that the inhibition of CE-OOH accumulation in AAPH-exposed plasma was also derived not only from free EC but also from EC metabolites or other phenomena.

Consumption of (-)-Epicatechin During Plasma Autoxidation

To verify which class of EC metabolites is the one engaged in antioxidative action against lipid peroxidation, total plasma (not diluted with PBS) from control and EC-treated (250 mg/kg) rats was exposed to air at 37°C in a shaking water bath for 4 d. The CE-OOH formation was significantly inhibited in the plasma from EC-treated rats compared with plasma from control rats (Fig. 18.4). Measurement of

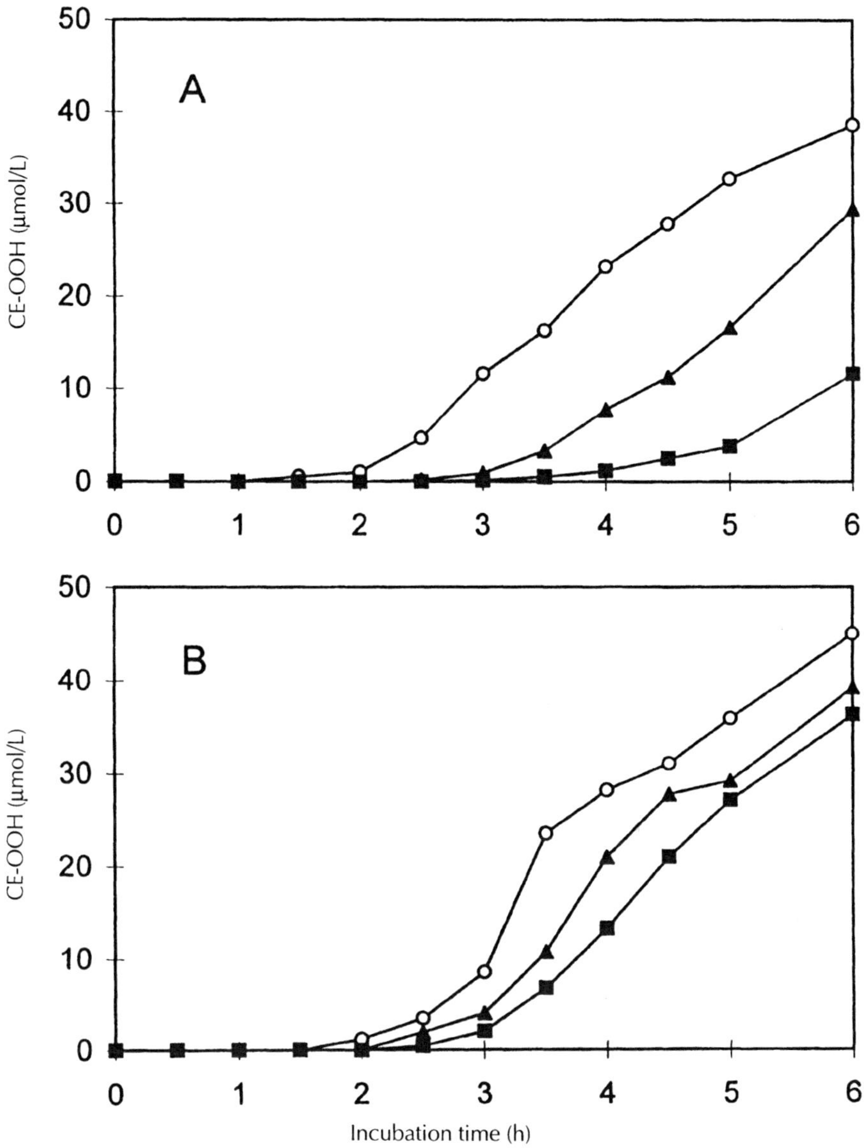

Fig. 18.3 Accumulation of cholesteryl esters hydroperoxides (CE-OOH) in radical generator [2,2′-azobis(2-amidinopropane) dihydrochloride (AAPH)]-induced rat plasma oxidation. (**A**) Plasma taken 1 h after administration of water (control) or (-)-epicatechin (EC) to rats. (B) Plasma taken 6 h after administration of water or EC to rats. Pooled plasma from four rats groups was diluted fourfold with phosphate-buffered saline (PBS) and incubated with 5 mmol/L AAPH at 37°C for 6 h. (○) Control group; (▲) group administered 50 mg EC/kg body weight; (■) group administered 250 EC mg/kg body weight. *Source*: de Silva *et al.* (1998).

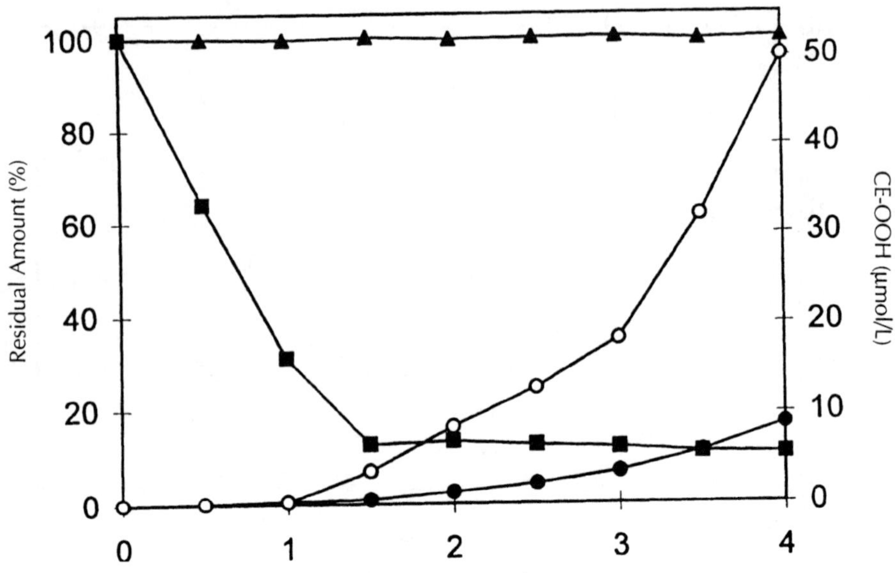

Fig. 18.4 Accumulation of cholesteryl esters hydroperoxides (CE-OOH) and consumption of (-)-epicatechin (EC) metabolites during autoxidation of rat plasma. Plasma taken 1 h after administration of water (control) or EC (250 mg/kg body weight) was pooled from four rats and autoxidized by incubation at 37°C for 4 d. (○) CE-OOH in plasma of control group; (●) CE- OOH in plasma of EC group; (■) percentage decrease of nonmethylated EC, (▲) percentage decrease of methylated EC. *Source*: de Silva *et al.* (1998).

the concentration of EC metabolites in plasma from EC-treated rats during this experiment showed that the level of nonmethylated EC metabolites decreased by ~90% in the first 1.5 d, whereas methyl-EC metabolites showed a remarkable stability. This indicates a relatively high activity of nonmethylated EC metabolites compared with the methylated metabolites in this oxidation model.

Discussion

There are a number of reports demonstrating *in vitro* catechin antioxidative activity in oxidation of plasma and its lipoproteins (da Silva *et al.* 1998, Frankel *et al.* 1993, Miura *et al.* 1995). Nevertheless, the bioavailability of catechins should be taken into account when their *in vivo* antioxidative action is estimated from *in vitro* model experiments.

Conjugation with glucuronic or sulfuric acid and methylation are the common steps in flavonoid metabolism (Cook and Samman 1996). In this study, most of EC absorbed by rats was found to be converted to conjugates (Fig. 18.1). Because the fivefold EC dose increase resulted in an approximately fivefold increase in free EC and only a twofold increase in EC conjugates in rat plasma 1 h after administration (Table 18.1), the presence of nonconjugated EC in rat plasma is most likely due to the overdosing of animals with EC. This also suggests that absorption of high amounts of EC is not dose dependent and that rats possess high but limited ability to conjugate orally administered EC.

The plasma oxidation experiments described here clearly show that oral administration of EC enhances its oxidative resistance and that the plasma level of nonmetabolized EC is too low to attribute this effect to the presence of free EC only. It suggests that EC exerts some antioxidative activity even after its metabolic conversion. Bors *et al.* (1990) reported that the dihydroxy structure at the 3′- and 4′-positions of the flavonoid molecule, including EC, is the radical target site for all flavonoids with a saturated C_2-C_3 bond. Moreover, it was reported that when (+)-catechin is metabolized, methylation of the hydroxyl group at the 3′-position takes place (Hackett *et al.* 1982). Therefore, it is unlikely that methyl-EC conjugates significantly contribute to the enhancement of oxidative resistance of EC-treated rat plasma. The fact that methyl-EC metabolites were stable during the autoxidation of EC-treated rat plasma (Fig. 18.4) also strongly supports the idea that 3′ methylation of EC is a crucial metabolic reaction, depressing EC antioxidative activity. It is likely that a nonmethylated EC conjugate or conjugates is responsible for the plasma oxidative resistance. In addition, a part of this resistance could also be derived from the metal-chelating ability of the EC nonmethylated derivatives (Cotelle *et al.* 1992).

The maximum concentration of methyl-EC metabolites was delayed for ~1 h compared with nonmethylated metabolites (Fig. 18.1), indicating that methylation takes place in the later stage of metabolism. Therefore, the demonstrated decrease in antioxidative potential of plasma from EC-treated rats with the lapse of time (Fig. 18.4) is due not only to the excretion of EC metabolites from the organism but also to the intensive EC methylation process.

Despite the effective metabolism of orally administered EC, a part of its antioxidative potential remains in its metabolites; this may increase the plasma antioxidative defense. It was reported that tea polyphenols administered orally to rats were effective in the suppression of plasma lipid peroxidation (Yoshino *et al.* 1994). Also, there is evidence that ingestion of tea by humans elevates their plasma antioxidative capacity (Serafini *et al.* 1996).

In conclusion, orally administered EC decreased the susceptibility of rat blood plasma to oxidation. Nevertheless, although EC is converted mainly to metabolites by glucuronization, sulfation, and methylation, it may still contribute to the epidemiologically discovered beneficial effects of tea consumption, at least in part, by its antioxidative action.

Acknowledgments

The research of M.K. Piskula and E.L. da Silva was done at the National Food Research Institute, Tsukuba, Japan, where they worked as Science and Technology Agency research fellows.

References

Afanas'ev, I.B., A.I. Dorozhko, A.V. Brodskii, V.A. Kostyuk, and A.I. Potapovitch, Chelating and Free Radical Scavenging Mechanism of Inhibitory Action of Rutin and Quercetin in Lipid Peroxidation, *Biochem. Pharmacol. 38*: 1763–1769 (1989).

Arai, H., J. Terao, D.S.P. Abdalla, T. Suzuki, and K. Takama, Coulometric Detection in High-Performance Liquid Chromatographic Analysis of Cholesteryl Ester Hydroperoxides, *Free Radic. Biol. Med. 20*: 365–371 (1996).

Bors, W., W. Heller, C. Michel, and M. Saran, Flavonoids as Antioxidants, Determination of Radical-Scavenging Efficiencies, *Methods Enzymol. 186:* 343–355.

Cook, N.C., and S. Samman, Flavonoids—Chemistry, Metabolism, Cardioprotective Effects, and Dietary Sources, *J. Nutr. Biochem. 7*: 66–76 (1996).

Cotelle, N., J.L. Bernier, J.P. Hénichart, J.P. Catteau, E. Gaydou, and J.C. Wallet, Scavenger and Antioxidant Properties of Ten Synthetic Flavones, *Free Radic. Biol. Med. 13*: 211–219 (1992).

da Silva, E.L., M.K. Piskula, and J. Terao, Enhancement of Antioxidative Ability of Rat Plasma by Oral Administration of (-)-Epicatechin, *Free Radic. Biol. Med. 24:* 1209–1216 (1998).

deRijke, Y.B., P.N.M. Demacker, N.A. Assen, L.A. Sloots, M.B. Katan, and A.F.H. Stalenhoef, (1996) Red Wine Consumption Does Not Affect Oxidizability of Low-Density Lipoproteins in Volunteers, *Am. J. Clin. Nutr. 63*: 329–334 (1996).

Eaton, S., K. Bartlett, and M. Pourfarzan, Mammalian Mitochondrial β-Oxidation, *Biochem. J. 320*: 345–357 (1996).

Frankel, E.N., J. Kanner, J.B. German, E. Parks, and J.E. Kinsella, Inhibition of Oxidation of Human Low-Density Lipoprotein by Phenolic Substances in Red Wine, *Lancet 341*: 454–457 (1993).

Fuhr, U., and A.L. Kummert, The Fate of Naringin in Humans: A Key to Grapefruit Juice-Drug Interactions? *Clin. Pharmacol. Ther. 58*: 365–373 (1995).

Fuhrman, B., A. Lavy, and M. Aviram, Consumption of Red Wine with Meals Reduces the Susceptibility of Human Plasma and Low-Density Lipoprotein to Lipid Peroxidation, *Am. J. Clin. Nutr. 61*: 549–554 (1995).

Hackett, A.M., and L.A. Griffiths, The Metabolism and Excretion of 3-Palmitoyl-(+)-Catechin in the Rat, *Xenobiotica 12*: 447–456 (1982).

Hackett, A.M., I.C. Shaw, and L.A. Griffiths, 3′-*O*-Methyl-(+)-Catechin Glucuronide and 3′-O-methyl-(+)-Catechin Sulphate: New Urinary Metabolites of (+)-Catechin in the Rat and the Marmoset, *Experientia 38*: 338–340 (1982).

Hertog, M., D. Kromhout, C. Aravanis, H. Blackburn, R. Buzina, F. Fidanza, S. Giampaoli, A. Jansen, A. Menotti, S. Nedeljkovic, M. Pekkarinen, B.S. Simic, H. Toshoma, E.J.M. Feskens, P.C.H. Hollman, and M. Katan, Flavonoid Intake and Long-Term Risk of Coronary Heart Disease and Cancer in the Seven Countries Study, *Arch. Intern. Med. 155*: 381–386 (1995).

Hollman, P.C.H., M. v.d. Gaag, M.J.B. Mengelers, J.M.P. van Trijp, J.H.M. deVries, and M.B. Katan, Absorption and Disposition Kinetics of the Dietary Antioxidant Quercetin in Man, *Free Radic. Biol. Med. 21*: 703–707 (1996).

Knekt, P., R. Järvinen, A. Reunanen, and J. Maatela, Flavonoid Intake and Coronary Mortality in Finland: A Cohort Study, *Br. Med. J. 312:* 478–481 (1996).

Manach, C., C. Morand, C. Demigné, O. Texier, F. Régérat, and C. Rémésy, Bioavailability of Rutin and Quercetin in Rats, *FEBS Lett. 409*: 12–16 (1997).

Miura, S., J. Watanabe, M. Sano, T. Tomita, T. Osawa, Y. Hara, and I. Tomita, Effects of Various Antioxidants on the Cu^{2+}-Mediated Oxidative Modification of Low Density Lipoprotein, *Biol. Pharm. Bull. 18*: 1–4 (1995).

Okushio, K., N. Matsumoto, T. Kohri, M. Suzuki, F. Nanjo, and Y. Hara, Absorption of Tea Catechins into Rat Portal Vein, *Biol. Pharm. Bull. 19*: 326–329 (1996).

Piskula, M.K., and J. Terao, Accumulation of (-)-Epicatechin Metabolites in Rat Plasma After Oral Administration and Distribution of Conjugation Enzymes in Rat Tissues, *J. Nutr. 128:* 1172–1178 (1998).

Renaud, S., and M. deLorgeril, Wine, Alcohol, Platelets, and the French Paradox for Coronary Heart Disease, *Lancet 339*: 1523–1526 (1992).

Rice-Evans, C.A., N.J. Miller, P.G. Bolwell, P.M. Bramley, and J.B. Pridham, The Relative Antioxidant Activities of Plant-Derived Polyphenolic Flavonoids, *Free Radic. Res. 22*: 375–383 (1995).

Serafini, M., A. Ghiselli, and A. Ferro-Luzzi, *In vivo* Antioxidant Effect of Green Tea and Black Tea in Man, *Eur. J. Clin. Nutr. 50*: 28–32 (1996).

Terao, J., and M.K. Piskula, Flavonoids as Inhibitors of Lipid Peroxidation in Membranes, in *Flavonoids in Health and Disease,* edited by C. Rice-Evans, and L. Packer, Marcel Dekker, New York, 1998, pp. 277–293.

Terao, J., M. Piskula, and Q. Yao, Protective Effect of Epicatechin, Epicatechin Gallate and Quercetin on Lipid Peroxidation in Phospholipid Bilayers, *Arch. Biochem. Biophys. 308*: 278–284 (1994).

Unno, T., K. Kondo, H. Itakura, and T. Takeo, Analysis of (-)-Epigallocatechin Gallate in Human Serum Obtained After Ingesting Green Tea, *Biosci. Biotechnol. Biochem. 60*: 2066–2068 (1996).

van Acker, S.A.B.E., D.-J. van den Berg, M.N.J.L. Tromp, D.H. Griffioen, W.P. van Bennekom, W.J.F. van der Vijgh, and A. Bast, Structural Aspects of Antioxidant Activity of Flavonoids, *Free Radic. Biol. Med. 20*: 331–342 (1996).

Yoshino, K., I. Tomita, M. Sano, I. Oguni, Y. Hara, and M. Nakano, Effects of Long-Term Dietary Supplement of Tea Polyphenols on Lipid Peroxide Level in Rats, *Age 17*: 79–85 (1994).

Chapter 19

Antibacterial Actions of Tea Polyphenols and Their Practical Applications in Humans

Yukihiko Hara

Food Research Laboratories, Mitsui Norin Company, Ltd., Fujieda, 426-01 Japan

Introduction

Bacteriostatic actions of tea brew have been known since ancient times, primarily on a folklore basis. Over the last 20 years, as a result of the introduction of high-performance liquid chromatography (HPLC) isolation techniques, it has become possible to fractionate and purify the components of the soluble solids of tea, particularly those of a polyphenolic nature. Crude catechins from green tea, crude theaflavins from black tea, and the individual components thereof that were obtained were tested *in vitro* on several bacterial species. As a result, the bacteriostatic/bactericidal potency of tea infusion was ascribed to these polyphenolic components in tea. However, no detailed experiments have been conducted on the bacteriostatic effects of tea polyphenols in human cohorts. By using oral administration of crude tea catechins to human carriers, we have been attempting the suppression of *Helicobacter pylori,* a bacterium that lives in the gastric mucosa and is associated with gastritis and ulcer formation. The influence of catechins on the intestinal microflora and the resultant fecal conditions were also tested in humans by administering tea catechins to bed-ridden in-patients who were all receiving liquid alimentation.

Minimum Inhibitory Concentrations (MIC) of Tea Polyphenols on Foodborne Pathogenic Bacteria

The MIC of tea polyphenols on representative foodborne pathogenic bacteria were determined by Hara and Ishigami (1989) and others as shown in Table 19.1. The concentration of tea polyphenols in a normal brew is ~500–1000 ppm. It is apparent from these results that tea polyphenols, at concentrations equivalent to or below that of daily brew, are bactericidal on *Staphylococcus aureus,* three *Vibrio* strains (*fluvialis, paraheamolyticus,* and *metschnikovii*), *Clostridium perfringens, Bacillus cereus, Plesiomonas shigelloides, Aeromonas sobria* and *A. hydrophila.* As can be seen from the table, other species such as *E. coli, Salmonellas, Campilobaters,* and *Yersinia* were affected only after hours of interaction with tea polyphenols; therefore the MIC showed higher values. Among catechins, gallated catechins (epicatechin gallate, epigallocatechin gallate) showed lower MIC than free catechins (epicatechin, epigallocatechin). There was little difference in the MIC among theaflavins.

TABLE 19.1
Minimum Inhibitory Concentrations (MIC) of Tea Polyphenols Against Pathogenic and Enteric Bacteria[a]

		MIC (ppm)									
Bacteria	MIC (ppm)	Crude catechins	EC	ECg	EGC	EGCg	Crude theaflavins	TF1	TF2A	TF2B	TF3
Staphylococcus aureus	IAM1011	450	>800	800	150	250	500	500	400	150	200
Vibrio fluvialis	JCM3752	200	800	300	300	200	400	500	500	400	500
V. parahaemolyticus	IFO12711	200	800	500	300	200	300	300	500	400	500
V. metshnikovii	IAM1039	500	>1000	>1000	500	1000	300	660	800	700	800
Clostridium perfringens	JCM3816	400	>1000	400	1000	300	200	500	600	500	400
C. botulinum	A,B mix	100	>1000	200	300	100	200	150	250	150	200
Bacillus cereus	JCM2152	600	>1000	600	>1000	600	500	600	1000	1000	1000
Plesiomonas shigelloides	IID No. 3	<100	700	100	200	<100	100	200	200	200	100
Aeromonas sobria	JCM2139	400	>1000	700	400	300	300	600	600	400	500
A. hydrophila subsp. *Hydrophila*	JCM1072	>1000	>1000	>1000	>1000	>1000	>1000	>1000	>1000	>1000	>1000
Escherichia coli	IID952	>1000	>1000	>1000	>1000	>1000	>1000	>1000	>1000	>1000	>1000
E. coli	IID954	>1000	>1000	>1000	>1000	>1000	>1000	>1000	>1000	>1000	>1000
Salmonella enteritidis	IFO3313	>1000	>1000	>1000	>1000	>1000	>1000	>1000	>1000	>1000	>1000
S. typhimurium	IFO12529	>1000	>1000	>1000	>1000	>1000	>1000	>1000	>1000	>1000	>1000
Campylobacter jejuni	JCM2013	>1000	>1000	>1000	>1000	>1000	>1000	>1000	>1000	>1000	>1000
C. coli	JCM2529	>1000	>1000	>1000	>1000	>1000	>1000	>1000	>1000	>1000	>1000
Yersinia enterocolitica	IID981	>1000	>1000	>1000	>1000	>1000	>1000	>1000	>1000	>1000	>1000
Lactobacillus brevis subsp. *Gravesensis*	JCM1102	>1000	>1000	>1000	>1000	>1000	>1000	>1000	>1000	>1000	>1000
L. brevis subsp. *brevis*	JCM1059	>1000	>1000	>1000	>1000	>1000	>1000	>1000	>1000	>1000	>1000
L. brevis subsp. *otakiensis*	JCM1183	>1000	>1000	>1000	>1000	>1000	>1000	>1000	>1000	>1000	>1000
Bifidobacterium bifidum	JCM1255	>1000	>1000	>1000	>1000	>1000	>1000	>1000	>1000	>1000	>1000
B. adolescentis	JCM1275	>1000	>1000	>1000	>1000	>1000	>1000	>1000	>1000	>1000	>1000
B. longum	JCM1217	>1000	>1000	>1000	>1000	>1000	>1000	>1000	>1000	>1000	>1000

[a]EC, (-)-epicatechin; ECg, (-)-epicatechin gallate; EGC, (-)-epigallocatechin; EGCg, (-)-epigallocatechin gallate; TF1, free theaflavin; TF2A, theaflavin monogallate A; TF2B, theaflavin monogallate B; TF3, theaflavin digallate.

The Effect of Catechin Administration on *H. Pylori* in the Human Stomach

The MIC of tea polyphenols on *H. pylori* is shown in Table 19.2. In this case too, gallated catechins exhibited a better bactericidal effect than free catechins (50 vs. 200 ppm). On the basis of these data, the following cohort study was conducted by Yamada *et al.* (1997).

Methods

Subjects and Catechin Dosage

In-patients and out-patients in Hamamatsu Clinical Center, who were confirmed to be carriers of *H. pylori*, were given orally 700 mg/d of catechins. Polyphenon 60 (155 mg), a crude catechin powder with 65% total catechin content containing 100 mg of catechins, was placed in a capsule. Subjects were requested to take two capsules after each meal plus one capsule before going to bed, a total of 7 capsules/d. Subjects (n = 34) continued this catechin intake for a period of 1 mo.

Determination of H. Pylori *in the Stomach*

^{13}C-Urea was administered to each subject before and after the completion of the catechin intake; the ^{13}C concentration in the exhaled CO_2 was determined by a conventional method. *H. pylori* is known to produce ammonia, metabolizing urea, in order to survive in the acidic conditions of the mucous membrane. Carbon dioxide thus produced, along with ammonia, will be absorbed into the body and exhaled eventually as $^{13}CO_2$.

As shown in Figure 19.1, the concentration of ^{13}C decreased in most of the 34 subjects, indicating a decrease in the number of *H. pylori.* In six subjects, ^{13}C was noted to be in the zone in which no survival of the bacterium is warranted. In these six cases, the elimination of *H. pylori* was confirmed, and even in the month after the termination of the catechin treatment, the breath test showed no recurrence of *H. pylori* activity. In the remaining subjects, the ^{13}C concentration, which had

TABLE 19.2
Minimum Inhibitory Concentration of (µg/mL) Tea Catechin Against *Helicobacter pylori*[a]

Strain	(-)-EC	(-)-EGC	(-)-ECg	(-)-EGCg	P-60	Crude TF
ATCC43526	>200	200	50	50	100	50
ATCC43629	>200	>200	50	50	100	100
ATCC43579	>200	200	50	50	100	50

[a](-)-EC, (-)-epicatechin; (-)-EGC, (-)-epigallocatechin; (-)-ECg, (-)-epicatechin gallate; (-)-EGCg, (-)-epigallocatechin gallate; P-60, Polyphenon 60; crude TF, crude theaflavin.

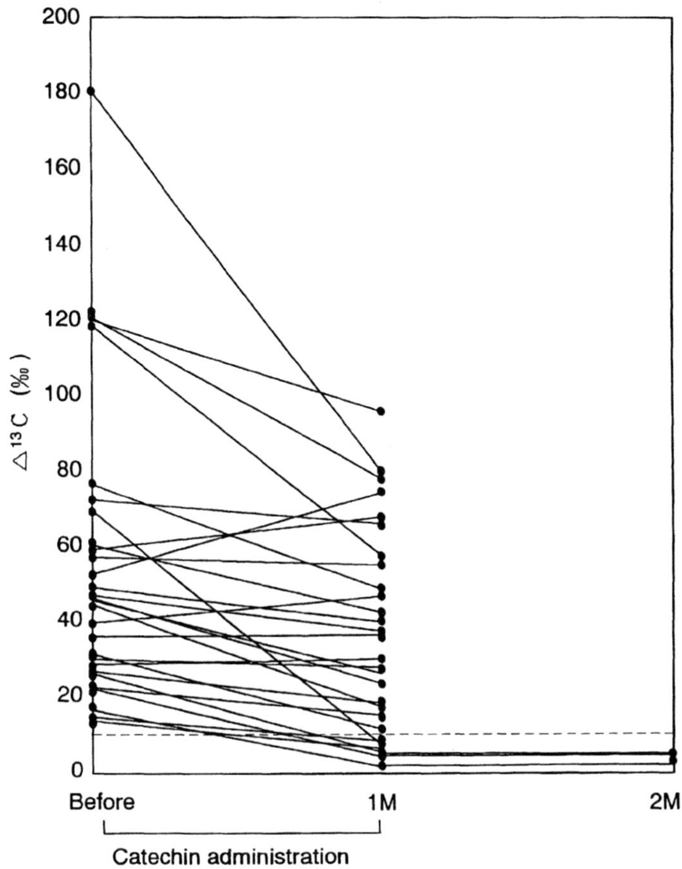

Fig. 19.1. Changes in ^{13}C-urea breath test results after 1 mo of catechin treatment (1M) and 1 mo (2M) after the termination of treatment.

decreased after a month of catechin treatment, increased again in the month after termination of the treatment, demonstrating the failure to eliminate the bacterium.

A wider cohort study is planned as a 5-y project, funded by the Ministry of Agriculture, Forestry and Fishery, covering >1000 subjects who will be classified in various ways. The project aims to investigate the efficacy of catechins in combination with antibiotics and to confirm the suppressive effect of catechins against the revival of the bacterium after elimination by antibiotics. The most effective forms of capsules and the timing of administration will be studied as well.

Improvement of Bowel Condition

Tea polyphenols were confirmed to suppress the growth of the foodborne pathogenic bacteria, whereas they do not tend to affect the lactic acid bacteria. These properties of tea polyphenols suggest the possibility that tea polyphenol ingestion will modulate bowel conditions in a very favorable way. Thus, the following experiments were conducted by Goto *et al.* (1998).

Elderly inpatients (n = 15) in health care centers, who were receiving the same gastroenteral liquid alimentation (1000 kcal/d) *via* nasogastric tubes, received a daily supplementation of 300 mg of tea catechins for a period of 3 wk.

Feces were collected on d 0 (before catechin administration), d 7, 14, and 21 of administration and d 7 postadministration. The following fecal parameters were measured immediately after the collection: microfloral population, pH, moisture content, ammonia, sulfide, other metabolic products, and organic acids.

Comparing the values before administration with those during tea catechin administration, the levels of lactobacilli and bifidobacteria increased significantly, whereas those of Enterobacteriaceae decreased significantly (Fig. 19.2). The detection rate of clostridia tended to decrease during the administration period. The levels of Bacteroidaceae and eubacteria decreased significantly. Fecal concentrations of ammonia, sulfide, phenol, cresol, ethylphenol, indole, and skatol decreased significantly. The pH values decreased during the administration. The content of fecal short-chain fatty acids increased significantly. Some of these data are graphed in Figures 19.3 and 19.4. On d 7 after the termination of catechin administration, the data tended to reverse

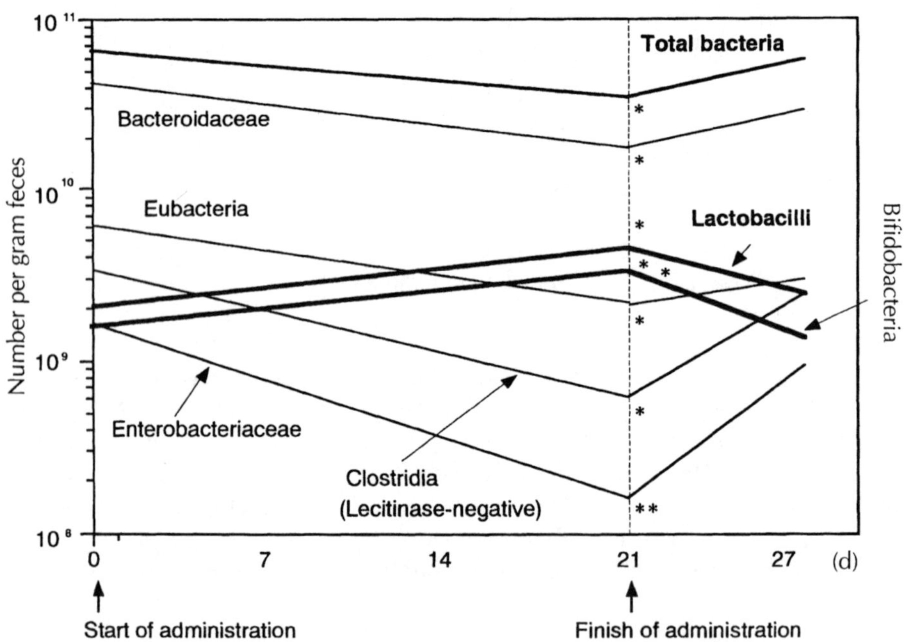

Fig. 19.2. Effect of green tea catechin (GTC) administration on fecal flora of 15 human volunteers (460 mg/d). Data are expressed as means ± SD. Asterisks indicate significant difference (*$p < 0.05$, **$p < 0.01$) from value at d 0 (before the administration).

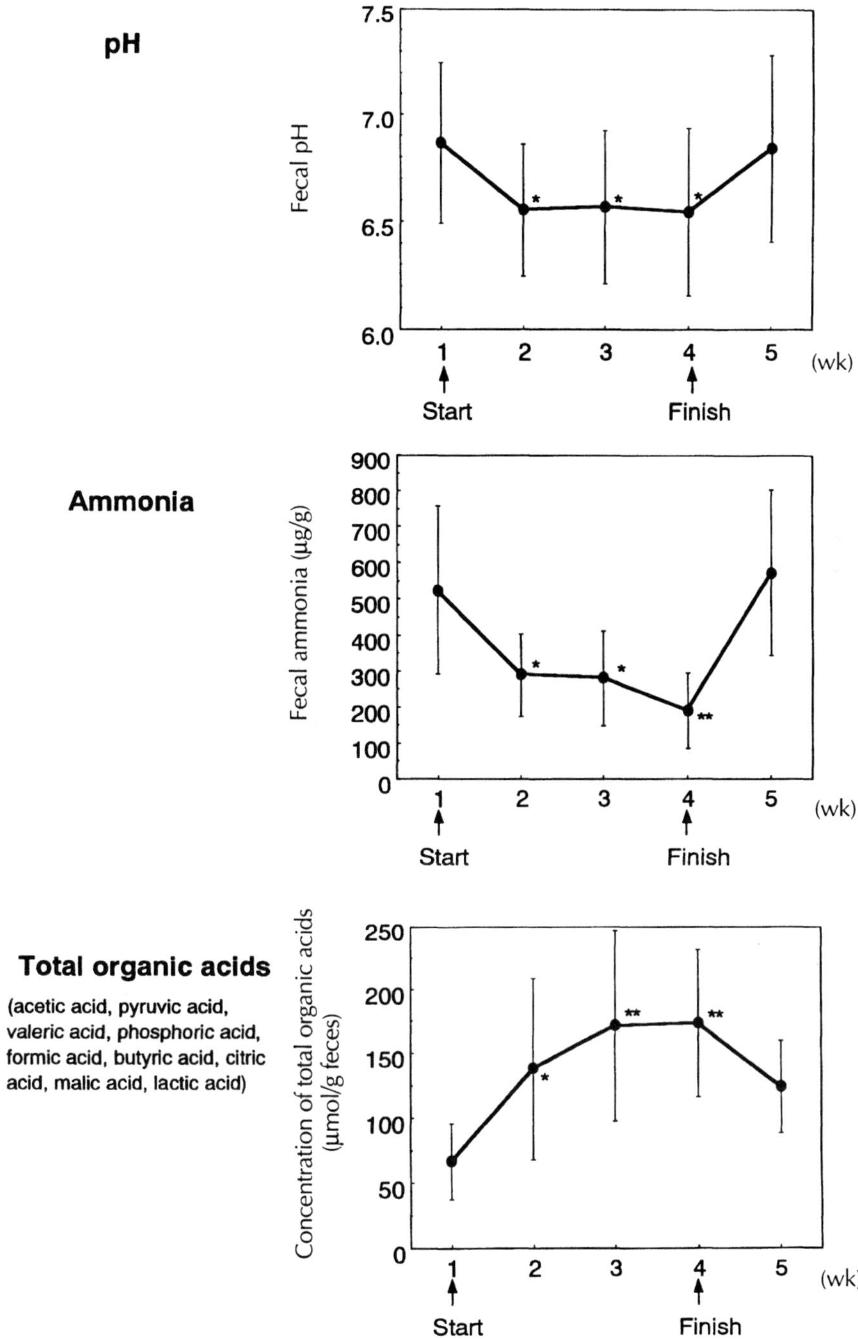

Fig. 19.3. Effects of green tea catechin (GTC; 300 mg catechins/d) on fecal composition.

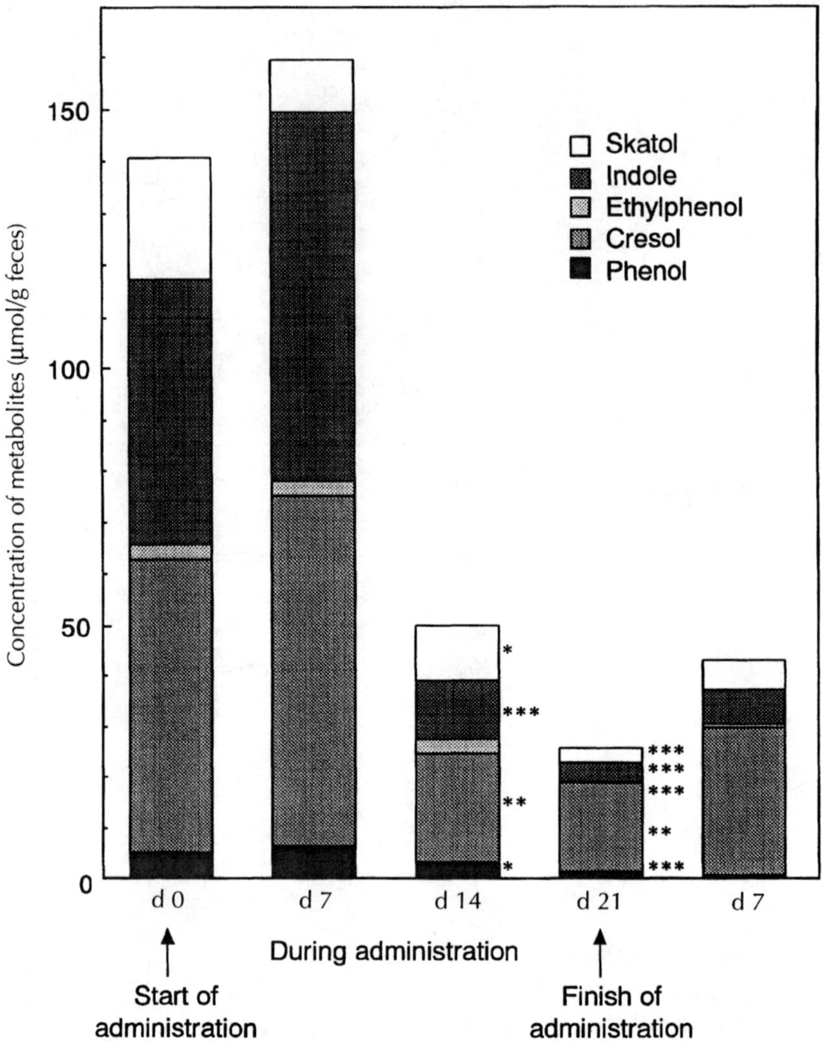

Fig. 19.4. Effect of tea catechin administration on fecal metabolites of human volunteers. Values are means ± SD. Asterisks indicate significant difference (*$p < 0.05$, **$p < 0.01$, ***$p < 0.001$) from the value at d 0 (before the administration).

toward the levels of d 0 (before catechin administration).

Thus, after administration of tea catechins equivalent to 5–6 cups of green tea brew a day to human subjects, significant changes in fecal flora and fecal odorous parameters were noted. These changes indicated marked improvement of fecal conditions and the reduction of fecal odors.

Discussion

Although tea polyphenols have a strong bactericidal effect against pathogenic bacteria, quite curiously their effect against the so-called good bacteria in the bowel, i.e., lactic acid bacteria, is only slight. These *in vitro* data prompted us to investigate how the oral intake of crude catechins will affect the bacteria in the digestive tract *in vivo*. In a preliminary cohort study, daily administration of 700 mg of catechins to the carriers of *H. pylori* tended to decrease the number of the bacteria in the stomach, and elimination of the bacteria was even observed in several cases. However, the majority of the bacteria survived in the stomach; thus, in the forthcoming cohort study and relevant experiments, we must investigate ways in which catechins will interact with the bacteria whose habitat is deep inside the mucous of the membrane. Because tea catechins modulate bowel conditions very favorably and decrease fecal malodor, they may be of great benefit in the near future when the number of bed-ridden aged people in nursing homes or other facilities will increase in most advanced countries. Some unpleasantness in the work of nursing staff may be mitigated if fecal odor is diminished. Further studies with black tea polyphenols and investigation into the ultimate conditions for the improvement of the microflora of ordinary people are warranted.

References

Goto, K., S. Kanaya, T. Nishikawa, H. Hara, A. Terada, T. Ishigami, and Y. Hara, The Influence of Tea Catechins on Fecal Flora of Elderly Residents in Long-Term Care Facilities, *Ann. Long-Term Care 6*: 1–6 (1998).

Hara, Y., and T. Ishigami, Antibacterial Activities of Tea Polyphenols Against Foodborne Pathogenic Bacteria, *Nippon Shokuhin Kogyo Gakkaishi 36*: 996–999 (1989a).

Hara, Y., and M. Watanabe, Antibacterial Activity of Tea Polyphenols Against *Clostridium botulinum*, *Nippon Shokuhin Kogyo Gakkaishi 36*: 951–955 (1989b).

Kono, K., I. Tatara, S. Takeda, K. Arakawa, and Y. Hara, Antibacterial Activity of Epigallocatechin Gallate Against Methicillin-Resistant *Staphylococcus aureus*, *J. Jpn. Assoc. Infect. Dis. 68*: 1518–1522 (1994).

Toda, M., S. Okubo, H. Ikigai, T. Suzuki, Y. Suzuki, Y. Hara, and T. Shimamura, The Protective Activity of Tea Catechins Against Experimental Infection by *Vibrio cholerae* O1, *Microbiol. Immunol. 36:* 999–1001 (1992).

Yamada, M., B. Murohisa, I. Oguni, N. Harada, and Y. Hara, Effect of Tea Polyphenols Against *Helicobacter pylori,* Abstract of Papers Part T, 213th ACS National Meeting, American Chemical Society, San Francisco, CA, April 13–17, 1997.

Chapter 20

Free Radical–Scavenging Action of Catechin and Related Compounds in Homogeneous and Micellar Solutions

Kazuo Mukai, Yu Kanesaki, Yoshifumi Egawa, and Shin-ichi Nagaoka

Department of Chemistry, Faculty of Science, Ehime University, Matsuyama 790-8577, Japan

Introduction

Flavonoids are well known as polyphenols widely distributed in foods and plants (Larson 1988, Shahidi and Wanasundara 1992). Catechins are a group of flavonoids that have attracted much attention because of their relatively high antioxidant capacity, reducing a variety of toxic effects caused by reactive oxygen metabolites (Leanderson *et al.* 1997, Lotito and Fraga 1998). Green and black teas contain considerable amounts of catechins (Graham 1992). Catechins are able to inhibit the formation of lipid peroxyl radical species and to act as inhibitors of low-density lipoprotein (LDL) oxidation (De Whalley *et al.* 1990, Jessup *et al.* 1990, Vinson *et al.* 1995). Green tea polyphenols, including catechins, inhibit oxidant-induced DNA strand breakage in cultured lung cells (Leanderson *et al.* 1997) and also prevent human plasma oxidation (Lotito and Fraga 1998). Further, there is now growing evidence that green tea and polyphenols may possess inhibitory effects against cancer (Jankun *et al.* 1997, Mukhtar *et al.* 1994, Yang and Wang 1993).

Several kinetic studies have been performed for the reaction of flavonoids with active free radicals, such as $N_3^{\bullet}$, $HO^{\bullet}$, $O_2^{-\bullet}$, *t*-BuO$^{\bullet}$, and LOO$^{\bullet}$, by using pulse radiolysis technique (Bors and Aaran 1987, Bors *et al.* 1990, Jovanovic *et al.* 1994 and 1995). It has been reported that flavonoids may act as efficient scavengers of these active free radicals. However, the above kinetic studies have been generally performed at relatively high pH (pH 10 and 11.5). Because of the various dissociable phenolic hydroxyl groups in flavonoids, it is expected that the reaction rates between flavonoids and active free radicals show notable pH dependence (Jovanovic *et al.* 1994 and 1995). Therefore, it is necessary to measure the free radical–scavenging rates of flavonoids at various pH in aqueous solution and in organic solvents in order to discuss the antioxidant activity of the flavonoids in foods and biological systems. It was reported that quercetin, epicatechin, and epicatechin gallate can act as chain-breaking antioxidants in the hydroperoxidation of methyl linoleate; their peroxyl radical–scavenging rates (k_{inh}) are 5–20 times smaller than that of α-tocopherol in *n*-hexane/2-propanol (1:1, vol/vol) solution

(Ioku *et al.* 1995, Terao *et al.* 1994). However, it remains unclear how these flavonoids contribute to the inhibition of lipid peroxidation.

Recently, Foti *et al.* (1994) measured the reaction rates between the phenoxyl radical and some natural phenolic antioxidants including α-, γ-, and δ-tocopherols [Reaction (1)]. The phenoxyl radical has been found to be ~100–300 times more reactive than peroxyl radicals (k_{inh}), i.e., $k_1 \sim [(1–3) \times 10^2]k_{inh}$ for the same phenolic antioxidants. The relative reactivities obtained for tocopherols were $\alpha:\gamma:\delta = 100:29:6$ in CH_3CN at 20°C. Such a high reactivity of the phenoxyl radical is interesting because the reactivity of the phenoxyl radical, for example, is expected to be similar to that of the tyrosyl radical as described (Foti *et al.* 1994).

$$PhO^{\bullet} + TocH \xrightarrow{k_1} PhOH + Toc^{\bullet} \tag{1}$$

In a previous work, detailed kinetic studies were performed for the reaction of 2,6-di-*tert*-butyl-4-(4-methoxyphenyl)phenoxyl [$ArO^{\bullet}$ (abbreviated to aroxyl, hereafter)] and 5,7-diisopropyltocopheroxyl ($Toc^{\bullet}$) radicals (see Fig. 20.1) with flavone and its derivatives (chrysin, flavonol, apigenin, rutin, and quercetin) in homogeneous and micellar solutions [Reactions (2) and (3)] (Mukai *et al.* 1997). The second-order rate constants (k_s and k_r) obtained in micellar solution showed notable pH dependence. For instance, rutin is a tetrabasic acid and can exist in five different molecular forms, depending on the pH value. By comparing the k_s values with the mole fraction (f) of each molecular form of rutin, the reaction rates k_{s1} for the undissociated form (RuH_4), k_{s2} for monoanion (RuH_3^-), k_{s3} for dianion (RuH_2^{2-}), and k_{s4} for the trianion (RuH^{3-}) were determined. The reaction rates (k_{si}) increased remarkably with the increasing anionic character of rutin, that is, the electron-donating capacity of rutin. It was found that quercetin and rutin have high activity in vitamin E regeneration.

$$ArO^{\bullet} + \text{flavonoid} \xrightarrow{k_s} ArOH + \text{flavonoid}^{\bullet} \tag{2}$$

$$Toc^{\bullet} + \text{flavonoid} \xrightarrow{k_r} TocH + \text{flavonoid}^{\bullet} \tag{3}$$

In this work, to clarify the structure-activity relationship in the scavenging reaction of free radicals by catechins, we measured the second-order rate constants (k_s and k_r) for the reaction of catechins (catechin and epicatechin) and related compounds (4-methylcatechol and 5-methoxyresorcinol) with $ArO^{\bullet}$ and $Toc^{\bullet}$ radicals in ethanol, 2-propanol/water (5:1, vol/vol), and aqueous Triton X-100 micellar solution (5.0 wt%) [Reactions (2) and (3)]. The rate constants obtained in micellar solution were pH dependent because of the dissociation of various phenolic hydroxyl groups in the catechins. The rates observed (k_s and k_r) were compared

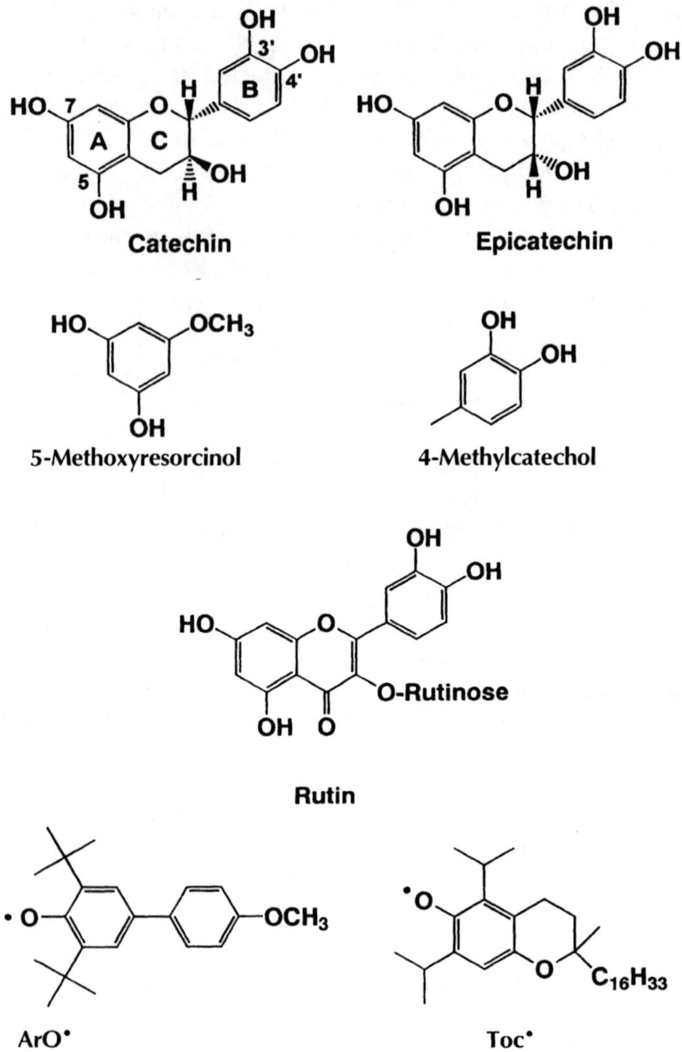

Fig. 20.1. Molecular structures of catechin, epicatechin, 4-methylcatechol, 5- methoxyresorcinol, rutin, the aroxyl radical (ArO•), and 5,7-diisopropyltocopheroxyl (Toc•).

with those of α-tocopherol, ubiquinol-10, and vitamin C (L-ascorbic acid), which are among the best known biological antioxidants.

Experimental Procedures

Catechin, epicatechin, 4-methylcatechol, and 5-methoxyresorcinol are commercially available. Aroxyl radical (ArO•) was prepared according to the method of Rieker and Scheffler (1965). The 5,7-diisopropyltocopheroxyl (Toc•) radical is fairly stable and was prepared by PbO_2 oxidation of the corresponding 5,7-diisopropyltocol in ethanol or in 2-propanol/water (5:1, vol/vol) solution under a nitrogen atmosphere (Mukai *et al.* 1989). In the case of the reaction in micellar solution, the Toc• radical was prepared by the reaction between the ArO• radical and 5,7-diisopropyltocol in aqueous Triton X-100 micellar solution (5.0 wt%) at 25°C, and was reacted immediately with a Triton X-100 micellar solution (5.0 wt%) of the antioxidants (Mukai *et al.* 1991 and 1992).

The kinetic data were obtained with a Unisoku Model RS-450 stopped-flow spectrophotometer (Osaka, Japan) by mixing equal volumes of ethanol [or 2-propanol/water (2-PrOH/H_2O) (5:1, vol/vol), or aqueous 5.0 wt% Triton X-100 micellar] solutions of antioxidants and ArO• (or Toc•) under nitrogen atmosphere. If the reaction rates (k_s and k_r) were slower than 1 $[(\text{mol} \cdot \text{s})/\text{L}]^{-1}$, measurements of the rate constants were performed using a Shimadzu UV-2100S spectrophotometer (Kyoto, Japan). All measurements were performed at 25.0 ± 0.5°C.

Results

Rates (k_r) of Vitamin E Regeneration Reaction with Catechins and Related Compounds in Ethanol and 2-Propanol/Water Solutions

Measurements of the rate constant (k_r) for the reaction of tocopheroxyl radical (Toc•) with catechins and related compounds were performed in ethanol and 2-PrOH/H_2O (5:1, vol/vol) solutions at 25.0°C. Figure 20.2 shows an example of the interaction between Toc• (~0.185 mmol/L) and catechin (1.75 mmol/L) in ethanol

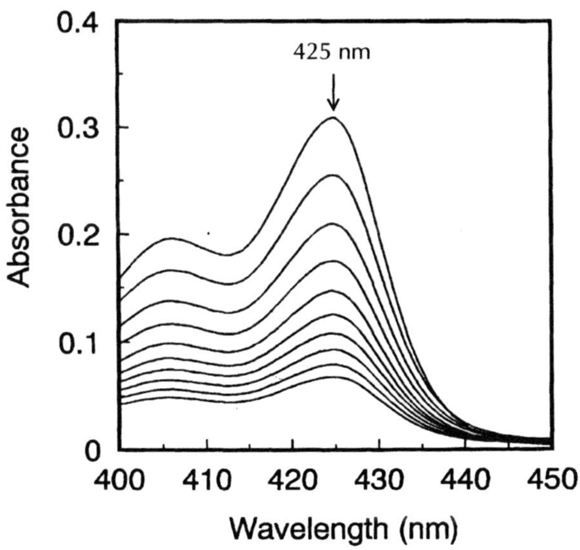

Fig. 20.2. Change in electronic absorption spectrum of tocopheroxyl radical (Toc•) during reaction of Toc• with catechin in ethanol solution at 25.0°C. $[\text{Toc}^\bullet]_{t=0}$ = ~ 0.185 mmol/L and $[\text{catechin}]_{t=0}$ = 1.75 mmol/L. The spectra were recorded at 995-ms intervals. The arrow indicates a decrease in absorbance with time.

solution. The pseudo-first-order rate constant (k_{obsd}) was obtained by following the decrease in absorbance of the Toc$^\bullet$ radical at 425 nm. The details of the experiments were reported previously (Mukai *et al.* 1989, 1991 and 1992). The Toc$^\bullet$ radical shows a slow natural decay in ethanolic solutions. Therefore, the rate constant (k_{obsd}) for Toc$^\bullet$ bleaching is given in Equation (4):

$$k_{obsd} = k_o + k_r\,[\text{Catechin}] \tag{4}$$

where k_o is the rate constant for the natural decay of Toc$^\bullet$ in the medium, and k_r is the second-order rate constant for the reaction of Toc$^\bullet$ with catechin. These parameters are obtained by plotting k_{obsd} against [Catechin] (see Fig. 20.3) (Mukai *et al.* 1997). The second-order rate constant (k_r) obtained for catechin is 1.20×10^2 $[(\text{mol} \cdot \text{s})/\text{L}]^{-1}$.

The k_r values obtained are summarized in Table 20.1, together with those obtained for ubiquinol-10 (Mukai *et al.* 1992) and the sodium ascorbate anion (Na^+AsH^-) (Mukai *et al.* 1997). The reactions of α-tocopheroxyl with vitamin C (ascorbate) (Mukai *et al.* 1991, Packer *et al.* 1979) and ubiquinol-10 (Bowry and Stocker 1993, Mukai *et al.* 1992) yield the normal tocopherol regeneration reaction seen in biomembrane systems. The rate constant of rutin, which has a catechol structure, is also listed in Table 20.1 (Mukai *et al.* 1997). The k_r values of catechin, epicatechin, and related compounds obtained in ethanol are similar to those obtained correspondingly in 2-PrOH/H_2O (5:1, vol/vol) solution. As is clear from the k_r values listed in Table 20.1, the rate of the regeneration reaction of Toc$^\bullet$ with the above antioxidants increases in the order of 5-methoxyresorcinol << rutin <

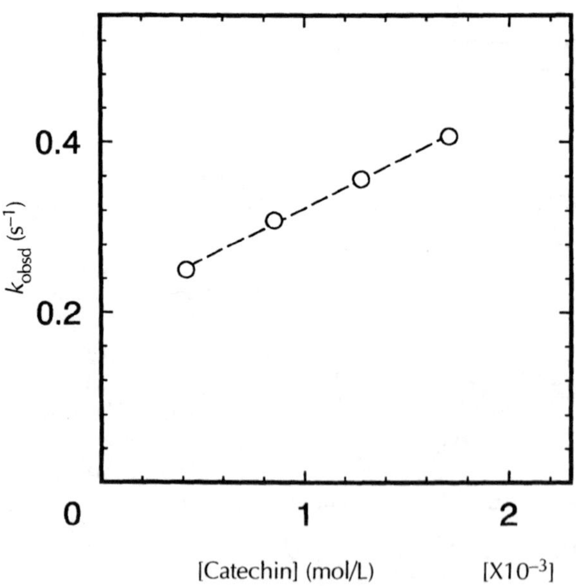

Fig. 20.3. Dependence of pseudo first-order rate constant (k_{obsd}) on concentration of catechin in ethanol.

TABLE 20.1
Second-Order Rate Constants (k_s and k_r) for the Reaction of ArO$^•$ and Toc$^•$ Radicals with Catechin, Epicatechin, and Related Antioxidants in Ethanol, 2-Propanol/Water (2-PrOH/H_2O), and Triton X-100 Micellar Solution[a]

Antioxidant	k_s / [(mol · s)/L]$^{-1}$ Ethanol	(ArO$^•$) pH 7.0	k_r / [(mol · s)/L]$^{-1}$ Ethanol	2-PrOH/H_2O (5:1, vol/vol)	(Toc$^•$) pH 7.0
Catechin	8.26×10	6.57×10^2	1.20×10^2	1.57×10^2	7.31×10^2
Epicatechin	1.34×10^2	7.78×10^2	1.52×10^2	1.91×10^2	1.27×10^3
4-Methyl catechol	2.67×10^2	2.22×10^3	4.41×10^2	8.07×10^2	4.41×10^3
5-Methoxy resorcinol	10^{-2}	2.17	10^{-2}	10^{-2}	4.93×10^{-1}
Rutin[b]	1.42×10	2.28×10^2	1.03×10	7.4	5.48×10^2
BHT[c]	3.50×10	1.08×10^4	1.26×10^2	3.28×10^2	8.62×10^3
α-Tocopherol[c]	5.12×10^3	4.53×10^5	—	—	—
Na^+AsH^-[b]	Insoluble	2.90×10^2	Insoluble	6.32×10^4	2.49×10^3
Ubiquinol-10[c]	5.19×10^3	1.25×10^5	3.64×10^4	5.33×10^4	9.24×10^5
Methyl linoleate[d]	—	—	1.26×10^{-2}	—	2.40×10^{-2}

[a]Conditions: Triton X-100 micellar solution, 5.0 wt%; pH, 7.0; temperature, 25.0°C.
[b]*Source*: Mukai *et al.* (1997).
[c]*Source*: Mukai *et al.* (1992).
[d]*Source*: Nagaoka *et al.* (1990).

catechin < epicatechin < 4-methylcatechol << ubiquinol-10 < sodium ascorbate in both solutions. The rate constants (k_r) of catechin and epicatechin are ~2 orders of magnitude smaller than those of sodium ascorbate and ubiquinol-10.

It was reported that the hydrogen abstraction reaction [Reaction (5)] from unsaturated lipids, such as methyl linoleate and methyl linolenate, by the α-tocopheroxyl radical may relate to the prooxidant effect of α-tocopherol (Mukai and Okauchi 1989, Nagaoka *et al.* 1990).

$$\text{Toc}^\bullet + \text{LH} \xrightarrow{k_p} \text{TocH} + \text{L}^\bullet \tag{5}$$

The reaction rate constant (k_p) between Toc$^•$ and methyl linoleate is 1.26×10^{-2} [(mol · s)/L]$^{-1}$ in ethanol. The k_r values of catechin and epicatechin are 1.20×10^2 and 1.52×10^2 [(mol · s)/L]$^{-1}$, respectively, and are about four orders of magnitude larger than that (k_p) of methyl linoleate. The result suggests that catechin and epicatechin may contribute to the tocopheroxyl radical–scavenging reaction in biological systems. In fact, suppression of the α-tocopherol consumption by flavonoids (morin, fisetin, and quercetin) was previously reported in the case of oxidative modification of human low-density lipoprotein treated with macrophage or metal ion (De Whalley *et al.* 1990, Jessup *et al.* 1990). Further, the flavonoids in wine are hypothesized to act synergistically with tocopherol to inhibit lipid peroxidation (Negre-Salvayre *et al.* 1991). Catechin prevents human plasma oxidation (Lotito *et al.* 1998).

Aroxyl Radical–Scavenging Rate (k_s) of Catechins and Related Compounds in Ethanol Solution

Measurements of the rate constant (k_s) for the reaction of ArO• with catechin, epicatechin, 4-methylcatechol, and 5-methoxyresorcinol were performed in ethanol solution at 25°C. The rate was measured by following the decrease in absorbance of the ArO• radical at 580 nm (Mukai *et al.* 1986 and 1989). The values of k_s obtained are listed in Table 20.1, together with those obtained for α-tocopherol, ubiquinol-10, and butylated hydroxytolulene (BHT), which are among the best-known lipid-soluble antioxidants. As is clear from the k_s values listed in Table 20.1, the rate of the scavenging reaction of ArO• with the above antioxidants increases in the order of 5-methoxyresorcinol << rutin < BHT < catechin < epicatechin < 4-methylcatechol << α-tocopherol ~ ubiquinol-10. The rate constant of epicatechin $\{k_s = 1.34 \times 10^2$ [(mol · s)/L]$^{-1}\}$ is ~9 times as large as that of rutin $\{k_s = 1.42 \times 10$ [(mol · s)/L]$^{-1}\}$. However, the k_s value of epicatechin is ~40 times smaller than that of α-tocopherol $\{k_s = 5.12 \times 10^3$ [(mol · s)/L]$^{-1}\}$, which has the highest free radical–scavenging activity among natural phenolic antioxidants (Burton *et al.* 1985, Mukai *et al.* 1989).

pH Dependencc of the Aroxyl and Tocopheroxyl Radical–Scavenging Rates (k_s and k_r) of Catechins and Related Compounds in Aqueous Triton X-100 Micellar Solution

Measurements of the rate constant (k_s) for the reaction of ArO• with catechins and related compounds were performed at various pH values in aqueous Triton X-100 micellar solution (5.0 wt%). The k_s values obtained are summarized in Tables 20.1 and 20.2, together with those obtained for α-tocopherol, ubiquinol-10, and BHT. In those cases in which 4 > pH or pH > 12, the ArO• radical and/or antioxidants are unstable, and measurement of the k_s value was unsuccessful. The k_s value of 5-methoxyresorcinol is 2.17 mol/(L · s) at pH 7.0. The aroxyl radical–scavenging rate of 5-methoxyresorcinol, which is considered to be a model of the A-ring in catechin and epicatechin, is very slow in micellar solution, as observed for the reaction in homogeneous solution. On the other hand, catechin, epicatechin, and 4-methylcatechol, which have a catechol structure in the molecule, show higher reactivity. The k_s values of these antioxidants are ~2–3 orders of magnitude larger than that of 5-methoxyresorcinol at pH 7.0. The pH dependence of the second-order rate constants (k_s) of catechin and 5-methoxyresorcinol is shown in Figures 20.4 and 20.5, respectively. By increasing pH values, the k_s of catechin remains constant $\{k_s = 5.5 \times 10^2$ [(mol · s)/L]$^{-1}\}$ between pH 4.0 and 6.0, increases rapidly at pH 8.25, and shows a maximum $\{k_s = 2.31 \times 10^3$ [(mol · s)/L]$^{-1}\}$ at pH 9.5. The rate constant k_s of epicatechin also shows pH dependence similar to that of catechin (see Table 20.2). The k_s values of 5-methoxyresorci-

TABLE 20.2
pH Dependence of the Second-Order Rate Constants (k_s) for the Reaction of Catechin, Epicatechin, and Related Antioxidants with $ArO^{\bullet}$ Radical in Triton X-100 Micellar Solution

	k_s /[(mol · s)/L]$^{-1}$ ($ArO^{\bullet}$)			
pH	Catechin	Epicatechin	4-Methylcatechol	5-Methoxyresorcinol
4	5.49×10^2	6.39×10^2	1.87×10^3	
5	5.45×10^2	7.50×10^2	1.41×10^3	
6	5.68×10^2	8.03×10^2	1.95×10^3	1
7	6.57×10^2	7.78×10^2	2.22×10^3	2.17
8	6.69×10^2	8.69×10^2	2.78×10^3	8
8.25	8.93×10^2			
8.5	1.22×10^3	1.28×10^3	3.20×10^3	
9	1.53×10^3	3.42×10^3	8.19×10^3	3.93×10
9.25	1.98×10^3			
9.4	2.27×10^3			
9.5	2.31×10^3	4.32×10^3	1.84×10^4	
9.6	2.21×10^3			
9.75	2.15×10^3			
10	2.16×10^3	3.80×10^3	3.29×10^4	5.05×10
10.25	1.70×10^3			
10.5	1.40×10^3			
11	1.01×10^3	1.18×10^3		3.61×10
12				2.58×10

[a]Conditions as in Table 20.1.

nol also showed notable pH dependence with a maximum at pH 10.0. The pH dependence reflects a complex mechanism that will be discussed later.

Similarly, the pH effect on the reaction of the $Toc^{\bullet}$ radical with catechins and related compounds was studied. As listed in Tables 20.1 and 20.3, the tocopheroxyl radical–scavenging rates (k_r) of 5-methoxyresorcinol obtained at pH 7.0 are very slow. On the other hand, the k_r values of catechin, epicatechin, and 4-methylcatechol are ~3–4 orders of magnitude larger than that of 5-methoxyresorcinol. As shown in Figure 20.6, the reaction rates (k_r) of catechin and epicatechin also increase with increasing pH value >6.0, and show a maximum at pH 9.5, as observed for the reaction of $ArO^{\bullet}$ with catechin and epicatechin. The k_r value of catechin at pH 9.5 {$k_r = 3.65 \times 10^3$ [(mol · s)/L]$^{-1}$} is 25 times as large as that at pH 5.0 {$k_r = 1.45 \times 10^2$ [(mol · s)/L]$^{-1}$}, as listed in Table 20.3.

Discussion

Structure-Activity Relationship of the Free Radical–Scavenging Reaction by Catechins and Related Compounds in Homogeneous Solution

As listed in Table 20.1, the reactivity of 5-methoxyresorcinol {$k_s = \sim 10^{-2}$ [(mol · s)/L]$^{-1}$} in ethanol is very weak and almost negligible. On the other hand, catechin

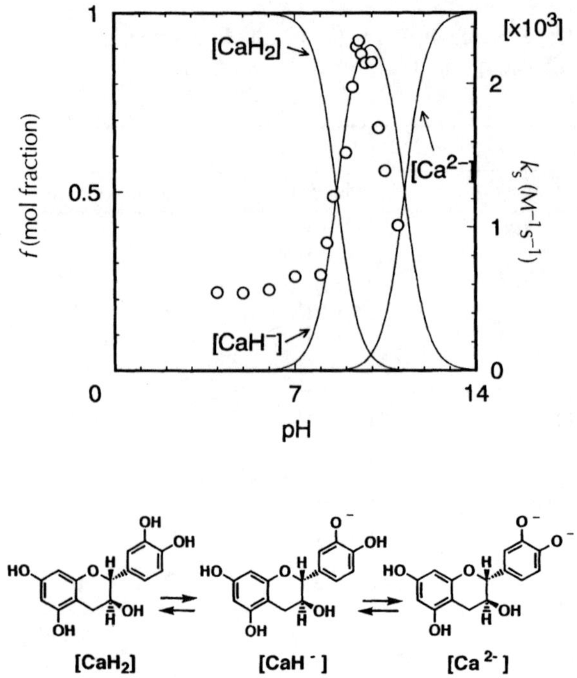

Fig. 20.4. Plots of second-order rate constant (k_s) for catechin (open circle) vs. pH and of mol fraction (*f*) of three catechin species (CaH_2, CaH^-, and Ca^{2-}) vs. pH (solid line).

and epicatechin, which have 3′- and 4′-OH groups at the B-ring, and 4-methylcatechol show higher reactivities, i.e., $k_s = 8.26 \times 10$ $[(\text{mol} \cdot \text{s})/\text{L}]^{-1}$ for catechin, $k_s = 1.34 \times 10^2$ $[(\text{mol} \cdot \text{s})/\text{L}]^{-1}$ for epicatechin, and $k_s = 2.67 \times 10^2$ $[(\text{mol} \cdot \text{s})/\text{L}]^{-1}$ for 4-methylcatechol. 5-Methoxyresorcinol and 4-methylcatechol are considered to be a

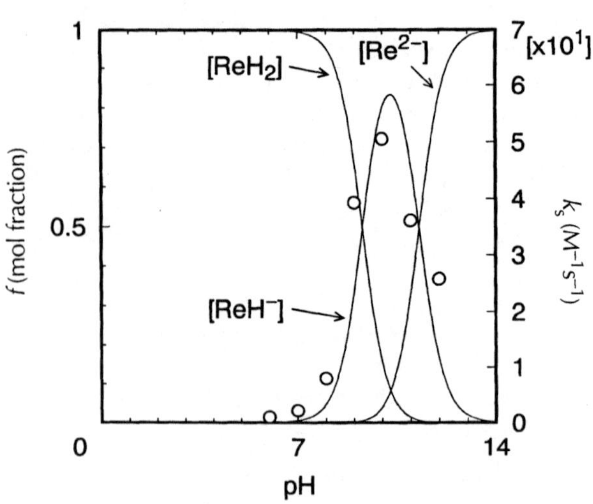

Fig. 20.5. Plots of second-order rate constant (k_s) for 5-methoxyresorcinol (open circle) vs. pH and of mol fraction (*f*) of three 5-methoxyresorcinol species (ReH_2, ReH^-, and Re^{2-}) vs. pH (solid line).

TABLE 20.3
pH Dependence of the Second-Order Rate Constants (k_r) for the Reaction of Catechin, Epicatechin, and Related Antioxidants with Toc• Radical in Triton X-100 Micellar Solution[a]

	k_r /[(mol · s)/L]$^{-1}$ (Toc•)				
pH	Catechin	Epicatechin	4-Methylcatechol	5-Methoxyresorcinol	Vitamin C[b]
3					1.72×10^2
4			2.31×10^3		7.98×10^2
5	1.45×10^2	5.00×10^2	2.36×10^3		1.56×10^3
5.5	1.59×10^2				
6	1.51×10^2	5.81×10^2	3.39×10^3	2.30×10^{-1}	2.26×10^3
6.5	4.75×10^2				
7	7.31×10^2	1.27×10^3	4.41×10^3	4.93×10^{-1}	2.49×10^3
8	1.16×10^3	1.76×10^3	5.93×10^3	8.05×10^{-1}	2.56×10^3
8.5			6.47×10^3		
9	2.47×10^3	2.96×10^3	8.18×10^3	8.28	2.47×10^3
9.5	3.65×10^3	3.75×10^3	1.51×10^4		
10	3.67×10^3	3.07×10^3	4.93×10^4		2.30×10^3
10.5	2.75×10^3				
11	1.69×10^3	1.31×10^3			

[a]Conditions as in Table 20.1.
[b]Source: Mukai *et al.* (1997).

model of A- and B-rings in catechin (and epicatechin), respectively. Consequently, the above results indicate that the *o*-dihydroxyl (catechol) structure in the B-ring is the obvious radical target site for catechin (and epicatechin).

Similar behavior was observed for the reaction of Toc• with catechins and related compounds in ethanol and 2-PrOH/H_2O (5:1, vol/vol) solutions (see Table 20.1),

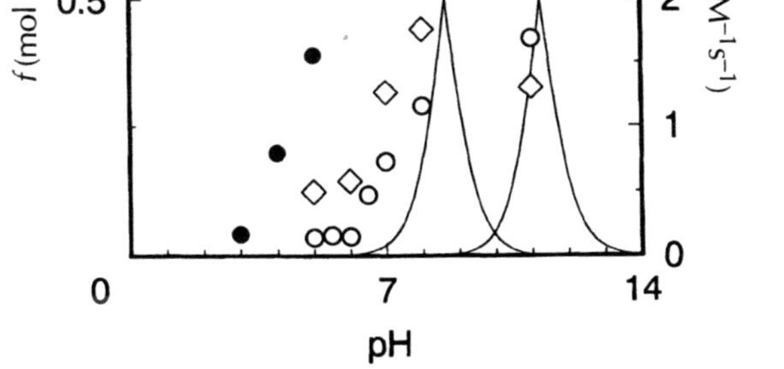

Fig. 20.6. Plots of second-order rate constant (k_r) for the reaction of 5,7-diisopropyltocopheroxyl with catechin (open circle), epicatechin (open square), and ascorbic acid (closed circle) (Mukai *et al.* 1997) vs. pH and of mol fraction (*f*) of three catechin (or epicatechin) species vs. pH (solid line).

indicating that the existence of the catechol structure in the B-ring is essential for the free radical–scavenging of the catechins.

Bors and co-workers (1987 and 1990) measured the second-order rate constants for the reaction of flavonoids with active free radicals, such as $HO^{\bullet}$, $N_3^{\bullet}$, and tBuO$^{\bullet}$, at pH 11.5, using pulse radiolysis technique, and reported that the following three structural groups in flavonoids are important determinants for radical scavenging: (i) the 3′- and 4′-OH groups (catechol structure) in the B-ring, which are the obvious radical target site for all flavonoids; (ii) the 2,3-double bond in conjugation with a 4-oxo function, which is responsible for electron delocalization from the B-ring; and (iii) the existence of both 3- and 5-OH groups for maximal radical-scavenging activity.

Because of the various dissociable phenolic hydroxyl groups in flavonoids, the active free radicals mentioned above react with phenolate anions at pH 11.5 rather than with undissociated phenols. The k_s and k_r values obtained by the measurements in micellar solution increase in the order of 5-methoxyresorcinol << rutin < catechin < epicatechin < 4-methylcatechol << ubiquinol-10 independent of pH value, as listed in Tables 20.1, 20.2, and 20.3. On the other hand, in ethanol, catechins and related compounds will take undissociated forms, differing from those in aqueous solution. However, the k_s and k_r values of flavonoids observed in ethanol solution also increase in the order of 5-methoxyresorcinol << rutin < catechin < epicatechin < 4-methylcatechol << ubiquinol-10. These results indicate that the above determinant (i) reported by Bors and co-workers (1987 and 1990) is important for free radical–scavenging in flavonoids independent of both pH value and solvent system. Catechin, epicatechin, and rutin have 5- and 7-OH groups at the A-ring and 3′- and 4′-OH groups at the B-ring. Further, only rutin has a 2,3-double bond, which is responsible for π-conjugation between B- and C-rings. However, the free radical–scavenging rates (k_s and k_r) of the catechin and epicatechin are larger than the corresponding rates of rutin. It seems that the existence of the 2,3-double bond in rutin is not important for radical scavenging. However, the reason is not clear at present why the k_s and k_r values of catechin and epicatechin are larger than the corresponding values of rutin. The existence of 5- and 7-OH groups at the A-ring is also not necessary for radical scavenging, as described above.

Catechin and epicatechin are stereoisomers; thus we can expect similar rate constants (k_s and k_r) for these antioxidants. However, the k_s and k_r values of epicatechin are ~1.2–1.7 times larger than those corresponding to catechin in homogeneous and micellar (pH 7.0) solutions, as listed in Table 20.1. It is clear from the structures of catechin and epicatechin (see Fig. 20.1) that the steric repulsion between the 6′-ring proton at the B-ring and the 3-OH group at the C-ring will differ in catechins and epicatechins. The difference in the Coulombic interaction between π-electrons at the B-ring and electrons at the C-ring in catechin and epicatechin may induce the change in the electronic state of the B-ring and thus in the reaction rates. Further, we can expect that the rate constant of 4-methylcatechol would be similar to those of catechins and epicatechins. However, the k_s and k_r values of 4-methylcatechol, for example, are 3.2 and 3.7 times larger than the corresponding values of catechins in ethanol solution, respectively, as shown in Table 20.1. Similarly, the k_s and k_r values of 4-methylcatechol are 3.4 and 6.0 times larger than the corresponding values of catechins in micellar solution (pH 7.0), respectively, as listed in Table 20.1. The cause of such a large difference in the rate constant is not clear at present.

Structure-Activity Relationship of the Free Radical–Scavenging Reaction by Catechins in Aqueous Micellar Solution

The rates of reaction (k_s and k_r) of catechins (catechin and epicatechin) and related compounds (5-methoxyresorcinol and 4-methylcatechol) with ArO• and Toc• in Triton X-100 micellar solutions have been measured by varying the pH value. The second-order rate constants (k_s and k_r) obtained increased with increasing pH value and showed a maximum at ~pH 9.5–10.0 (see Figs. 20.4–20.6, and Tables 20.2 and 20.3).

5-Methoxyresorcinol is dibasic and can exist in three different molecular forms, i.e., the undissociated form (ReH_2), monoanion (ReH^-), and dianion (Re^{2-}), depending on the pH value. The equilibrium reactions of 5-methoxyresorcinol have the following form:

$$ReH_2 \overset{K_{a1}}{\leftrightharpoons} ReH^- \overset{K_{a2}}{\leftrightharpoons} Re^{2-} \tag{6}$$

The pK_{a1} and pK_{a2} values of 5-methoxyresorcinol were reported by Jovanovic *et al.* (1995); the values are 9.30 and 11.3, respectively. Therefore, the mole fractions (*f*) present as the ReH_2 molecule and the ReH^- and Re^{2-} ions were calculated as a function of pH (Mukai *et al.* 1991). The analytical concentration (C_a) is given in Equation (7).

$$C_a = [ReH_2] + [ReH^-] + [Re^{2-}] \tag{7}$$

Mole fractions (*f*) present as ReH_2, ReH^- and Re^{2-} are shown as functions of pH in Figure 20.5.

If we assume that the k_{s1}, k_{s2}, and k_{s3} are the reaction rates for undissociated (ReH_2), monoanion (ReH^-), and dianion (Re^{2-}) forms of 5-methoxyresorcinol, respectively, the total rate k_s will be expressed as Equation (8).

$$k_s = k_{s1} f(ReH_2) + k_{s2} f(ReH^-) + k_{s3} f(Re^{2-}) \tag{8}$$

By comparing the observed pH dependence of k_s with the pH dependence of the mole fraction, the values of k_{si} (i = 1–3) were determined. For instance, at pH 6, only the undissociated form of 5-methoxyresorcinol exists in solution, that is, $f(ReH_2) = 1$, and we can immediately determine the k_{s1} value. At $7 < pH < 9$, both the undissociated and monoanion forms exist in solution. Consequently, we can determine the k_{s2} value, using Equation (8). The k_{s3} value is considered to be small (i.e., negligible) compared

TABLE 20.4
The Reaction Rates (k_{si} and k_{ri}, $i = 1$–3) for Undissociated, Monoanion, and Dianion Forms of Catechin, Epicatechin, and Related Antioxidants in Micellar Solution

Antioxidant	k_{si} /[(mol · s)/L]$^{-1}$ (ArO$^\bullet$)		
	k_{s1}	k_{s2}	k_{s3}
Catechin	5.5×10^2	2.5×10^3	5.1×10
Epicatechin	7.4×10^2	4.7×10^3	5.1×10
4-Methylcatechol	1.8×10^3	4.4×10^4	0
5-Methoxyresorcinol	1	5.1×10	0
	k_{ri} /[(mol · s)/L]$^{-1}$ (Toc$^\bullet$)		
Antioxidant	k_{r1}	k_{r2}	k_{r3}
Catechin	1.5×10^2	4.0×10^3	10
Epicatechin	5.5×10^2	4.2×10^3	10
4-Methylcatechol	2.7×10^3	6.6×10^4	0
5-Methoxyresorcinol	2.3×10^{-1}	10	0
Vitamin C[a]	1.7×10^2	2.6×10^3	0

[a]*Source*: Mukai *et al.* (1997).

with k_{s2}. The k_{s1}, k_{s2}, and k_{s3} values obtained for three molecular forms of 5-methoxyresorcinol are listed in Table 20.4. The k_{s2} value is 51 times as large as the k_{s1} value. The result indicates that the reaction rate k_{si} increases by increasing the anionic character of 5-methoxyresorcinol, that is, the electron-donating capacity of 5-methoxyresorcinol.

4-Methylcatechol is also dibasic and can exist in three different molecular forms, depending on the pH value. The pK_{a1} and pK_{a2} values of 4-methylcatechol are 9.56 and 14.00, respectively (Slabbert 1977). By comparing the observed pH dependence of k_s with the pH dependence of the mole fraction, the reaction rates k_{s1} (undissociated form), k_{s2} (monoanion), and k_{s3} (dianion) of 4-methylcatechol were determined. The k_{s2} value is 24 times larger than the k_{s1} value. Similarly, the k_{ri} ($i = 1$–3) values of 5-methoxyresorcinol and 4-methylcatechol were determined. The k_{ri} values obtained are listed in Table 20.4.

On the other hand, catechin is tetrabasic and can exist in five different molecular forms in micellar solution, depending on the pH value. The pK_{a1}, pK_{a2}, pK_{a3}, and pK_{a4} values of catechin have been reported by Jovanovic *et al.* (1994); the values are 8.64, 9.41, 11.26, and 13.26, respectively.

As described in a previous section and listed in Table 20.2, the aroxyl radical–scavenging rates (k_s) of 5-methoxyresorcinol are very slow and are ~2–3 orders of magnitude smaller than those of catechin in all pH regions. The result suggests that the 5- and 7-OH groups at the A-ring of catechin do not contribute to the scavenging of the aroxyl radical. The pH dependence of the rate constants (k_s) in catechin and epicatechin will be explained by considering only the dissociation of protons of 3′- and 4′-OH groups at the B-ring.

By comparing the observed pH dependence of k_s with the pH dependence of the mole fraction calculated, two kinds of pK_a values for the B-ring of catechin were determined. If we used the values $pK_{a1} = 8.64$ and $pK_{a3} = 11.26$ for the two OH groups of the B-ring of catechin, a good correlation between the rate constants (k_s) and the mole fraction (*f*) of monoanion (CaH^-) of catechin was obtained, as shown in Figure 20.4. Any other sets of pK_{ai} values for the B-ring of catechin could not explain the pH dependence of the k_s value. The result shown in Figure 20.4 shows that the monoanion form (CaH^-) of the B-ring of catechin contributes mainly to the scavenging of free radicals at a pH region of 8–11. On the other hand, at a lower pH region, both the undissociated form (CaH_2) and the monoanion (CaH^-) contribute to the scavenging of free radicals.

By comparing the observed pH dependence of k_s with the pH dependence of the mole fraction, the values of k_{si} were determined. The values are listed in Table 20.4. By using these k_{s1}, k_{s2}, and k_{s3} values, we simulated the experimental data. As shown in Figure 20.7, good agreement between the observed rate constants (k_s) and theoretical curve was obtained, suggesting that each reaction rate (k_{si}) estimated is reasonable.

Similarly, the reaction rates (k_{si}) for epicatechin and the reaction rates (k_{ri}) for catechin and epicatechin were estimated; they are summarized in Table 20.4. The k_{r2} values for the monoanion form of catechin and epicatechin are 27 and 7.6 times as large as the k_{r1} values for the undissociated form, respectively.

Comparison Between Rates of Vitamin E Regeneration Reaction with Catechins and Vitamin C in Aqueous Triton X-100 Micellar Solution

The pH dependence of the tocopheroxyl radical–scavenging rate (k_r) of vitamin C (L-ascorbic acid) was studied previously in aqueous Triton X-100 micellar solution

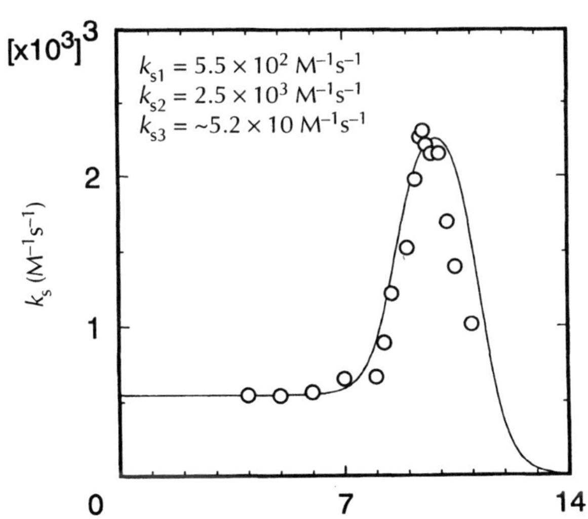

Fig. 20.7. Plots of second-order rate constant (k_s) for catechin (open circle) vs. pH and simulation curve (solid line).

(5.0 wt%) (Mukai *et al.* 1991 and 1997). The second-order rate constants (k_r) obtained showed notable pH dependence with a broad maximum around pH 8, as shown in Figure 20.6. For example, the k_r values obtained are 1.72×10^2 [(mol · s)/L]$^{-1}$ at pH 3.0, 2.49×10^3 [(mol · s)/L]$^{-1}$ at pH 7.0, and 2.56×10^3 [(mol · s)/L]$^{-1}$ at pH 8.0. As shown in Figure 20.6 and listed in Table 20.3, the rates of vitamin E regeneration reaction with catechin and epicatechin are slower than the corresponding rates of ascorbic acid at a low pH region (pH 5–7). However, the rates are similar to each other at ~pH 8–9. Further, the rate constants (k_r) obtained for catechin and epicatechin are 1.6 and 1.3 times larger, respectively, than that of ascorbic acid at pH 10.0. The reaction rate (k_p) between Toc• and methyl linoleate [Reaction (5)] is 2.40×10^{-2} [(mol · s)/L]$^{-1}$ in micellar solution. The k_r values of catechin {7.31×10^2 [(mol · s)/L]$^{-1}$} and epicatechin {1.27×10^3 [(mol · s)/L]$^{-1}$} are much larger than that (k_p) of methyl linoleate. However, the tocopheroxyl radical–scavenging rates (k_r) of catechin and epicatechin are smaller than that {$k_r = 9.24 \times 10^5$ [(mol · s)/L]$^{-1}$} at pH 7.0] of ubiquinol-10, which is also well known as a vitamin E regeneration compound in biomembrane systems (Bowry *et al.* 1993, Mukai *et al.* 1992).

Green and black teas contain considerable amounts of catechins (Graham 1992). In a Dutch epidemiologic study, dietary flavonoid intake, the majority from tea, was found to significantly reduce death from heart disease (Hertog *et al.* 1993 and 1995). Of special importance to heart disease is LDL oxidation. Several flavonoids have been found to inhibit LDL oxidation *in vitro* by cupric ion and by macropharges (De Whalley *et al.* 1990, Jessup *et al.* 1990, Vinson *et al.* 1995). The results of this kinetic study suggest that catechin and epicatechin may function as vitamin E regeneration compounds in biological systems. In fact, suppression of the α-tocopherol consumption by flavonoids was previously reported in the case of oxidative modification of human LDL treated with macrophage or metal ion (De Whalley *et al.* 1990, Jessup *et al.* 1990).

Summary

Detailed kinetic studies have been performed for the reaction of aroxyl (ArO•) and 5,7-diisopropyltocopheroxyl (Toc•) radicals with catechin, epicatechin, and related antioxidants in homogeneous and micellar solutions. The structure-activity relationship in the scavenging reaction of free radicals by catechin and epicatechin in solution was discussed. It has been found that catechin and epicatechin have high activity in vitamin E regeneration. The results of this kinetic study should provide a foundation for the interpretation of reactions of catechins with active free radicals, such as lipid peroxyl and α-tocopheroxyl radicals in more complex biological systems.

References

Bors, W., and M. Saran, Radical Scavenging by Flavonoid Antioxidants, *Free Radic. Res. Commun.* 2: 289–294 (1987).

Bors, W., W. Heller, C. Michel, and M. Saran, Flavonoids as Antioxidants: Determination of Radical-Scavenging Efficiencies, *Methods Enzymol.* 186: 343–355 (1990).

Bowry, V.W., and R. Stocker, Tocopherol-Mediated Peroxidation. The Prooxidant Effect of Vitamin E on the Radical-Initiated Oxidation of Human Low-Density Lipoprotein, *J. Am. Chem. Soc. 115*: 6029–6044 (1993), and references cited therein.

Burton, G.W., T. Doba, E.J. Gabe, L. Hughes, F.L. Lee, L. Prasad, and K.U. Ingold, Autoxidation of Biological Molecules. 4. Maximizing the Antioxidant Activity of Phenols, *J. Am. Chem. Soc. 107*: 7053–7065 (1985) and references cited therein.

De Whalley, C.V., S.M. Rankin, R.S. Hoult, W. Jessup, and D.S. Leake, Flavonoids Inhibit the Oxidative Modification of Low Density Lipoproteins by Macrophages, *Biochem. Pharmacol. 39*: 1743–1750 (1990).

Foti, M., K.U. Ingold, and J. Lusztyk, The Surprisingly High Reactivity of Phenoxyl Radicals, *J. Am. Chem. Soc. 116*: 9440–9447 (1994).

Graham, H.N., Green Tea Composition, Consumption and Polyphenol Chemistry, *Prev. Med. 21*: 334–350 (1992).

Hertog, M.G.L., E.G.M. Feskens, P.C.H. Lollman, M.B. Katan, and D. Kromhout, Dietary Antioxidant Flavonoids and Risk of Coronary Heart Disease: The Zutphen Elderly Study, *The Lancet 342*: 1007–1011 (1993).

Hertog, M.G.L., D. Kromhout, C. Aravans, H. Blackburn, R. Buzina, F. Fidanza, S. Giampaoli, A. Jansen, A. Menotti, S. Nedeljkovic, M. Pekkarinen, B.S. Simic, H. Toshima, E.J.M. Feskens, P.C.H. Hollman, and M. Katan, Flavonoid Intake and Long-Term Risk of Coronary Heart Disease and Cancer in the Seven Countries Study, *Arch. Intern. Med. 155*: 381–386 (1995).

Ioku, K., T. Tsushida, Y. Takei, N. Nakatani, and J. Terao, Antioxidant Activity of Quercetin and Quercetin Monoglucosides in Solution and Phospholipid Bilayers, *Biochim. Biophys. Acta 1234*: 99–104 (1995).

Jankun, J., S.H. Selman, R. Swiercz, and E. Skrzypczak-Jankun, Why Drinking Green Tea Could Prevent Cancer, *Nature 387*: 561 (1997).

Jessup, W., S.M. Rankin, C.V. De Walley, J.R.S. Hoult, J. Scott, and D.S. Leake, α-Tocopherol Consumption During Low-Density-Lipoprotein Oxidation, *Biochem. J. 265*: 399–405 (1990).

Jovanovic, S.V., Y. Hara, S. Steenken, and M.G. Simic, Antioxidant Potential of Gallocatechins. A Pulse Radiolysis and Laser Photolysis Study, *J. Am. Chem. Soc. 117*: 9881–9888 (1995).

Jovanovic., S.V., S. Steenken, M. Tosic, B. Marjanovic, and M.G. Simic, Flavonoids as Antioxidants, *J. Am. Chem. Soc. 116*: 4846–4851 (1994).

Larson, R.A., The Antioxidants of Higher Plants, *Phytochemistry 27*: 969–978 (1988).

Leanderson, P., A.O. Faresjo, and C. Tagesson, Green Tea Polyphenols Inhibit Oxidant-Induced DNA Strand Breakage in Cultured Lung Cells, *Free Radic. Biol. Med. 23*: 235–242 (1997).

Lotito, S.B., and C.G. Fraga, (+)-Catechin Prevents Human Plasma Oxidation, *Free Radic. Biol. Med. 24*: 435–441 (1998).

Mukai, K., S. Itoh, and H. Morimoto, Stopped-Flow Kinetic Study of Vitamin E Regeneration Reaction with Biological Hydroquinones (Reduced Forms of Ubiquinone, Vitamin K, and Tocopherolquinone) in Solution, *J. Biol. Chem. 267*: 22277–22281 (1992).

Mukai, K., Y. Kageyama, T. Ishida, and K. Fukuda, Synthesis and Kinetic Study of Antioxidant Activity of New Tocopherol (Vitamin E) Compounds, *J. Org. Chem. 54*: 552–556 (1989).

Mukai, K., M. Nishimura, and S. Kikuchi, Stopped-Flow Investigation of the Reaction of Vitamin C with Tocopheroxyl Radical in Aqueous Triton X-100 Micellar Solutions, *J. Biol. Chem. 266*: 274–278 (1991).

Mukai, K., W. Oka, K. Watanabe, Y. Egawa, and S. Nagaoka, Kinetic Study of Free-Radical-Scavenging Action of Flavonoids in Homogeneous and Aqueous Triton X-100 Micellar Solutions, *J. Phys. Chem. A. 101*: 3746–3753 (1997).

Mukai, K., and Y. Okauchi, Kinetic Study of the Reaction Between Tocopheroxyl Radical and Unsaturated Fatty Acid Esters in Benzene, *Lipids 24*: 936–939 (1989).

Mukai, K., Y. Watanabe, Y. Uemoto, and K. Ishizu, Stopped-Flow Investigation of Antioxidant Activity of Tocopherols, *Bull. Chem. Soc. Jpn. 59*: 3113–3116 (1986).

Mukhtar, H., S.K. Katiyar, and R. Agarwal, Cancer Chemoprevention by Green Tea Components, in *Diet and Cancer, Markers, Prevention, and Treatment,* edited by M.M. Jacobs, Plenum Press, New York, 1994.

Nagaoka, S., Y. Okauchi, S. Urano, U. Nagashima, and K. Mukai, Kinetic and *ab initio* Study of the Prooxidant Effect of Vitamin E, Hydrogen Abstraction from Fatty Acid Esters and Egg Yolk Lecithin, *J. Am. Chem. Soc. 112*: 8921–8924 (1990), and references cited therein.

Negre-Salvayre, A., A. Alfany, C. Hariton, and R. Salvayre, Additional Antilipoperoxidant Activities of Alpha-Tocopherol and Ascorbic Acid on Membrane-Like Systems Are Potentiated by Rutin, *Pharmacology 42*: 262 (1991).

Packer, J.E., T.F. Slater, and R.L. Willson, Direct Observation of a Free Radical Interaction Between Vitamin E and Vitamin C, *Nature 278*: 737–738 (1979).

Rieker, A., and K. Scheffler, Die Beteiligung von Phenylresten an der Aroxylmesomerie, *Liebigs Ann. Chem. 689*: 78–92 (1965).

Shahidi, F., and P.K.J.P.D. Wanasundara, Phenolic Antioxidants, *Crit. Rev. Food Sci. Nutr. 32*: 67–103 (1992).

Slabbert, N.P., Ionization of Some Flavanols and Dihydroflavonols, *Tetrahedron 33*: 821–824 (1977).

Terao, J., M. Piskula, and Q. Yao, Protective Effect of Epicatechin, Epicatechin Gallate, and Quercetin on Lipid Peroxidation in Phospholipid Bilayers, *Arch. Biochem. Biophys. 308*: 278–284 (1994).

Vinson, J.A., J. Jang, Y.A. Dabbagh, M.M. Serry, and S. Cai, Plant Polyphenols Exhibit Lipoprotein-Bound Antioxidant Activity Using an *in vitro* Oxidation Model for Heart Disease, *J. Agric. Food Chem. 43*: 2798–2799 (1995).

Yang, C.S., and Z.-Y. Wang, Tea and Cancer, *J. Natl. Cancer Inst. 85*: 1038–1049 (1993).

Chapter 21

Modulating Prooxidative Mitotic Signaling as Action Mechanisms of Cancer Chemopreventive Agents Including Curcumin, Epigallocatechin Gallate, and Other Phytopholyphenols

Jen-Kun Lin, Yu-Li Lin, Yu-Chih Liang, Yen-Chou Chen, and Shoei-Yn Lin-Shiau

Laboratory for Cancer Research, Institutes of Biochemistry and Toxicology, College of Medicine, National Taiwan University, Taipei, Taiwan

Introduction

To date, many phytopolyphenols with antioxidant activity have been discovered in numerous plant sources such as tea (Rice-Evans *et al.* 1995, Vinson *et al.* 1995), celery, parsley, grapefruit, buckwheat plants, soybeans, bananas, grapes, rice hulls, green barley leaves (Shibamoto 1997), and tumeric (Osawa *et al.* 1992). Among these phytopolyphenols, certain representative compounds have been demonstrated to be potential cancer chemopreventive agents (Table 21.1). A number of antioxidant compounds have been found in several plant materials such as oilseeds, crops, vegetables, fruits, leaves, leaf wax, bark, and roots. Spices, herbs, and crude drugs are also important sources of natural antioxidants, although the chemical properties and biological roles of the active principles are not fully understood and remain to be explored. Recently, much attention has been given to the cancer chemopreventive effects of these phytopolyphenols.

During the past few years, many phytochemicals have been postulated to prevent cancer through their ability to act as antioxidants, i.e., they intercept potential DNA-damaging molecules such as direct-acting carcinogens, reactive xenobiotic electrophiles, and reactive oxygen species (ROS), thereby preventing mutation and tumor promotion and progression. These include carotenoids, tocopherols, ascorbate, antioxidant enzymes, and polyphenols such as flavonoids (Rice-Evans *et al.* 1995, Vinson *et al.* 1995).

Phytopolyphenols with Cancer Chemopreventive Potential

One of the most exciting findings in the field of cancer chemoprevention is the very large number of compounds that have been demonstrated to prevent the occurrence of cancer; compounds belonging to >20 different classes of chemicals have been shown to have cancer chemopreventive capacities (Morse and Stoner 1993, Wattenberg 1985). During the last decade, several naturally occurring phytopolyphenols have

TABLE 21.1
Phytopolyphenols with Antioxidant Activity and Cancer Chemoprevention Potential[a]

	General chemical name	Source	Antioxidant activity
Curcumin	1,7-Bis(4-hydroxy-3-methoxyphenyl)-1,6-hepadiene-3,5-dione	Tumeric	+++
Desmethoxycurcumin		Tumeric	++
Bisdesmethoxycurcumin		Tumeric	+
Tetrahydrocurcumin		Synthetic	++++
Tetrahydromethoxycurcumin		Synthetic	+++
Tetrahydrobisdesmethoxycurcumin		Synthetic	++
Tetrahydroxycurcumin (Bisdesmethylcurcumin)		Synthetic	+++
Tetrahydrotetrahydroxycurcumin (Bisdesmethyltetrahydrocurcumin)		Synthetic	++++
Apigenin	5,7-Dihydroxy-2-(4-hydroxyphenyl)-4H-1-benzopyran-4-one	Celery and parsley	+
Quercetin	2-(3,4-Dihydroxyphenyl)-3,5,7-trihydroxy-4H-1-benzopyran-4-one	Plant kingdom	++++

(continued)

TABLE 21.1
(Continued)

	General chemical name	Source	Antioxidant activity
Kaempferol	2-(4-Hydroxyphenyl)-3,5,7-trihydroxy-4H-1-benzopyran-4-one	Grapefruit	++
Rutin	3,5,7,3′-Pentahydroxyflavone-3-rutinoside	Buckwheat plants	+++
Genistein	5-7-Dihydroxy-3-(4-hydroxyphenyl)-4H-1-benzopyran-4-one	Soy beans and forage plants	+
Cyanidin chloride	2-(3,4-Dihydroxyphenyl)-3,5,7-trihydroxy-1-benzopyrylium chloride	Bananas	++++
Resveratrol	4,3′,5′-Trihydroxystilbene	Grapes	++
Isovitexin		Rice hulls	++++++
2″-O-Glycosyl-isovitexin		Barley leaves	++++++
Catechin		Tea	++++
Epicatechin		Tea	+++
Epigallocatechin		Tea	++++++
Epicatechin gallate		Tea	+++++
Epigallocatechin gallate		Tea	+++++++
α-Tocopherol (as control, 100% activity)			++++++

[a]Data for the antioxidant activity of flavonoids are from Rice-Evans *et al.* (1995) and Vinson *et al.* (1995); those for isovitexins are from Shibamoto (1997), and those for curcumin derivatives are from Osawa *et al.* (1992).

been demonstrated to be effective in chemoprevention of cancer in animal and human experiments (Kelloff *et al.* 1996). The common and general chemical names of some well-known phytopolyphenols belonging to this category are listed in Table 21.1. The cancer chemopreventive effects of these compounds are briefly described below.

Curcumin

Curcumin is the major yellow pigment in tumeric and curry; it is obtained from the rhizome of the plant *Curcuma longa.* Synthetic hydrocurcumin derivatives were prepared by hydrogenation, whereas hydroxy derivatives were made by demethylation of curcumin. Curcumin demonstrated chemopreventive activity in mouse skin, colon (Huang *et al.* 1988), and rat mammary models (Steele *et al.* 1994).

Curcumin has multiple mechanisms of action. It is a potent anti-inflammatory agent (Srimal and Dhawan 1973). It inhibits arachidonic acid metabolism by blocking both the lipoxygenase and cyclooxygenase pathways and possibly, phospholipase A_2. Curcumin exhibits strong antioxidant activity (Sharma 1976), i.e., it is an effective superoxide scavenger (Srivastava 1985). It inhibits 12-*O*-tetradecanoyl-13-acetate (TPA)-induced DNA synthesis, demonstrating an inhibitory effect on proliferation (Huang *et al.* 1988). It also modifies cytochrome p450 and enhances glutathione-S-transferase activity, and it may modify the metabolic activation and DNA-binding of *p*-aminohippuric acid (PAH) carcinogens by this mechanism. Furthermore, curcumin decreases expression of c-jun (Huang *et al.* 1991), c-fos, and c-myc (Kakar and Roy 1994), possibly through inhibition of protein kinases (Liu *et al.* 1993). It inhibits ornithine decarboxylase activity and epidermal growth factor (EGF) receptor function (Koruta *et al.* 1995).

Apigenin

Apigenin is a low-toxic, nonmutagenic plant flavonoid (Kuo *et al.* 1992) and protein kinase inhibitor. Apigenin exhibits antiproliferating effects on human breast cancer cells and enhances gap junctional intercellular communication in liver cells (Chaumontent *et al.* 1994). It inhibits the ornithine decarboxylase activity and tumor-promoting effects of TPA on mouse skin initiated with dimethylbenzanthracene (DMBA) (Wei *et al.* 1990). Our recent studies demonstrated that apigenin has the ability to inhibit protein kinase C activity and suppress c-jun and c-fos protoncogene expression (Huang *et al.* 1996).

Genistein

The isoflavone genistein occurs naturally in soybeans and some forage plants (Naim *et al.* 1973) as the glucoside and has been identified in soy food products and beer. Epidemiologic studies have shown an association between decreased hormone-dependent cancer (e.g., in breast and prostrate) incidence/mortality and a traditional soy-rich Asian diet (Wang and Murphy 1994). Individuals consuming this diet have 7- to 110-fold higher plasma and 30-fold higher urinary genistein concentrations than individuals consuming a typical Western diet. The potential chemopreventive efficacy of genistein appears to be related to its phytoestrogenic effects. Genistein competes with estradiol for estrogen receptors (Shutt and Cox 1972); the complex translocates to the nucleus, stimulating estrogen-related cellular events (Cheng *et al.* 1953).

Genistein inhibits tyrosine-specific protein kinase activity (Akiyama *et al.* 1987); this may, in turn, inhibit cell proliferation, growth-factor stimulated responses [EGF-R, insulin-like growth factor receptor (IGF-R), platelet-derived growth factor receptor (PDGF-R), or oncogene expression (c-ras, c-fos, c-jun)] (Akiyama *et al.* 1987). Genistein may induce the cellular differentiation of HL-60 cells (Katagiri *et al.* 1992).

Resveratrol

Resveratrol, a phytoalexin found in grapes and other food products, was purified and shown to have cancer chemopreventive activity in assays representing three major stages of carcinogenesis, including tumor initiation, promotion, and progression (Jang *et al.* 1997). Resveratrol was found to act as an antioxidant and antimutagen and to induce phase II xenobiotic-metabolizing enzymes (these properties display its anti-initiation activity); it mediated anti-inflammatory effects and inhibited cyclooxygenase and hydroperoxidase functions (these properties manifest its antipromotion activity); and it induced human promyelocytic leukemia cell differentiation (antiprogression activity). In addition, it inhibited the development of preneoplastic lesions in carcinogen-treated mouse mammary gland in culture and inhibited tumorigenesis in a mouse skin cancer model. These data suggest that resveratrol, a common constituent of the human diet in foods such as grape raisin and red grape wine, merits investigation as a potential cancer chemopreventive agent in humans (Jang *et al.* 1997).

Isovitexin and 2″-O-Glycosyl-Isovitexin

These two compounds are isoflavonoids. Isovitexin is found in rice hulls (Ramarathnam *et al.* 1989), whereas 2″-*O*-glycosyl-isovitexin is isolated from an ethanol extract of young green barley leaves (Shibamoto 1997). The antioxidative activity of 2″-*O*-glycosyl-isovitexin toward n-3 fatty acids such as eicosapentaenoic acid (EPA) was higher than that of the known natural antioxidant α-tocopherol (Table 21.1). These two isoflavonoids demonstrated significant antioxidative activity toward lipid peroxidation; furthermore, they can be obtained in large quantities from a natural source at low cost. The cancer chemopreventive effects of these two compounds deserve further studies.

Tea Polyphenols

Tea polyphenols have demonstrated chemopreventive activities in animal cancer models in the colon and large intestine, duodenum, esophagus, forestomach, liver, lung, mammary gland, and skin (Kelloff *et al.* 1996). Multiple mechanisms may be responsible for the chemopreventive properties of tea and tea polyphenols (Stoner and Mukhtar 1995); these include inhibition of lipid peroxidation and free radical formation (antioxidant activity), ornithine decarboxylase, lipoxygenase and cyclooxygenase, and protein kinase C and cellular proliferation.

Tea polyphenols also inhibit carcinogen-DNA binding and adduct formation by modulation of cytochrome p450 enzymes and enhancement of phase II enzymes such as glutathione-S-transferase, as well as glutathione peroxidase, catalase, and NAD(P)H:quinone reductase (Lee *et al.* 1995).

Recent Investigations on the Action Mechanisms of Curcumin

Induction of Apoptosis in Immortalized NIH 3T3 and Cancer Cells by Curcumin

Many antitumor agents are capable of inducing cell death by the pathway of apoptosis; we found that curcumin could stop the growth of the immortalized mouse embryo fibroblast NIH 3T3 cell, the human colon cancer cell HT-29, the human kidney cancer cell 293, and human hepatoma Hep G2 cells, demonstrating various common features of apoptosis (Jiang *et al.* 1996b). Under defined experimental conditions, curcumin could induce cell shrinkage, chromatin condensation, and DNA-fragmentation characteristics of apoptosis. The apoptotic changes were not detected in the primary culture of mouse embryonic fibroblast C3H 10T 1/2, rat embryonic fibroblast, and human foreskin fibroblast cells (Jiang *et al.* 1996b).

Molecular Mechanism of Apoptosis Induced by Curcumin

Apoptosis of Hep-G2 cells triggered by curcumin and other agents was characterized in an attempt to delineate the common apoptosis signaling pathway in human hepatoma cells. Several hallmarks of apoptosis, including DNA laddering, chromatin condensation, DNA fragmentation, and an apoptosis-specific cleavage of 28S and 18S ribosomal RNA were observed after treatment with curcumin (Jiang *et al.* 1996a). Curcumin treatment, however, did not alter the expression levels of Bcl-2 and Bax proteins. The p53 protein accumulated slowly and decreased abruptly after reaching a maximum. Conversely, the c-Myc protein decreased initially and subsequently increased prior to the onset of apoptosis (Jiang *et al.* 1996a).

When COLO205 colorectal carcinoma cells were treated with curcumin (60 μmol/L), the appearance of apoptotic DNA ladders was delayed by ~5 h, and G1 arrest was detected (Chen *et al.* 1996). The reduction of p53 gene expression was accompanied by the induction of HSP 70 gene expression in the curcumin-treated cells. The intracellular Ca^{++} concentration was depleted by curcumin in a dose-dependent manner. These findings suggest that curcumin may induce the expression of the HSP70 gene through the initial depletion of intracellular Ca^{++}, followed by the suppression of p53 gene function in the target cells (Chen *et al.* 1996).

Recent Investigations on the Action Mechanisms of Tea Polyphenols

Suppression of Extracellular Signals and Cell Proliferation Through EGF Receptor Modulation by EGCG in Cancer Cells

Tea polyphenols are known to inhibit a wide variety of enzymatic activities associated with cell proliferation and tumor progression (Lin *et al.* 1997). The effects of the major tea polyphenol (-)-epigallocatechin-3-gallate (EGCG) on the proliferation of human epidermoid cell line A431 were investigated. With the use of a [^{3}H]-thymidine incorporation assay, EGCG could significantly inhibit the DNA synthesis of A431 cells. In an *in vitro* assay, EGCG strongly inhibited the protein tyrosine kinase (PTK) activities of EGF-R, PDGF-R and fibroblast growth factor response (FGF-R), and exhibited an IC_{50} (concentration required for 50% inhibition) value of 0.5–1.0 μg/mL (Liang *et al.* 1997). But EGCG scarcely inhibited the protein kinase activities of pp60$^{v\text{-}src}$, PKC, and PKA (IC_{50} > 10 μg/mL). In an *in vivo* assay, EGCG could reduce the autophosphorylation level of EGF-R by EGF. Phosphoamino acid analysis of the EGF-R revealed that EGCG inhibited the EGF-stimulated increase in phosphotyrosine level in A431 cells. In addition, we showed that EGCG blocked EGF binding to its receptor. The results of further studies suggested that the inhibition of proliferation and suppression of the EGF signaling by EGCG might mediate mainly dose-dependent blocking of ligand binding to its receptor, and subsequently through inhibition of EGF-R kinase activity (Liang *et al.* 1997).

EGCG Blocks the Induction of Nitric Oxide Synthase by Down-Regulating Nuclear Factor B (NFκB)

Nitric oxide (NO) plays an important role in inflammation and in multiple stages of carcinogenesis. We investigated the effect of various tea polyphenols on the induction of NO synthase (NOS) in thioglycollate-elicited and lipopolysaccharide (LPS)-activated peritoneal macrophages. Gallic acid (GA), (-)-epigallocatechin (EGC), and EGCG, the major tea catechins, were found to inhibit inducible NOS (iNOS) protein in the activated macrophage. EGCG, a potent antitumor agent with anti-inflammatory and antioxidant properties, inhibited NO generation, as measured by the amount of nitrite released into the culture medium (Lin and Lin 1997). An electrophoretic mobility shift assay indicated the EGCG blocked the activation of NFκB, a transcription factor necessary for iNOS induction. EGCG also blocked the disappearance of the inhibitor κB (IκB) from the cytosolic fraction. These results suggest that EGCG decreases the activity and protein levels of iNOS by reducing the expression of iNOS mRNA, and the reduction could occur through prevention of the binding of NFκB to the iNOS promoter, thereby inhibiting the induction of iNOS transcription (Lin and Lin 1997).

Discussion of the Action Mechanisms of Phytopolyphenols

Function of Protooxidative Mitotic Signaling

Reactive oxygen species (ROS) such as superoxide ($O^{\bullet}$), hydrogen peroxide (H_2O_2), the hydroxyl radical ($^{\bullet}OH$), and lipid hydroperoxide (LOOH) are now recognized as signaling molecules under subtoxic conditions. Nitric oxide (NO) is also an example of a redox mediator. ROS induce various biological processes such as gene expression by stimulating signal transduction components such as Ca^{++}-signaling and protein phosphorylation. Various oxidants increase cytosolic Ca^{++} and stimulate tyrosine as well as serine/threonine phosphorylation; direct stimulation of protein kinases and inhibition of protein phosphatases by oxidants have been proposed as mechanisms (Suzuki *et al.* 1997).

The oxidant stimulation of the effector molecules such as phospholipase A_2, as well as the activation of oxidative stress responsive transcription factors, may also depend on the oxidant-mediated activation of Ca^{++}-signaling and/or protein phosphorylation. In addition to the stimulation by oxidants, the observation that ligand-receptor interactions produce ROS and that antioxidants block receptor-mediated signal transduction led to a proposal that ROS may be second messengers for transcription factor activation, apoptosis, proliferation, and other related functions (Suzuki *et al.* 1997), as well as oncogene expression (Lee *et al.* 1996).

It has been demonstrated that the oxidant status of the cytosol increases the phosphorylation and degradation of IκB. NFκB action, however, requires a thioredoxin-dependent reduced status in the nucleus. Upstream kinases and phosphatases prone to thiolation or oxidation of a vicinal sulfhydryl (SH)-group are considered at present the best candidates for mediation of the redox regulation of NFκB (Flohe *et al.* 1997).

Most Phytopolyphenols Are Strong Antioxidants

As described in Table 21.1, most phytopolyphenols with cancer chemopreventive potential are strong antioxidants. These compounds may act through their intimate removal of ROS at their sites of generation or sequential blocking of the signal transduction pathways at the near or remote sites (Lin *et al.* 1997). Many phytopolyphenols have long been postulated to prevent cancer through their ability to act as antioxidants that intercept potential DNA-damaging molecules, thereby preventing tumor initiation, promotion, and progression.

Although the antioxidant properties of the flavonoids are associated with their chemopreventive properties, structural characteristics such as the absence of conjugation in the C-ring and the presence of orthopositioned functional groups in the B-ring (e.g., catechol structure) appear to be more important to their function as cancer chemopreventive agents (Franke *et al.* 1995).

Modulating Prooxidative Mitotic Signaling as Action Mechanisms

In aerobic life, oxidative stress arises from both endogenous and exogenous sources. Despite antioxidant defense mechanisms, cell damage from ROS is ubiquitous. ROS-related lesions that do not cause cell death (apoptosis or necrosis) can stimulate the development of cancer (Dreher and Junod 1996). The pathogenesis of cancer comprises a multistep carcinogenesis including tumor initiation, promotion, and progression. It seems that the oxidative stress induced by ROS may be involved in all stages of carcinogenesis. A large body of evidence suggests an important role for ROS in the expansion of tumor clones (tumor promotion) and the acquisition of malignant properties (tumor progression).

An accumulation of experimental results exists, indicating that the functioning of many growth-signal transducing proteins are significantly dependent on their redox state. Such proteins include growth factor receptors (EGF-R, PDGF-R, FGF-R, IGF-R, and others), protein kinases (PKC, PKA, PTK, and others), protein phosphatases (PP-1, PPIIa, and others), as well as a number of important transcription factors including AP-1 and NFκB. It has been speculated that the adjustment of the redox states of individual signal transduction proteins within cells is a prerequisite for their optimal functioning in the transmission of growth responses (Burdon 1996).

On the basis of the aforementioned facts, it would appear that ROS play important roles in the development of cancer and other disease states in humans. The role of antioxidant defenses, including endogenous and exogenous systems, in preventing such diseases is becoming better understood (Anderson 1996). Dietary antioxidants such as ascorbic acid, tocopherols, and carotenes may be especially important in protecting against human diseases associated with ROS damage to cellular DNA, lipids, and proteins (Jacob and Burri 1996). The antioxidant roles of phytopolyphenols in diet and their cancer chemopreventive effects have been intensively investigated (Cooney *et al.* 1997, Lin *et al.* 1997); more research is required to determine the occurrence and biological significance of these compounds *in vivo*. However, a tentative scheme is provided to describe the possible action sites of these phytopolyphenols in the pathway of prooxidative mitotic signaling (Fig. 21.1).

Acknowledgments

This study was supported by grants from the National Science Council (NSC-86-2314-B-002-042) and a National Health Research Institute Grant (DOH 86-HR-403).

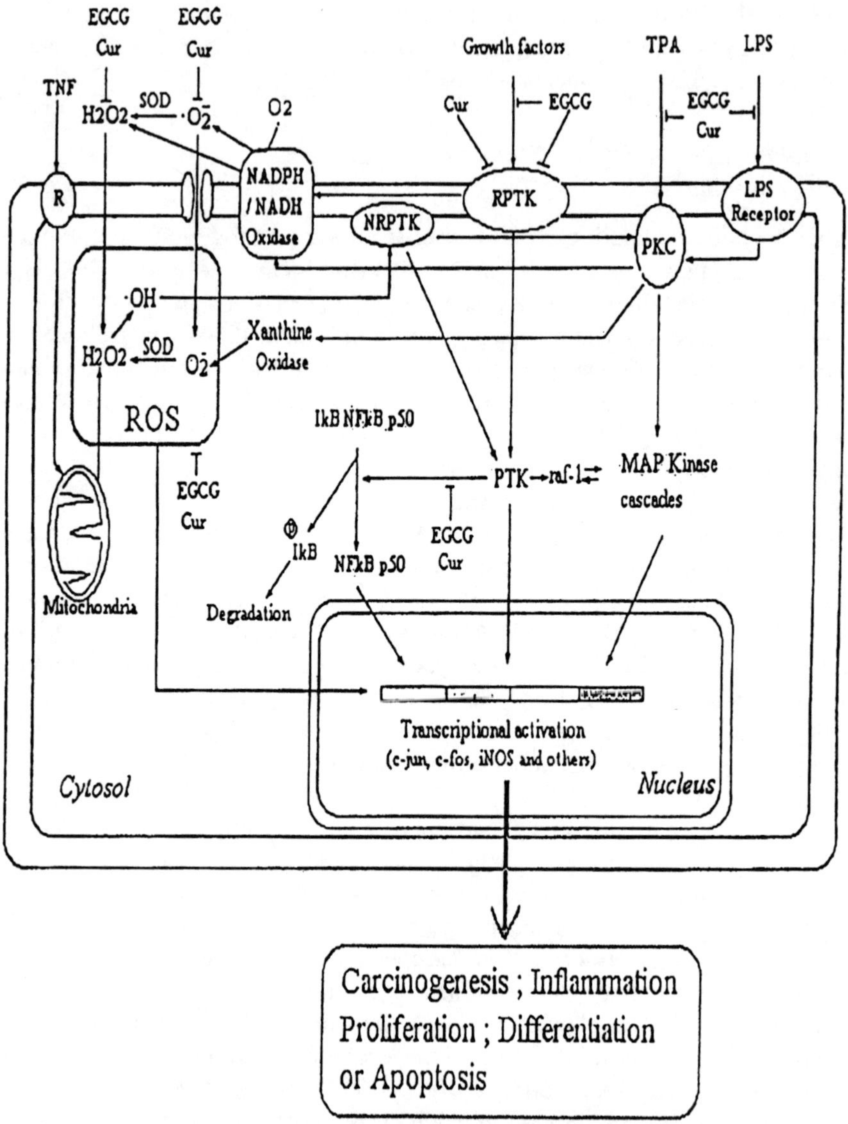

Fig. 21.1. Tentative prooxidative mitotic signal transduction pathways and their modulations by phytopolyphenolic antioxidants such as curcumin and (-)-epigallocatechin-3-gallate. Both exogenous stimulating factors (TPA and LPS) and endogenous stimulating factors (TNF, EGF, PDGF, FGF, IGF, and so on) bind to their intimate receptors. Autophosphorylation triggers the kinase activity of the cytoplasmic domain of their corresponding receptors. The target protein may be recognized by an SH-2 domain. The signal may be passed subsequently along a cascade of kinases. A common signal transduction cascade passes from a receptor tyrosine kinase through an adaptor (such as Grb2 and SOS) to Ras, which triggers a series of Ser/Thr phosphorylation events. Finally, the activat-

ed MAP kinases enter the nucleus and phosphorylate transcription factors. A signal transduction cascade passes to the nucleus by translocation of a component of the pathway or of a transcription factor (Lewin 1997). The factor may translocate directly as a result of phosphorylation (such as ERK2) or may be released when an inhibitor is phosphorylated (such as IκB). The activated PKC and RPTK are the important signal transducers that may deliver the signals to the MAP kinase cascade; meanwhile, they may also activate NADPH/NADH oxidase (or xanthine oxidase) to generate the prooxidative mitotic signal, i.e., ROS, which may in turn activate NRPTK, PKC, PTK, MAP kinases, and NFκB. These activated transducers (kinases and transcription activators) then turn on the nuclear vital genes (c-myc, c-jun, c-fos, and so on) that are actively participating in the processes of carcinogenesis, inflammation, proliferation, differentiation, and/or apoptosis. The possible sites that can be inhibited by Cur and EGCG on the signal transduction pathways are indicated. More experimental data are required to elucidate the molecular mechanisms of these inhibitions. Abbreviations: Cur, curcumin; EGCG, (-)-epigallocatechin-3-gallate; TNF, tumor necrosis factor; R, receptor for TNF; NRPTK, non-receptor protein tyrosine kinase; RPTK, receptor protein tyrosine kinase; PKC, protein kinase C; TPA, 12-*O*-tetradecanoyl-13-acetate; LPS, lipopolysaccharide; SOD, superoxide dismutase; ROS, reactive oxygen species; MAP kinase, mitotic activation protein kinase; PTK, protein tyrosine kinase; raf-1, ras activating factor-1; NFB, nuclear factor κB; IκB, inhibitor κB; iNOS, inducible nitric oxide synthase.

References

Akiyama, T., J. Ishida, and S. Nakagawa, Genistein, a Specific Inhibitor of Tyrosine Specific Protein Kinase, *J. Biol. Chem. 262:* 5592–5595 (1987).

Anderson, D., Antioxidant Defenses Against Reactive Oxygen Species Causing Genetic and Other Damage, *Mutat. Res. 350:* 103–108 (1996).

Burdon, R.H., Control of Cell Proliferation by Reactive Oxygen Species, *Biochem. Soc. Trans. 24:* 1028–1032 (1996).

Chaumontent, C., V. Bex, and I. Gallard-Sanchez, Apigenin and Tangeretin Enhance Gap Junction Intercellular Communication in Rat Liver Cells, *Carcinogenesis 15:* 2325–2330 (1994).

Chen, Y.C., T.C. Kuo, S.Y. Lin-Shiau, and J.K. Lin, Induction of HSP70 Gene Expression by Modulation of Ca^{++} Ion and Cellular p53 Protein by Curcumin in Colorectal Carcinoma Cells, *Mol. Carcinog. 17:* 224–234 (1996).

Cheng, E., C.D. Story, and L. Yoder, Estrogenic Activity of Isoflavone Derivatives Extracted and Prepared from Soybean Oil Meal, *Science 118:* 164–165 (1953).

Cooney, R.A., L.J. Mordan, and A. Franke, Multiple Mechanisms of Cancer Prevention by Photochemicals: Interaction Between Cellular Proliferation and Endogenous Mutagens, in *Food Factors for Cancer Prevention*, edited by H. Ohigashi, T. Osawa, J. Terao; S. Watanabe; and T. Yoshikawa, Springer, Tokyo, 1997, pp. 26–29.

Dreher, D., and A.F. Junod, Role of Oxygen Free Radicals in Cancer Development, *Eur. J. Cancer 32A:* 30–38 (1996).

Flohe, L., R. Brigelius-Flohe, and C. Saliou, Redox Regulation of NF-κB Activation, *Free Radic. Biol. Med. 21:* 437–448 (1997).

Franke, A.A., L.J. Mordan, and R.V. Cooney, Dietary Phenolic Agents Inhibit Neoplastic Transformation and Trap Toxic $N0_2$, Proceedings 86th Annual Meeting of the AACR, Toronto, ON, Canada, 1995, pp. 36, 125.

Huang, M.T., R.C. Smart, C.Q. Wong, and A.H. Conney, Inhibitory Effect of Curcumin, Chlorogenic Acid, Caffeic Acid and Ferulic Acid on Tumor Promotion in Mouse Skin by TPA, *Cancer Res. 48:* 5941–5946 (1988).

Huang, T.S., S.C. Lee, and J.K. Lin, Suppression of c–jun/Ap–1 Activation by an Inhibitor of Tumor Promotion in Mouse Fibroblast Cells, *Proc. Natl. Acad. Sci. USA 88:* 5292–5296 (1991).

Huang, Y.T., M.L. Kuo, and J.Y. Liu, Inhibition of Protein Kinase C and Proto-Oncogene Expression in NIH 3T3 cells by Apigenin, *Eur. J. Cancer 32A:* 146–151 (1996).

Jacob, R.A., and B.J. Burri, Oxidative Damage and Defense, *Am. J. Clin. Nutr. 63:* 985S–990S (1996).

Jang, M., L. Cai, and G.O. Udeani, Cancer Chemopreventive Activity of Resveratrol, a Natural Product Derived from Grapes, *Science 275:* 219–222 (1997).

Jiang, M.C, H.F. Yang-Yen, and J.K. Lin, Differential Regulation of p53, c-Myc, Bcl-2 and Bax Protein Expression During Apoptosis Induced by Widely Divergent Stimuli in Human Hepatopblastoma Cells, *Oncogene 13:* 609–616 (1996a).

Jiang, M.C., H.F. Yang-Yen, J.J.Y. Yen, and J.K. Lin, Curcumin Induces Apoptosis in Immortalized NIH 3T3 and Malignant Cancer Cell Lines, *Nutr. Cancer 26:* 111–120 (1996b).

Kakar, S.S., and D. Roy, Curcumin Inhibits TPA-Induced Expression of c-fos, c-jun and c-myc Protooncogenes Messenger RNAs in Mouse Skin, *Cancer Lett. 87:* 85–89 (1994).

Katagiri, K., T. Katagiri, and K. Kajiyama, Modulation of Monocytic Differentiation of HL-60 Cells by Inhibitors of Protein Tyrosine Kinases, *Cell Immunol. 140:* 282–294 (1992).

Kelloff, G.J., C.W. Boone, and J.A. Crowell, New Agents for Cancer Chemoprevention, *J. Cell. Biochem. 26S:* 1–28 (1996).

Koruta, L., Y.J. Cheung, J. Mendelsoln, and R. Kumar, Inhibition of Ligand-Induced Activation of Epidermal Growth Factor Receptor Tyrosine Phosphorylation by Curcumin, *Carcinogenesis 16:* 1741–1745 (1995).

Kuo, M.L., K.C. Lee, and J.K. Lin, Genotoxicity of Nitropyrnes and Their Modification by Apigenin, Tannic Acid, Ellagic Acid and Indole-3-Carbinol in the Salmonella and CHO Systems, *Mutat. Res. 270:* 87–95 (1992).

Lee, S.F., Y.T. Huang, W.S. Wu, and J.K. Lin, Induction of c-jun Protooncogene Expression by Hydrogen Peroxide Through Hydroxy Radical Generation and $p60^{src}$ Tyrosine Kinase Activation, *Free Radic. Biol. Med. 21:* 437–448 (1996).

Lee, S.F., Y.C. Liang, and J.K. Lin, Inhibition of 1,2,4-Benzenetriol-Generated Active Oxygen Species and Induction of Phase II Enzymes by Green Tea Polyphenols, *Chem. Biol. Interact. 98:* 283–301 (1995).

Lewin, B., *Genes, VI*, Oxford University Press, Oxford, 1997, pp. 1071–1084.

Liang, Y.C., S.Y. Lin-Shiau, C.F. Chen, and J.K. Lin, Suppression of Extracellular Signals and Cell Proliferation Through EGF Receptor Binding by (-)-Epigallocatechin Gallate in Human A431 Epidermoid Carcinoma Cells, *J. Cell. Biochem. 67:* 55–65 (1997).

Lin, J.K., Y.C. Liang, Y.C. Chen, I.M. Juan, and S.Y. Lin-Shiau, Anti-Carcinogenesis of Tea Polyphenols, in *Food Factor for Cancer Prevention*, edited by H. Ohigashi, T. Osawa, J. Terao; S. Watanabe; and T. Yoshikawa. Springer, Tokyo, 1997, pp. 122–126.

Lin, Y.L., and J.K. Lin, (-)Epigallocatechin-3-Gallate Blocks the Induction of Nitric Oxide Synthase by Down-Regulating Lipopolysaccharide-Induced Activity of Transcription Factor NF-κB, *Mol. Pharmacol. 52:* 465–472 (1997).

Liu, J.Y., S.J. Lin, and J.K. Lin, Inhibitory Effects of Curcumin on Protein Kinase C Induced by TPA in NIH 3T3 Cells, *Carcinogenesis 14:* 857–861 (1993).

Morse, M.A., and G.D. Stoner, Cancer Chemoprevention: Principle and Prospects, *Carcinogenesis 14:* 1737–1746 (1993).

Naim, M., B. Gestetner, and I. Kirson, A New Isoflavone from Soya Beans, *Phytochemistry 12:* 169–170 (1973).

Osawa, T., N. Ramarathnam, S. Kawahishi, and M. Namiki, Antioxidative Defense Systems Generated by Phenolic Plant Constituents, in *Phenolic Compounds in Food and Their Effects on Health II, Antioxidant and Cancer Prevention*, edited by M.T. Huang, C.T. Ho, and C.Y. Lee, ACS Symposium Series 507, American Chemical Society, Washington, DC 1992, pp. 122–134.

Ramarathnam, N., T. Osawa, M. Namiki, and S. Kawakishi, Chemical Studies on Novel Rice Hull Antioxidants II, Identification of Isovitexin, AC-Glycosyl Flavonoid, *J. Agric. Food Chem. 37:* 316–319 (1989).

Rice-Evans, C., N.J. Miller, and P.G. Bolwell, The Relative Antioxidant Activities of Plant Derived Polyphenolic Flavonoids, *Free Radic. Res. 22:* 375–383 (1995).

Sharma, O.P., Antioxidant Activity of Curcumin and Related Compounds, *Biochem. Pharmacol. 25:* 1811–1812 (1976).

Shibamoto, T., A Strong Antioxidant Found in Young Green Barley Leaves, in *Food Factors for Cancer Prevention*, edited by H. Ohigashi, T. Osawa, J. Terao, S. Watanabe; and T. Yoshikawa, Springer, Tokyo, 1997, pp. 304–308.

Shutt, D.A., and R.I. Cox, Steroid and Phytoestrogen Binding to Sheep Uterine Receptors *in vitro*, *J. Endocrinol. 52:* 299–310 (1972).

Srimal, R.C., and B.N. Dhawan, Pharmacology of Diferuloyl Methane (Curcumin), a Non-Steroidal Anti-Inflammatory Agent, *J. Pharm. Pharmacol. 25:* 447–452 (1973).

Srivastava, R., Inhibition of Neutrophil Response by Curcumin, *Agents Actions 28:* 298–303 (1985).

Steele, V.E., R.C. Moon, and R.A. Lubert, Preclinical Efficacy Evaluation of Potential Chemopreventive Agents in Animal Carcinogenesis Models: Methods and Results from the NCI Chemoprevention Drug Development Program, *J. Cell. Biochem. 20:* 32–54 (1994).

Stoner, G.D., and H. Mukhtar, Polyphenols as Cancer Chemopreventive Agents, *J. Cell. Biochem. 22:* 169–180 (1995).

Suzuki, Y.J., H.J. Forman, and A. Sevanian, Oxidants as Stimulators of Signal Transduction, *Free Radic. Biol. Med. 22:* 269–285 (1997).

Vinson, J.A., Y.A. Dabbagh, J.M. Serry, and J. Jang, Plant Flavonoids, Especially Tea Flavonoids, Are Powerful Oxidants Using a Nitrite Oxidation Model for Heart Disease, *J. Agric. Food Chem. 43:* 2800–2802 (1995).

Wang, H., and P.A. Murphy, Isoflavone Content in Commercial Soybean Foods, *J. Agric. Food Chem. 42:* 1666–1673 (1994).

Wattenberg, L.W., Chemoprevention of Cancer, *Cancer Res. 45:* 1–8 (1985).

Wei, H., L. Tye, E. Bresnick, and D.F. Brit, Inhibitory Effect of Apigenin, a Plant Flavonoid, on Epidermal Ornithine Decarboxylase and Skin Tumor Formation in Mice, *Cancer Res. 50:* 499–502 (1990).

Chapter 22

Radical-Scavenging Activity of Maillard Reaction Substances

M. Kawane[a], Y. Yoshiki[a], K. Okubo[a], and M. Tsunakawa[b]

[a]Department of Environmental Bioremediation, Graduate School of Agriculture, Tohoku University, 1-1 Tsutsumidori, Amamiyamachi, Aoba-ku, Sendai, Miyagi, 981 Japan

[b]Takano Foods Company, Noda, Ogawamachi, Higashiibaraki, Ibaraki, Japan 311-34

Introduction

Maillard reaction substances (MRS) are produced from the interaction of sugars and amino acids during the aging, cooking, and processing of foodstuffs. MRS, which involve the formation of Amadori products and Schiff bases, are responsible for mutagenesis, carcinogenesis, and DNA breaking (Hiramoto *et al.* 1994, Mihara and Shibamoto 1980). Earlier investigations showed that the generation of free radicals, such as dialkylpyrazine radicals and singlet oxygen from intermediate or secondary compounds, occurred during the early stages of the Maillard reaction (Namiki *et al.* 1973 and 1993). Further, a weak chemiluminescence (CL) was observed at an early stage of various amino-carbonyl reaction systems associated with the formation of free radicals. Namiki *et al.* (1993) suggested that the CL observed in Maillard reaction is caused by the reaction of oxygen with some reactive products such as free radicals or highly oxidizable compounds based on the effect of oxygen on CL occurrence of MRS.

On the other hand, we demonstrated previously that the photon emissions arose from the mixture of reactive oxygen species (X), catalytic species (Y), and receptive species (Z) at room temperature and under neutral conditions (Yoshiki *et al.* 1995). When phenolic compounds used as Y in the presence of reactive oxygen species (X) and acetaldehyde (Z), which acted as the typical Z, their CL occurrence and intensity were in agreement with the particle structures of phenolic compounds and the radical-scavenging activities identified with enzymatic (Puppo 1992) or high-performance liquid chromatography (HPLC)-electrochemical methods (Husain *et al.* 1987). Soy sauce, which is a fermented soybean-derived food, rich in MRS, is used widely as a seasoning in China and Japan. Recently, we found that a desalted soy sauce fraction exhibited CL in the presence of reactive oxygen species, with gallic acid as Y. This study examines the CL properties and superoxide and hydroxyl radical-scavenging activities of the soy sauce fraction. Further, we describe the inferred constituents of CL substances of soy sauce by comparison with the properties and activities of MRS obtained from glucose-amino acid model systems.

Materials and Methods

Preparation of Soy Sauce Chemiluminescence Fraction and the Maillard Reaction Substances from Glucose-Amino Acid Systems

Soy sauce was eluted with 50% MeOH after desalting, using a charcoal-activated column (50 × 50 mm) from coconut shells obtained from Nakarai Tesque (Kyoto, Japan). MeOH eluate (50%) was concentrated with an evaporator under reduced pressure to remove MeOH. The crude soy sauce fraction was applied to an ion-exchange column (Dowex 50W × 4, H^+ form, 25 × 250 mm) (Omaha, NE). The chemiluminescence substance of soy sauce was identified according to the "measurement of chemiluminescence." Maillard reaction substances (MRS) from glucose-amino acid systems were obtained from a mixture of 100 mg of glucose and 100 mg each of alanine, arginine, citrulline, glycine, histidine, homoserine, isoleucine, methionine, ornithine, phenylalanine, threonine, tryptophan, and valine solubilized in 5.0 mL of a 50 mmol/L phosphate buffer (pH 7.0). The solution was placed in a glass tube with a screw cap and heated at 100°C for 2 h. The absorbance of the MRS solution was measured at 420 nm after heating.

Measurement of Chemiluminescence

The chemiluminescence of H_2O_2/$FeCl_2$/gallic acid/MRS or soy sauce fraction was measured according to a single photon counting method, using a CL Detector Model CLD-110 (Tohoku Electronic Industrial) (Sendai, Japan) connected to a Waters Model 510 pump (Milford, MA) and a U6K injector. The wavelength was set in the range of 300–650 nm. The photons in the wavelength range of 300–650 nm were counted as total intensities. A 50 mmol/L phosphate buffer (pH 7.0) was used as the mobile phase; the flow rate was 1 mL/min, and a temperature of 23°C was used as described earlier (Yoshiki *et al.* 1995). The chemiluminescence of the soy sauce fraction or the MRS solution from the glucose-amino acid system, in the presence of hydroxyl radical and gallic acid, was measured as follows: 10 μL each of 176 mmol/L H_2O_2, 25 mmol/L $FeCl_2$, 5 mmol/L gallic acid, and soy sauce fraction or MRS solution (total 40 μL) were injected directly into the sample injector. The hydroxyl radical was generated in a system consisting of H_2O_2 and $FeCl_2$ (Fenton reaction).

Determination of Emission Spectra

The spectra of chemiluminescence were measured by a Simultaneous Multiwavelength Analyzer Model CLA-SP2 (Tohoku Electronic Industrial), equipped with a diffraction grating, a built-in spectroscope, and a two-dimensional photomultiplier tube. The wavelength range of the spectroscope was 270–900 nm. The reaction mixture (1.2 mL) contained a 50 mmol/L phosphate buffer (pH 7.0), 0.4 mL of 176 mmol/L H_2O_2, 0.4 mL of 5 mmol/L gallic acid, and soy sauce in a quartz cell. The light emission was determined for 180 s.

Measurement of Superoxide and Hydroxyl Radical-Scavenging Activities

The electron spin resonance (ESR) spectra were recorded on a JEOL JES-RE1X spectrometer (Tokyo, Japan) using an aqueous quartz flat cell (60 mm × 10 mm × 0.31 mm, effective volume 160 μL). The optical absorption spectra were measured by an Otsuka Electronics MCPD-100 multichannel photodetector (Otsuka Electronics Ltd., Osaka, Japan). Sample solutions [30 μL of soy sauce fraction and MRS obtained from glucose-amino acid system (see above)] were added to the mixture consisting of 50 μL of 2 mmol/L hypoxanthine, 50 μL of 0.4 unit/mL xanthine oxidase, and 20 μL of 9.2 mol/L 5,5-dimethyl-1-pyrroline-*N*-oxide (DMPO) for superoxide radical-scavenging activity. The superoxide-scavenging activity was calculated using the index of the DMPO-Ospin adduct by comparison with that of bovine erythrocyte superoxide dismutase as described by Misuta *et al.* (1990). Hydroxyl radicals were generated by the Fenton reaction. Sample solutions (30 μL) were added to the mixture consisting of 60 μL of 1 mmol/L H_2O_2, 60 μL of 0.1 mmol/L $FeCl_2$, 10 μL of 5.5 mmol/L diethylenetriamine-*N,N,N′,N″,N″*-pentaacetic acid, and 15 μL of 9.2 mol/L DMPO. The hydroxyl radical inhibitory ratio was calculated from the DMPO-OH signal.

Periodate Oxidation

Periodate oxidation of MRS was conducted as follows: Maillard products (1.0 mL) were treated with 50 mmol/L of periodic acid dihydrate ($HIO_4 \cdot 2H_2O$) for 4 d. After neutralization by barium carbonate ($BaCO_3$), the reaction mixture was centrifuged at 16,000 × *g* for 1 min. The supernatant solution was used as the periodate oxidation product.

Detection of Maillard Reaction Substances on Thin-Layer Chromatography by Chemiluminescence

The soy sauce fraction and MRS obtained from the glucose–amino acid system were detected on thin-layer chromatography (TLC) plates (TLC, Kieselgel 60, Merck, Darmstadt, Germany) using a solvent development system of $CHCl_3$/MeOH/H_2O (65:35:10, by vol, lower layer). The spots were visualized by spraying with 10% H_2SO_4 and heating at 120°C for 10 min. Chemiluminescence detection was performed by spraying with a mixture of 220 mmol/L H_2O_2 and 852 mmol/L MeCHO to catch the photon as Y, and with a mixture of 220 mmol/L H_2O_2 and 0.66 mmol/L gallic acid to catch the photon as Z, based on a XYZ system. The CL exhibited on TLC was counted for 1 min using a CCD camera (C2400-77; Hamamatsu Photonics, Hamamatsu, Japan) connected with an image processor (Argus-10; Hamamatsu Photonics).

Results and Discussion

A short burst of chemiluminescence (CL) from soy sauce was enhanced by the addition of H_2O_2. With the use of the CL method, we showed previously that gallic acid and MeCHO are typical Y (catalytic species) and Z (receptive species), respectively, in a XYZ system. The CL of soy sauce was examined to identify the activity as Y in the presence of 176 mmol/L H_2O_2 and 356 mmol/L MeCHO, and as Z in the presence of 176 mmol/L H_2O_2 and 5 mmol/L gallic acid. The CL analysis suggested that soy sauce acted as Z because the increase in CL was dependent on the fermentation period under the H_2O_2 and MeCHO condition compared with that under the H_2O_2 and gallic acid condition (Fig. 22.1). The fractionation and isolation of CL substances from soy sauce were performed using Sephadex G-100 and Sepharose 4B or 2B open columns (25 × 640 mm), (Pharmacia Biotech, Sweden), after desalting. However, CL substances were eluted in void fractions from any column, although they passed through the dialysis membrane. It is known that the sugar-amino acid browning reaction substances, commonly known as Maillard reaction substances (MRS), contain a large variety of high molecular weight compounds. Although earlier investigation of MRS analysis suggested the partial structure such as ketons, aldehydes, and ester moieties, the entire chemical structure of MRS was unclear. The fractionation results for CL substances from soy sauce had properties had properties similar to those for the MRS analysis. Further, soy sauce made from soybean fermentation is also a rich source of MRS. Therefore, the CL intensity of the desalted soy sauce fraction was compared with that of the solution of MRS from the glucose-amino acid system. The productive ratio of MRS was measured at 420 nm after heating using ultraviolet-visible (UV-vis) spectrophotometry. The solution of MRS gave absorptions from 0.35 to 0.60 for the glucose (Glc)-alanine (Ala), -arginine (Arg), -citrulline (Cit), -glycine (Gly), -histidine (His), homoserine (Hse), -isoleucine (Ile), -methionine (Met), -ornithine (Orn), -phenylalanine (Phe), -threonine (Thr), -tryptophan (Trp), and -valine (Val) systems, although MRS solution was not produced from Glc-cystine (Cys), -cysteine (Cys-Cys), -glutamic acid (Glu), and -asparagine (Asn) systems. CL observed with the Glc-amino system was due to the

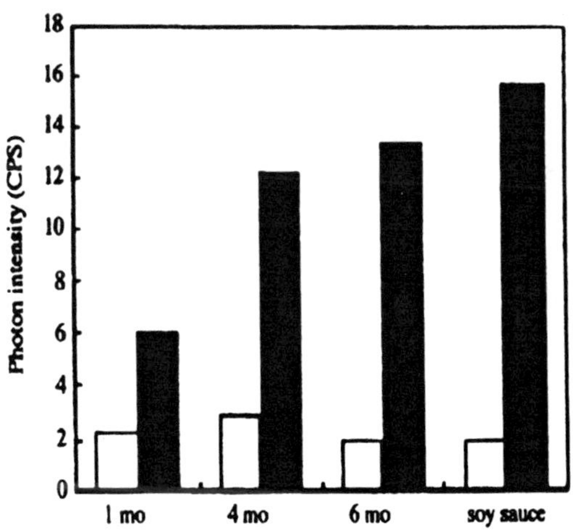

Fig. 22.1. Effects of fermentation period on the photon intensity [counts/s (CPS)]. □, photon intensity for activity as Y measured with hydroxyl radical (X) and acetaldehyde (Z); ■, photon intensity for activity as Z measured with X and gallic acid (Y).

MRS in the solution because solutions of Glc-Cys, -Cys-Cys, -Glu, and -Asn did not emit any CL. The CL intensity of the MRS solution showed a stronger enhancement in the presence of H_2O_2 and gallic acid than the CL in the presence of H_2O_2 and MeCHO. A similar tendency of CL with the soy sauce fraction indicated that the solution of MRS from the Glc-amino acid system acted as Z in the XYZ CL system.

To investigate the effect of the reactive oxygen species (X) on CL, the CL intensity of a 10% soy sauce fraction (10 μL) was compared with that of 10 μL of 176 mmol/L H_2O_2, 10 μL of 25 mmol/L HO•, 10 μL of 8 mol/L *tert*-BuOOH, 10 μL of 8 mol/L *tert*-BuO•, or 10 μL of 8 mol/L *tert*-BuOO• in the presence of 10 μL of 5 mmol/L gallic acid. The CL intensities were obtained at 38, 1124, 17, 71, and 52 counts/s for H_2O_2, HO•, *tert*-BuOOH, *tert*-BuO•, and *tert*-BuOO•, respectively. Because of the high CL intensities in the HO•, the comparison of activity as Y and Z of the soy sauce fraction and MRS from Glu-amino acid system was measured using 25 mmol/L HO• and 5 mmol/L gallic acid for Z activity (Table 22.1). Each CL intensity of HO•/gallic acid/sample and HO•/sample/MeCHO was calculated, with the CL intensity obtained from HO•/gallic acid/MeCHO mixtures arbitrarily set at 100. These results suggested that the main activity of the soy sauce fraction and the solution of MRS from the Glc-amino acid system was Z, although they had activities as both Y and Z in this CL system. In particular, the solutions of Glc-His, -Ile and -Val gave 5.1-, 5.2- and 4.1-fold higher CL intensity as Z compared with that of Y.

It has been reported that the CL observed in the Maillard reaction is caused by the reaction of oxygen associated with the formation of free radicals or oxidizable

TABLE 22.1
Photon Intensities of Soy Sauce and Maillard Reaction Substances Obtained from Glucose-Amino Acid System in the Presence of Hydroxyl Radical[a]

Substance	Y	Z
Control	100	100
Soy sauce	192	673
Ala	240	619
Arg	166	449
Cit	192	454
Gly	320	641
His	110	566
Hse	208	306
Ile	266	1377
Met	74	363
Orn	560	811
Phe	486	1239
Thr	134	256
Trp	147	235
Val	299	1239

[a]The photons of reactive oxygen (0.025 mol/L HO•), 5 mmol/L gallic acid (typical Y) and 0.35 mol/L MeCHO (typical Z) were arbitrarily set at 100 (control).

compounds (Namiki *et al.* 1993). On the other hand, there have been extensive studies on the mechanisms of CL in biological and chemical systems. According to the evidence, CL arises from singlet oxygen, dioxetane, and an excited carbonyl state. To elucidate the CL mechanism of soy sauce in the XYZ system, spectral analysis was carried out using a multiwavelength analyzer. The spectral analysis of the CL showed a distribution entirely in the visible region with main peaks at 660, 650, 660, and 680 nm for the soy sauce, H_2O_2/soy sauce, H_2O_2/soy sauce/MeCHO, and H_2O_2/gallic acid/soy sauce systems, respectively. A typical spectrum pattern is shown in Figure 22.2. These emission spectra were actually recorded during the XYZ system and were distinguishable from that of singlet oxygen molecules as observed by Slawlnska and Slawlnski (1975). Although it is difficult to elucidate the origin of CL in soy sauce, spectral analysis indicated that an excited carbonyl state caused the CL because a main peak was observed at 660 nm in the H_2O_2/gallic acid/MeCHO system.

The superoxide ($O_2^{-\bullet}$)- and hydroxyl radical ($HO^\bullet$)-scavenging activities were investigated using an ESR spin-trapping method. Table 22.2 shows the $O_2^{-\bullet}$- and $HO^\bullet$-scavenging activity of 30 μL of the soy sauce fraction and MRS solution from the Glc-amino acid system. Almost all of the MRS solution showed a slight $O_2^{-\bullet}$-scavenging activity, except for the soy sauce fraction, which was equal to 21.08 superoxide dismutase (SOD) unit/mL activity. On the other hand, the MRS solution from the Glc-Cit, -Met, -Phe, and -Trp systems had higher $HO^\bullet$ inhibition at

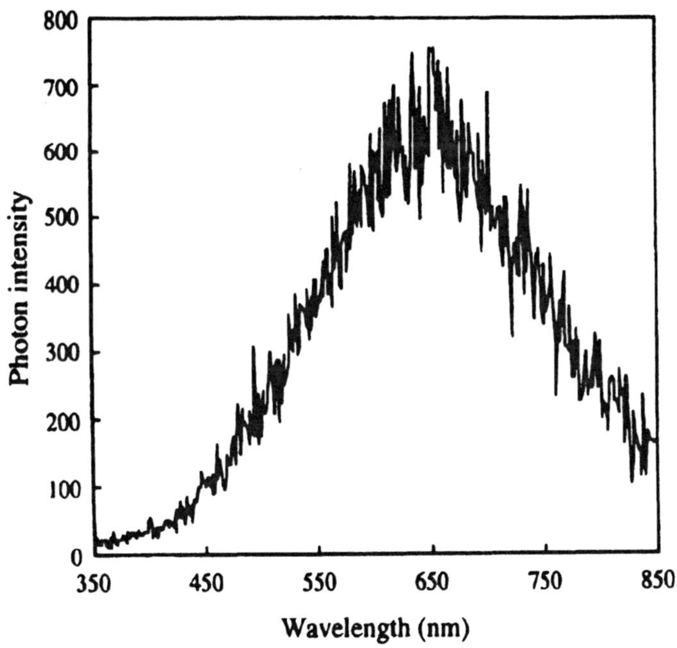

Fig. 22.2. Emission spectrum observed in H_2O_2/gallic acid/soy sauce system.

TABLE 22.2
Effects of Soy Sauce and Maillard Reaction Substances on Superoxide- and Hydroxyl Radical–Scavenging Activities[a]

Substance	$O_2^{-\bullet}$ inhibition (SOD unit/mL)	$HO^\bullet$ inhibition (%)
Soy sauce	21.08	10.2
Ala	0.24	28.6
Arg	0.38	29.3
Cit	0.53	51.6
Gly	1.01	21.8
His	2.86	25.8
Hse	0.31	37.9
Ile	0.54	31.5
Met	0.85	49.8
Orn	1.00	18.6
Phe	0.54	49.2
Thr	0.41	37.5
Trp	0.97	63.0
Val	0.38	38.5

[a]SOD, superoxide dismutase.

50–60% than other MRS solutions. Dimerization of amino acid radicals occurs easily from aromatic amino acids such as Phe, Trp, and tyrosine attached by active oxygen (Ushijima *et al.* 1985). Therefore, it is thought that a high $HO^\bullet$ inhibition of the MRS solution from Glc-Phe, and -Trp was caused by the aromatic moiety of MRS. In the ESR method, the MRS solution from Glc-His gave a result similar to that of the soy sauce fraction, i.e., they had a high $O_2^{-\bullet}$-scavenging activity and a low $HO^\bullet$ inhibition.

The decomposition was performed by periodate oxidation because the chemiluminescent constituents from soy sauce and the MRS solution had a high hydrophilicity and a large molecular weight. After periodate oxidation, the soy sauce fraction and MRS solution did not exhibit CL either as Y or as Z (Table 22.3).

These results suggest that the CL of the MRS solution was responsible for the adjoining hydroxyl moieties because the periodate oxidation decomposes the CHOH-CHOH bond. The MRS solution was fractionated using a TLC method based on the XYZ CL system (Fig. 22.3). This method, which is able to add photon emissions, depends on the reaction time and is different from the chemiluminescent flow system. CL was detected for 1 min in this study. The CL of the MRS solution, before and after periodate oxidation, was detected by spraying with a mixture of 220 mmol/L H_2O_2 and 852 mmol/L MeCHO to catch the photon as Y, and with a mixture of 220 mmol/L H_2O_2 and 0.66 mmol/L gallic acid to catch the photon as Z. After periodate oxidation, the disappearance of CL as Y was observed in soy sauce fractions, Glc-Phe, -Ile, and -His. The evidence from TLC suggested that the CL as Y and Z of MRS occurred *via* different constituents. Further, it sug-

TABLE 22.3
The Effects of Periodate Oxidation on the Photon Intensities of Soy Sauce and Maillard Reaction Substances Obtained from a Glucose-Amino Acid System and Soy Sauce in the Presence of Hydroxyl Radicals[a]

Substance	Before periodate oxidation		After periodate oxidation	
	Y	Z	Y	Z
Control	100	100	100	100
Soy sauce	614	790	8	7
Ala	192	673	16	7
Arg	166	449	6	7
Cit	208	363	7	7
Gly	192	454	11	8
His	560	811	14	7
Hse	74	363.	9	7
Ile	266	1377	9	8
Met	240	619	12	8
Orn	320	641	19	6
Phe	486	1239	10	7
Thr	134	256	16	7
Trp	147	235	27	5
Val	299	1239	8	7

[a]Maillard reaction substances were treated with 50 mmol/L HIO_4 (after periodate oxidation) and without 50 mmol/L HIO_4 (before periodate oxidation).

gested that the CL substances as Y resembled MRS from Glc-Ile or His because they have the same R_f values. This result is also supported by the behavior of CL, and of the $O_2^{-\bullet}$- and $HO^\bullet$-scavenging activity of the soy sauce fraction and MRS from Glc-His.

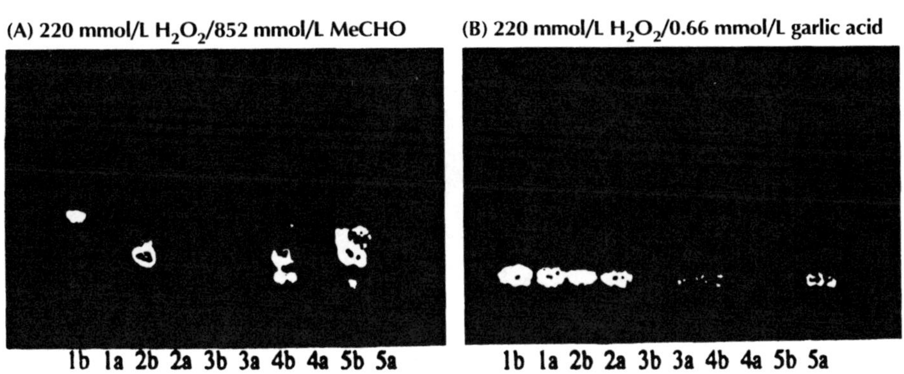

Fig. 22.3. Detection of Maillard reaction substances on thin-layer chromatography (TLC) by chemiluminescence. Maillard reaction substances treated (A) with 50 mmol/L HIO_4 and (B) without 50 mmol/L HIO_4. MRS obtained from (1) Glc-Phe, (2) -Ile, (3) -Val, and (4) -His systems and (5) soy sauce fraction.

It has been demonstrated that the free radical formation in the early stage of the reaction, before browning, is due to the formation of melanoidin (Namiki *et al.* 1973). The CL study in the XYZ system, however, indicated that MRS from the Glc-amino acid system and soy sauce had reactive oxygen-scavenging activity. The ESR method strongly supported the CL results. It is suggested that the responses differ in the two investigations because they arise from different mechanisms between the formation process of MRS and MRS properties.

References

Hiramoto, K., K. Kido, and K. Kikugawa, DNA Breaking by Maillard Products of Glucose-Amino Acid Mixtures Formed in an Aqueous System, *J. Agric. Food Chem. 42*: 689–694 (1994).

Husain, S.R., J. Cilard, and P. Cilard, Hydroxyl Radical Scavenging Activity of Flavonoids, *Phytochemistry 26*: 2489–2491 (1987).

Inatani, R., N. Nakatani, and H. Fuwa, Antioxidative Effect of the Constituents of Rosemary (*Rosmarinus officianalis* L.) and Their Derivatives, *Agric. Biol. Chem. 47*: 521–528 (1983).

Mihara, S., and T. Shibamoto, Mutagenicity of Products Obtained from Cysteamine-Glucose Browning Model Systems, *J. Agric. Food Chem. 28*: 62–66 (1980).

Mitsua, K., Y. Misuta, M. Kohno, M. Hiramatsu, and A. Mori, The Application of ESR Spin-Trapping Technique to the Evaluation of SOD-Like Activity of Biological Substances, *Bull. Chem. Soc. Jpn. 63*: 187–191 (1990).

Namiki, M., T. Hayashi, and S. Kawakishi, Free Radicals Developed in the Amino-Carbonyl Reaction of Sugars with Amino Acids, *Agric. Biol. Chem. 37*: 2935–2936 (1973).

Namiki, M., M. Oka, T. Miyazawa, K. Fujimoto, and K. Namiki, Weak Chemiluminescence at an Early Stage of the Maillard Reaction, *J. Agric. Food Chem. 41*: 1704–1709 (1993).

Puppo, A., Effect of Flavonoids on Hydroxyl Radical Formation by Fenton-Type Reactions; Influence of Iron Chelator, *Phytochemistry 31*: 85–88 (1992).

Slawlnska, D., and J. Slawlnski, Chemiluminescent Flow Method for Determination of Formaldehyde, *Anal. Chem. 47*: 2101–2109 (1975).

Ushijima, Y., M. Nakano, C. Takyu, and H. Inaba, Chemiluminescence in L-Tyrosine-H_2O_2-Horseradish Peroxidase System: Possible Formation of Tyrosine Cation Radical, *Biochem. Biophys. Res. Commun. 128*: 936–941 (1985).

Yoshiki, Y., Okubo, K., M. Onuma, and K. Igarashi, Chemiluminescence of Benzoic and Cinnamic Acids, and Flavonoids in the Presence of Aldehyderate and Hydrogen Peroxide or Hydroxyl Radical by Fenton Reaction, *Phytochemistry 39*: 225–229 (1995).

Chapter 23

Reactive Oxygen–Scavenging Compound Isolated from Adzuki Beans

T. Iida, Y. Yoshiki, M. Kawane and K. Okubo

Department of Environmental Bioremediation, Graduate School of Agriculture, Tohoku University, 1-1, Tsutsumidori, Amamiyamachi, Aoba-ku, Sendai, Miyagi 981, Japan

Introduction

Adzuki beans (*Vigna angularis*) are widely used for food and as medicinal plants in China and Japan. However, there have been few reports on their biological effects on human health. Recently, radical scavengers have received extensive attention because they can provide protection from oxidative cell injuries, which cause aging and many ailments, including carcinogenesis, inflammation, and mutagenesis. Superoxide is particularly involved in one of the crucial steps for controlling the reaction rate in oxidative damage to lipids and proteins. Therefore, inhibition or elimination of superoxide production can be effective in the promotion of antioxidative properties. Thus, we demonstrated the isolation of a superoxide-scavenging compound from Adzuki beans.

The flavonoids are one of the most numerous and widespread groups of natural compounds in the plant kingdom, i.e., they are present in seeds, fruits, nuts, vegetables, leaves, and flowers. Flavonoids have been reported to have antioxidative effects and the ability to inhibit free radicals (Cao *et al.* 1997, Cotelle *et al.* 1992, Vinson *et al.* 1995). Well-known examples include kaempferol, quercetin, myricetin, cyanidin, delphinidin, catechin, and epigallocatechin. These flavonoids also exhibit chemiluminescence (CL) in the presence of reactive oxygen species, such as hydrogen peroxide and the hydroxyl radical, and acetaldehyde (Yoshiki *et al.* 1995, 1996a, and 1996b). Chemiluminescence reflects the number and the position of the hydroxyl groups on the B-ring. This phenomenon suggests that the light emission occurs from the reaction of reactive oxygen species (X), catalytic species (Y), and receptive species (Z), caused by the radical-scavenging activity of Y. In this study, we describe the radical-scavenging and antioxidative activities of Glc-catechin from Adzuki beans. Further, we evaluate the correlation between radical-scavenging activity and the CL of Glc-catechin, using flavan-3-ols and gallic acid as standards (Fig. 23.1).

Materials and Methods

Materials

Adzuki beans were purchased from Marutane (Sendai, Japan). These seeds were cultivated in Hokkaido in 1995, cleaned and stored at 4°C, and then used. Gallic

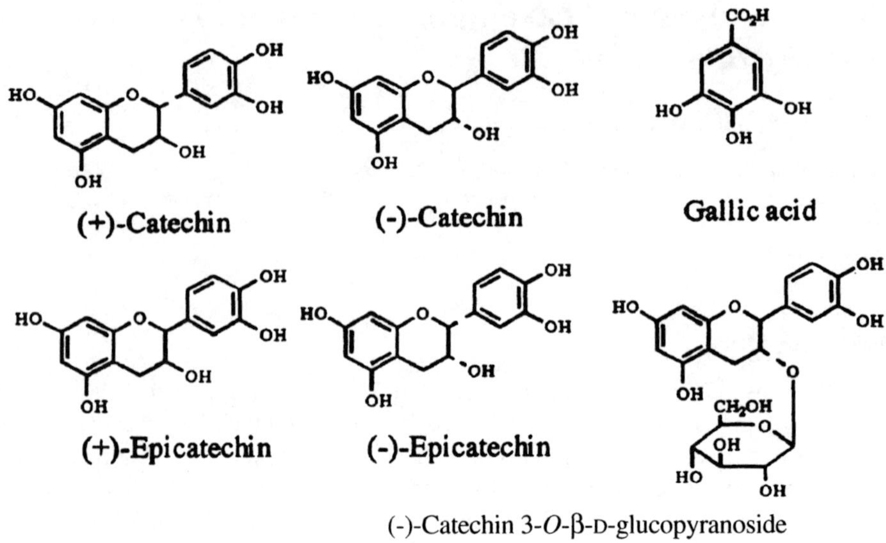

Fig. 23. 1. Comparative structures of Glc-catechin and other compounds.

acid, L-ascorbic acid, and 1,1-diphenyl-2-picrylhydrazyl (DPPH) were purchased from Nakarai (Kyoto, Japan). Bovine erythrocyte superoxide dismutase (SOD) and 5,5-dimethyl-1-pyrroline-*N*-oxide (DMPO) were from Sigma Chemical (St. Louis, MO) and Mitsui Toatsu Chemical (Tokyo, Japan), respectively. Hypoxanthine was from Kojin (Tokyo, Japan), and xanthine oxidase was from Behringer (Mannheim, Germany). Acetaldehyde (MeCHO) was from Merck (Darmstadt, Germany). Hydrogen peroxide (H_2O_2; 30%) and linoleic acid were from Santoku Chemical (Tokyo, Japan) and Wako Pure Chemicals (Tokyo, Japan), respectively. All of the flavan-3-ols were purchased from Kurita (Tokyo, Japan). All of the test solutions were freshly prepared before use.

Detection and Isolation

Glc-catechin was detected by the maximum absorbance at 280 nm. High-performance liquid chromatography (HPLC) was carried out by a CAPSELPACK C_{18} (Shiseido, (Tokyo, Japan) 4.6 mm × 250 mm) column using $MeCN/MeCO_2C_2H_5/H_2O$ containing a 0.05% H_3PO_4 (40:6:954) mixture as the mobile phase with a flow rate of 0.9 mL/min. To isolate Glc-catechin, whole seeds (1 kg) of Adzuki beans (*Vigna angularis*) were extracted with 70% EtOH and filtered using a Buchner funnel. The extract was evaporated and dissolved in water/*n*-BuOH (1:1). The H_2O layer was evaporated. The crude Glc-catechin fraction (30 g) was chromatographed on a ODS open column (25 mm × 640 mm) (YMC Co. Ltd., Kyoto, Japan) and eluted with 0, 10, and 40% MeCN solution. The 10% fraction was loaded onto the same column with a linear gradient elution from 0 to 10% aqueous MeCN. The eluate was further purified by a YMC-pack ODS-AM-323 (10 mm × 250 mm) column, (YMC Co. Ltd., Kyoto, Japan) using $MeCN/H_2O$ (3:97) as the mobile phase with a flow rate of 2.0 mL/min. This purification was replicated three times. The superoxide-scavenging compound, Glc-catechin, was isolated in the amount of 150 mg from the crude 10% fraction.

Analysis

Glc-catechin (5 mg) was dissolved in 5 mL of 2N HCl/50% dioxane and was hydrolyzed at 90°C for 2 h. After being cooled to room temperature, the reaction mixture was concentrated. After the hydrolysate was suspended in 300 μL of 50% MeOH, the sugar mixture was subjected to thin-layer chromatography (TLC) analysis (TLC, Kieselgel 60, Merck) using a solvent system of *n*-PrOH-acetone-H_2O (5:3:1). The spots were visualized by spraying with water-saturated *n*-BuOH containing 0.93% aniline and 1.66% phthalic acid and heating at 120°C for 10 min. ^1H nuclear magnetic resonance (NMR) and ^{13}C NMR spectra were recorded on a JEOL GSX spectrometer (JEOL Ltd., Tokyo, Japan) at 300 and at 75 MHz, respectively, in CD_3OD with tetramethylsilane (TMS) as an internal standard. Fast atomic bombardment (FAB)-mass spectra, using glycerol as a matrix, were obtained using a JEOL JMS HX-105.

Measurement of SOD-Like Activity

Superoxide was generated by the xanthine/xanthine oxidase system and measured by ultraviolet-visible (UV-vis) spectrophotometry (Hitachi U-1100, Tokyo, Japan) according to Shinohara (1985) (XOD-NH_2OH method) and ESR (JEOL JES-REIX) spectrometry. The activity of samples (Glc-catechin, gallic acid, and L-ascorbic acid) was measured in 1 mL of each concentration, 4 μmol/L–4 mmol/L (final concentration) by the XOD-NH_2OH method. SOD-like activity was indicated from the inhibitory ratio [100 – (ES – EB/EC – EB) × 100, where EB is a blank, EC is the control, and ES is the sample] by absorption at 550 nm. Electron spin resonance (ESR) spectrometry was carried out by the spin-trapping method using 5,5-dimethyl-1-pyrroline-*N*-oxide (DMPO) to bind the superoxide. Each sample (50 μL, 5 mmol/L Glc-catechin, flavan-3-ols, and gallic acid), was added to a solution mixture containing 50 μL of 2 mmol/L hypoxanthine, 50 μL of 0.4 unit/mL xanthine oxidase, and 20 μL of 9.2 mol/L DMPO. The superoxide-scavenging activity was calculated using the index of the DMPO-O spin adducts by comparison with that of SOD as described by Mitsuta *et al.* (1990).

Measurement of Chemiluminescence

Gallic acid and catechins were solubilized in 50% MeOH by stirring before use. The CL of gallic acid (final concentration, 2 mmol/L) and catechins (final concentration, 4 mmol/L) (Y) was measured in the presence of hydrogen peroxide (final concentration, 78 mmol/L), hydroxyl radical (final concentration, 4.2 mmol/L) (X), and acetaldehyde (final concentration, 72 mmol/L) (Z).The hydroxyl radical was produced by the Fenton reaction. The measurement of CL was carried out under the flow condition, using 50 mmol/L phosphate buffer with a flow rate of 1.0 mL/min as the mobile phase. The CL intensity was measured according to Yoshiki *et al.* (1995). The instrument used in this study was a filter-equipped photon counting-type spectrometer (CLD-110, Tohoku Electric), connected to a Waters Model 510 pump and a U6K injector (Milford, MA). The dispersed light at the grating was simultaneously detected on the photocathode with the image sensor. The photon counted between 300 and 650 nm was computed as the total spectral intensity.

Measurement of DPPH Radical-Scavenging Activity

Two milliliters of a 0.05 mmol/L ethanolic solution of samples and 1 mL of a 1.0 mmol/L ethanolic solution of 1,1-diphenyl-2-picryl hydrazyl (DPPH) radical were added to 2 mL of 0.1 mol/L acetate buffer (pH 5.5). The mixture was shacken vigorously and left to stand for 30 min. The absorbance of the mixture was measured at 517 nm using a spectrophotometer (Yoshikawa *et al.* 1994).

Measurement of Antioxidative Activity

The autoxidation of linoleic acid in aqueous ethanol solution was measured using the ferric thiocyanate method. Reaction mixtures containing 0.1 mmol/L samples for the autoxidation were prepared according to the procedure described by Inatani *et al.* (1987).

Results and Discussion

The crude fraction of Adzuki bean extract was obtained from the bottom phase, after being dispersed in water/*n*-BuOH (1:1). The fraction eluted with MeCN by a stepwise increase in the volume of H_2O on an ODS open column (25 mm × 640 mm). The SOD-like activity was measured by the XOD-NH_2OH method. The 10% eluate exhibited superoxide inhibition; it was loaded onto the same column with linear gradient elution from 0 to 10% aqueous MeCN (Fig. 23.2). The eluate with fraction numbers from 68 to 78 exhibited the strongest inhibition and absorbance at 280 nm. The eluate was further purified by a YMC-pack ODS-AM-323 (10 mm × 250 mm) column.

With the use of fast atom bombardment mass spectrometry (FAB-MS), the molecular formula of Glc-catechin was found to be $C_{21}H_{24}O_{11}$ (MW 452). The FAB-mass spectrum (positive-ion mode) gave quasimolecular ions at *m/z* 453 $[M+H]^+$, and fragment ions at 291 $[M+H\text{-}Glc]^+$. The FAB-mass spectrum (negative-ion mode) gave quasimolecular ions at *m/z* 451 $[M\text{-}H]^-$, and fragment ions at 289 $[M\text{-}H\text{-}Glc]^-$. The assignment of the ^1H and ^{13}C NMR signals in Glc-catechin was established by ^{13}C-^1H correlated spectroscopy (COSY) spectra and ^1H-^1H COSY spectra compared with the spectral data of (+)-epicatechin. The ^1H NMR of Glc-catechin indicated the presence of one anomeric proton in the sugar (δ 4.70, $1H^{23}{}_D$, *d*, $J = 7.7$ Hz). The C-3 signal of Glc-catechin (δ 76.6) in the ^{13}C NMR shifted downfield by δ 11.7 due to the glycosydation (Table 23.1). These results suggest that the sugar moiety in Glc-catechin is

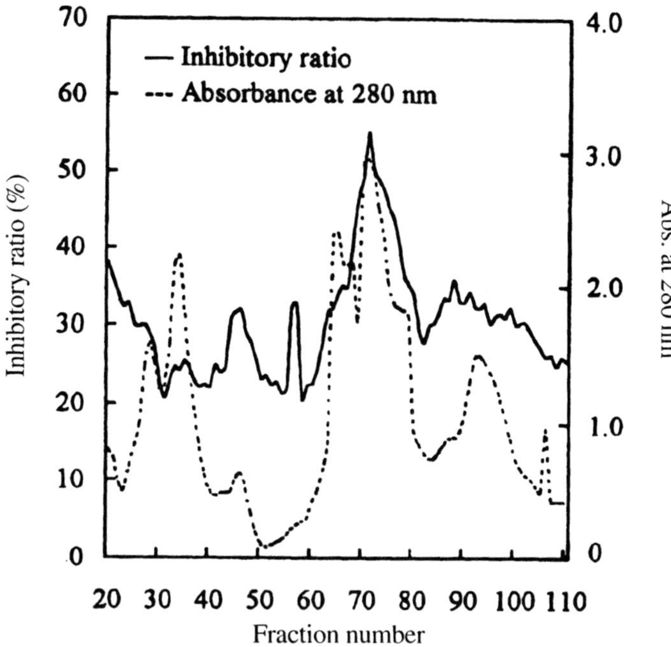

Fig. 23. 2. Chromatogram and superoxide dismutase (SOD)-like activity of 10% fraction on an ODS column. Sample eluate (0.4 mL) was measured for superoxide inhibition by the XOD-NH_2OH method.

linked to the oxygen at C-3 of the aglycone. The TLC analysis of sugar after hydrolysis indicated the presence of glucose. An anomeric proton with a large J value (7.7 Hz) showed the configuration at the glucose C-1 position to be in the β-form. The appearance of a doublet at δ 4.55 with a large J value (7.1 Hz), ascribable to the flavan 2-H and optical rotation suggested aglycone as (-)-catechin. The above spectral data identified Glc-catechin as (-)-catechin 3-*O*-β-D-glucopyranoside. Glc-catechin [α] = –52.3° (MeOH; c 0.4). ^{1}H NMR (CD_3OD): δ 2.37–2.70 (2H, *dd*, J = 16.2, 5.2 Hz, H-4), 4.70 (1H, *d*, J = 7.7 Hz, glc. H-1), 4.55 (1H, *d*, J = 7.1 Hz, H-2), 6.09 (1H, *d*, J = 2.2 Hz, H-6), 5.99 (1H, *d*, J = 2.2 Hz, H-8), 6.59 (1H, *d*, J = 1.7 Hz, H-6), 6.72 (1H, *d*, J = 1.9 Hz, H-5′), 6.71 (1H, *s*, J = 2.2 Hz, H-2′).

A comparison of the SOD-like activity with various concentrations (4 μmol/L, 40 μmol/L, 400 μmol/L, and 4 mmol/L) of Glc-catechin and standard samples (gallic acid and L-ascorbic acid) is shown in Figure 23.3. Glc-catechin was capable of scavenging superoxide in a concentration-dependent fashion. The calculations of the 50% inhibitory concentration (IC_{50}) of Glc-catechin, gallic acid, and L-ascorbic acid were 176.8, 32.8, and 39.8 μmol/L, respectively. Glc-catechin showed about one fifth the superoxide-scavenging activity compared with gallic acid and L-ascorbic acid.

Samples (4 μmol/L–4 mmol/L) were measured by the XOD-NH_2OH method. Table 23.2 shows the CL for Glc-catechin in the presence of hydrogen peroxide

TABLE 23.1
^{13}C Nuclear Magnetic Resonance (NMR) Spectral Assignments for Glc-Catechin from Adzuki Beans[a]

(+)-Epicatechin		Glc-catechin	
A- and C-ring			
C-2	78.1	2	81.2
C-3	64.9	3	76.6
C-4	28.1	4	27.5
C-4a	98.6	4a	102.0
C-5	155.9	5	155.4
C-6	95.2	6	96.4
C-7	156.4	7	156.4
C-8	94.2	8	94.7
C-8a	156.7	8a	157.0
B-ring			
C-1′	130.7	1′	130.5
C-2′	114.8	2′	115.3
C-3′	144.6	3′	145.2
C-4′	144.6	4′	145.2
C-5′	115.0	5′	114.6
C-6′	118.1	6′	118.4
		3-*O*-β-D-Glucopyranosyl	
		C-1″	100.6
		C-2″	73.3
		C-3″	77.0
		C-4″	69.7
		C-5″	66.1
		C-6″	60.7

[a]In CD_3OD with tetramethylsilane (TMS) as an internal standard.

and hydroxyl radical (X), and acetaldehyde (Z). To investigate the correlation between the structure and the CL intensity of Glc-catechin, four stereoisomers of flavan-3-ols and gallic acid were also used as Y standards. Only flavan-3-ols with (-)-configuration gave CL comparable to that of gallic acid in the presence of hydrogen peroxide as X. (-)-Epicatechin showed the highest photon intensity (22.0 counts/s) in flavan-3-ols; (-)-catechin gave a CL intensity about half a magnitude lower than that of (-)-epicatechin. However, no light emission was observed, when Glc-catechin was used as Y. Because Glc-catechin has the (-)-configuration, photon emission was expected to be observed in the presence of Y and Z. This observation suggests that the hydroxyl group at the C-3 position of the C-ring is probably needed for photon emission, and that attachment of glucose to the C-3 position on the C-ring of the flavan-3-ol skeleton is responsible for the lack of emission. Moreover, the same results were obtained using hydrogen peroxide as X. These results indicate that the CL intensity in the presence of hydrogen peroxide and the hydroxyl radical as X is related not only to the steric factor of the flavan-3-ol

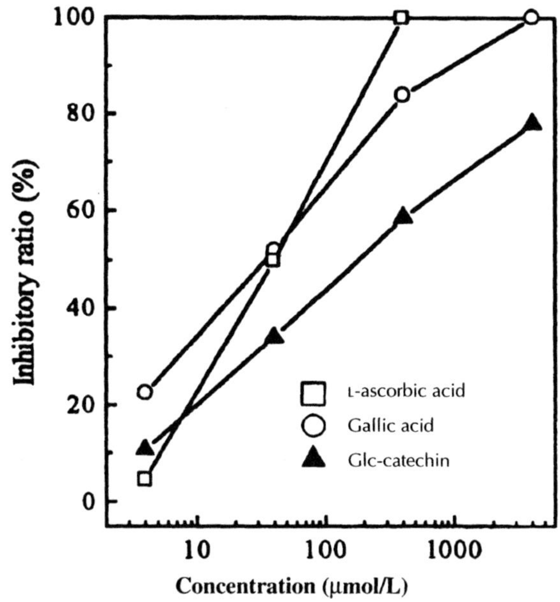

Fig. 23. 3. Effects of Glc-catechin and other compounds on superoxide inhibition. Samples (4 μmol/L–4 mmol/L) were measured by the XOD-NH_2OH method.

skeleton, but also to the substitution of the C-3 position on the C-ring. With respect to the steric factor, we suspect that substances with chirality and stereoselectivity substances may play a role in the course of the reaction.

The scavenging activity of Glc-catechin on superoxide and the DPPH radical, compared with flavan-3-ols and gallic acid, is shown in Table 23.3. The SOD-like

TABLE 23.2
Photon Intensity of Glc-Catechin and Other Compounds as Y in the Presence of Hydrogen Peroxide or Hydroxyl Radical as X and Acetaldehyde as Z[a]

	Photon intensity (counts/s)	
	Hydrogen peroxide	Hydroxyl radical
(+)-Catechin	—	—
(-)-Catechin	13.3	5.3
(+)-Epicatechin	—	—
(-)-Epicatechin	22.0	16.6
Glc-catechin	—	—
Gallic acid	4.7	13.3

[a]Conditions: Buffer, 50 mmol/L phosphate buffer (pH 7); reaction mixture containing 10 μL of 195 mmol/L reactive oxygen species (X), 10 μL of catalytic species (Glc-catechin and flavan-3-ols; 10 mmol/L, gallic acid; 5 mmol/L) (Y), and 5 μL of 360 mmol/L acetadehyde (Z) was injected with a 50-μL syringe. The hydroxyl radical was produced by Fenton reaction (5 μL of 25 mmol/L $FeCl_2$ and 10 μL of 195 mmol/L hydrogen peroxide).

TABLE 23.3
Superoxide and 1,1-Diphenyl-2-picrylhydrazyl (DPPH) Radical-Scavenging Activities of Glc-Catechin and Other Compounds[a]

Compounds	SOD activity (unit/mL)	DPPH inhibitory ratio (%)
(+)-Catechin	6.5	46.6
(-)-Catechin	8.7	52.4
(+)-Epicatechin	6.8	54.8
(-)-Epicatechin	7.4	57.2
Glc-catechin	6.2	37.7
Gallic acid	10.3	70.4

[a]Superoxide generated by hypoxanthine and XOD. Superoxide dismutase (SOD)-like activity was measured by the electron spin resonance (ESR) spin-trapping method. Modulation amplitude, 0.079 G; time constant, 0.3 s; sweep time, 5 min. The concentration of the sample was 5 mmol/L for SOD activity and 0.05 mmol/L for DPPH inhibition.

activities of (+)-catechin, (-)-catechin, (+)-epicatechin, (-)-epicatechin, and Glc-catechin measured by ESR were 6.5, 8.7, 6.8, 7.4, and 6.2 unit/mL, respectively. The DPPH radical-scavenging activities of (+)-catechin, (-)-catechin, (+)-epicatechin, (-)-epicatechin, and Glc-catechin were 46.6, 52.4, 54.8, 57.2, and 37.7%, respectively. Slight differences of SOD-like activities among samples were thought to be influenced only marginally by the substituent at the C-3 position of the C-ring. It has been reported that the hydroxyl group at the C-3 position of the C-ring in flavonols is the first point of attack for superoxide. In fact, flavonol glucosides, such as rutin and quercitrin, have a lower SOD-like activity than their aglycone (Takahama 1983 and 1987). These results suggest that C-3 substituents on the C-ring of flavanol are less important for the reaction with superoxide than flavonol. It is predicted that there are different mechanisms for reactions with superoxide between flavanol and flavonol. The DPPH radical-scavenging activity of Glc-catechin was obviously lower than that of other flavan-3-ols. The hydroxyl group at the C-3 position of the C-ring is thought to be related to the reduction of DPPH, which indicates its hydrogen-donating ability. However, the hydroxyl group of the A- or B-ring is related to the hydrogen donor to the DPPH radical because Glc-catechin also has reduction power.

The antioxidative activity of 0.1 mmol/L Glc-catechin was compared with 0.1 mmol/L flavan-3-ols and 0.1 mmol/L gallic acid, measured by the ferric thiocyanate method (Fig. 23.4). Glc-catechin was the most efficient among the samples in retarding oxidative deterioration of linoleic acid. Epicatechins have prooxidative activity because they are easily polymerized or decomposed; they are probably photosensitized by oxidation caused by singlet oxygen (which might have occurred). A hydrogen donor can break the chain reaction, but Glc-catechin has a lower donating activity than flavan-3-ols, hence glucose attached to the C-3 position on the C-ring of the aglycone may affect the stability of the molecule or may become the target of the hydroxyl radical, which initiates lipid peroxidation.

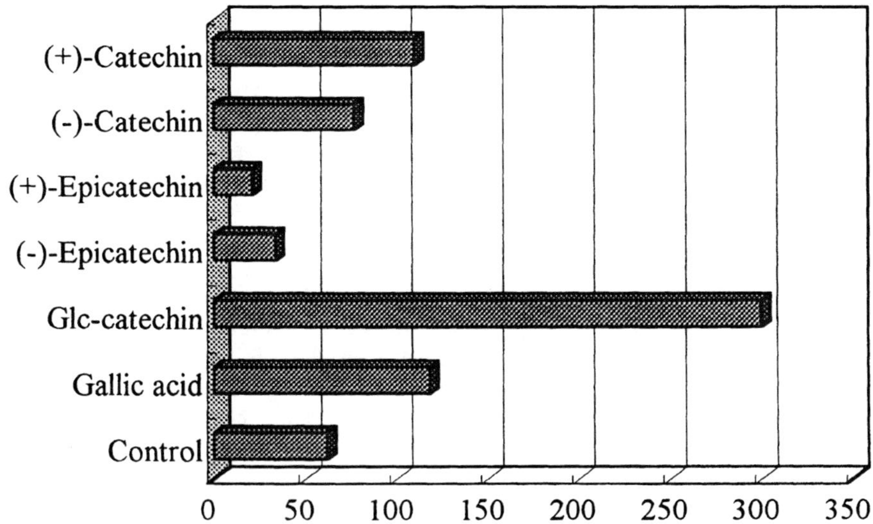

Fig. 23. 4. Antioxidative activity of Glc-catechin and other compounds. Incubation periods were values at which absorbance reached 0.3. Each sample of 0.1 mmol/L at the final concentration was added to a reaction mixture of linoleic acid (final concentration, 0.52%)/ethanol-phosphate buffer (pH 7.0) and incubated at 37°C.

In conclusion, we indicated that the SOD-like activity of 5 mmol/L Glc-catechin measured by ESR was 6.2 unit/mL and that the DPPH radical inhibitory ratio of 0.02 mmol/L Glc-catechin was 37.7%. Furthermore, it was observed that Glc-catechin from Adzuki beans acted as an effective and stable antioxidant for a long period. However, the presence of any correlation between radical-scavenging activity and photon intensity was not clarified in this experiment.

References

Cao, G., E. Sofic, and R.L. Prior, Antioxidative and Prooxidant Behavior of Flavonoids: Structure-Activity Relationships, *Free Radic. Biol. Med. 22*: 749–760 (1997).

Cotelle, N., J.L. Bernier, J.P. Henichart, J.P. Catteau, E. Gaydou, and J.C. Wallet, Scavenger and Antioxidant Properties of Ten Synthetic Flavones, *Free Radic. Biol. Med. 13*: 211–219 (1992).

Inatani, R., N. Nakatani, and H. Fuwa, Antioxidative Effect of the Constituents of Rosemary (*Rosemarinus officinalis* L.) and Their Derivatives, *Agric. Biol. Chem. 47*: 521–528 (1983).

Mitsuta, K., Y. Mizuta, M. Kohno, M. Hiramatsu, and A. Mori, The Application of ESR Spin-Trapping Technique to the Evaluation of SOD-Like Activity of Biological Substances, *Bull. Chem. Soc. Jpn. 63*: 187–191 (1990).

Shinohara, R., Superoxide Dismutase, *Rinsyokensa 29*: 1447–1452 (1985).

Takahama, U., Redox Reactions Between Kaempferol and Illuminated Chloroplasts, *Plant Physiol. 71*: 598–601 (1983).

Takahama, U., Oxidation Products of Kaempferol by Superoxide Anion Radical, *Plant Cell Physiol. 28*: 953–957 (1987).

Vinson, J.A., Y.A. Dabbagh, M.M. Serry, and J. Jang, Plant Flavonoids, Especially Tea Flavonols, Are Powerful Antioxidants Using an *in vitro* Oxidation Model for Heart Disease, *J. Agric. Food Chem. 43*: 2800–2802 (1995).

Yoshikawa, M., E. Harada, A. Miki, K. Tsukamoto, S.Q. Liang, J. Yamahara, and N. Murakami, Antioxidant Constituents from the Fruit Hulls of Mangosteen (*Garcinia mangostana* L.) Originating in Vietnam, *Yakugaku Zasshi 114*: 129–133 (1994).

Yoshiki, Y., T. Kahara, K. Okubo, K. Igarashi, and K. Yotsuhashi, Mechanism of Catechin Chemiluminescence in the Presence of Active Oxygen, *J. Biolumin. Chemilumin. 11*: 131–136 (1996a).

Yoshiki, Y., M. Kinumi, T. Kahara, and K. Okubo, Chemiluminescence of Soybean Saponins in the Presence of Active Oxygen Species, *Plant Sci. 116*: 125–129 (1996b).

Yoshiki, Y., and K. Okubo, Chemiluminescence of Anthocyanins in the Presence of Acetaldehyde and *tert*-Butyl Hydroperoxide, *J. Biolumin. Chemilumin. 10*: 1–4 (1995).

Chapter 24

Antioxidative and Cytotoxic Components of Highbush Blueberry (*Vaccinium corymbosum* L.)

Mingfu Wang[a], Jiangang Li[a], Yu Shao[a,b], Tzou-Chi Huang[c], Mou-Tuan Huang[d], Chee-Kok Chin[b], Robert T. Rosen[e], and Chi-Tang Ho[a]

[a]Department of Food Science, [b]Department of Plant Science and [e]Center for Advanced Food Technology, Rutgers University, New Brunswick, NJ 08901–8520

[c]Department of Food Science and Technology, National Pingtung Polytechnic Institute, Pingtung, Taiwan

[d]Department of Chemical Biology, Laboratory for Cancer Research, College of Pharmacy, Rutgers University, Piscataway, NJ 08855–0789

Introduction

Cancer was reported to cause 23% of the premature loss of life and about 530,000 deaths in the United States in 1993. It is well known that consumption of adequate amounts of fruits and vegetables is associated with a lowered risk of degenerative diseases such as cancer (Ames *et al.* 1995). Blueberry is a fruit produced commercially in North America, Europe, and Japan. It is consumed fresh or as processed products. The reported per capita consumption of blueberry in the United Stated in 1985 was 6 ounces per person (Eck 1988). The chemical components of blueberry, other than the anthocyanins, have not been extensively studied. The major anthocyanins in highbush blueberry were identified as malvidin 3-galactoside, delphinidin 3-galactoside, delphinidin 3-arabinoside, petunidin 3-galactoside, petunidin 3-arabinoside, and malvidin 3-arabinoside (Gao and Mazza 1994). There are also some reports on the determination of flavonols and hydroxybenzoic acid derivatives in blueberry (Bilyk and Sapers 1986, Gao and Mazza 1994, Teeling *et al.* 1971).

It has been reported recently that the extracts of the fruits of the *Vaccinium* species showed anticancer activities *in vitro* (Bomser *et al.* 1996). It was observed that the crude extracts, the anthocyanin and proanthocyanidin fractions, were not highly active in QR (Phase II xenobiotic enzyme, quinone reductase) induction, whereas the ethyl acetate extracts were active QR inducers. This contribution reports on a systematic study of the chemical components of the ethyl acetate–extracted fraction of highbush blueberry and their antioxidative and cytotoxic activities.

Materials and Methods

General Procedures

Electron-impact mass spectrometry (EI-MS) spectra were obtained on a Finnigan Mat 8320 mass spectrometer. The nuclear magnetic resonance (NMR) spectra were recorded on a Bruker AMX-400 spectrometer.

Materials

Dried fruits of blueberry were a gift from Tru-Blu Cooperative Association (New Lisbon, NJ). The silica gel (130–270 mesh) and thin-layer chromatography (TLC) plates were purchased from Aldrich Chemical (Milwaukee, WI); the Sephadex LH-20 was purchased from Sigma Chemical (St. Louis, MO). All organic solvents and other chemicals were obtained from Fisher Scientific (Springfield, NJ).

Extraction

The dried fruits of highbush blueberry (4.5 kg) were extracted with 95% EtOH at room temperature. The extract was evaporated to dryness under reduced pressure, and the residue was dissolved in water; the aqueous solution was extracted with hexane, ethyl acetate, and *n*-butanol to give fractions H (25 g), E (17.3 g), and B (150 g), respectively.

Part E (17.3 g) was subjected to column chromatography (CC) on silica gel, eluted with hexane/ethyl acetate (10:1, 9:1, 8:1, 7:1, 6:1, 5:1, 4:1, 3:1, 2:1, 1:1, vol/vol, each 3500 mL) and separated into 15 fractions (determined by TLC). Fraction 7 (Fr 7) was rechromatographed on Sephadex LH-20, eluted with MeOH, then purified with CC (chloroform/methanol, 40:1, vol/vol) to give 180 mg of compound (**1**) (Fig. 24.1). Fraction 8 (Fr 8) was rechromatographed on Sephadex LH-20, eluted with methanol, and then subjected to CC on silica gel (hexane/ethyl acetate, 3:1, vol/vol) to yield 24 mg of compound (**2**). Fraction 13 (Fr 13) was first subjected to CC on silica gel (chloroform/methanol/acetic acid, 20:1:1, by vol) and then purified by Sephadex LH-20 column to afford 70 mg of compound (**3**). Fraction 14 (Fr 14) was first rechromatographed on Sephadex LH-20 column, then applied to Silica gel column, eluted with hexane/acetone/acetic acid (2:1:0.1, v/v/v), hexane/ethyl acetate/acetic acid (2:1:0.1, v/v/v) and chloroform/methanol/acetic acid (20:1:0.5, v/v/v) to give 40 mg of compound (**4**) and 80 mg of compound (**5**). Fraction 15 (Fr 15) was also first separated by Sephadex LH-20 column, then subjected to silica gel, eluted with acetone/hexane/acetic acid (1:2:0.1, v/v/v) and chloroform/acetone/acetic acid (5:1: 0.1, by vol), respectively, to yield 50 mg of compound (**6**).

Identification and Structure Elucidation of the Isolated Compounds

Compound (**1**) was isolated as a white powder; melting point: 292°C; ^{1}H NMR [in dimethyl sulfoxide (DMSO)-d_6, 400 MHz] exhibited δ 0.65, 0.71, 0.84, 0.86, 1.02 (each, 3H, *s*), 0.78 (3H, *d*, $J = 5$ Hz), 3.1 (1H, *m*), 5.14 (1H, *m*). ^{13}C NMR (DMSO-d_6, 100 MHz) showed signals at 33.0 (*t*, C-1), 27.1 (*t*, C-2), 77.3 (*d*, C-3), 38.4 (*s*, C-4), 55.1 (*d*, C-5), 18.2 (*t*, C-6), 33.0 (*t*, C-7), 38.6 (*s*, C-8), 47.3 (*d*, C-9), 36.8 (*s*, C-10), 23.1 (*t*, C-11), 124.9 (*d*, C-12), 138.4 (*s*, C-13), 41.9 (*s*, C-14), 27.8 (*t*, C-15), 24.1 (*t*, C-16), 47.1 (*s*, C-17), 52.7 (*d*, C-18), 38.9 (*d*, C-19), 38.5 (*d*, C-20), 30.4 (*t*, C-21), 36.6 (*t*, C-22), 28.5 (*q*, C-23), 16.3 (*q*, C-24), 16.3 (*q*, C-25), 17.2 (*q*, C-26), 23.5 (*q*, C-27), 178.7 (*s*, C-28), 17.2 (*q*, C-29), 21.3 (*q*, C-30). These

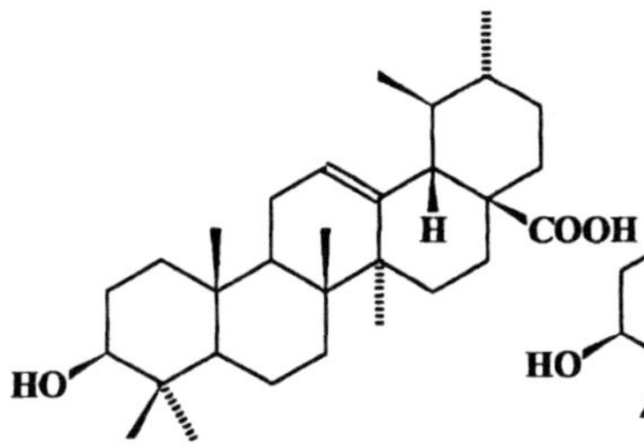

(1)

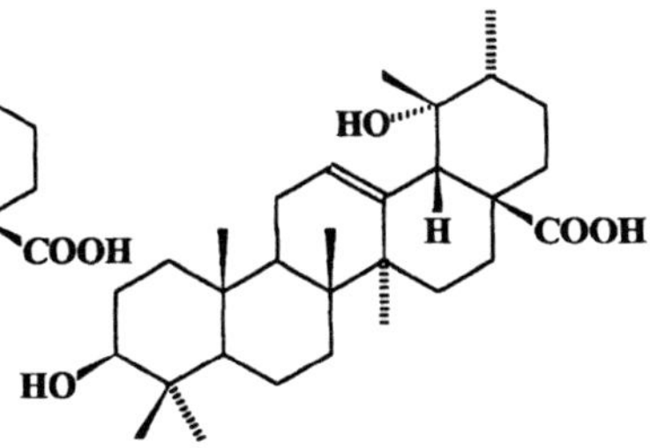

(2)

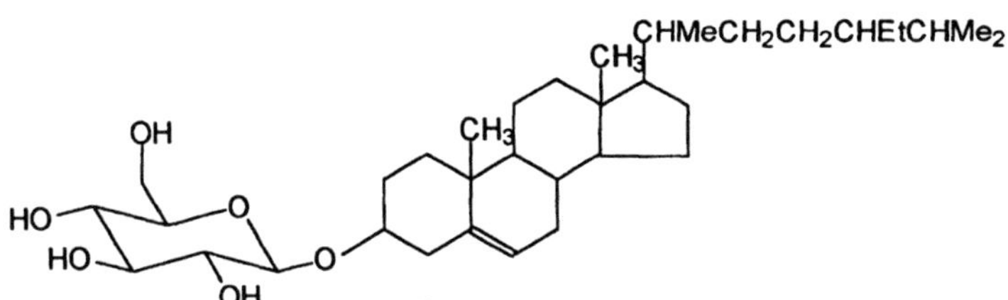

(3)

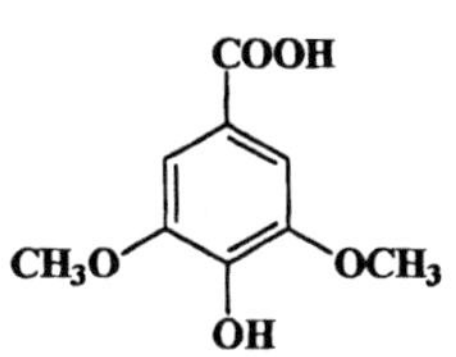

(4)

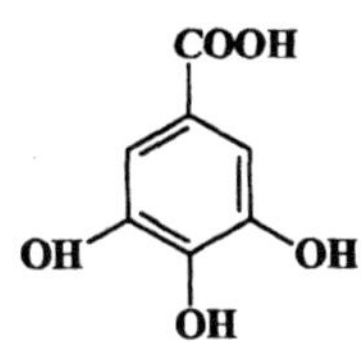

(5)

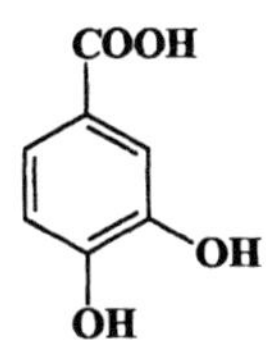

(6)

Fig. 24.1. Chemical structures of compounds identified in the ethyl acetate extract of highbush blueberry: (**1**), ursolic acid; (**2**), 3β,19α-dihydroxy-urs-12-en-28-oic acid; (**3**), β-sitosterol-β-D-glucoside; (**4**), syringic acid; (**5**), gallic acid; (**6**), protocatechuic acid.

data were identical to those of ursolic acid; therefore, compound (**1**) was identified as ursolic acid.

Compound (**2**) was isolated as a white powder; its molecular formula, $C_{30}H_{48}O_4$, was deduced from EI-MS and ^{13}C NMR. Compound (**2**) showed ^1H NMR and ^{13}C NMR spectra similar to those of (**1**), allowed the identification of (**2**) as an urs-12-ene type triterpene with a 3-OH. According to the molecular weight, two hydroxyl groups were present in (**2**). The second hydroxyl group was attributed to the 19-position, according to the chemical shifts of protons and carbons in 19-hydroxy-urs-12-ene derivatives (Terreaux *et al.* 1996). This was confirmed by the EI-MS, which showed a peak at *m/z* 264 as a result of retro-Diels-Alder fragmentation of ring C. This peak indicated the presence of a hydroxyl group on ring D or E (Fullas *et al.* 1996, Terreaux *et al.* 1996); therefore, (**2**) was identified as 3β,19α-dihydroxy-urs-12-en-28-oic acid. Its ^{13}C NMR (in DMSO, 100 MHz) was assigned according to literature data (Fullas *et al.* 1996, Seto *et al.* 1984, Tommasi *et al.* 1996) as follows: 32.9 (*t*, C-1), 27.1 (*t*, C-2), 77.3 (*d*, C-3), 38.9 (*s*, C-4), 55.2 (*d*, C-5), 18.4 (*t*, C-6), 33.0 (*t*, C-7), 38.6 (*s*, C-8), 47.0 (*d*, C-9), 36.8 (*s*, C-10), 23.4 (*t*, C-11), 127.1 (*d*, C-12), 138.8 (*s*, C-13), 41.3 (*s*, C-14), 27.1 (*t*, C-15), 23.4 (*t*, C-16), 47.2 (*s*, C- 17), 53.5 (*d*, C-18), 72.0 (*s*, C-19), 41.5 (*d*, C-20), 26.2 (*t*, C-21), 37.5 (*t*, C-22), 28.5 (*q*, C-23), 16.3 (*q*, C-24), 16.5 (*q*, C-25), 16.9 (*q*, C-26), 24.2 (*q*, C-27), 179.3 (*s*, C-28), 15.3 (*q*, C-29), 26.7 (*q*, C-30); ^{1}H NMR (in DMSO, 400 MHz): 0.65, 0.66, 0.82, 0.82, 0.86, 1.21 (each, 3H, *s*), 3.1 (1H, *m*), 5.14 (1H, *m*); EI-MS *m/z:* 472 (M^+), 454, 426, 408, 354, 264, 246, 218, 207, 146.

Other compounds were determined as, β-sitosterol-β-D-glucoside (**3**), syringic acid (**4**), gallic acid (**5**), and protocatechuic acid (**6**) by comparing MS spectra with the literature (Schuster and Herrmann 1985) and by comparing with authentic compounds using the TLC method.

Cytotoxic Activity Test With Human Leukemia Cell HL-60

The HL-60 cells were cultivated in the 1.62% RPMI 1640 medium supplemented with 10% fetal calf serum and 1% penicillin-streptomycin. The concentration of the cells in the medium was kept between 2 and 10×10^5 counts/mL.(**AQ3**) At the beginning of the experiment, the cell concentration was counted with a microscope. The medium with antibiotics was discarded after centrifugation. The cells were washed by the RPMI medium without antibiotics and collected by centrifuge. The cells were then diluted to $\sim 3 \times 10^5$ counts/mL with a cold medium without antibiotics. ^{3}H-Thymidine was then mixed into the cold medium at a concentration of 150 Ci/mmol.

The samples were dissolved in DMSO at concentrations between 0.1 g/L and 150 μg/L. A 2-μL sample of each concentration was added into an empty 6-mL glass tube. Each testing point had three parallel tubes to ensure the accuracy and reproducibility of the result. A 1-mL sample of the HL-60 cell culture containing the ^{3}H isotope was added into each tube. The sample was diluted 500-fold. After thorough mixing, all of the tubes were kept in the culture oven at 37°C and 5% CO_2 except for the three zero controls. The zero controls were treated with 2 mL of cold phosphate buffer saline (PBS) to terminate the cell growth immediately after the mixing of the sample and the cell culture.

The test tubes were kept in the oven for 90 min and then cooled with ice water. Cold PBS (2 mL) was added into each tube. All of the tubes, including the zero controls, were centrifuged at 1000 rpm for 5 min. The supernatant was discarded after centrifugation. Distilled water (2 mL) was mixed with the cells to swallow up and break the cell plasma system, releasing into the water both the free ^{3}H-thymidine and the ^{3}H-thymidine residue on the DNA chain. Trichloroacetic acid in water solution (2 mL of 10% TCA) was added into each tube to make a final concentration of 5% TCA. The free and residual ^{3}H-thymidines were separated by vacuum glass wool filtration. To ensure complete separation, the filter glass wool was washed three times with 5 mL of 5% TCA and then once with 5 mL of acetone. The residual ^{3}H-thymidine on the filtration glass wool was synthesized during incubation with the liquid. The filter glass wool was put into a scintillation vial with 5 mL scintillation fluid. At the end, the ^3H concentration was measured using a Beckman LS 1701 scintillation counter (Fullerton, CA).

Evaluation of the Inhibitory Effect on Lipid Oxidation by the Rancimat Method

Pure lard was used as the lipid substrate to evaluate the lipid oxidation of compounds isolated from blueberry. A Metrohm 679 Rancimat instrument (Herisan, Switzerland) was used. These compounds (2 mg) were mixed with 2.5 g lard in different glass cylinders and the control was 2.5 g pure lard. The air supply was maintained at 20 mL/min and the temperature was kept at 110°C.

Results and Discussion

Many dietary constituents have been shown to inhibit chemical carcinogenesis in experimental models (Bomser *et al.* 1996). Recently, it was shown that specific fractions of the *Vaccinium* species had the potential to inhibit the initiation and promotion of chemical carcinogenesis; however, only the bioactivity of crude extracts was reported. The authors related the anticarcinogenic components to lipid, sterol, carotenoid, chlorophyll, and additional nonpolar components or proanthocyanidin (Bomser *et al.* 1996). In our studies, we systematically studied the chemical components of the ethyl acetate fraction of highbush blueberry. We found that the main components were triterpenoids and phenolic acids. The seven main components were determined as ursolic acid, 3β,19α-dihydroxy-urs-12-en-28-oic acid, β-sitosterol-β-D-glucoside, gallic acid, syringic acid, and protocatechuic acid (Fig. 24.1).

Human leukemia cell HL-60 is a strain with a high tolerance to penicillin-streptomycin. The anticarcinogenic activity of chemicals is represented by the inhibitory effect of chemicals on the growth of the HL-60 cell. The stronger the anticarcinogenic activity of the chemical compound, the more it inhibits DNA synthesis and the lower is the isotope counting result. In this test, two compounds, ursolic acid and 3β,19α-dihydroxy-urs-12-en-28-oic acid, effectively inhibited DNA synthesis in HL-60 cells (Fig. 24.2). The 50% inhibitory concentrations (IC_{50}) were determined to be 1.5 and 1 ppm, respectively. Ursolic acid, a compound isolated from apples, pears, and other fruits, was the main component of the ethyl acetate fraction and was found to have significant cytotoxicity in the lymphocytic leukemia cells p-388 and L-1210 as well as in the human lung carcinoma cell A-549. It also demonstrated marginal cytotoxicity in the KB and human colon (HCT-8) and mammary (MCF-7) tumor cells (Lee *et al.* 1988). 3β,19α-Dihydroxy-urs-12-en-28-oic acid, a 19α-OH derivative of ursolic acid, showed even better antitumor activity than ursolic acid. This is the first report on the anticancer activity of this compound.

The Rancimat method has been a common method used to measure the potential antioxidative abilities of synthetic and natural antioxidants. The mechanism of this method is reported to be based on measuring the changes of electrical conduc-

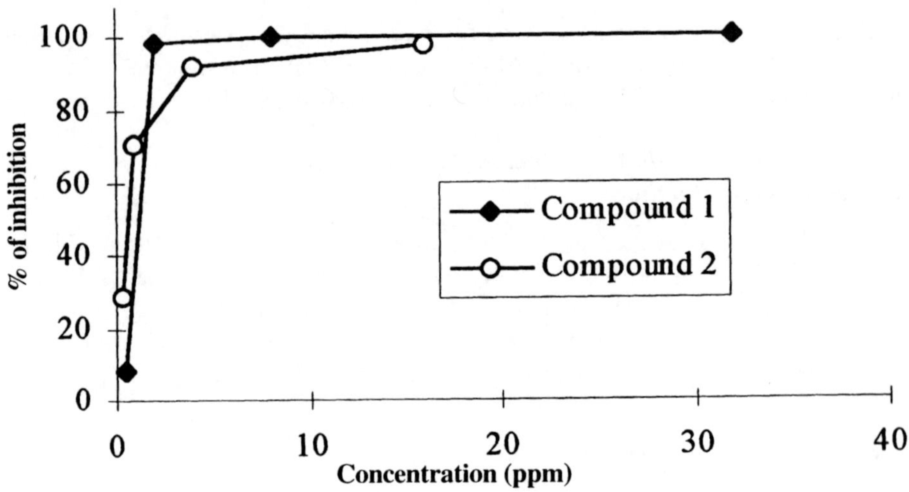

Fig. 24.2. Inhibitory effects of compounds (**1**) and (**2**) on DNA synthesis in HL-60 cells. Results represent the average values for three experiments each performed in triplicate.

tivity of water caused by the formation of short-chain compounds when fats and oils are oxidized under elevated temperature and accelerated aeration.

The effects of 2-mg samples in 2.5 g lard on the oxidative stability of lard are shown in Table 24.1. Two compounds, protocatechuic acid and gallic acid, showed high oxidation-inhibition activity, with induction times of 28.1 and 18.4 h, respectively. Phenolic acids are widely distributed in the plant kingdom, and many of these types of compounds have been reported as antioxidants. Their antioxidative activities are generally ascribed to their hydroxyl groups (Chen and Ho 1997).

In conclusion, this study shows that the anticarcinogenic activity of highbush blueberry is related at least in part to its triterpenoid components. The antioxidative activity of this plant may be ascribed to the phenolic acids. Highbush blueberry is a healthy food with anticancer and antioxidative activities.

TABLE 24.1
Antioxidative Activity (Rancimat test) of Compounds Isolated from Blueberry

Compound	Induction time (h)	Antioxidant index
BHT	5.8	2.76
1	2.2	1.05
2	2.2	1.05
3	2.4	1.14
4	3.1	1.48
5	28.1	13.38
6	18.5	8.81
Lard control	2.1	1.00

References

Ames, B.N., L.S. Gold, and W.C. Willett, The Cause and Prevention of Cancer, *Proc. Nat. Acad. Sci. USA 92*: 558–565 (1995).

Bilyk, A., and G.M. Sapers, Varietal Differences in the Quercetin, Kaempferol, and Myricetin Contents of Highbush Blueberry, Cranberry, and Thornless Blackberry, *J. Agric. Food Chem. 34*: 585–588 (1986).

Bomser, J., D.L. Madhavi, K. Singletary, and M.A.L. Smith, *In vitro* Anticancer Activity of Fruit Extracts from *Vaccinium* Species, *Planta Med. 62*: 212–216 (1996).

Chen, J.H., and C.-T. Ho, Antioxidant Activities of Caffeic Acid and Its Related Hydroxycinnamic Acid Compounds, *J. Agric. Food Chem. 45*: 2374–2378 (1997).

Eck, P., *Blueberry Science,* Rutgers University Press, New Brunswick, NJ, 1988, pp. 3–21.

Fullas, F., M.C. Wani, M.E. Wall, J.C. Tucker, C.W.W. Beecher, and A.D. Kinghorn, Triterpenes from the Combined Leaf and Stem of *Lithospermum caroliniense*, *Phytochemistry 43*: 1303–1305 (1996).

Gao, L., and G. Mazza, Quantitation and Distribution of Simple and Acylated Anthocyanins and Other Phenolics in Blueberries, *J. Food Sci. 59*: 1057–1059 (1994).

Lee, K.H., Y.M. Lin, T.S. Wu, D.C. Zhang, T. Yamagishi, T. Hayashi, I.H. Hall, J.J. Chang, R.Y. Wu, and T.H. Yang, The Cytotoxic Principles of *Prunella vulgaris, Psychotria serpens*, and *Hyptis caspitata*: Ursolic Acid and Related Derivatives, *Planta Med.,* 308–311 (1988).

Schuster, B., and K. Herrmann, Hydroxybenzoic and Hydroxycinnamic Acid Derivatives in Soft Fruits, *Phytochemistry 24*: 2761–2764 (1985).

Seto, T., T. Tanaka, O. Tanaka, and N. Naruhashi, β-Glucosyl Esters of 19α-Hydroxyursolic Acid Derivatives in Leaves of *Rubus* Species, *Phytochemistry 23*: 2829–2834 (1984).

Teeling, C.G.V., P.E. Cansfield, and R.A. Gallop, An Anthocyanin Complex Isolated from the Syrup of Canned Blueberries, *J. Food Sci. 36*: 1061–1063 (1971).

Terreaux, C., M.P. Maillard, M.P. Gupta, and K. Hostettmann, Triterpenes and Triterpene Glycosides from *Paradrymonia macrophylla*, *Phytochemistry 42*: 495–499 (1996).

Tommasi, N.D., L. Rastrelli, J. Cumanda, G. Speranza, and C. Pizza, (1996) Aryl and Triterpenic Glycosides from *Margyricarpus setosus*, *Phytochemistry 42*: 163–167.

Chapter 25

Antioxidants of Evening Primrose

F. Shahidi, M. Wettasinghe, R. Amarowicz, and M.A. Khan

Department of Biochemistry, Memorial University of Newfoundland, St. John's, NF, A1B 3X9 Canada

Introduction

Lipids are heterogeneous compounds that are primarily a source of fuel (9 kcal/g) for both plants and animals. In foods, lipids contribute to sensory characteristics such as flavor and aroma, color, texture, and mouthfeel. In human nutrition, food lipids not only provide the essential fatty acids such as linoleic and linolenic acids, but also serve as a source and carrier of fat-soluble vitamins such as A, D, E, and K (St. Angelo 1996). Like many other food constituents, lipids are not immune to deterioration with time. In fact, the presence of unsaturated bonds in their fatty acid chains makes them susceptible to deterioration due to oxidation. During production, distribution, and storage preceding actual consumption, foods undergo various deterioration processes that involve physical, chemical, and microbial pathways. Lipid oxidation, a chemical process, causes undesirable changes in the flavor, color, and nutritional value of foods. Deoxygenation, airtight packaging, and several other techniques have solved some of these problems, but the role of antioxidants cannot be overlooked (Namiki 1990).

In a peroxide-free system, lipid peroxidation is initiated when a hydrogen atom is abstracted from a diallylic methylene ($-CH_2^-$) group of an unsaturated fatty acid (Halliwell and Gutteridge 1985). Oxygen-derived free radicals such as the superoxide radical ($O_2^{\bullet-}$), hydroxyl radical ($^{\bullet}OH$), hydroperoxyl radical ($HO_2^{\bullet}$), and nitric oxide radical ($^{\bullet}NO$) may abstract a hydrogen atom from the fatty acid chains of lipid molecules (Packer and Glazer 1990). Singlet oxygen ($^1O_2$), an electronically excited molecule, and hydrogen peroxide (H_2O_2), another reactive oxygen species, can initiate lipid oxidation through the "ene" reaction in which olefenic groups of fatty acids are converted to their corresponding allyl hydroperoxides and $^{\bullet}OH$ generation (Halliwell and Gutteridge 1985), respectively. Phenolic antioxidants are reported to quench oxygen-derived free radicals as well as those that are substrate-derived by donating a hydrogen atom or an electron to the free radical (Yuting *et al.* 1990). Extracts of plants have been shown to neutralize free radicals in various model systems (Ruch *et al.* 1989, Silva *et al.* 1991, Zhang *et al.* 1996). Several authors have reported that the extracts of various oilseed meals possess antioxidant properties, which in some cases exerted better antioxidant properties than those observed for synthetic antioxidants under the same conditions and concentrations (Amarowicz *et al.* 1993, Fukuda *et al.* 1985, Shahidi *et al.* 1994, Wanasundara *et al.* 1994).

Evening primrose (*Oenothera biennis*) is a major source of γ-linoleic acid (GLA, 18:3n-3) (Rahmatullah *et al.* 1994a and 1994b). The potential beneficial health effects of evening primrose oil as an anti-inflammatory agent are well recognized (Gunstone 1992, Horrobin 1992). However, the meal that remains after oil extraction is largely unutilized. Therefore, it is prudent to look for ways to develop value-added components from primrose meal after oil extraction. Similar to other plant materials, evening primrose meal might be used as a potential source of phenolic antioxidants for use in food and non-food applications. Phenolic compounds present in oilseed meals are chemically diverse, but generally belong to the phenolic acid, flavonoid, and related compound groups. The antioxidant activity of the meals, as such, or their extracts depends heavily on the chemical nature of the active components, the relative concentration of the active ingredients, as well as the system to which they are added. The antioxidant activity of the meals of soybean (Hammerschmidt and Pratt 1978), canola (Wanasundara *et al.* 1994), cottonseed (Whittern *et al.* 1984), mustard (Saleemi *et al.* 1993, Shahidi *et al.* 1994), sesame (Fukuda *et al.* 1985, Shahidi *et al.* 1997), flaxseed (Amarowicz *et al.* 1993), and other oilseeds (Shukla *et al.* 1997) has been reported. The objectives of this study were to examine the effect of endogenous antioxidants in evening primrose oil on its stability and also to extract and evaluate the antioxidant components of evening primrose meal.

Materials and Methods

Oxidative Stability of Nonstripped and Stripped Evening Primrose Oil

Cold-pressed evening primrose oil was obtained from Efamol (Kentville, Canada). All chemicals used in this study were of ACS grade or better. Evening primrose oil (100 g), dissolved in 100 mL hexane, was passed through a chromatographic column (3.4-cm internal diameter × 40 cm) connected to a suction apparatus. The column was packed with three layers of four adsorbants. The top layer consisted of activated silicic acid (40 g); immediately beneath it was a layer of 1:2 (w/w) mixture of Celite 545 and activated charcoal (20 g), followed by a layer of 1:2 (w/w) mixture of Celite 545 and powdered sucrose (80 g) (Lampi *et al.* 1992). The silicic acid was activated as described by Min (1973). Silicic acid (100 g) was rinsed three times with 1 L of distilled water; each time, the slurry was allowed to settle for 30 min and suspended silicic acid was decanted. Water-washed silicic acid was further washed with methanol and the slurry was suction-filtered. Washed silicic acid was activated by heating in an oven at 200°C for 22 h. Stripped evening primrose oil was recovered by removing hexane under vacuum at 30°C.

Oxidative stability of both nonstripped and stripped oils was determined under Schaal oven conditions. Oil samples (5 g) in 25-mL Erlenmeyer flasks were allowed to oxidize for 7 d in a forced-air oven at 60°C. Peroxide value (PV) of oil on d 0, 1, 3, 5, and 7 was determined according to the AOCS (1990) method. In addition, gas chromatographic headspace analysis of samples was performed according to the method of Shahidi and Pegg (1993).

Preparation of the Meal

Evening primrose seeds were obtained from Efamol. Seeds (100 g) and hexane (500 mL) were blended for ~15 min in a Waring blender (Waring Product Division, Dynamics Corporation of America, New Hartford, CT). The slurry was filtered through a Whatman No. 3 filter paper using a Buchner funnel attached to a suction aspirator. This procedure was repeated three times, and the defatted seed meals were air-dried under a fumehood at room temperature for 12 h.

Extraction of Antioxidants from the Meal

The meal (6 g) was mixed with 100 mL of ethanol in a 250-mL round-bottomed flask and heated under reflux condition for 20 min in a thermostated water bath at 80°C. The slurry was centrifuged at 4000 × *g* for 5 min, and the supernatant containing antioxidant compounds was collected. The residue was reextracted two more times and the supernatants from each extraction were combined. A dried extract was prepared by evaporating the supernatant to dryness under vacuum at 40°C (Amarowicz *et al.* 1995b).

Column Chromatographic Fractionation

The ethanol extract (350 g) was dissolved in 5 mL methanol, applied to a chromatographic column (2.5-cm internal diameter × 45 cm) packed with Sephadex LH-20 (Pharmacia, Upsala, Sweden) and eluted with methanol (Amarowicz *et al.* 1992). The eluates were collected in 20-mL tubes with the use of a fraction collector; their absorbance was read at 280 nm. On the basis of their absorbance data, eluates were pooled into 10 major fractions, followed by desolventization and weighing. The content of total phenolics (as catechin equivalents) in each fraction was determined using the Folin-Denis reagent (Naczk and Shahidi 1989). Ultraviolet (UV) spectra of individual fractions were recorded using a Hewlett-Packard 8452 A (Palo Alto, CA) diode array spectrophotometer.

Determination of Antioxidant Activity

Antioxidant activity of the crude extract and its fractions was assessed in a β-carotene/linoleate model system (Miller 1971). β-Carotene (5 mg) was dissolved in 10 mL of chloroform, and 2 mL of the solution was evaporated (at 40°C) to dryness under vacuum. Linoleic acid (40 mg), Tween 40 (400 mg), and 10 mL of aerated distilled water were added to the flask and sonicated in the dark for 1 min. A methanolic solution (1 mg in 0.2 mL) of each of the dried fractions was added to a series of test tubes containing 4.8 mL of the prepared emulsion. Butylated hydroxytoluene (BHT, 0.3 mg in 0.2 mL methanol) was added into one of the tubes and used as a reference to evaluate the antioxidant activity of the fractions. After the initial absorbance values (at 470 nm) of systems were measured, they were kept in a thermostated water bath (50°C), and subsequent absorbance values were recorded at 15-min intervals during a 120-min period.

Assessment of Hydrogen Peroxide–Scavenging Properties of the Crude Extract

The effect of the crude extract on the scavenging of H_2O_2 was assessed using the spectrophotometric method of Ruch *et al.* (1989). The crude extract of evening primrose (at 10, 100, and 200 ppm, based on total phenolics as catechin equivalents) was dissolved in 3.4 mL of 0.1 mol/L phosphate buffer (pH 7.4) and mixed with 600 μL of a 43 mmol/L solution of H_2O_2 (prepared in the same buffer). The absorbance value (at 230 nm) of the reaction mixture was recorded at zero time and then every 10 min for up to 40 min. For each concentration, a separate blank sample (devoid of H_2O_2) was used for background subtraction.

Assessment of Free Radical–Scavenging Properties of the Crude Extract

The hydroxyl radical was selected for assessing the free radical–scavenging properties of the crude extract because of its importance in the initiation of lipid autoxidation. It was generated *via* iron-catalyzed Haber-Weiss reaction (Fenton driven Haber-Weiss reaction) and trapped with 5,5-dimethyl-1-pyrroline *N*-oxide (DMPO). The resultant DMPO-OH adduct was detected using an electron paramagnetic resonance (EPR) spectrometer (Bruker ESP-300, Bruker Instruments, Billerica, MA). The crude extract was dissolved in a sufficient amount of 0.1 mol/L phosphate buffer (pH 7.4) so that a 200 μL volume of the aliquot contained 200 ppm of the extract phenolics in the final assay medium (final volume was 800 μL). For 10 and 100 ppm concentrations, 10 and 100 μL, respectively, of the same stock solution were used, but the volume was adjusted to 200 μL by adding the buffer. Extract (200 μL) was mixed with 200 μL of 100 mmol/L DMPO, 200 μL of 10 mmol/L $FeSO_4$, and 200 μL of 10 mmol/L H_2O_2. All solutions were prepared in 0.1 mol/L phosphate buffer (pH 7.4). After 3 min, 10 μL of the mixture were drawn into a syringe and transferred into a quartz capillary tube. The spectrum was recorded in an EPR spectrometer set at 2×10^5 receiver gain, 1.0 G modulation amplitude, 200 s scan time, 3460 G center field, 100 G sweep width, and 0.5 s time constant (Shi *et al.* 1991).

Isolation and Identification of Antioxidant Compounds—Analytical Thin-Layer Chromatography (TLC)

Fractions were examined on silica gel 60 TLC plates with a layer thickness of 0.2 mm (Merck, Darmstad, Germany) using petroleum ether/diethyl ether/acetic acid (80:20:1, by vol) (Amarowicz *et al.* 1995b) and chloroform/methanol/water (65:35:10, by vol, lower phase) (Amarowicz *et al.* 1994) as mobile phases. Compounds with antioxidant activity were visualized after spraying a second developed plate with a solution of β-carotene and linoleic acid (Philip 1974). Each fraction and commer-cial standards of (-)-epicatechin, (-)-epigallocatechin, (-)-epicatechin-3-gallate, and (-)-epigallocatechin-3-gallate were also developed on silica gel 60 TLC plates with toluene/acetone/formic acid (30:30:10, by vol) (Lea 1978) and sprayed with a vanillin-hydrochloric acid reagent (Karchesy *et al.* 1989).

Preparative TLC

Two pure compounds were separated from fraction IV using preparative silica gel 60 TLC plates with a layer thickness of 0.5 mm (Merck) and chloroform/methanol/water (65:35:10, by vol, lower phase) as the mobile phase.

Results and Discussion

Figure 25.1 shows the effect of the stripping of evening primrose oil from its endogenous antioxidants on oxidative stability of the stripped oil. The peroxide value (PV) of the nonstripped oil was markedly lower than that of the stripped oil throughout the entire storage period. The length of the induction period for nonstripped oil was longer compared with that of the stripped oil. The presence of γ-tocopherol (211 ppm) in the nonstripped oil might be responsible, at least in part, for this difference. Tocopherols can quench both oxygen-derived and lipid-derived free radical species by donating a hydrogen atom to radicals, eventually reducing the amount of free radicals available for the initiation reaction (Yamauchi *et al.* 1995). Because the tocopherol reserve was depleted over time, even the nonstripped oil became more susceptible to free radical attack, resulting in higher PV toward the end of the storage period.

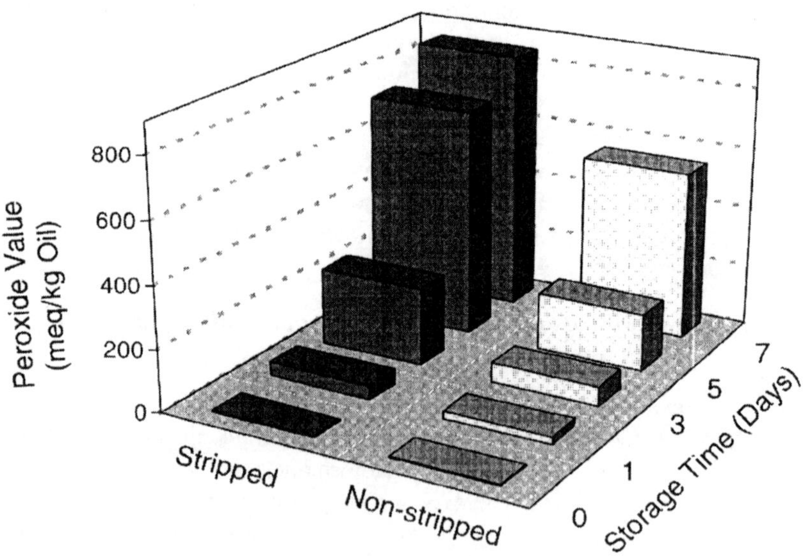

Fig. 25.1. Effect of stripping of evening primrose oil from its antioxidants on peroxide formation in the oil.

Linoleic acid was the major fatty acid present in nonstripped (59.46%) and stripped (58.89%) evening primrose oils. According to Satue *et al.* (1995), the major decomposition product formed during the oxidation of linoleic acid is hexanal. Therefore, hexanal may be used to monitor lipid oxidation. Initially, the stripped evening primrose oil had a significantly ($P \leq 0.05$) lower content of hexanal compared with the nonstripped evening primrose oil (Table 25.1). This can be attributed to the effectiveness of the stripping technique employed in this study in removing any of the oxidation products originally present in the oil. Several authors have reported that stripped vegetable oils do not contain any significant amount of volatiles, and in most cases, the oils are devoid of volatiles (Huang *et al.* 1995, Lampi *et al.* 1997, Yoshida *et al.* 1993). However, the hexanal content of the stripped oil was significantly ($P \leq 0.05$) higher than that of the nonstripped oil during the entire storage time. The higher stability of the nonstripped oil compared with the stripped oil can be attributed to the presence of γ-tocopherol in the former (Lampi *et al.* 1997, Shahidi *et al.* 1997). Tocopherols play an important role in retarding lipid oxidation in edible oils (Wanasundara *et al.* 1997, Yoshida *et al.* 1993). Satue (1995) showed that α-tocopherol (100 and 500 ppm) is an efficient inhibitor of hexanal formation pathways. Yoshida *et al.* (1993) reported that the antioxidant activity of tocopherols decreased in the order $\alpha < \beta = \gamma < \delta$ in an ethyl linoleate system and tocopherol-stripped oil. The contribution of α-tocopherol to the stability of turnip and sesame oil was reported by Lampi *et al.* (1997) and Shahidi *et al.* (1997), respectively.

Although a large proportion of seed tocopherol reserve was co-extracted along with the oil, other phenolic antioxidants were retained in the meal. A multitude of phenolic antioxidants ranging from simple phenolic acids to complex procyanidins have been isolated from various oilseed meals (Wanasundara *et al.* 1994). The extracts of evening primrose meal contained ~300 mg of total phenolics as catechin equivalents per gram. In an attempt to characterize these compounds, the extracts were subjected to a series of chromatographic separation techniques. Sephadex LH-20 column chromatography of ethanolic extract of the meal yielded ten fractions (I–X) (Fig. 25.2).

TABLE 25.1
Comparison of the Hexanal Contents in Nonstripped and Stripped Evening Primrose Oils Stored at 60°C[a]

	Hexanal content (mg/kg oil)	
Storage time (d)	Nonstripped	Stripped
0	4.0 ± 0.05[c]	0.0 ± 0.00[b]
1	21.7 ± 0.46[b]	35.6 ± 0.78[c]
3	23.2 ± 0.41[b]	82.8 ± 0.58[c]
5	69.7 ± 0.82[b]	619.3 ± 24.96[c]
7	517 ± 22.5[b]	1300 ± 7.3[c]

[a]Results are mean values of three determinations ± SD. Means in a row are significantly ($P \leq 0.05$) different from one another.

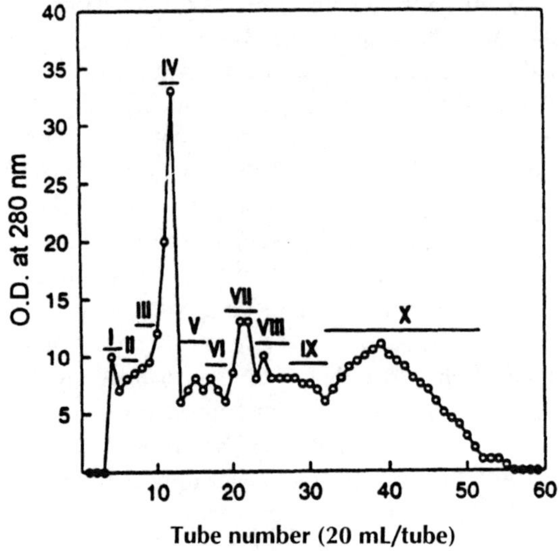

Fig. 25.2. Column chromatographic fraction profile of ethanol extract.

The fraction profile was characterized by one sharp intense peak (IV), one broad peak (X), one medium-sized peak (VII), and four smaller peaks (I, V, VI, VIII). Fractions II and IX did not exhibit typical peaks.

The relative contribution of fraction IV in the extracts was the highest (26.8%), whereas those of fractions I, II, and X varied from 10.2 to 16.9%. The remaining fractions were present in amounts <10% (Fig. 25.3). The total phenolic contents in fractions IV–X were very similar and ranged from 132 to 165 mg catechin equivalents/g. Fractions I–III contained lower amounts of total phenolics (Fig. 25.4).

As depicted in Figure 25.5, UV spectra of fractions I, II, V, VI, IX, and X showed absorption maxima at ~280 nm. For fractions III, IV, and VIII, UV absorption maxima occurred at 286, 274, and 278 nm, respectively. The UV spectrum of fraction VII showed two maxima at 256 and 364 nm. With the exception of fraction I, all other fractions showed shoulders between 266 and 248 nm. UV spectral data of fractions I–X are given in Table 25.2.

Figure 25.6 shows the antioxidant activity of each fraction, as evidenced in their ability to inhibit the bleaching of β-carotene, compared with that of butylated hydroxyanisole (BHA). The antioxidant efficacy of fractions I–III was low because they contained small quantities of total phenolics (Fig. 25.4). Antioxidant properties of fraction V decreased during the second hour of incubation. Meanwhile, antioxidant activity of fractions IV and VI–X was comparatively strong because ~50% of β-carotene remained unoxidized after 2 h of incubation in the systems containing these fractions. Antioxidant activity of fractions IV and VI–X was comparable to those of extracts from rapeseed (Amarowicz *et al.* 1995b), crude catechins from green tea (Amarowicz and Shahidi 1995), and the most active fractions from extracts of flaxseed (Amarowicz *et al.* 1993) and mustard (Amarowicz *et al.* 1996, Shahidi *et al.* 1994).

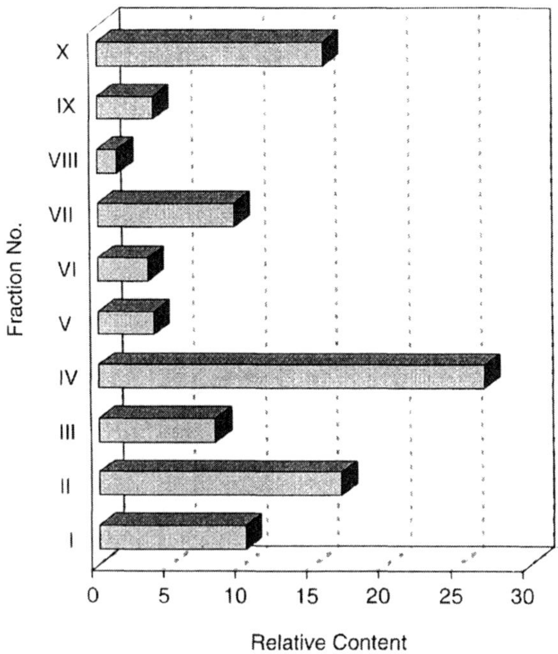

Fig. 25.3. Relative contribution of column chromatographic fractions in ethanol extract.

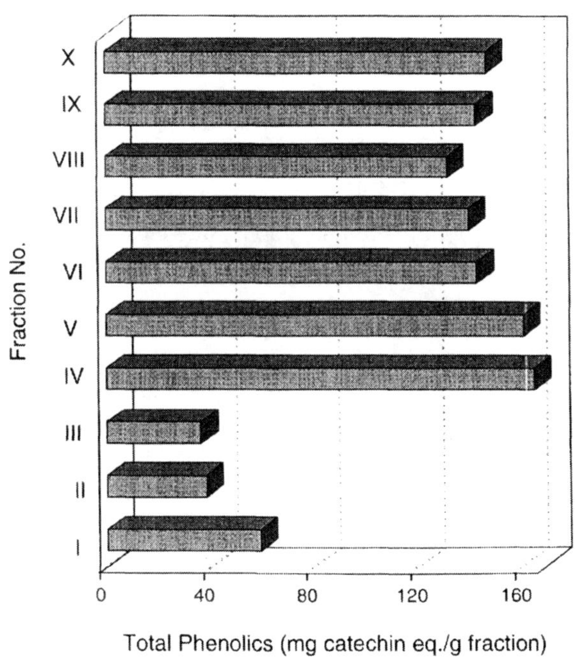

Fig. 25.4. Content of total phenolics in the column chromatographic fractions of ethanol extract.

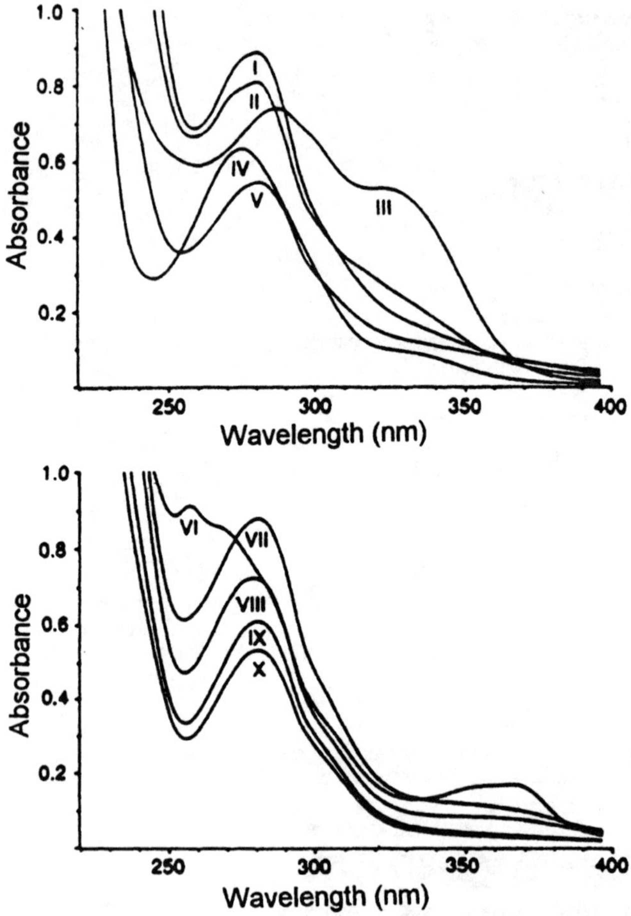

Fig. 25.5. Ultraviolet (UV) spectra of column chromatographic fractions of ethanolic extract of evening primrose meal.

Thin-layer chromatographic analysis of fractions separated from a Sephadex LH-20 column indicated the presence of several phenolic compounds (Figs. 25.7 and 25.8). TLC plates, developed with a polar chloroform/methanol/water solvent system, displayed more compounds than those separated using a nonpolar petroleum ether/diethyl ether/acetic acid solvent system. Spots from all fractions that were located on the baselines of plates, using both developing solvent systems, possessed antioxidant activity. Furthermore, plates developed with nonpolar systems had one spot with antioxidant activity as detected for fractions I, II, IV, and V, and three spots for fraction III (Fig. 25.7). Application of a more polar solvent system (chloroform/methanol/water) (Fig. 25.8) allowed separation and subsequent detection of more spots ($R_f > 0$) from different fractions.

TABLE 25.2
Ultraviolet (UV) Spectral Data of Primrose Phenolic Fractions Separated by Sephadex LH-20 Column Chromatography

Fraction	λ max (nm)	Shoulder (nm)
I	280	—
II	280	272
III	286	322
IV	274	328
V	280	304, 314
VI	280	306
VII	256, 364	266, 282
VIII	278	304, 348
IX	280	304
X	280	304

The presence of vanillin-positive compounds was noted for fractions VI–X on plates developed with toluene/acetone/formic acid (30:30:10, by vol) (Fig. 25.9). The compound isolated from fraction VI with an R_f value of 0.63 was tentatively identified as (-)-epicatechin. In fractions VII–X, compounds that were more polar than (-)-epigallocatechin-3-gallate were present. The R_f values for these compounds, most probably dimers and trimers of proanthocyanidins, were 0.27 and 0.25 (fraction VII), 0.38, 0.31, 0.27, and 0.17 (fraction VIII) and 0.33 and 0.28 (fraction IX). For fractions VIII–X, red spots were noted on or close to the baseline; these compounds might be due to polymerization of polyphenols.

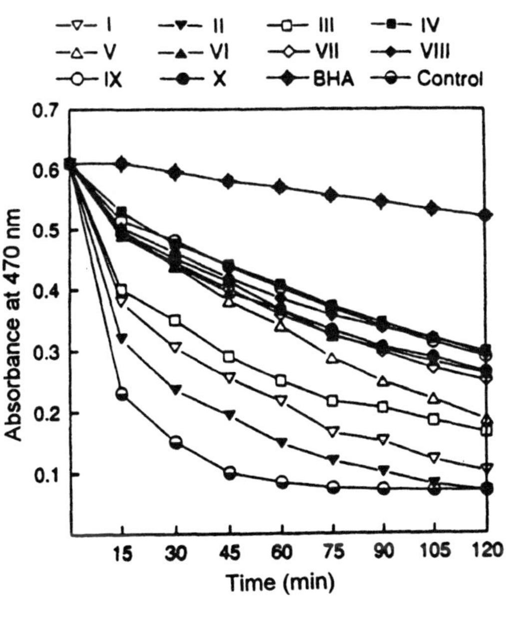

Fig. 25.6. Antioxidant activity of column chromatographic fractions in a β-carotene-linoleate model system.

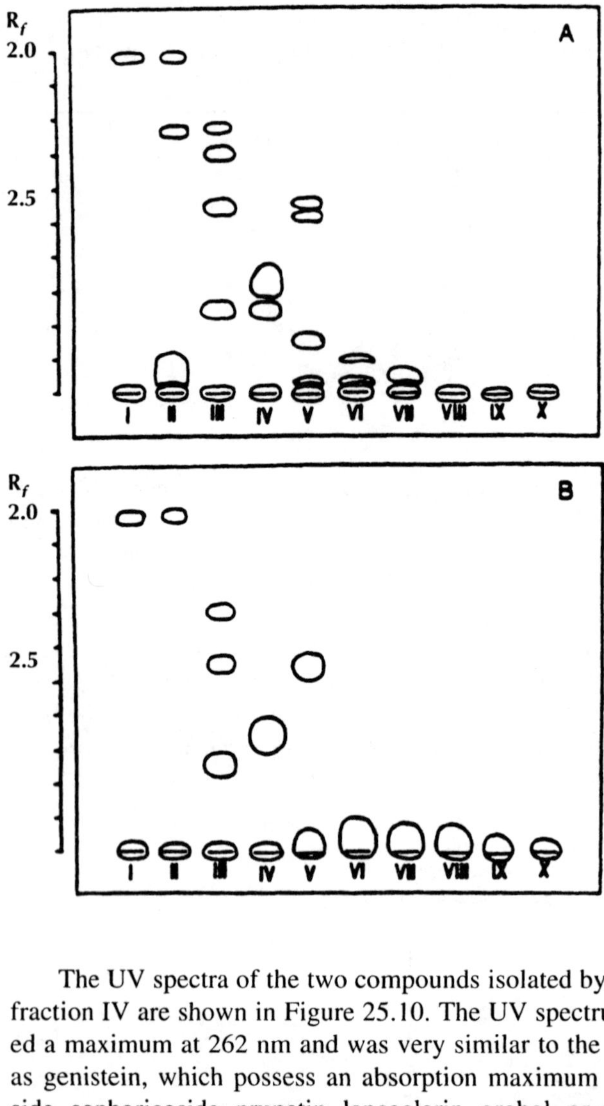

Fig. 25.7. Thin-layer chromatograms of column chromatographic fractions. Mobile phase used was petroleum ether/diethyl ether/acetic acid (80:20:1, by vol). (A) Plate was sprayed with a solution of ferric chloride to locate phenolic compounds; (B) plate was sprayed with β-carotene-linoleate emulsion to evaluate the antioxidant activity of spots.

The UV spectra of the two compounds isolated by semipreparative TLC from fraction IV are shown in Figure 25.10. The UV spectrum of compound A exhibited a maximum at 262 nm and was very similar to the spectra of isoflavones such as genistein, which possess an absorption maximum at 261 nm, and sphaerobiside, sophoricoside, prunetin, lanceolarin, orobol, or pratensein (all isoflavones), which display an absorption maximum at 262 nm (Marbry *et al.* 1970). The maximum absorbance of compound B was observed at 276 nm, which is similar to that of 2-hydroxychalcone (Marbry *et al.* 1970).

As depicted in Figure 25.11, evening primrose crude extract exerted a concentration-dependent scavenging of hydrogen peroxide (H_2O_2). The concentration of (H_2O_2) in the systems containing crude extracts dropped very sharply during the initial 10 min of the assay. During this period, the reduction in hydrogen peroxide concentration in the systems containing 0, 10, 100, and 200 ppm of crude extract

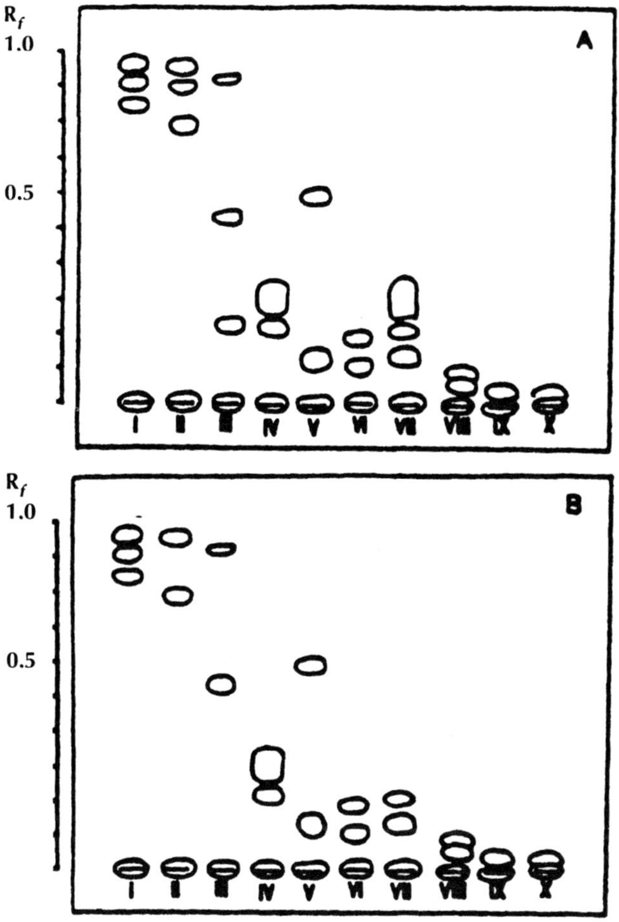

Fig. 25.8. Thin-layer chromatograms of column chromatographic fractions. Mobile phase used was chloroform methanol/water (65:35:10, by vol). (A) and (B) refer to Fig. 25.7.

was 3, 15, 23, and 86%, respectively. The decomposition of H_2O_2 into water may occur according to the following reaction:

$$H_2O_2 + 2H^+ + 2e^- \rightarrow 2H_2O$$

Because phenolic compounds present in the crude extract are good electron donors, they may accelerate the conversion of H_2O_2 into H_2O. However, H_2O_2 in the presence of iron ions may generate hydroxyl radicals, which are extremely reactive. This reaction is sometimes called the iron-catalyzed Haber-Weiss reaction as given below.

$$Fe^{3+} + O_2^{\bullet-} \rightarrow Fe^{2+} + O_2 \text{ (metal reduction)}$$

$$Fe^{2+} + H_2O_2 \rightarrow Fe^{3+} + {}^{\bullet}OH + OH^- \text{ (Fenton reaction)}$$

$$O_2^{\bullet-} + H_2O_2 \rightarrow O_2 + {}^{\bullet}OH + OH^- \text{ (Haber-Weiss reaction)}$$

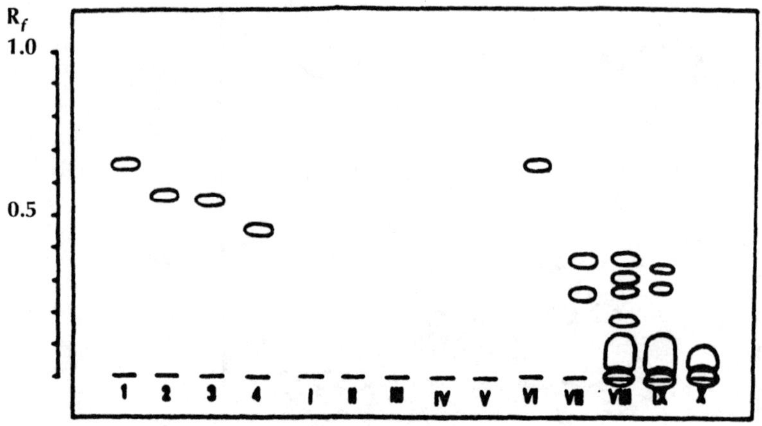

Fig. 25.9. Thin-layer chromatograms of the column chromatographic fractions. Mobile phase used was toluene/acetone/formic acid (30:30:10, by vol). Plate was sprayed with vanillin-hydrochloric acid reagent. I–X: Column chromatographic fractions. (-)-epicatechin (1), (-)-epicatechin-3-gallate (2), (-)-epigallocatechin (3), and (-)-epigallocatechin-3-gallate (4) were used as standards.

In this study, •OH generated *via* these reactions was spin trapped with DMPO, which resulted in the DMPO-OH adduct, a relatively stable free radical (Halliwell and Guteridge 1985, Ruch *et al.* 1984). This was done because the detection of •OH, as such, is extremely difficult due to its very short lifetime (Ruch *et al.* 1984). As depicted in Figure 25.12, the DMPO-OH adduct generated a 1:2:2:1 quartet with hyperfine coupling parameters of 14.9 G (Yen and Chen 1995). The intensity of this signal was reduced, but did not completely disappear, when 100 ppm (as catechin equivalents) of evening primrose crude extract was present in the assay medium. At 200 ppm, a 100% quenching of the signal was evident. There is some evidence that the phenolic compounds found in various plant materials possess free radical–scavenging properties. Shi *et al.* (1991) demonstrated the ability of caffeine, a phenolic compound, to effectively scavenge •OH. Husain *et al.* (1987) and van Acker *et al.* (1996) reported that flavonoids such as myricetin, quercetin, and rhamnetin were •OH scavengers. They also noted that the effectiveness of such compounds increases with an increasing number of hydroxyl groups attached to the aromatic B-ring of flavonoids. Thus, the phenolic compounds present in the crude extracts had the ability to donate a hydrogen atom to •OH, resulting in the quenching of the EPR signal.

Conclusions

The oxidative stability of cold-pressed evening primrose oil is greatly influenced by the endogenous antioxidant compounds, and the stripping of oil from these antioxi-

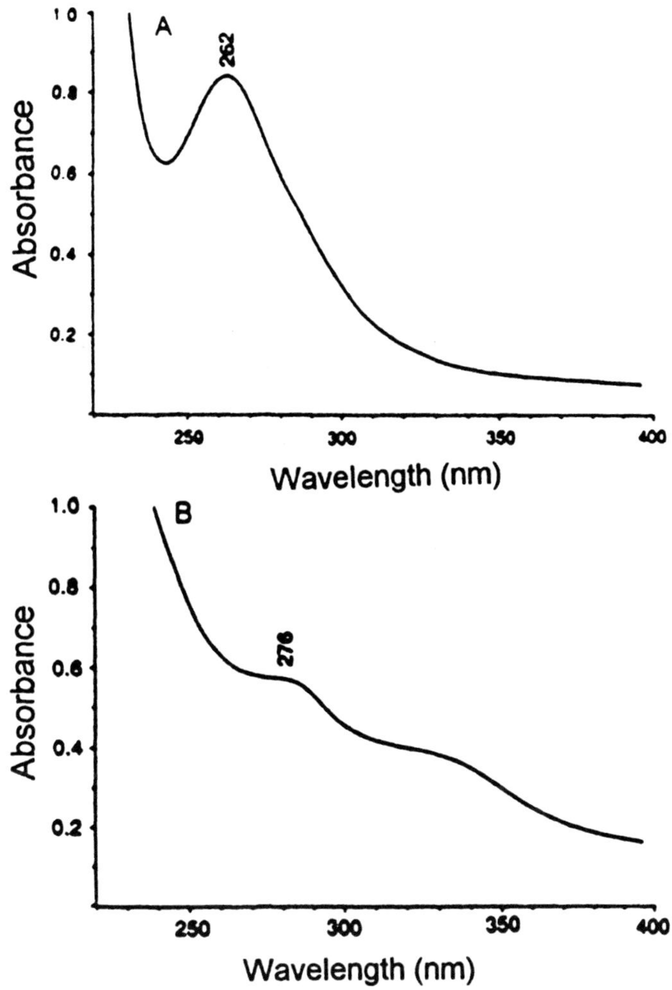

Fig. 25.10. Ultraviolet (UV) spectra of the compounds A and B isolated from fraction IV by preparative thin-layer chromatography.

dants caused its increased susceptibility to oxidation. The extracts of evening primrose reduced the coupled oxidation of β-carotene and linoleate and scavenged hydrogen peroxide, indicating its antioxidant activity. EPR experiments involving crude extract revealed the quenching of the EPR signal of the DMPO-OH adduct, suggesting that the compounds present in the crude extract participate in free radical neutralization. UV spectral data suggest that the isolated compound may belong to the proanthocyanidin family.

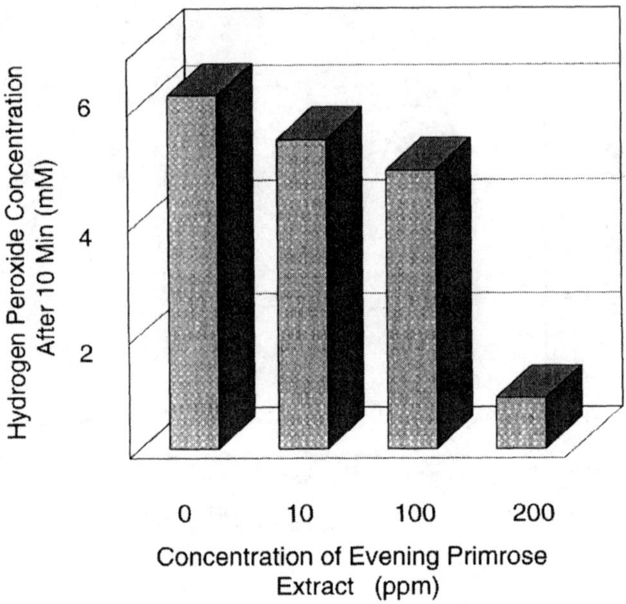

Fig. 25.11. Effect of ethanol extract of evening primrose meal on hydrogen peroxide decomposition.

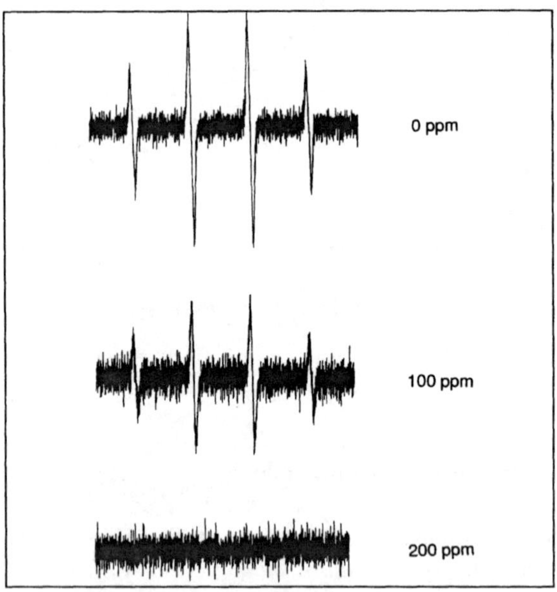

Fig. 25.12. Effect of ethanol extract of evening primrose meal on quenching of the hydroxyl radical as represented by reduction in the intensity of electron paramagnetic resonance (EPR) signal of the 5,5-dimethyl-1-pyrroline *N*-oxide (DMPO)-OH adduct.

References

Amarowicz, R., J. Fornal, and M. Karamac, Effect of Seed Moisture on Phenolic Acids in Rapeseed Oil Cake, *Grasas Aceites 46*: 354–356 (1995a).

Amarowicz, R., H. Kozlowska, M. Shimoyamada, and K. Okubo, Chromatographic Analysis of Rapeseed Glucoside Fractions, *Pol. J. Food Nutr. Sci. 42*: 89–93 (1992).

Amarowicz, R., M. Piskula, J. Honke, B. Rudnicka, A. Troszynska, and H. Kolzlowska, Extraction of Phenolic Compounds from Lentil Seeds (*Lens culinaris*) with Various Solvents, *Pol. J. Food Nutr. Sci. 45*: 53–62 (1995b).

Amarowicz, R., and F. Shahidi, Antioxidant Activity of Green Tea Catechins in a β-Carotene-Linoleate Model System, *J. Food Lipids 2*: 47–56 (1995).

Amarowicz, R., and F. Shahidi, A Rapid Chromatographic Method for Separation of Individual Catechins from Green Tea, *Food Res. Int. 29*: 71–76 (1996).

Amarowicz, R., U. Wanasundara, M. Karamac, and F. Shahidi, Antioxidant Activity of Mustard Extract, *Nahrung 40*: 261–263 (1996).

Amarowicz, R., J. Wanasundara, and F. Shahidi, Chromatographic Separation of Flaxseed Phenolics, *Nahrung 38*: 520–526 (1994).

Amarowicz, R., U. Wanasundara, J. Wanasundara, and F. Shahidi, Antioxidant Activity of Ethanolic Extracts of Flaxseed in a β-Carotene-Linoleate Model System, *J. Food Lipids 1*: 111–117 (1993).

AOCS, *Official Methods and Recommended Practices of the American Oil Chemists' Society*, 4th edn., American Oil Chemists' Society, Champaign, IL, 1990.

Bisby, R.H., M. Coleman, R.B. Cundall, G.W. Gould, and K.E. Tomaszewski, Lipid Oxidation Initiated by Free Radicals and by Lactoperoxidase, in *Oxy Radicals and Their Scavenger Systems, Vol. I, Molecular Aspects,* edited by G. Cohen, and R.A. Greenwald, Elsevier Biomedical, New York, 1982, pp. 50–55.

da Silva, J.M.R., N. Darmon, Y. Fernandez, and S. Mitjavila, Oxygen Free Radical Scavenger Capacity in Aqueous Models of Different Procyanidins from Grape Seeds, *J. Agric. Food Chem. 39*: 1549–1552 (1991).

Fukuda, Y., T. Osawa, M. Namiki, and T. Ozaki, Studies on Antioxidative Substances in Sesame Seeds, *Agric. Biol. Chem. 49*: 301–306 (1985).

Gunstone, F.D., Gamma Linoleic Acid—Occurrence and Physical and Chemical Properties, *Prog. Lipid Res. 31*: 145–161 (1992).

Halliwell, B., and J.M.C. Gutteridge, *Free Radicals in Biology and Medicine*, Oxford Science Publications, Oxford University Press, Oxford, 1985.

Hammerscmidt, P.A., and D.E. Pratt, Phenolic Antioxidants of Dried Soybeans, *J. Am. Oil Chem. Soc. 48*: 556–559 (1978).

Horrobin, D.F., Nutritional and Medicinal Importance of Gamma-Linoleic Acid, *Prog. Lipid Res. 31*: 163–194 (1992).

Huang, S.-W., E.N. Frankel, and J.B. German, Effects of Individual Tocopherols and Tocopherol Mixtures on the Oxidative Stability of Corn Oil Triacylglycerols, *J. Agric. Food Chem. 43*: 2345– 2350 (1995).

Husian, S.R., J. Cillard, and P. Cillard, Hydroxyl Radical Scavenging Activity of Flavonoids, *Phytochemistry 26*: 2489–2493 (1987).

Karchesy, J.J., Y. Bae, L. Chalker-Scott, R.F. Helm, and L.Y. Foo, Chromatography of Proanthocianidins, in *Chemistry and Significance of Condensed Tannins,* edited by R.W. Hemingway, and J.J. Karchesy, Plenum Press, New York, 1989, pp. 139–151.

Lampi, A.-M., H. Anu, and P. Vieno, Antioxidant Activity of Minor Amounts of γ-Tocopherol in Natural Triacylglycerol, *J. Am. Oil Chem. Soc. 74*: 549–555 (1997).

Lampi, A.-M., A. Hopia, P. Ekholm, and V. Piironen, Method for the Preparation of Triacylglycerol Fractions from Rapeseed and Other Oils for Autoxidation Studies, *Food Sci. Technol. 25*: 386–388 (1992).

Lea, A.G.H., The Phenolics of Ciders: Oligomeric and Polymeric Procyanidins, *J. Sci. Food Agric. 29*: 471–478 (1978).

Marbry, T.J., K.R. Markham, and M.B. Thomas, *The Systematic Identification of Flavonoids*, Springer-Verlag, Berlin, 1970.

Miller, H.E., A Simplified Method for the Evaluation of Antioxidants, *J. Am. Oil Chem. Soc. 45*: 91 (1971).

Min, D.B., A Study of Color Stability and Red Color Development During Epoxidation of Tall Oil Fatty Acids Through the Isolation and Characterization of Minor Constituents, Ph.D. Thesis, Rutgers University, New Brunswick, NJ, 1973.

Naczk, M., and F. Shahidi, The Effect of Methanol-Ammonia-Water Treatment on the Content of Phenolic Acids of Canola, *Food Chem. 31*: 159–164 (1989).

Namiki, M., Antioxidants/Antimutagens in Food, *Crit. Rev. Food Sci. Nutr. 29*: 273–300 (1990).

Packer, L., and A.N. Glazer, *Methods in Enzymology, Vol. 186, Part B: Oxygen Radicals and Antioxidants*, Academic Press, San Diego, CA 1990.

Philip, F., An Investigation of the Antioxidants in a Textured Vegetable Protein Product from Soy Products, Ph.D. Thesis, Purdue University, W. Lafayette, IN, 1974.

Rahmatullah, M.S.K.S., V.K.S. Shukla, and K.D. Mukherjee, γ-Linolenic Acid Concentrates from Borage and Evening Primrose Oil Fatty Acids *via* Lipase-Catalyzed Esterification, *J. Am. Oil Chem. Soc. 71*: 563–568 (1994a).

Rahmatullah, M.S.K.S., V.K.S. Shukla, and K.D. Mukherjee, Enrichment of γ-Linolenic Acid from Evening Primrose Oil and Borage Oil *via* Lipase-Catalyzed Hydrolysis, *J. Am. Oil. Chem. Soc. 71*: 569–573 (1994b).

Ruch, R.J., S.-J. Cheng, and E. Klaunig, Prevention of Cytotoxicity and Inhibition of Intercellular Communication by Antioxidant Catechins Isolated from Chinese Green Tea, *Carcinogenesis 10*: 1003–1008 (1989).

Ruch, R.J., S.U. Chug, and J.E. Klaunig, Spin Trapping of Superoxide and Hydroxyl Radicals, *Methods Enzymol. 105*: 198–209 (1984).

Saleemi, Z.O., P.K.J.P.D. Wanasundara, and F. Shahidi, Effects of Low-Pungency Ground Mustard Seed on Oxidative Stability, Cooking Yield and Color Characteristics of Comminuted Pork, *J. Agric Food Chem. 41*: 641–643 (1993).

Satue, M.T., S.W. Huang, and E.N. Frankel, Effect of Natural Antioxidants in Virgin Olive Oil on Oxidative Stability of Refined, Bleached and Deodorized Olive Oil, *J. Am. Oil Chem. Soc. 72*: 1131–1137 (1995).

Shahidi, F., R. Amarowicz, H.A. Abou-Gharbia, and A.Y. Shehata, Endogenous Antioxidants and Stability of Sesame Oil as Affected by Processing and Storage, *J. Am. Oil Chem. Soc. 74*: 143–148 (1997).

Shahidi, F., and R.B. Pegg, Hexanal as an Indicator of Meat Flavor Deterioration, *J. Food Lipids 1*: 177–186 (1993).

Shahidi, F., U. Wanasundara, and R. Amarowicz, Natural Antioxidants from Low-Pungency Mustard Flour, *Food Res. Int. 27*: 489–493 (1994).

Shi, X., N.S. Dalal, and A.C. Jain, Antioxidant Behavior of Caffeine: Efficient Scavenging of Hydroxyl Radicals, *Food Chem. Toxicol. 29*: 1–6 (1991).

Shukla, V.K.S., P.K.J.P.D. Wanasundara, and F. Shahidi, Natural Antioxidants from Oilseeds, in *Natural Antioxidants: Chemistry, Health Effects and Application,* edited by F. Shahidi, AOCS Press, Champaign, IL, 1997, pp. 97–132.

St. Angelo, A.J., Lipid Oxidation in Foods, *Crit. Rev. Food Sci. Nutr. 36*: 175–224 (1996).

van Acker, S.A.B.E., D.-J. van den Berg, M.N.J.L. Tromp, D.H. Griffioen, W.P. van Bennekom, W.J.F. van der Vijgh, and A. Bast, Structural Aspects of Antioxidant Activity of Flavonoids, *Free Radic. Biol. Med. 20*: 331–342 (1996).

Wanasundara, U., R. Amarowicz, and F. Shahidi, Isolation and Identification of an Antioxidative Component in Canola Meal, *J. Agric. Food Chem. 42*: 1285–1290 (1994).

Wanasundara, P.K.J.P.D., F. Shahidi, and V.K.S. Shukla, Endogenous Antioxidants from Oilseeds and Edible Oils, *Food Res. Int. 13*: 225–292 (1997).

Whittern, C.C., E.E. Miller, and D.E. Pratt, Cottonseed Flavonoids as Lipid Antioxidants, *J. Am. Oil Chem Soc. 61*: 1075–1078 (1984).

Yamauchi, R., K. Kato, and Y. Ueno, Free-Radical Scavenging Reactions of α-Tocopherol During the Autoxidation of Methyl Linoleate in Bulk Phase, *J. Agric. Food Chem. 43*: 1455–1461 (1995).

Yen, G.-C., and H.-Y. Chen, Antioxidant Activity of Various Tea Extracts in Relation to Their Antimutagenicity, *J. Agric. Food Chem. 43:* 27–32 (1991).

Yoshida, H., K. Goro, and E. Shiuji, Antioxidant Effects of δ-Tocopherols at Different Concentrations in Oil During Microwave Heating, *J. Am. Oil Chem. Soc. 70*: 989–995 (1993).

Yuting, C., Z. Rongliang, J. Zhongjian, and J. Yong, Flavonoids as Superoxide Scavengers and Antioxidants, *Free Radic. Biol. Med. 9*: 19–21 (1990).

Zhang, D., T. Yasuda, Y. Yu, P. Zheng, T. Kawabata, Y. Ma, and S. Okada, Ginseng Extract Scavenges Hydroxyl Radical and Protects Unsaturated Fatty Acids from Decomposition Caused by Iron-Mediated Lipid Peroxidation, *Free Radic. Biol. Med. 20*: 145–150 (1996).

Chapter 26

Antioxidant and Antitumor Activity of Rosemary Leaves

Chi-Tang Ho[a], Mou-Tuan Huang[b], You-Rong Lou[b], Wei Ma[b], Yu Shao[a,c], Guor-Jien Wei[a], Mingfu Wang[a], and Chee-Kok Chin[c]

[a]Department of Food Science, [b]Laboratory for Cancer Research, and [c]Department of Plant Science, Rutgers University, New Brunswick, NJ 08901

Introduction

Rosemary (*Rosmarinus officinalis* L.) and sage (*Salvia officinalis*) leaves are commonly used as spices and flavoring agents. Rosemary (in the form of dried leaves) is among the most widely used spices in processed foods because it has a desirable flavor and possesses high antioxidant activity (Ho *et al.* 1994, Nakatani 1997, Wu *et al.* 1982). Crude and refined extracts of rosemary are now commercially available for application to food stabilization (Offord *et al.* 1997).

Antioxidant Activity of Phenolic Diterpene in Rosemary

Many antioxidants in rosemary are nonvolatile, and the antioxidant activity of rosemary in lard is comparable to that of butylated hydroxyanisole (BHA) and butylated hydroxytoluene (BHT) (Chang *et al.* 1997, Nakatani and Inatani 1981). Twelve diterpenes, as shown in Figure 26.1, have been isolated from rosemary (Cuvelier *et al.* 1996, Houlihan *et al.* 1984 and 1985, Inatani *et al.* 1982, Nakatani and Inatani 1983, Nakatani and Inatani 1984, Wu *et al.* 1982). Several of these, such as carnosol, carnosic acid, rosmanol, epirosmanol, isorosmanol, rosmaridiphenol, rosmadial, and miltirone, have been shown to have antioxidant activities. Carnosol and carnosic acid were found to account for >90% of the antioxidative activity of rosemary (Offord *et al.* 1997). The antioxidative activity of carnosol and carnosic acid was reported as early as the 1960s (Brieskorn and Domling 1969). The antioxidant activities of carnosol were confirmed in a later study by Wu *et al.* (1982). The studies by Nakatani's group showed that rosmanol, epirosmanol, isorosmanol, and carnosol had remarkably high activity when evaluated in lard using the Active Oxygen Method (Inatani *et al.* 1983, Nakatani and Inatani 1984). Among them, rosmanol, epirosmanol, and isorosmanol were more than four times as effective as BHT, and carnosol was twice as active (Cuvelier *et al.* 1983, Nakatani and Inatani 1984). In an aqueous ethanolic solution, the results by the ferric thiocyanate method indicated that the antioxidative activity of rosmanol and carnosol was stronger than that of α-tocopherol, but weaker than that of BHT (Nakatani and Inatani 1984). Rosmaridiphenol, isolated

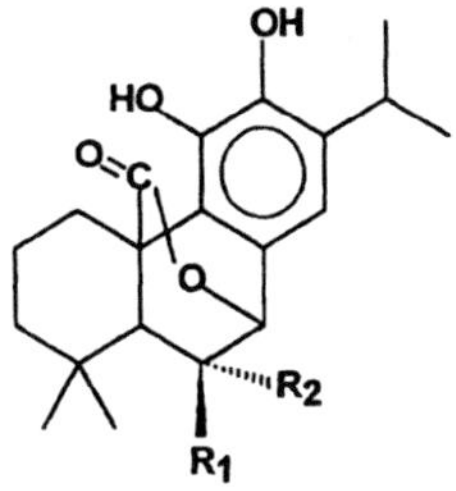

	R_1	R_2
Carnosol	H	H
Epiisorosmanol	OH	H
Epiisorosmanol ethyl ether	OCH_2CH_3	H

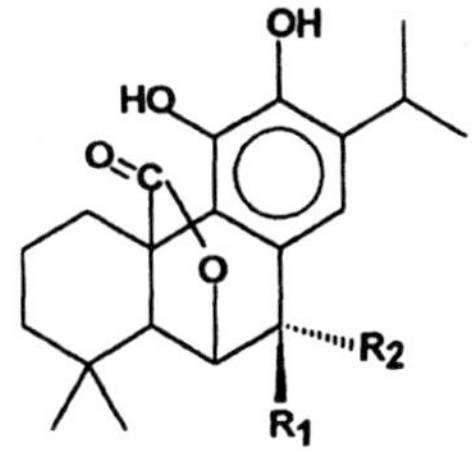

	R_1	R_2
Rosmanol	H	OH
Epirosmanol	OH	H
Epirosmanol methyl ether	OCH_3	H

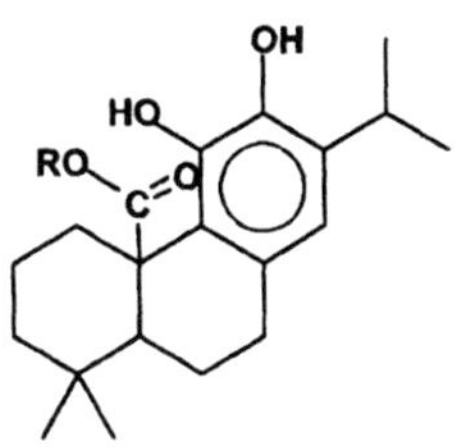

Carnosic acid R=H
Methyl carnosate R=CH_3

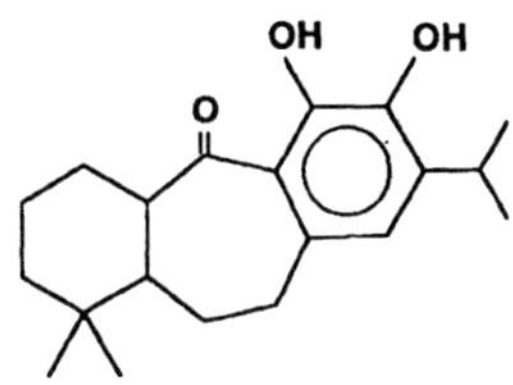

Rosmaridiphenol

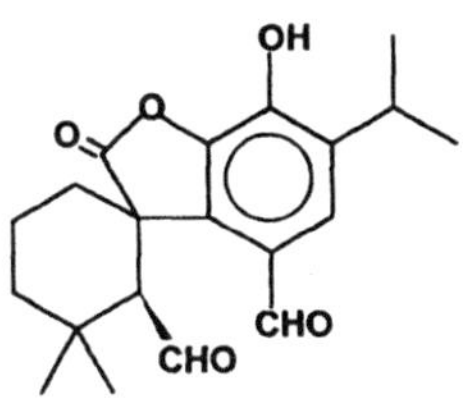

Rosmadial

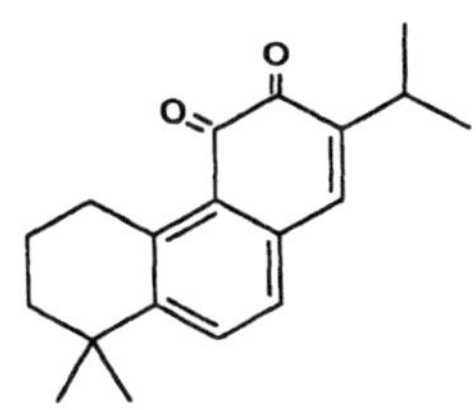

Miltirone

Fig. 26.1. Phenolic diterpenes identified in rosemary.

by Houlihan *et al.* (1984), has been shown to have antioxidative activity superior to that of BHA in lard.

Miltirone belongs to a family of compounds known as phenanthrenediones; it was first isolated from the roots of tanshen (*Salvia miltiorrhiza* Bunge) (Hayashi *et al.* 1970). This compound has also been found to have antioxidant activity (Hall *et al.* 1994, Houlihan *et al.* 1984, Weng and Gordon 1992). A chain-breaking acceptor mechanism has been suggested to be responsible for the antioxidant activity of miltirone (Weng and Gordon 1992).

Antioxidant Activity of Flavonoids in Rosemary

Generally, flavonoids are a major group of antioxidants in plants and their antioxidative activities have been widely reported (Ho 1997, Kandaswami and Middleton 1997). Figures 26.2 and 26.3 show the flavones, flavanones, and flavonoid glucuronides isolated and identified from rosemary (Cuvelier *et al.* 1996, Okamura *et al.* 1994, Shao *et al.*, unpublished results). 5-Hydroxy-7,4′-dimethoxyflavone (also known as 4′-methoxytectochrysin) isolated by Inatani *et al.* (1983) has been shown to possess antioxidative activity. However, its antioxidative activity is weaker than that of rosmanol. Compared with synthetic antioxidants, BHA and BHT, the antioxidative activity of flavones and flavanones may be relatively weak. The lack of or low antioxidative activity of these groups of compounds may be attributed to the fact that there is no hydroxyl group on the C_3 position of the molecule (Hayashi *et al.* 1970). Recently, Okamura *et al.* (1994) isolated four flavonoid glycosides, luteolin 3′-*O*-β-D-glucuronide, luteolin 3′-*O*-(4″-*O*-acetyl)-β-D-glucuronide, luteolin 3′-*O*-(3″-*O*-acetyl)-β-D-glucuronide, and hesperitin (Fig. 26.3). The antioxidative activities of these flavonoids were evaluated by a ferric thiocyanate method. All four compounds were active, and the most active compound was hesperitin.

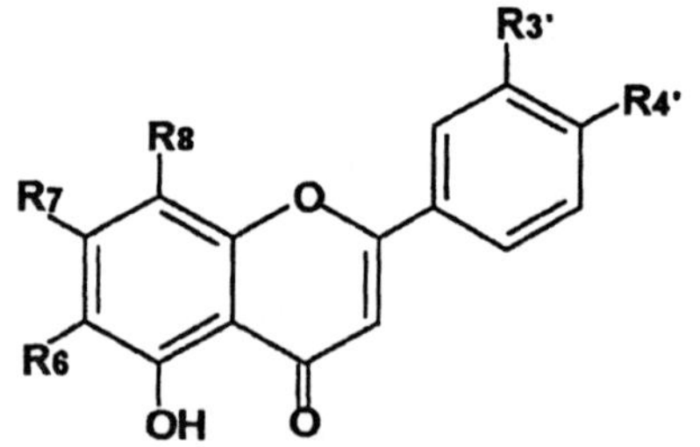

Flavones	R_6	R_7	R_8	$R_{3'}$	$R_{4'}$
Apigenin	H	OH	H	H	OH
Genkwanin	H	OCH_3	H	H	OH
4′-Methoxytectochrysin	H	OCH_3	H	H	OCH_3
Cirsimaritin	OCH_3	OCH_3	H	H	OH
Scutellarein	OH	OH	H	H	OH
4′,5,7,8-Tetrahydroxyflavone	H	OH	OH	H	OH
Homoplantaginin	OCH_3	O-Glu	H	H	OH
6-Hydroxyluteolin 7-glucoside	OH	O-Glu	H	OH	OH
Hispiduloside	OCH_3	O-Glu	H	H	OH
Nepitrin	OCH_3	O-Glu	H	OH	OH
Linaroside	OCH_3	O-Glu	H	H	OCH_3

Fig. 26.2. Flavones identified in rosemary.

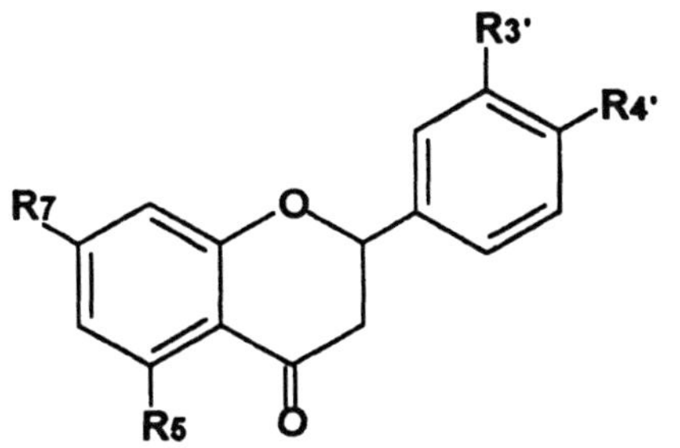

Flavanones	R_5	R_7	$R_{3'}$	$R_{4'}$
Hesperetin	OH	OH	OH	OCH_3
Hesperidin	OH	*O*-rutinose	OH	OCH_3

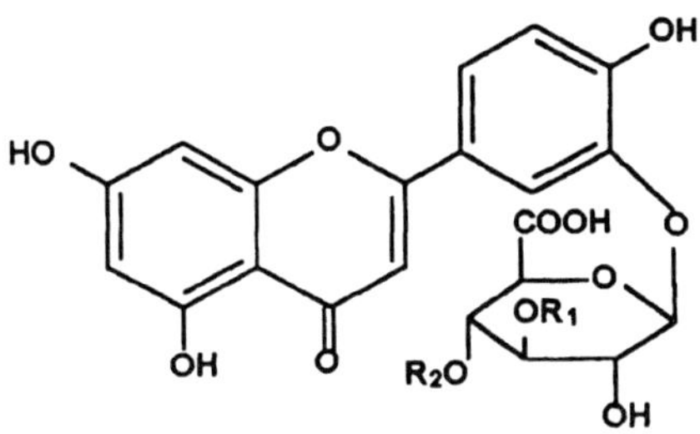

	R_1	R_2
Luteolin 3′-*O*-β-D-glucuronide	H	H
Luteolin 3′-*O*-(4″-*O*-acetyl)-β-D-glucuronide	H	Ac
Luteolin 3′-*O*-(3″-*O*-acetyl)-β-D-glucuronide	Ac	H

Fig. 26.3. Flavanones and flavonoid glucuronides identified in rosemary.

Other Compounds in Rosemary

Rosemary is known to contain a significant quantity of rosmarinic acid, which is a dimer of caffeic acid, as well as minor quantities of other phenolic acids such as vanillic acid, ferulic acid, and caffeic acid (Cuvelier *et al.* 1996). Ursolic acid, a triterpene acid, is also a well-known component of rosemary. Ursolic acid is not an active antioxidant (Wu *et al.* 1982), but its biological activity has been explored (Huang *et al.* 1994). Recently, we isolated a new lingan glycoside, rosmarinoside A (Fig. 26.4) from the more polar fraction of rosemary leaves extract (Shao *et al.*, unpublished results). The antioxidant and biological activities of the lignan glycoside is under investigation. The chemical structures of phenolic acids, ursolic acid, and rosmarinoside A are shown in Figure 26.4.

Cancer Preventive Activity of Rosemary

Many compounds that possess antioxidant activity inhibit tumor initiation and tumor promotion in different organs of several animal models (Smart *et al.* 1987, Huang *et al.* 1988, 1992, 1994, 1995, and 1996). Because rosemary extract is a naturally occurring antioxidant, we were interested in evaluating the effect of rose-

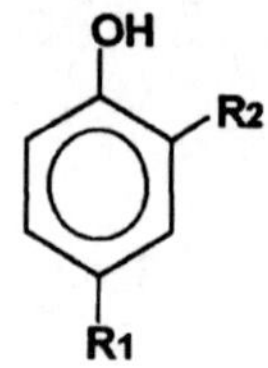

	R_1	R_2
Vanillic acid	COOH	OCH_3
Ferulic acid	-CH=CH-COOH	OCH_3
Caffeic acid	-CH=CH-COOH	OH

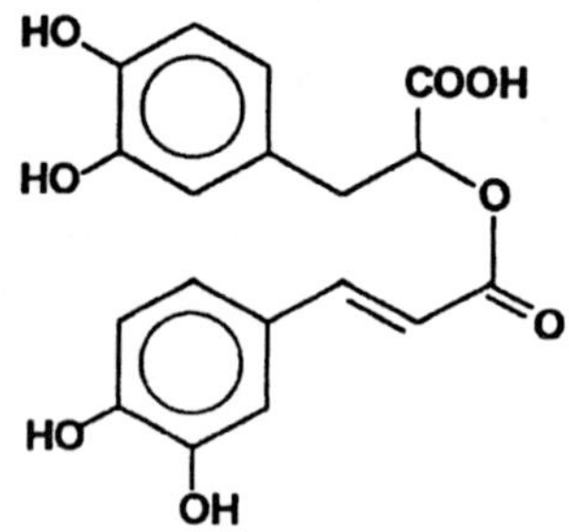

Rosmarinic acid

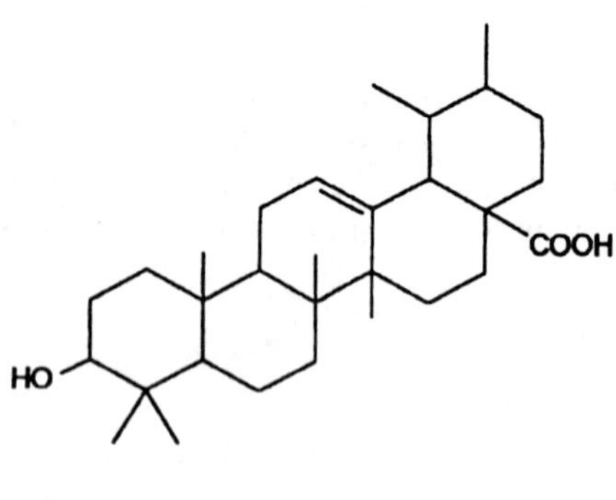

Ursolic acid

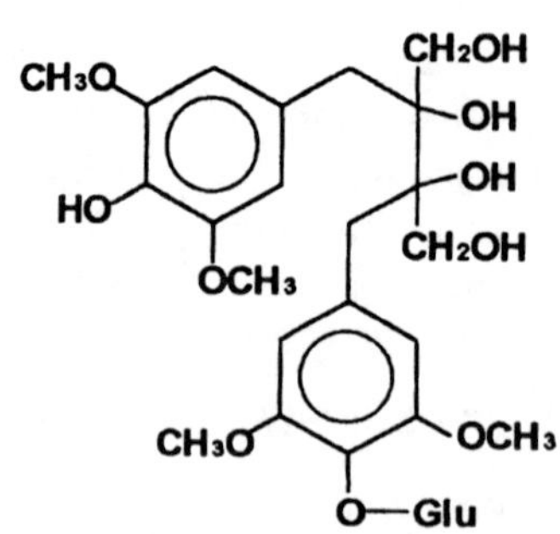

Rosmarinoside A

Fig. 26.4. Phenolic acids, ursolic acid, and rosmarinoside A identified in rosemary.

mary on tumorigenesis in several animal models. We have prepared a methanolic extract from the leaves of *R. officinalis* L. and have also studied the cancer chemopreventive effects of rosemary on tumorigenesis in several animal models as described below.

Preparation of Methanol Extract of Ground Dried Leaves of **Rosmarinus officinalis** *L.*

A methanol extract of the ground dried leaves of the plant *R. officinalis* L. (rosemary extract; rosemary) was prepared according to a previously reported procedure (Huang *et al.* 1994). Rosemary leaves (3 kg), which had been ground to a fine powder, were extracted with 18 L methanol at 60°C for 2 h. The extraction was performed in a custom-made 30-L steam-jacketed, stainless extractor equipped with a Lightning mixer (Lightning Group, Rochester, NY). After extraction, the mixture was filtered, and the residue was extracted again with 12 L fresh methanol. The combined filtrate was bleached with 600 g activated charcoal and then filtered to yield a light-brown filtrate. The methanol solution was concentrated to a final volume of 2 L with a vacuum rotary evaporator and then filtered to remove precipitates. Water (3 L) was added to the filtrate. The precipitate that was formed after the addition of water was filtered and air-dried to yield a rosemary extract (rosemary) that was used for our studies. Approximately 200–300 g dry powdered rosemary extract was obtained from 3 kg of ground dried rosemary leaves. This preparation of rosemary was used for all of the experiments as indicated below.

High-Performance Liquid Chromatographic (HPLC) Analysis of Rosemary Extract

HPLC analysis was performed using a Varian (Palo Alto, CA) 5000 Liquid Chromatograph with an ACS model 750/14 mass detector from Peris Industries (State College, PA). In this detector, the flow of the column moves through a nebulizer and then passes through a tube held at a temperature at which the solvent vaporizes and the solute forms small droplets. The light scattered by the droplets is measured by a photomultiplier. The detector was operated at a temperature at 50°C and 110.3 KPa air pressure. The column was a Whatman (Clifton, NJ) Parti Sphere C column (12.5 cm × 0.46 cm internal diameter). The following ternary solvent system was utilized for column elution: Solvent A, 1% acetic acid in water; Solvent B, acetonitrile; and solvent C, 1% acetic acid in methanol. The mobile phase was programmed linearly from 30% A/70% C at the start of elution to 5% A/90% B over a 30-min interval, followed by 5% A/5% B/90% C for 5 min and then 100% C for another 15 min. The flow rate was 0.7 mL/min. The retention times of carnosol, miltirone, carnosic acid, and ursolic acid were 7.9, 9.2, 15.7, and 30.8 min, respectively. HPLC analysis of several preparations of the rosemary extract indicated that they contained 16.5–19.2% ursolic acid, 3.8–4.6% carnosol, 0.1–0.5% carnosic acid, and a trace to 0.1% miltirone. Each preparation of dry, powdered rosemary extract was examined by HPLC and tested for its ability to inhibit 12-*O*-tetradecanoylphorbol-13-acetate (TPA)-induced increases in epidermal ornithine decarboxylase activity and TPA-induced mouse ear edema.

Inhibitory Effects of Rosemary on the Covalent Binding of [^{3}H]Benzo[a]pyrene (B[a]P) to Epidermal DNA

Table 26.1 shows the inhibitory effect of rosemary on the formation of carcinogen DNA adducts and TPA-induced short-term biochemical marker changes. We have previously reported the inhibitory effect of topical application of rosemary on the covalent binding of [^{3}H]B[a]P to epidermal DNA (Huang *et al.* 1994). Topical application of 20 nmol [^{3}H]B[a]P resulted in the covalent binding of 0.74 pmol [^{3}H]B[a]P metabolites/mg epidermal DNA at 15 h after application. Topical application of 1.2 or 3.6 mg rosemary 5 min before the application of [^{3}H]B[a]P inhibited covalent binding to DNA by 30 and 54%, respectively. Singletary and Nelshoppen (1991) showed that feeding 0.5–1% rosemary in the diet inhibited the formation of 7,12-dimethylbenz[a]anthracene (DMBA)-DNA adducts in mammary glands of rats (Singletary and Nelshoppen 1991). These results indicate that rosemary may inhibit B[a]P-induced skin tumor initiation and DMBA-induced mammary tumor initiation.

Inhibitory Effects of Rosemary on TPA-Induced Short-Term Biochemical Marker Changes

Topical application of rosemary to the backs of CD-1 mice markedly inhibited TPA-induced increases in ornithine decarboxylase activity, epidermal thickness, the number of epidermal cell layers, and inflammatory cell infiltration into dermis (Huang *et al.* 1994).

TABLE 26.1
Inhibitory Effect of Rosemary on Formation of Carcinogen-DNA Adducts and TPA-Induced Short-Term Biochemical Marker Changes[a,b]

Observed effects	Inhibition
Tumor initiation marker changes	
Rosemary inhibited the formation of [^{3}H]B[a]P-DNA adducts in mouse epidermis	+++
Rosemary inhibited the formation of [^{3}H]DMBA-DNA adducts in mammary gland of rats	+++
Tumor promotion marker changes	
Rosemary inhibited TPA-induced increases in ornithine decarboxylase activity	+++
Rosemary inhibited TPA-induced skin inflammation (edema of mouse ear)	+++
Rosemary inhibited TPA-induced increases in epidermal thickness	+++
Rosemary inhibited TPA-induced increases in the number of epidermal cell layers	+++
Rosemary inhibited TPA-induced leukocyte infiltration into dermis	++

[a] The amount of inhibition is indicated by + (weak), ++ (moderate), or +++ (strong).
[b] Abbreviations: TPA, 12-*O*-tetradecanoylphorbol-13-acetate; B[a]P, [^{3}H]benzo[a]pyrene; DMBA, 7,12-dimethylbenz[a]anthracene.

TABLE 26.2
Some Studies on the Inhibitory Effect of Rosemary on Tumorigenesis in Several Animal Models[a]

Organ	Species	Administration of rosemary	Carcinogen or promoter treatment	Inhibitory effect
Skin	Mouse	Topical application of rosemary	B[a]P and DMBA	Inhibited B[a]P- or DMBA-induced skin tumor initiation
			TPA	Inhibited TPA-induced skin tumor promotion in DMBA-treated mice
			Mirex or okadaic acid	Inhibited non-TPA tumor promoter, mirex, or okadaic acid–induced tumor promotion in DMBA-treated mice
Lung	Mouse	Feeding rosemary diet during the initiation period	B[a]P	Inhibited B[a]P-induced lung tumorigenesis during the initiation period in A/J mice
		Feeding rosemary diet during postinitiation period	B[a]P	Inhibited B[a]P-induced lung tumorigenesis during the postinitiation period
Forestomach	Mouse	Feeding rosemary diet during the initiation period	B[a]P	Inhibited B[a]P-induced forestomach tumorigenesis during the initiation period
		Feeding rosemary diet during postinitiation period	B[a]P	Inhibited B[a]P-induced forestomach tumorigenesis during postinitiation period
Breast	Mouse	Feeding rosemary diet during the initiation period	DMBA	Inhibited DMBA-induced breast tumorigenesis during the initiation period
		Feeding rosemary diet during postinitiation period	DMBA	Had little or no effect on DMBA-induced breast tumorigenesis during the postinitiation period

Continued

TABLE 26.2
(Continued)[a]

Organ	Species	Administration of rosemary	Carcinogen or promoter treatment	Inhibitory effect
	Rats	Feeding rosemary diet	DMBA	Inhibited DMBA-induced breast during the initiation period tumorigenesis during the initiation period
Lymphoma/ Leukemia	Mouse	Feeding rosemary diet during the initiation period	DMBA	Inhibited DMBA-induced formation of lymphomas/leukemias during the initiation period
		Feeding rosemary diet during postinitiation period	DMBA	Inhibited DMBA-induced formation of lymphomas/leukemias during the postinitiation period
Ovary	Mouse	Feeding rosemary diet during the initiation period.	DMBA	Inhibited DMBA-induced formation of ovary tumors in Sencar mice
Uterus	Mouse	Feeding rosemary diet during the initiation period	DMBA	Inhibited DMBA-induced formation of uterine tumors in Sencar mice during the initiation period
Colon	Mouse	Feeding rosemary diet during the initiation period	AOM	Inhibited AOM-induced colon tumorigenesis during the initiation period
		Feeding rosemary diet during postinitiation period	AOM	Had little or no effect on AOM-induced colon tumorigenesis during the postinitiation period

[a]Abbreviations: B[a]P, [^{3}H]benzo[a]pyrene; DMBA, 7,12-dimethyl-benz[a]anthracene; TPA, 12-*O*-tetradecanoylphorbol-13-acetate; AOM, azoxymethane.

Topical application of rosemary to the ears of mice also markedly inhibited TPA-induced edema of mouse ear (skin inflammation) (Huang *et al.* 1994). The results were summarized in Table 26.1. These results indicate that rosemary may inhibit tumor promotion in mice previously initiated with B[a]P or DMBA.

Inhibitory Effects of Rosemary on Tumorigenesis in Several Animal Models

Preliminary studies in our laboratory indicated that rosemary strongly inhibits skin tumor initiation and skin tumor promotion in DMBA-treated mice (Table 26.2). Topical application of rosemary to the backs of CD-1 mice markedly inhibited skin tumor initiation by polycyclic aromatic hydrocarbons, B[a]P and DMBA, and inhibited TPA-induced tumor promotion in DMBA-initiated mice (Huang *et al.* 1994). Topical application of rosemary to the backs of mice also inhibited non-TPA type tumor promoter mirex or okadaic acid–induced skin tumor promotion in DMBA-initiated mice.

Feeding 0.5–2% rosemary in AIN 76A diet of mice during the initiation period or during the postinitiation period markedly inhibited B[a]P-induced forestomach tumorigenesis and lung tumorigenesis in A/J mice. Feeding 0.5–2% rosemary in the AIN 76A diet during the initiation period or during the entire period (the initiation period + the postinitiation period) to mice strongly inhibited azoxymethane (AOM)-colon tumorigenesis in CF-1 mice and strongly inhibited DMBA-induced breast tumorigenesis in Sencar mice. Feeding 0.52% rosemary in the AIN 76A diet during the postinitiation period had little or no effect on AOM-induced colon tumorigenesis and DMBA-induced breast tumorigenesis in Sencar mice (Table 26.2). In this study, dietary rosemary did not inhibit AOM-induced colon tumorigenesis and DMBA-induced breast tumorigenesis in mice. It is possible in these studies that the carcinogen doses used were too high and that tumor appearing time (latency of tumor development) was too short and too rapid. It is necessary to reevaluate the effect of dietary rosemary on AOM-induced colon tumorigenesis and DMBA-induced breast tumorigenesis by using a low dose of carcinogens in this model. Feeding 0.5–1.0% rosemary in the AIN 76A diet to Sprague-Dawley rats during the initiation period also inhibited DMBA-induced breast tumorigenesis (Singletary and Nelshoppen 1991). In addition, feeding 0.5–2% rosemary inhibited DMBA-induced formation of lymphoma/leukemia, ovarian tumors, and uterine tumors in Sencar mice (Table 26.2). Therefore, dietary rosemary serves as a potent cancer chemopreventive agent.

References

Brieskorn, C.H., and H.J.Z. Domling, Carnosic Acid as an Antioxidant in Rosemary and Sage Leaves, *Lebensm. Unters. Forsch. 141*: 10–16 (1969).

Chang, S.S., B. Ostric-Matijasvic, O.A.L. Hsieh, and C.-L. Huang, Natural Antioxidants from Rosemary and Sage, *J. Food Sci. 42*: 1102–1106 (1977).

Cuvelier, M.-E., H. Richard, and C. Berset, Antioxidant Activity and Phenolic Composition of Pilot-Plant and Commercial Extracts of Sage and Rosemary, *J. Am. Oil Chem. Soc. 73*: 645–652 (1996).

Hall, C., III, S. Cuppett, D. Wheeler, and X. Fu, Effects of Bleached and Unbleached Rosemary Oleoresin and Rosmariquinone on Light-Sensitized Oxidation of Soybean Oil, *J. Am. Oil Chem. Soc. 71*, 533–535 (1994).

Hayashi, T., H. Kakisawa, H.Y. Hsu, and Y.P. Chen, Structure of Miltirone; a New Diterpenoid Quinone, *Chem. Commun. 1*: 299–300 (1970).

Ho, C.-T., Antioxidant Properties of Plant Flavonoids, in *Food Factors for Cancer Prevention*, edited by H. Ohigashi, T. Osawa, S. Terao, S. Watanabe, and T. Yoshikswa, Springer-Verlag, Tokyo, 1997, pp. 593–597.

Ho, C.-T., T. Ferraro, Q. Chen, R.T. Rosen, and M.T. Huang, Phytochemicals in Teas and Rosemary and Their Cancer-Preventive Properties, in *Food Phytochemicals for Cancer Prevention II, Teas, Spices, and Herb,* edited by C.-T. Ho, T. Osawa, M.T. Huang, and R.T. Rosen, ACS Symposium, Series No. 547, American Chemical Society, Washington, DC, 1994, pp. 2–19.

Houlihan, C.M., C.-T. Ho, and S.S. Chang, Elucidation of the Chemical Structure of a Novel Antioxidant, Rosmaridiphenol, Isolated from Rosemary, *J. Am. Oil. Chem. Soc. 61*: 1036–1039 (1984).

Houlihan, C.M., C.-T. Ho, and S.S. Chang, The Structure of Rosmariquinone, a New Antioxidant Isolated from *Rosmarinus officinalis* L., *J. Am. Oil. Chem. Soc. 62*: 96–98 (1985).

Huang, M.T., C.-T. Ho, Z.Y. Wang, T. Ferraro, Y.R. Lou, K. Stauber, W. Ma, C. Georgiadis, J.D. Laskin, and A.H. Conney, Inhibition of Skin Tumorigenesis by Rosemary and Its Constituents Carnosol and Ursolic Acid, *Cancer Res. 54*: 701–708 (1994).

Huang, M-T., W. Ma, Y.-P. Lu, R.L. Chang, C. Fisher, P.S. Manchand, H.L. Newmark, and A.H. Conney, Effects of Curcumin, Demethoxycurcumin, Bisdemethoxycurcumin, and Tetrahydrocurcumin on 12-*O*-Tetradecanoylphorbol-13-acetate-Induced Tumor Promotion, *Carcinogenesis 16*: 2493–2497 (1995).

Huang, M-T., W. Ma, P. Yen, J.-G. Xie, J. Han, K. Frenkel, D. Grunberger, and A.H. Conney, Inhibitory Effects of Caffeic Acid Phenethyl Ester (CAPE) on 12-*O*-Tetradecanoylphorbol-13-acetate-Induced Tumor Promotion in Mouse Skin and the Synthesis of DNA, RNA and Protein in HeLa Cells, *Carcinogenesis 17*: 761–765 (1996).

Huang, M-T., R.C. Smart, C.-Q. Wong, and A.H. Conney, Inhibitory Effect of Curcumin, Chlorogenic Acid, Caffeic Acid, and Ferulic Acid on Tumor Promotion in Mouse Skin by 12-*O*-Tetradecanoylphorbol-13-acetate, *Cancer Res. 48*: 5941–5946 (1988).

Huang, M-T., Z.Y. Wang, C.A. Georgiadis, J.D. Laskin, and A.H. Conney, Inhibitory Effect of Curcumin on Tumor Initiation by Benzo[a]pyrene and 7,12-Dimethylbenz[a]anthracene, *Carcinogenesis (Lond.) 13*: 947–954 (1992).

Inatani, R., N. Nakatani, and H. Fuwa, Antioxidative Effect of the Constituents of Rosemary (*Rosmarinus officinalis* L.) and Their Derivatives, *Agric. Biol. Chem. 47*: 521–528 (1983).

Inatani, R., N. Nakatani, H. Fuwa, and H. Seto, Structure of a New Antioxidative Phenolic Diterpene Isolated from Rosemary (*Rosmarinus officinalis* L.), *Agric. Biol. Chem. 46*: 1661–1666 (1982).

Kandaswami, C., and E. Middleton, Flavonoids as Antioxidants, in *Natural Antioxidants: Chemistry, Health Effects, and Application*, edited by F. Shahidi, AOCS Press, Champaign, IL, 1997, pp. 174–203.

Nakatani, N., Antioxidants from Spices and Herbs, in *Natural Antioxidants: Chemistry, Health Effects, and Application*, edited by F. Shahidi, AOCS Press, Champaign, IL, 1997, pp. 64–75.

Nakatani, N., and R. Inatani, Structure of Rosmanol, a New Antioxidant from Rosemary (*Rosmarinus officinalis* L.), *Agric. Biol. Chem. 45*: 2385–2386 (1981).

Nakatani, N., and R. Inatani, A New Diterpene Lactone, Rosmadiol, from Rosemary (*Rosmarinus officinalis* L.), Agric. Biol. Chem. 47: 353–358 (1983).

Nakatani, N., and R. Inatani, Two Antioxidative Diterpenes from Rosemary (*Rosmarinus officinalis* L.) and a Revised Structure of Rosmanol, *Agric. Biol. Chem. 48*: 2081–2085 (1984).

Offord, E.A., F. Guillot, R. Aeschbach, J. Loliger, and A.M.A. Pfeifer, Antioxidant and Biological Properties of Rosemary Components: Implications for Food and Health, in *Natural Antioxidants: Chemistry, Health Effects, and Application*, edited by F. Shahidi, AOCS Press, Champaign, IL, 1997, pp. 88–96.

Okamura, N., H. Haraguchi, K. Hashimoto, and A. Yagi, Flavonoids in *Rosmarinus* Leaves, *Phytochemistry 37*: 1463–1466 (1994).

Singletary, K.W., and J.M. Nelshoppen, Inhibition of 7,12-Dimethylbenz[a]anthracene (DMBA)-Induced Mammary Tumorigenesis and of *in vivo* Formation of Mammary DMBA DNA Adducts by Rosemary Extract, *Cancer Lett. 60*: 169–175 (1991).

Smart, R.C., M.-T. Huang, Z.T. Han, M.C. Kaplan, A. Focella, and A.H. Conney, Inhibition of 12-O-Tetradecanoylphorbol-13-acetate Induction of Ornithine Decarboxylase Activity, DNA Synthesis, and Tumor Promotion in Mouse Skin by Ascorbic Acid and Ascorbyl Palmitate, *Cancer Res. 47*: 6633–6638 (1987).

Weng, X.C., and M.H. Gordon, Antioxidant Activity of Quinones Extracted from Tanshen (*Salvia miltiorrhiza* Bunge), *J. Agric. Food Chem. 40*: 1331–1336 (1992).

Wu, J.W., M.-H. Lee, C.-T. Ho, and S.S. Chang, Elucidation of the Chemical Structures of Natural Antioxidants Isolated from Rosemary, *J. Am. Oil Chen. Soc. 59*: 339–345 (1982).

Chapter 27

Antitumor Activity of an Extract of the Gum Exudate from the Tree *Boswellia serrata*

Mou-Tan Huang[a], Wei Ma[a], Vladimir Badmaev[b], Yu Shao[c,d], Chee-Kok Chin[c], and Chi-Tang Ho[d]

[a]Department of Chemical Biology, Laboratory for Cancer Research, College of Pharmacy, Rutgers University, Piscataway, NJ, 08854–8020

[b]121 Ethel Road, Piscataway, NJ, 08854

[c]Department of Plant Science, Cook College, Rutgers University, New Brunswick, NJ, 08901

[d]Department of Food Science, Cook College, Rutgers University, New Brunswick, NJ, 08901–8520

Introduction

The gum resin exudate from portions of the stem of the tree, *Boswellia serrata*, has been used as a traditional medicine in India and China for the treatment of many ailments such as inflammation, arthritic pain, and wounds (Gupta *et al.* 1987, Kakrani 1981). The crude alcohol extract of the gum resin exudate from *B. serrata* and its major constituent, boswellic acid, have been reported to have potent anti-inflammatory and antiarthritic properties (Gupta *et al.* 1987, Kakrani 1981); they are known to inhibit 5-lipoxygenase activity but have little or no effect on cyclooxygenase activity (Ammon *et al.* 1991, Singh *et al.* 1993). The metabolites of arachidonic acid such as 5-hydroxyeicosatatraenoic acid, leukotriene C_4, and leukotriene D_4 have been shown to play important roles in the enhancement and stimulation of the growth of certain tumor cell lines *in vitro* (Ondrey *et al.* 1989) as well as tumor promotion and progression *in vivo* (Kato *et al.* 1983, Nakadate *et al.* 1984). It was observed that certain inhibitors of 5-lipoxygenase inhibit DNA synthesis of several cancer lines (Avis *et al.* 1996, Ondrey *et al.* 1989, Tsukada *et al.* 1986). Tsukata *et al.* (1986) reported that arachidonic 5-lipoxygenase inhibitors had potent antiproliferative effects on human leukemic cell lines (HL-60). On the basis of the above observations, we hypothesize that a crude extract of gum resin exudate from *B. serrata* or boswellic acid will inhibit the growth and synthesis of DNA in HL-60 cells and inhibit 12-*O*-tetradecanoylphorbol-13-acetate (TPA)-induced tumor promotion on the skin of mice previously initiated with 7,12-dimethylbenz[a]anthracene (DMBA). In this chapter, we report that an alcohol extract of gum resin exudate from *Boswellia serrata* (BSE) markedly inhibits the growth and synthesis of DNA, RNA, and protein in HL-60 cells.

Materials and Methods

Materials

[^{3}H]Thymidine (50 Ci/mol), [^{3}H]uridine (55 Ci/mol), and [^{3}H]leucine (150 Ci/mol) were purchased from the Du Pont Chemical Company, NEN Research Products (Boston, MA). HL-60 cells were purchased from the American Type Cell Cultures (Rockville, MD). RPMI medium, fetal calf serum, trypsin, penicillin, and streptomycin were purchased from GIBCO BRL (Grand Island, NY). Glass microfiber filters (2.5 cm in diameter), ScientiVerse, trichloroacetic acid (TCA), and high-performance liquid chromatography (HPLC)-grade dimethyl sulfoxide (DMSO), ethyl acetate, methanol, and *n*-butanol were purchased from Fisher Scientific (Springfield, NJ).

Cultured HL-60 Cells

The HL-60 human promyelocytic leukemia cell line was purchased from American Type Culture Collection. HL-60 cells (5×10^5 cells/mL) were grown in a 75-cm^3 canted flask in RPMI medium supplemented with 10% fetal calf serum and 1% penicillin-streptomycin in a humidified atmosphere containing 5% CO_2 at 37°C. Under these conditions, HL-60 cells had a doubling time of ~24–30 h, depending on the medium and calf serum used. To maintain normal cell growth, cells were usually passed every 2–3 d. To measure the effect of an alcohol extract of gum resin exudate from *B. serrata* on the growth of HL-60 cells, different concentrations of an alcohol extract of gum resin exudate from *B. serrata* were added to HL-60 cells (5×10^5 cells/mL) in a flask containing 50 mL of RPMI medium supplemented with 10% fetal calf serum and 1% penicillin-streptomycin. Cultures were counted and recorded daily using a hemacytometer under a microscope with 10 × magnification every 24 h.

Measurement of DNA, RNA, and Protein Synthesis in HL-60 Cells

HL-60 cells were harvested by centrifugation for 5 min at $1000 \times g$. Cells were resuspended in RPMI without fetal calf serum and placed in a series of 13 × 100 mm test tubes [5×10^5 cells/(mL · tube)]. For measurement of DNA, RNA, or protein synthesis, 3 μL [^{3}H]thymidine (50 Ci/mol), 5 μL [^{3}H]uridine (55 Ci/mol), or 10 μL [^{3}H]leucine (150 Ci/mol) respectively, was added to each individual tube followed by 2 μL DMSO or inhibitor in DMSO. Tubes were incubated at 37°C for 120 min. The reactions were terminated by adding 3 mL of cold phosphate-buffered saline (PBS) to the tubes, which were subsequently rinsed twice with 3 mL cold PBS. The tubes were then centrifuged for 5 min at $1000 \times g$ and the supernatant was discarded. Finally, cells were lysed in 2 mL distilled water followed by 2 mL of an ice-cold 10% TCA solution and kept in an ice bath. The precipitates were collected on glass-fiber filters and washed three times with a cold 5% TCA solution. The dry, glass-fiber filters were placed in scintillation vials with

10 mL of ScientVerse fluid, and radioactivity was determined in a Beckman LS 1701 scintillation counter (Beckman Instruments Inc., Palo Alto, CA).

Results and Discussion

Major Constituents of an Alcohol Extract of the Gum Exudate of **Boswellia serrata** *(BSE)*

An ethanol extract of the gum exudate from the stem of the plant *B. serrata* contains ~60% boswellic acid and its derivatives, namely, 3-*O*-acetyl-β-boswellic acid, 11-keto-β-boswellic acid, and 3-*O*-acetyl-11-keto-β-boswellic acid (see Fig. 27.1).

Inhibitory Effect of an Alcohol Extract of Gum Resin Exudate from **Boswellia serrata** *on the Growth of HL-60 Cells*

An alcohol extract of gum resin exudate from *B. serrata* (BSE) inhibited the growth of HL-60 cells in a dose-dependent fashion. Addition of BSE of the cell culture to the final concentrations at 4, 12, or 50 μg/mL and incubation at 37°C for 48 h inhibited the growth by 11, 30, or 97%, respectively, at 37°C for 72 h inhibited the growth by 9, 40, or 100%, respectively, and incubation at 37°C for 96 h inhibited the growth of HL-60 cells by 5, 45, or 96%, respectively (Table 27.1). BSE at 50 μg/mL completely inhibited HL-60 cell growth. To test whether BSE inhibition was the result of killing the cells (cytocide) or stopping the cell division (cytostate), the effects of BSE on HL-60 cells viability were examined using the trypan blue exclusion method. More than 95% of cells were not stained with trypan

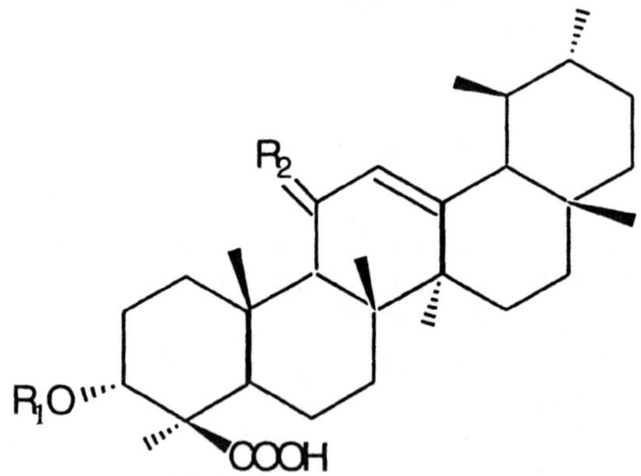

1.	β-Boswellic acid	$R_1 = H$	$R_2 = H_2$
2.	3-*O*-Acetyl-β-boswellic acid	$R_1 = Ac$	$R_2 = H_2$
3.	11-Keto-β-boswellic acid	$R_1 = H$	$R_2 = O$
4.	3-*O*-Acetyl-11-keto-β-boswellic acid	$R_1 = Ac$	$R_2 = O$

Fig. 27.1. Structure of β-boswellic acid and its derivatives.

TABLE 27.1
Inhibitory Effect of an Alcohol Extract of Exudate of *Boswellia serrata* (BSE) on the Growth of HL-60 Cells in Culture[a]

	Concentration of BSE (μmol/L)			
Incubation time (h)	0	4	12	50
0	25 ± 2.3	25 ± 2.3	25 ± 2.3	25 ± 2.3
24	45 ± 2.1	40 ± 4.2 (25%)	39 ± 3.7 (30%)	30 ± 4.0 (75%)
48	71 ± 1.5	66 ± 4.7 (11%)	57 ± 3.5 30%)	27 ± 2.0 (97%)
72	102 ± 2.1	95 ± 2.9 (9%)	72 ± 7.8 (40%)	25 ± 1.2 (100%)
96	166 ± 16.6	159 ± 11 (5%)	102 ± 2.6 (45%)	31 ± 2.2 (96%)

[a]HL-60 cells (2.5×10^5 cells/mL) were grown in a 75-cm^3 canted flask in RPMI medium supplemented with 10% fetal calf serum and 1% penicillin-streptomycin in a humidified atmosphere containing 10% CO_2 at 37°C. Various concentrations of BSE were added to the cultures, and cultures were counted daily using a hemacytometer under a microscope with 10 × magnification every 24 h. Data are expressed as the mean ± SEM from triplicate tests. Data in parentheses are the percentages of inhibition of cell growth.

blue, indicating that most cells were alive. BSE, at concentrations that significantly inhibited cell growth, did not kill the cells.

Inhibitory Effect of BSE on the Synthesis of DNA, RNA, and Protein in HL-60 Cells

The inhibitory effects of BSE on the synthesis of DNA, RNA, and protein were studied in HL-60 cells. The initial rates of incorporation of [^{3}H]thymidine, [^{3}H]uridine, or [^{3}H]leucine into TCA-insoluble materials were utilized to estimate the rates of DNA, RNA, and protein synthesis, respectively, in HL-60 cells. Inhibition of BSE on DNA synthesis in HL-60 cells occurred rapidly. Table 27.2 shows the inhibitory effect of BSE on DNA synthesis in HL-60 cells. Incubation of HL-60 cells with BSE at 4, 12, or 50 μg/mL at 37°C for 120 min inhibited [^{3}H]thymidine incorporation into DNA in HL-60 cells by 69, 77, or 86%, respectively. Studies on the inhibitory effect of BSE on DNA, RNA, and protein synthesis in HL-60 cells revealed that BSE strongly inhibited DNA, RNA, or protein synthesis in HL-60 cells (Table 27.3). BSE at 0.75, 1.5, 3.0, 6.0, 12.0, or 25.0 μmol/L inhibited DNA synthesis by 20, 55, 65, 77, 81, or 82%, respectively, inhibited RNA synthesis by 9, 36, 38, 80, 90, at 92%, respectively, and inhibited protein synthesis by 30, 48, 74, 88, 91, or 92%, respectively.

Irreversible Inhibitory Effect of BSE on the Synthesis of DNA in HL-60 Cells

HL-60 cells (5×10^5 cells/mL) were incubated with BSE at 50 μg/mL at 37°C for 30 min. The cells were harvested and washed three times with 3 mL PBS each

TABLE 27.2
Inhibitory Effect of Various Concentrations of an Alcohol Extract of Exudate of *Boswellia serrata* (BSE) on the Incorporation of [^{3}H]Thymidine into DNA of HL-60 Cells[a]

	Concentration of BSE (μmol/L)			
	0	4	12	50
Incubation time (min)	(cpm/5 × 10^5 cells)			
0	279 ± 76	252 ± 114	312 ± 54	225 ± 15
120	11,112 ± 1897	4039 ± 737 (69%)	2794 ± 306 (77%)	1893 ± 505 (86%)

[a]HL-60 cells (2.5 × 10^5 cells/mL) were grown in a 75-cm^3 canted flask in RPMI medium supplemented with 10% fetal calf serum and 1% penicillin-streptomycin in a humidified atmosphere containing 10% CO_2 at 37°C. [^{3}H]Thymidine (3 μL; 50 Ci/mol) vehicle or various concentrations of BSE in vehicle were added to 1 mL of HL-60 cells (5 × 10^5 cells/mL) in 100 × 13 mm test tubes; cultures were incubated at 37°C for 120 min. Incorporation of [^{3}H]thymidine into DNA was determined as indicated in the Materials and Methods section. Data are expressed as the mean ± SEM from triplicate tests. Data in parentheses are the percentages of inhibition of [^{3}H]thymidine incorporation into DNA of HL-60 cells.

time, and then resuspended in the fresh medium for the determination of [^{3}H]thymidine incorporation into DNA. It was essentially an irreversible inhibitory effect (data not shown). These results indicated that inhibition of cell growth was due to the interference with cell proliferation rather than with cytotoxicity. The inhibition of cell proliferation can be explained by the interference with the biosynthesis of DNA, RNA, and protein that is required for cell proliferation. Further isolation, purification, and identification of β-boswellic acid, 3-acetyl-β-boswellic acid, 11-keto-β-boswellic acid, and 3-acetyl-11-keo-β-boswellic acid or other

TABLE 27.3
Inhibitory Effect of an Alcohol Extract of Exudate of *Boswellia serrata* (BSE) on the Synthesis of DNA, RNA, and Protein in HL-60 Cells[a]

	DNA synthesis		RNA synthesis		Protein syn thesis	
BSE added (μmol/L)	% Control	% Inhibition	% Control	% Inhibition	% Control	% Inhibition
0	100	0	100	0	100	0
0.75	80	20	91	9	70	30
1.5	45	55	64	36	52	48
3.0	35	65	62	38	26	74
6.0	23	77	20	80	12	88
12.0	19	81	10	90	9	91
25.0	18	82	8	92	8	92

[a]An extract of various concentrations of exudate of *Boswellia serrata* was added at the indicated final concentrations to 1 mL of HL-60 cells (5 × 10^5 cells) suspended in RPMI medium without supplementation with 10% fetal calf serum. [^{3}H]Thymidine (50 Ci/mol; 3 mL), [^{3}H]uridine (55 Ci/mol; 5 μL), or [^{3}H]leucine (200 Ci/mol; 10 μL) was added to the cell suspension and incubated at 37°C for 120 min. Reactions were terminated by addition of 3 mL of cold phosphate-buffered saline (PBS) and the rates of DNA, RNA, and protein synthesis were determined according to the procedures as described in Materials and Methods section.

derivatives of β-boswellic acid are required to determine which compound is responsible for the inhibition of growth and synthesis of DNA in HL-60 cells.

References

Ammon, H.P.T., T. Mack, G.B. Singh, and H. Safayhi, Inhibition of Leukotriene B Formation in Rat Peritoneal Neutrophils by an Ethanolic Extract of the Gum Resin Exudate of *Boswellia serrata, Planta Med. 57*: 203–205 (1991).

Ammon, H.P.T., H. Safayhi, T. Mack, and J. Sabieraj, Mechanism of Anti-Inflammatory Actions of Curcumin and Boswellic Acids, *J. Ethnopharmacol. 38*: 113–119 (1993).

Avis, I.M., M. Jett, T. Boyle, M.D. Vos, T. Moody, A.M. Treston, A. Martinez, and J.L. Mulshine, Growth Control of Lung Cancer by Interruption of 5-Lipoxygenase-Mediated Growth Factor Signaling, *J. Clin. Investig. 97*: 806–813 (1996).

Gupta, V.N., D.S. Yadav, M.P. Jain, and C.K. Atal, Chemistry and Pharmacology of the Gum Resin of *Boswellia serrata, Indian Drugs 24*: 221–231 (1987).

Kakrani, H.K., Guggul—A Review, *Indian Drugs 18:* 417–421 (1981).

Kato, R., T. Nakadata, S. Yammoto, and T. Sugimura, Inhibition of 12-*O*-Tetradecanoylphorbol-13-acetate-Induced Tumor Promotion and Ornithine Decarboxylase Activity by Quercetin: Possible Involvement of Lipoxygenase Inhibition, *Carcinogenesis 4*: 1301–1305 (1983).

Nakadate, T., S. Yammamoto, E. Aizu, and R. Kato, Effects of Flavonoids and Antioxidants on 12-*O*-Tetradecanoylphorbol-13-acetate-Caused Epidermal Ornithine Decarboxylase Induction and Tumor Promotion in Relation to Lipoxygenase Inhibition by These Compounds, *Gann 75*: 214–222 (1984).

Ondrey, F., J. Harris, and K. Anderson, Inhibition of U937 Eicosanoid and DNA Synthesis by 5,8,11,14-Eicosatetraenoic Acid, an Inhibitor of Arachidonic Acid Metabolism and Its Partial Reversal by Leukotriene C_4, *Cancer Res. 49*: 1138–1142 (1989).

Singh, G.B., S. Singh, and S. Bani, Boswellic Acids, in *Drugs of the Future*, *Indian Drugs 18*: 307–309 (1993).

Tsukata, T., K. Nakashima, and S. Shirakawa, Arachidonic 5-Lipoxygenase Inhibitors Show Potent Antiproliferative Effects on Human Leukemic Cells Lines, *Biochem. Biophys. Res. Commun. 140*: 812–816 (1986).

Chapter 28

Inhibitory Effect of Resveratrol and Related Compounds on the Macromolecular Synthesis in HL-60 Cells and the Metabolism of 7,12-Dimethylbenz[a]anthracene by Mouse Liver Microsomes

Benaifer D. Dubash[a], Bo Lin Zheng[b], Calvin Hyungchan Kim[b], Kan He[b], Yu Shao[b], Qun Yi Zheng[b], Chi-Tang Ho[a], and Mou-Tuan Huang[a]

[a]Laboratory for Cancer Research, College of Pharmacy and Department of Food Science, Cook College, Rutgers University, Piscataway, NJ 08854–8020

[b]Department of Research and Development, Madis Botanicals, South Hackensack, NJ 07606

Introduction

Resveratrol (3,5,4′-trihydroxystilbene) and its glucoside, *cis* and *trans* forms, are found in grapes, mulberries, peanuts, and wine (Creasy and Coffee 1988, Dercks and Creasy 1989, Ingham 1976, Jeandet *et al.* 1991, Soleas *et al.* 1997). Resveratrol and its related compounds are also found in other plants, including a traditional Chinese medicinal plant (the dried roots of *Polygonum cuspidatum* Sieb. et Zucc.) (Tang and Eisenbrand 1992). Resveratrol possesses many biological effects that include antioxidant, antimutagenic, anti-inflammatory, and antitumor-promoting activities, and demonstrates a preventive effect on cancer (Jang *et al.* 1997). It is also an inhibitor of 12-*O*-tetradecanoylphorbol-13-acetate (TPA)-induced expression of the cyclooxygenase-2 gene (Subbaramaih *et al.* 1998). Considerable amounts of resveratrol are present in red wine (Goldberg *et al.* 1995a), and it has been found that moderate consumption of red wine may reduce the risk of cardiovascular disease (Goldberg *et al.* 1995b). Much evidence has been accumulated in support of the fact that resveratrol may have preventive effects on cancer and heart disease. Therefore, resveratrol and related compounds may soon become one of the most popular food supplements. Commercial sources of resveratrol, extracted from roots of the plant *Polygonum cuspidatum*, contain *cis*- and *trans*-resveratrols and their corresponding glucosides. It has been the focus of our studies to uncover the biological activity of *cis*- and *trans*-resveratrol as well as the activity of its glucosides, also in the *cis* and *trans* forms. We have extracted and isolated the following three fractions: resveratrol, piceid, and stilbenes. We have also evaluated the inhibitory effects of the three fractions on the synthesis of DNA and RNA in HL-60 cells and the metabolism of 7,12-dimethylbenz[a]anthracene (DMBA) by mouse liver microsomes *in vitro*.

Materials and Methods

Plant Material

Roots of the *Polygonum cuspidatum* Sieb. et Zucc. sample used in this study were acquired from China in October 1997. A Voucher specimen was preserved at Madis Botanicals (South Hackensack, NJ). ResveraPure was an extract of *P. cuspidatum* from Madis Botanicals.

Isolation and Purification

Roots (4 kg) of the plant *P. cuspidatum* were repeatedly extracted with 75% MeOH (10 × 4 L) at 45°C. The ethanol extract was filtered and evaporated under reduced pressure to obtain a dark gummy solid (1.38 kg). The solids were pretreated with Amberlite resin and then loaded onto a polyamide column (10.16 × 124.46 cm), gradient-eluted with MeOH. About 117 g of 45% *trans*-piceid (ResveraPure-45P) product was obtained from the earlier fractions, and ~108 g of 30% *trans*- and *cis*-resveratrol product (ResveraPure-30R) was obtained from the later fractions (16.13% *trans*-resveratrol and 13.86% *cis*-resveratrol; Fig. 28.1, compounds 1 and 3, respectively). Finally, ~225 g of 40% stilbenes (ResveraPure-40S) were obtained from the mixture of the aforementioned products (21.56% *trans*-piceid, 2.15% *cis*-piceid, 7.74% *trans*-resveratrol, and 8.38% *cis*-resveratrol; Fig. 28.1, compounds 2, 3, 5, 6, 1 and 4, respectively). The final preparations were as follows: 30% resveratrol containing 16.13% *trans*-resveratrol and 13.86% *cis*-resveratrol; 45% piceid containing 45% *trans*-piceid; 40% stilbenes containing 21.56% *trans*-piceid, 2.15% *cis*-piceid, 7.74% *trans*-resveratrol, and 8.38% *cis*-resveratrol. Chemical structures of *cis*- and *trans*-resveratrol, its glucosides, *cis* and *trans* forms, are shown in Figure 28.1.

HL-60 Cell Cultured and Measurement of DNA and RNA Synthesis in HL-60 Cells

HL-60 cells (2×10^5) cells/mL were grown in 100-mm diameter culture dishes containing RPMI 1640 medium and supplemented with 10% fetal calf serum (FCS), 100 units/ml penicillin, and 100 μg/mL streptomycin sulfate. Cells were incubated at 37°C in a 5% CO_2 humidified atmosphere. HL-60 cells had a doubling time of ~24–30 h under these conditions. Cells were inoculated every 3 d to maintain their normal growth.

HL-60 cells were harvested by centrifugation for 5 min at 1000 × *g*. Pellets were resuspended in RPMI-1640 medium without FCS and penicillin-streptomycin, and diluted to 5×10^5 cells/mL. [^{3}H]Labeled thymidine (for monitoring DNA synthesis), or uridine (for monitoring RNA synthesis) was added. A series of 13 × 100 mm culture tubes containing 1 mL of the above solution were incubated at 37°C for 90 min, and reactions were terminated by addition of 2 mL ice-cold (0–4°C) phosphate-buffered saline (PBS) solution (1.37 mol/L sodium chloride in 0.1 mol/L sodium phosphate). The solution was then centrifuged for 5 min at 1000 × *g*. The super-

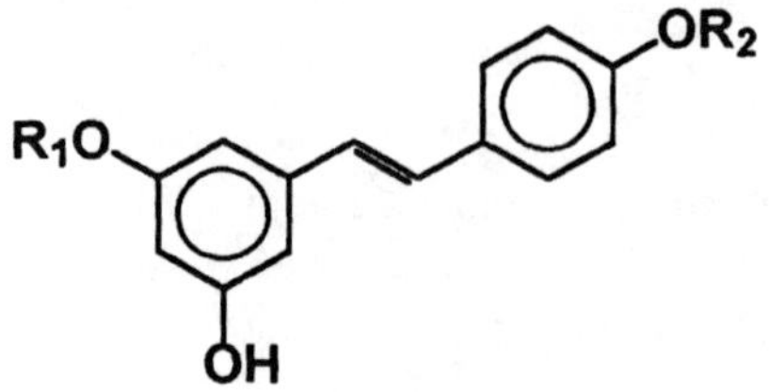

***trans*-resveratrol or *trans*-piceid**

1 $R_1=R_2=H$

2 $R_1=\beta$-glc, $R_2=H$

3 $R_1=H$, $R_2=\beta$-glc

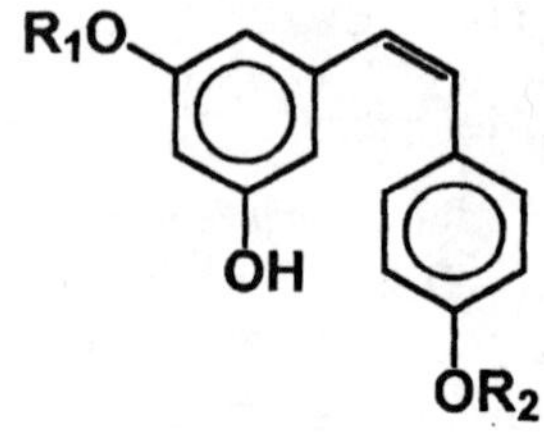

***cis*-resveratrol or *cis*-piceid**

4 $R_1=R_2=H$

5 $R_1=\beta$-glc, $R_2=H$

6 $R_1=H$, $R_2=\beta$-glc

Fig. 28.1. Structures of resveratrols and piceids.

natant was discarded, cells were lysed in 2 mL ice-cold deionized water, and 2 mL 10% trichloroacetic acid (TCA) solution was added. The macromolecular substance was collected on glass-fiber filters using a Millipore Sampling Manifold (Bedfold, MA) and washed with 5% ice-cold TCA solution three times. The dry glass fibers were put into scintillation vials with 5 mL of ScientVase fluid; radioactivity was determined in a Beckman LS1701 scintillation counter (Beckman, Palo Alto, CA).

Measurement of DMBA Metabolism by Liver Microsomes

Resveratrol, stilbenes, and piceid were diluted to concentrations of 20, 10, and 50 µmol/L, respectively. Glucose-6 phosphate, glucose-6-phosphate dehydrogenase, 100 mmol/L NADPH, liver microsome, and double-distilled water were prepared. The compound of interest (2 µL) was dissolved in 1 mL of the stock solution. DMSO (2 µL) in 1 mL of stock solution was used as the control, and the blank was comprised solely of stock solution. Cold DMBA (0.5 mL of a 0.1 mol/L solution) was then added to 10 µL of radioactively labeled DMBA; 10 L of the solution was placed into each of the tubes except the blank. All tubes were kept in an ice bath until after the radiolabeled DMBA was added. Tubes were then placed in a 37°C water bath for 1 h. After the incubation, tubes were transferred back into the ice bath and 200 µL of 5 mol/L HCl was added to each tube, including the blank. The tubes were subsequently vortexed. [^{3}H]DMBA (10 µL) was then added to the blank and the tubes were vortexed again. Activated charcoal (~20–30 mg) was added to each tube, including the blank, and all tubes were vortexed and spun in a centrifuge at 1000 × *g* for 7–10 min. The supernatant (1 mL) was then passed through glass-wool filters and 200 μL was taken to count.

Results and Discussions

Inhibitory Effect of Resveratrol, Piceid, and Stilbenes Fractions on the Synthesis of DNA and RNA in HL-60 Cells

The resveratrol, piceid, and stilbenes fractions markedly inhibited the [^{3}H]thymidine incorporation into DNA and [^{3}H]uridine incorporation into RNA in HL-60 cells in a dose-dependent manner (Figs. 28.2 and 28.3). The figures show that increasing dosages of these fractions reduced the amount of radioactive thymidine and uridine incorporation into the DNA and RNA of HL-60 cells, respectively. At concentrations of 2.5, 5, 10, 20, 40 and 80 μg/mL, resveratrol inhibited [^{3}H]thymidine incorporation into the DNA of HL-60 cells by 30, 56, 67, 81, 83, and 87%, respectively. Identical concentrations of stilbenes inhibited [3H]thymidine incorporation into the DNA of HL-60 cells by 45, 53, 57, 77, 84, and 89%, respectively.

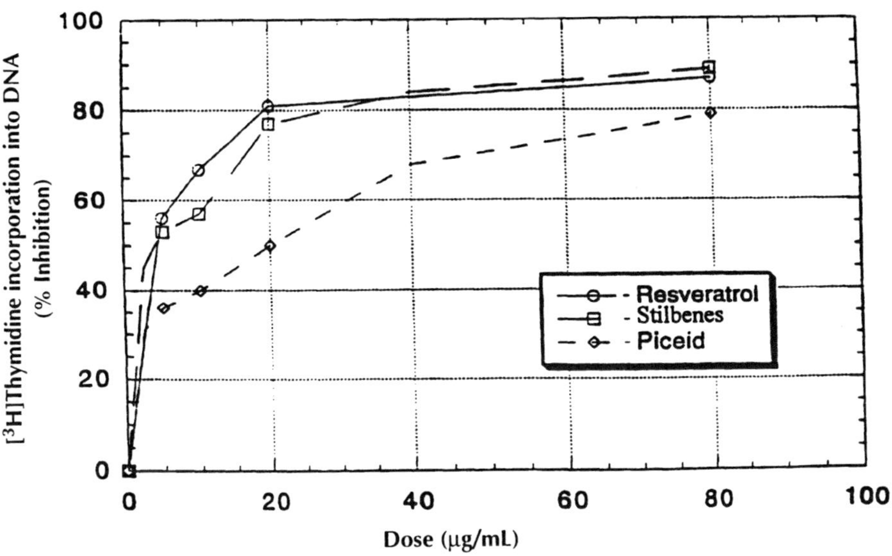

Fig. 28.2. Effect of resveratrol and its derivatives on the synthesis of DNA in HL-60 cells. [^{3}H]Thymine incorporation is inhibited in HL-60 cells in a dose-dependent manner. [^{3}H]Thymine is incorporated into the cells, which are subsequently treated with varying concentrations of resveratrol and its glucoside derivatives. Cells are incubated for 90 min at 37°C under 5% CO_2 atmosphere. Results indicate a negative correlation between increasing dosage of the compounds and amount of [^{3}H]thymine incorporation detected.

Identical concentrations of piceid also inhibited [3H]thymidine incorporation into the DNA of HL-60 cells by 31, 36, 40, 50, 68, and 79%, respectively (Fig. 28.2). All three fractions also demonstrated an inhibitory effect on the [^{3}H]uridine incorporation into the RNA of HL-60 cells. With the dosages mentioned above, resveratrol demonstrated inhibitory rates of 43, 54, 72, 85, 90, and 93%, respectively; stilbenes inhibited [^{3}H]uridine incorporation into RNA of HL-60 cells by 24, 45, 60, 76, 85, and 91%, respectively; and piceid produced inhibitory rates of 10, 10, 27, 50, 67, and 81%, respectively (Fig. 28.3).

Inhibitory Effect of Resveratrol, Piceid, and Stilbenes Fractions on the Metabolism of DMBA by Mouse Liver Microsomes

Resveratrol, piceid, and stilbenes inhibited the DMBA metabolism by liver microsomes *in vitro* in a dose-dependent manner. As shown in Figure 28.4, at 10, 20, 40, and 80 μg/mL, resveratrol inhibited DMBA metabolism by 37, 48, 61, and 69%, respectively. Stilbenes inhibited the rates of DMBA metabolism by 31, 42, 61, and 70%, respectively, at identical concentrations, and piceid demonstrated inhibitory effects of 30, 35, 52, and 58%, respectively, also at identical concentrations. Figure 28.4 shows the inhibition of DMBA metabolism by mouse liver microsomes *in*

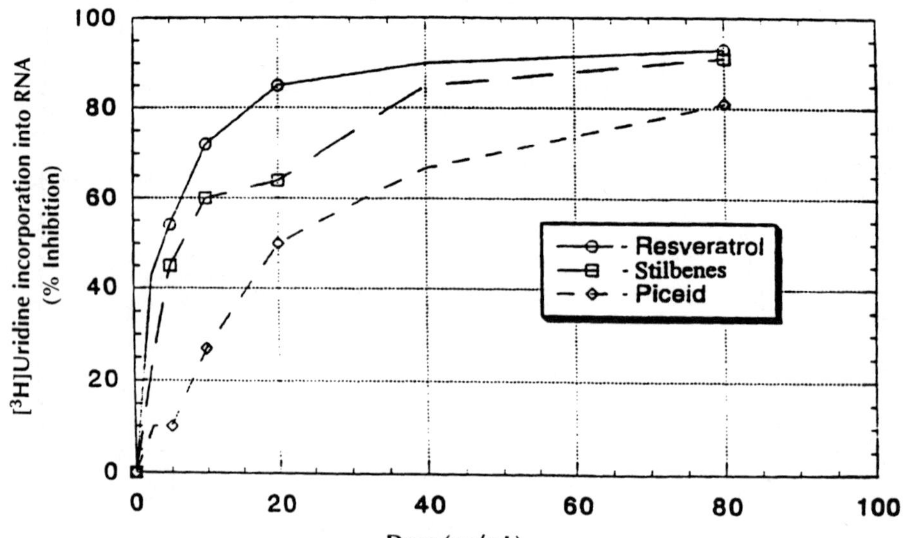

Fig. 28.3. Effect of resveratrol and its derivatives on the synthesis of DNA in HL-60 cells. [^{3}H]Uridine incorporation is inhibited in HL-60 cells in a dose-dependent manner. [^{3}H]Uridine is incorporated into the cells, which are subsequently treated with varying concentrations of resveratrol and its glucoside derivatives. Cells are incubated for 90 min at 37°C under 5% CO_2 atmosphere. Results indicate a negative correlation between increasing dosage of the compounds and amount of [^{3}H]uridine incorporation detected.

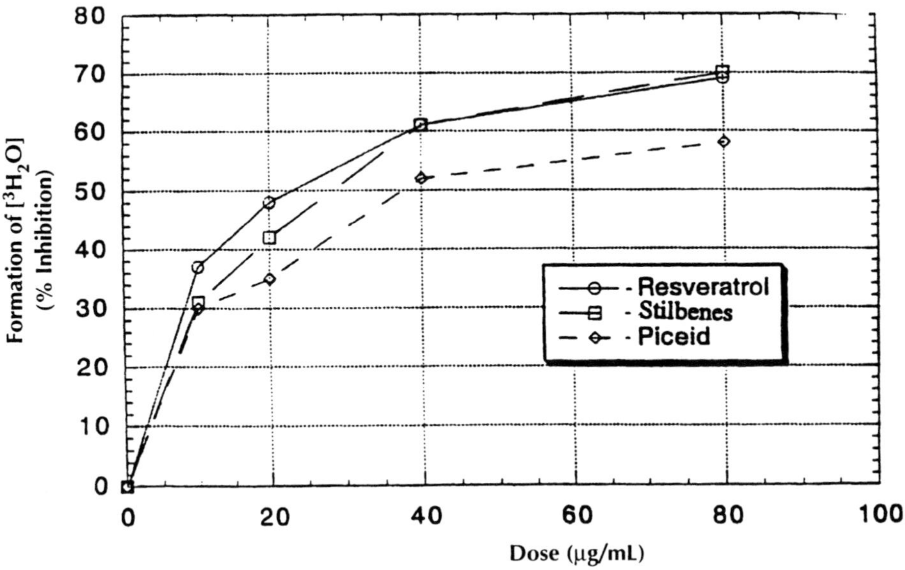

Fig. 28.4. Effect of resveratrol and its derivatives on the synthesis of DNA in HL-60 cells. DMBA metabolism by rat liver microsomes is inhibited by resveratrol and its glucoside derivatives. The graph indicates increasing inhibition of levels of radiolabeled H_2O, which emerged as a by-product of the metabolism of 7,12-dimethylbenz[a]anthracene (DMBA) by liver microsomes *in vitro,* with respect to increasing dosages of the compounds.

vitro by all three compounds. It is reasonable to expect that the resveratrol, piceid, and stilbenes fractions may inhibit DMBA activation to an ultimate carcinogen. Therefore, the resveratrol, piceid and stilbenes fractions may inhibit tumor initiation and tumor promotion *in vivo.*

References

Creasy, L.L., and M. Coffee, Phytoalexin Production Potential of Grape Berries, *J. Am. Soc. Hort. Sci. 113*: 230–234 (1988).

Dercks, W., and L.L. Creasy, The Significance of Stilbene Phytoalexins in the *Plasmopara viticola*–Grapevine Interaction, *Physiol. Mol. Plant Pathol. 34*: 189–202 (1989).

Goldberg, D.M., J. Yan, E. Ng, E.P. Diamandis, A. Karumanchiri, G. Soleas, and A.L. Waterhouse, A Global Survey of *trans*-Resveratrol Concentrations in Commercial Wines, *Am. J. Enol. Vitic. 46*: 159–165 (1995a).

Goldberg, D.M., S.E. Hahn, and J.G. Parkes, Beyond Alcohol: Beverage Consumption and Cardiovascular Mortality, *Clin. Chim. Acta. 237*: 155–187 (1995b).

Ingham, J.L., 3,5,4′-Trihydroxystilbene as a Phytoalexin from Groundnut (*Arachis hypogaea), Phytochemistry 15*: 1791–1793 (1976).

Jang, M., L. Cai, G.O. Udeani, K.V. Slowing, C.F. Thomas, C.W.W. Beecher, H.H.S. Fong, N.R. Farnsworth, A.D. Kinghorn, R.G. Mehta, R.C. Moon, and J.M. Pezzuto, Cancer

Chemopreventive Activity of Resvertrol, a Natural Product Derived from Grapes, *Science 275*: 218–220 (1997).

Jeandet, P., R. Bessis, and B. Gautheron, The Production of Resveratrol (3,5,4′)-Trihydroxystilbene by Grape Berries in Different Development Stages, *Am. J. Enol. Vitic. 42*: 41–46 (1991).

Soleas, G.J., E.P. Diamandis, and D.M. Goldberg, Resveratrol: a Molecule Whose Time Has Come? And Gone? *Clin. Biochem. 30*: 91–113 (1997).

Subbaramaih, K., W.J. Chung, P. Michaluart, N. Telang, T. Tanabe, H. Inoue, M. Jang, J.M. Pezzuto, A.J. Dannenberg, Resveratrol Inhibits Cyclooxygenase-2 Transcription and Activity in Phorbol Ester-Treated Human Mammary Epithelial Cells, *J. Biol. Chem. 273*: 21875–21882 (1998).

Tang, W., and G. Eisenbrand, *Chinese Drugs of Plant Origin*, Springer-Verlag, Berlin, 1992, pp. 787–791.

Chapter 29

Chemiluminescence of a Hydrogen Peroxide/Gallic Acid/Horseradish Peroxidase System

Y. Yoshiki[a], H. Yuan[a], M. Kawane[a], T. Iida[a], K. Okubo[a], H. Ohrui[a], T. Ishizawa[b], and S. Kawabata[b]

[a]Department of Environmental Bioremediation, Graduate School of Agriculture, Tohoku University, 1-1 Tsutsumidori, Amamiyamachi, Aoba-ku, Sendai, 981 Japan

[b]Tohoku Electronic Industrial Company, Rifu Office, Rifu-Cho, Miyagi, 981-01, Japan

Introduction

Peroxidase decomposes hydrogen peroxide to oxygen and water in the form of an intermediate compound I, which then reacts with an array of oxidizable organic and inorganic substrates in consecutive one-electron steps. The catalytic cycles of peroxidase, such as horseradish peroxidase (HRP) or catalase, have been reviewed extensively (Araiso *et al.* 1976, Hanson *et al.* 1981). The reaction between peroxidase and hydroperoxides was detected by low-level chemiluminescence (Kanofsky 1983, Kobayashi *et al.* 1980). An earlier investigation suggested that the chemiluminescence (CL) arose from singlet molecular oxygen produced in the reaction of H_2O_2 with high-oxidation state iron complexes and hydroperoxyl radical reactions, and/or secondary reaction of the porphyrin or protein moieties. However, Campa *et al.* (1984) and Nassi and Cilento (1985), using an aldehyde/HRP/O_2/chlorophyll system, reported that light emission of peroxidase-oxidase was caused by an electronically excited carbonyl state (triplet state) leading to subsequent transfer of their excitation energy to several acceptors. On the other hand, it was demonstrated earlier that the photon emission arose from the mixture of reactive oxygen species (X), catalytic species (Y), and receptive species (Z) at room temperature and under neutral conditions. The CL occurrence and intensity of phenolic compounds as Y were closely related to their particle structures and radical-scavenging activities (Yoshiki *et al.* 1995 and 1996). In the organic system, various enzymes, especially heme-protein, are associated with reactive oxygen metabolism and/or electron translation. Such heme-protein, including cytochrome c and HRP, act as Z in this CL system.

This study shows the results obtained after the generation of CL from an H_2O_2/gallic acid/HRP system and the effect of gallic acid and catechins as Y upon CL behavior. Further, we describe the roles of gallic acid and HRP in this CL system on the basis of the examination of lipid autoxidation and the electron spin resonance (ESR) spin trapping method in order to better understand the CL mechanism.

Materials and Methods

Reagents

Horseradish peroxidase (HRP) was obtained from Sigma Chemical (St. Louis, MO). The concentration of HRP was determined by measuring the absorption at 403 nm {ε; 1.02×10^5 [(mol · cm)/L]$^{-1}$}. The apoprotein was obtained by treating HRP with acid-butanone as described by Yonetani *et al.* (1967). Gallic acid and linoleic acid were from Nakarai Tesque (Tokyo, Japan). Hydrogen peroxide (H_2O_2; 30%) and acetaldehyde (MeCHO) were from Santoku Chemical (Tokyo, Japan) and Merck (Darmstadt, Germany), respectively. Hypoxanthine was from Kojin (Tokyo, Japan); xanthine oxidase from cow milk was obtained from Behringer (Mannheim, Germany); superoxide dismutase (SOD) from bovine erythrocytes was from Merck, and 5,5-dimethyl-1-pyrroline-*N*-oxide (DMPO, nitrone spin trap) was from Mitsui Toatsu Chemical (Tokyo, Japan). Diethylenetriamine-*N,N,N′,N″,N″*-pentaacetic acid (DETAPAC, a chelator for trace metal impurities) was obtained from Wako Pure Chemical (Tokyo, Japan). (-)-Catechin, (-)-epicatechin, (-)-gallocatechin, (-)-epigallocatechin, (-)-catechin gallate, (-)-epicatechin gallate, (-)-gallocatechin gallate, and (-)-epigallocatechin gallate from *Camellia sinensis* were procured from Kurita (Tokyo, Japan).

Measurement of Chemiluminescence

The chemiluminescence of H_2O_2/gallic acid/HRP was measured using a CL Detector Model CLD-110 (Tohoku Electronic, Miyagi, Japan; single-photon counting method) connected to a Waters Model 510 pump (Milford, MA) and a U6K injector. The wavelength was set in the range of 300–650 nm. The photons were counted as total intensities. The measurement was carried out using 50 mmol/L phosphate buffer as the mobile phase with a flow rate of 1 mL/min and a temperature of 23°C as described previously (Yoshiki *et al.* 1995). Ten microliters of various concentrations of H_2O_2, gallic acid or catechins and MeCHO or HRP (total 30 μL) were injected directly into the sample injector. Gallic acid and catechins were dissolved in 50% MeOH. To compare the photon intensity of the H_2O_2/gallic acid/MeCHO and H_2O_2/catechin/HRP systems, photon intensities expressed the vale (count/s) at 5 mmol/L concentration of gallic acid and catechins, respectively.

Determination of Emission Spectra

The spectra of chemiluminescence were measured by a Simultaneous Multiwavelength Analyzer Model CLA-SP2 (Tohoku Electronic) equipped with a built-in diffraction grating, a spectroscope, and a two-dimensional photomultiplier tube. The wavelength range of the spectroscope was 400–850 nm. Hydrogen peroxide, HRP, and MeCHO were dissolved in a 50 mmol/L phosphate buffer (pH 7.0); gallic acid and catechins were in 50% MeOH. The reaction mixture contained 50 mmol/L phosphate buffer (pH 7.0), 0.4 mL of 176 mmol/L H_2O_2, 0.4 mL of 5 mmol/L gallic acid, 10.2 mmol/L (-)-epicatechin, 8.2 mmol/L (-)-gallocatechin, 9.6 mmol/L (-)-epicatechin gallate, 6.7 mmol/L (-)-gallocatechin gallate, or 6.6 mmol/L (-)-epigallocatechin gallate and 0.4 mL of 354 mmol/L MeCHO or 3.48 mol/L HRP in a quartz cell (total volume, 1.2 mL). The light emission was determined for 180 s.

Measurement of Superoxide and Hydroxyl Radical–Scavenging Activities

The ESR spectra were recorded on a JEOL JES-RE1X spectrometer (JEOL Ltd., Tokyo, Japan) using an aqueous quartz flat cell (60 mm × 10 mm × 0.31 m, effective volume 160 μL). The optical absorption spectra were measured using an Otsuka Electronics MCPD-100 multichannel photodetector (Otsuka Electronics Ltd., Otsuka, Japan). Fifty microliters of sample solution (various concentrations of gallic acid and HRP) were added to the mixture consisting of hypoxanthine (2 mmol/L, 50 μL), xanthine oxidase (0.4 unit/mL, 50 μL) and DMPO (9.2 mol/L, 20 μL) for measuring superoxide radical–scavenging activity. The superoxide-scavenging activity was calculated using the index of the DMPO-O_2^- spin adduct and by comparison with that of bovine erythrocyte superoxide dismutase as described by Mitsuta *et al.* (1990). Hydroxyl radicals were generated by the Fenton reaction. Sample solutions (50 μL) were added to the mixture consisting of H_2O_2 (1 mmol/L, 60 μL), $FeCl_2$ (0.1 mmol/L, 60 μL), DETAPAC (5.5 mmol/L, 10 μL) and DMPO (9.2 mol/L, 15 μL). The hydroxyl radical inhibitory ratio was calculated from the DMPO-OH signal.

Measurement of Antioxidative Activity

Different amounts of samples dissolved in 100 μL of 50% MeOH were added to a mixture of 1.026 mL of 99% ethanol containing 2.51% linoleic acid and 2 mL of a 50 mmol/L phosphate buffer (pH 7.0). The total volume was made up to 5 mL with distilled water. The mixture (100 μL) was used in the antioxidative assay with the ferric thiocyanate method as described by Inatani *et al.* (1983).

Results and Discussion

The addition of 158 mmol/L H_2O_2 (10 μL) to 627 μmol/L HRP (10 L) gave a short burst of chemiluminescence (17 counts/s, CPS). We have previously shown through the chemiluminescence (CL) method that gallic acid and MeCHO are typical Y and Z, respectively, in an XYZ system. Based on this system, the CL of HRP as Y or Z is detected by combining it with gallic acid and MeCHO. Although equal CL intensity (28 CPS) to the H_2O_2/HRP system was observed from a mixture of 158 mmol/L H_2O_2 (10 μL), 627 μmol/L HRP (10 μL), and 354 mmol/L MeCHO (10 μL), the 11 mmol/L H_2O_2 (10 μL), 5 mmol/L gallic acid (10 μL), and 6.27 μmol/L HRP (10 μL) mixture afforded a high intensity (1.1×10^3 CPS). The intensity of CL suggested that HRP acted as Z in the XYZ system. When HRP was inactivated by heating, the maximum intensity decreased and was dependent upon the heating time. Further, it is thought that heme iron and peripheral amino residues played an important role in the CL of heme-protein in a H_2O_2/gallic acid/HRP system because the apoprotein obtained with the acid-butanone method did not emit CL.

To identify the cause of CL in a H_2O_2/gallic acid/HRP system, CL intensities were measured using various concentrations of H_2O_2, gallic acid, and HRP. The CL intensity was affected by H_2O_2, gallic acid, or HRP concentrations. In particular, variation of H_2O_2 and gallic acid concentrations was markedly affected under a certain HRP concentration. The behavior of CL in this system, however, was complicated because the emission integral indicated that the concentration was dependent on two factors (Fig. 29.1). The variation of H_2O_2 and HRP concentration in 0.1 mmol/L gallic acid gave a CL curve similar to that from the variation of H_2O_2 and gallic acid concentrations in 34.8 nmol/L HRP. The saturation curve of the CL intensity at high H_2O_2 concentration has been reported for H_2O_2/cytochrome c (Cadenas *et al.* 1980) and H_2O_2/hemoglobin (Lissi *et al.* 1994). A tendency for saturation was observed at high H_2O_2 concentrations (>8.8 mmol/L H_2O_2), which was dependent upon gallic acid concentration in the presence of HRP as well as HRP concentration in the presence of gallic acid. In the cytochrome c/H_2O_2 system, the CL intensity was linearly related to the cytochrome c concentration, up to 40 μmol/L, a concentration considerably higher than those shown in Figure 29.1. The evidence of effects of H_2O_2 and HRP concentrations (Fig. 29.1A) and H_2O_2 and gallic acid concentrations (Fig. 29.1B) suggested that the linearity of the present data could be obtained by the intensity plot against gallic aid and HRP concentrations in the presence of H_2O_2 at concentrations >8.8 mmol/L. In fact, CL intensity increased linearly depending upon the increase of HRP and gallic acid concentrations in the presence of 8.8 mmol/L H_2O_2 (Fig. 29.1C).

It is known that phenyl compounds such as aromatic amino acids and phenols act as electron donors. To assess the role of Y on the intensity of CL, a comparison was carried out by using eight catechin compounds and gallic acid. CL intensities and wavelengths were measured in the H_2O_2/catechin/MeCHO and H_2O_2/catechin/HRP systems (Table 29.1). In the H_2O_2/catechin/MeCHO system, the CL intensity was affected markedly by the structures of catechins. Although a similar observation was made in the H_2O_2/catechin/HRP system, the difference in intensities was slight in comparison with that of the H_2O_2/catechin/MeCHO system. Interesting results were obtained when using gallic acid as Y; it showed a high intensity in the H_2O_2/catechin/HRP system, despite having the lowest intensity in the MeCHO system. The different intensities of both systems could indicate the importance of Y peculiarities, i.e., molecular size, substrate dependency, stereospecificity in interacting with the active site of Z, and especially high-molecular-weight compounds such as heme-protein.

The spectral analysis of the CL showed a distribution entirely in the visible region with main peaks at 620, 610, 610, 610, 660, 650, 640, 640, and 620 nm for the (-)-catechin, (-)-epicatechin, (-)-gallocatechin, (-)-epigallocatechin, (-)-catechin

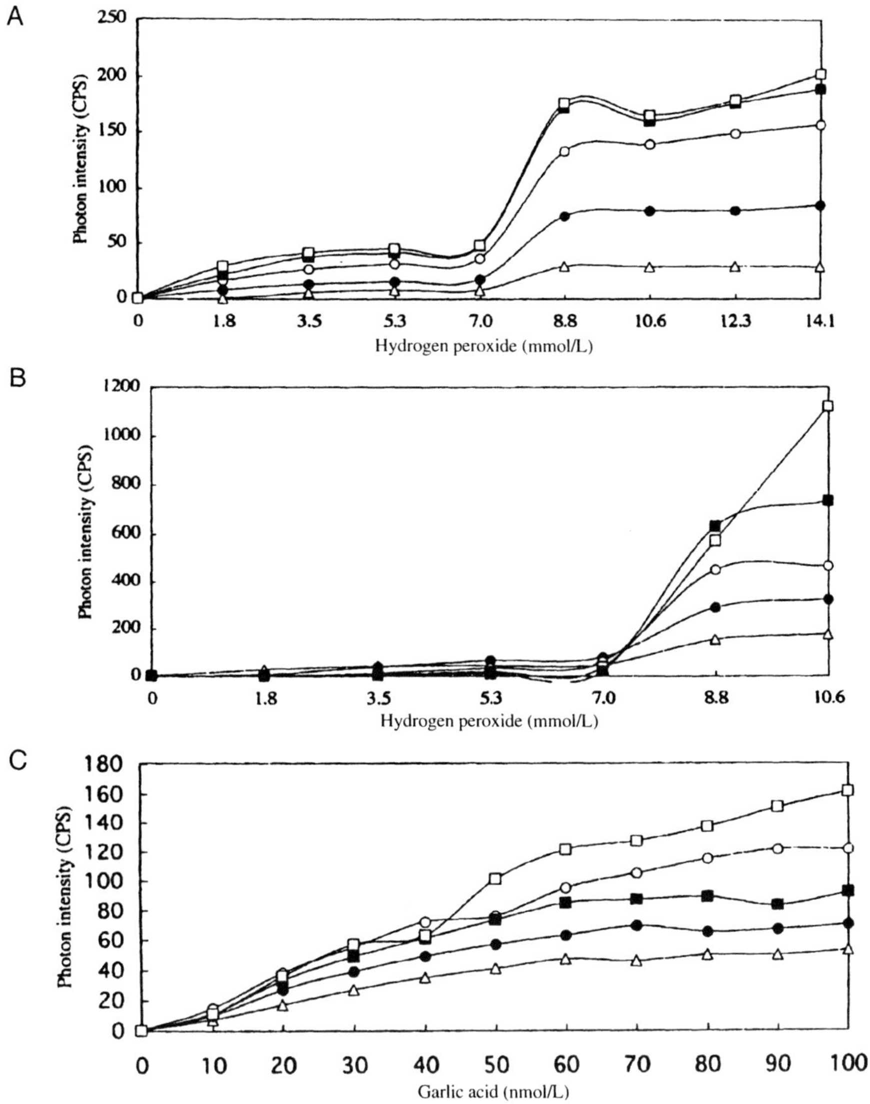

Fig. 29.1. Effects of (A) H_2O_2 and horseradish peroxidase (HRP) concentrations, (B) H_2O_2 and gallic acid (GA) concentrations, and (C) GA and HRP concentrations on photon intensity measured in counts/s (CPS). Conditions: (A) Chemiluminescence was measured in the presence of 0.1 mmol/L GA; (△) 34.8 nmol/L HRP; (●) 69.6 nmol/L HRP; (○) 104.4 nmol/L HRP; (■) 139.2 mmol/L HRP; (□) 17.0 nmol/L HRP. (B) Chemiluminescence was measured in the presence of 34.8 μmol/L HRP; (△) 0.1 mmol/L GA; (●) 0.2 mmol/L GA; (○) 0.3 mmol/L GA; (■) 0.4 mmol/L GA; (□) 0.5 mmol/L GA. (C) Chemiluminescence was measured in the presence of 8.8 mmol/L H_2O_2; (△) 52.0 nmol/L HRP; (●) 69.6 nmol/L HRP; (○) 87.0 nmol/L HRP; (■) 104.4 nmol/L HRP; (□) 139.2 nmol/L HRP.

gallate, (-)-epicatechin gallate, (-)-gallocatechin gallate, (-)-epigallocatechin gallate, and gallic acid, respectively (Table 29.1). These emission spectra, which were actually recorded from the XYZ system, were distinguishable from that of singlet oxygen molecules as observed by Slawlnska and Slawlnski (1975) (Emax: 643, 702, 672, 1070, 1270 nm). The earlier evidence suggested that the singlet oxygen molecules were the main products of oxidized heme-protein that caused CL in heme-protein, including enzymes such as peroxidase (Cadenas *et al.* 1980). However, the spectral analysis indicated predominance of excited carbonyl compounds in the XYZ system. The CL of the H_2O_2/catechin/HRP system was strongly affected by catechins that acted as Y because (-)-catechin gallate, (-)-epicatechin gallate, (-)-gallocatechin gallate, and (-)-epigallocatechin gallate, which have galloyl moieties, showed different spectra compared with those of other catechins, i.e., they had shoulder wavelengths at 550, 550, 500, and 490 nm, respectively. Figure 29.2 shows the typical patterns of spectra obtained from H_2O_2/(-)-epigallocatechin/HRP and H_2O_2/(-)-epigallocatechin gallate/HRP systems. On the other hand, catechin compounds were not affected as compared with the spectra in the H_2O_2/catechin/MeCHO system. Furthermore, the spectra that were observed in H_2O_2/catechin or gallic acid/MeCHO and H_2O_2/catechin or gallic acid/HRP systems did not correspond to Y compounds except for (-)-epicatechin and (-)-epigallocatechin. Thus, the spectral analysis indicated that the Y and Z complex caused CL in the XYZ system because spectra did not correspond for either the Y or the Z compounds.

The scope of hydrogen atom transfer and electron transfer mechanisms has been studied for metalloporphyrin model compounds (Groves and Han 1995). Hydrogen atom transfer or electron transfer is affected by both electron-donation activity of oxidized compounds and environmental conditions, including solvent. The effects of gallic acid and HRP on lipid autoxidation were measured to investi-

TABLE 29.1
Photon Intensity and Spectra of Catechins or Gallic Acid as Y in the Presence of Hydrogen Peroxide as X and Horseradish Peroxidase (HRP) as Z[a]

Substance	MeCHO Intensity (CPS)	MeCHO Wavelength (nm)	HRP Intensity (CPS)	HRP Wavelength (nm)
(-)-Catechin	1.3×10^3	610	1.0×10^3	620
(-)-Epicatechin	5.3×10^4	610	0.6×10^3	610
(-)-Gallocatechin	2.3×10^4	610	2.2×10^3	610
(-)-Epigallocatechin	4.6×10^6	610	7.0×10^3	610
(-)-Catechin gallate	1.3×10^3	ND	1.9×10^3	660 (550)
(-)-Epicatechin gallate	3.2×10^3	610	0.7×10^3	650 (550)
(-)-Gallocatechin gallate	6.8×10^3	610	1.8×10^3	640 (500)
(-)-Epigallocatechin gallate	3.6×10^5	610	4.9×10^3	640 (490)
Gallic acid	0.8×10^3	660	2.6×10^5	620

[a]Photon intensities and spectra are expressed as counts/s (CPS) and maximum wavelength (nm), respectively. The parentheses show the shoulder peak. ND, not determined.

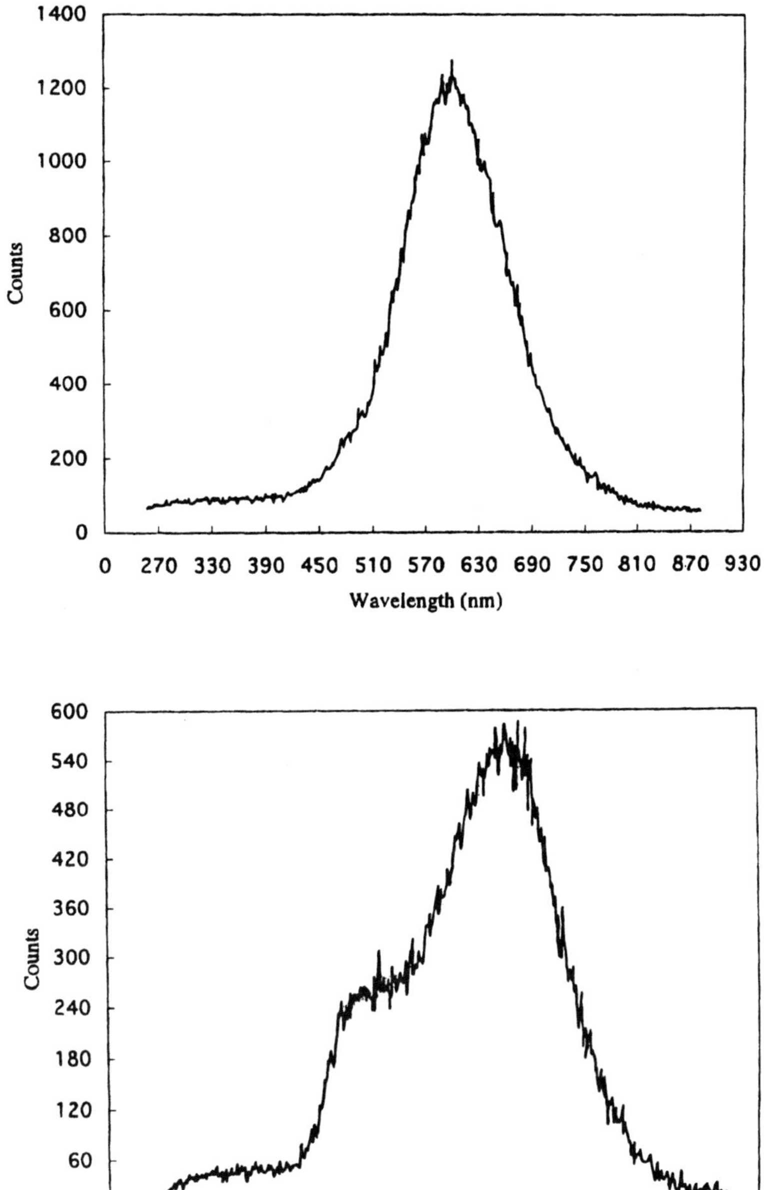

Fig. 29.2. Emission spectra observed in (A) H_2O_2/(-)-epigallocatechin/horseradish peroxidase (HRP) and (B) H_2O_2/(-)-epigallocatechin gallate/HRP systems.

gate hydrogen-abstraction activities of gallic acid, HRP, and gallic acid and HRP mixture.

When linoleic acid was incubated with HRP, prooxidative activity, which was dependent on the HRP concentrations, was observed. Gallic acid, which is generally thought to be a radical scavenger, acted as an antioxidant. An interesting activity was observed in the linoleic acid/gallic acid/HRP system; that is, the linoleic acid/gallic acid/HRP system prolonged the induction period of lipid autoxidation to a greater extent than did the linoleic acid/gallic acid system (Fig. 29.3). The inhibition of hydrogen abstraction or the scavenging of linoleic hydroperoxide must be responsible for such strong antioxidative activity of the linoleic acid/gallic acid/HRP system.

Superoxide and hydroxyl radical–scavenging activities were measured using the ESR spin-trapping method. It was confirmed that superoxide-scavenging activities of 0.5 mmol/L gallic acid and 174 nmol/L HRP were equal to 1.41 and 1.54 SOD unit/mL activities, respectively (Table 29.2). The hydroxyl radical–scavenging activities of 87 nmol/L HRP and 1 mmol/L gallic acid were 10.7 and 9.3%, respectively. Addition of gallic acid to HRP caused a dramatic synergistic inhibition of the hydroxyl radical; that is, the 87 nmol/L HRP and 1 mmol/L gallic acid mixture led to complete inhibition of the hydroxyl radical, despite an inhibition of only 20% from calculation. This evidence on the effects on lipid autoxidation and the use of the ESR method indicated that HRP had electron-donating and hydrogen-abstracting activities different from gallic acid, which is an electron donor.

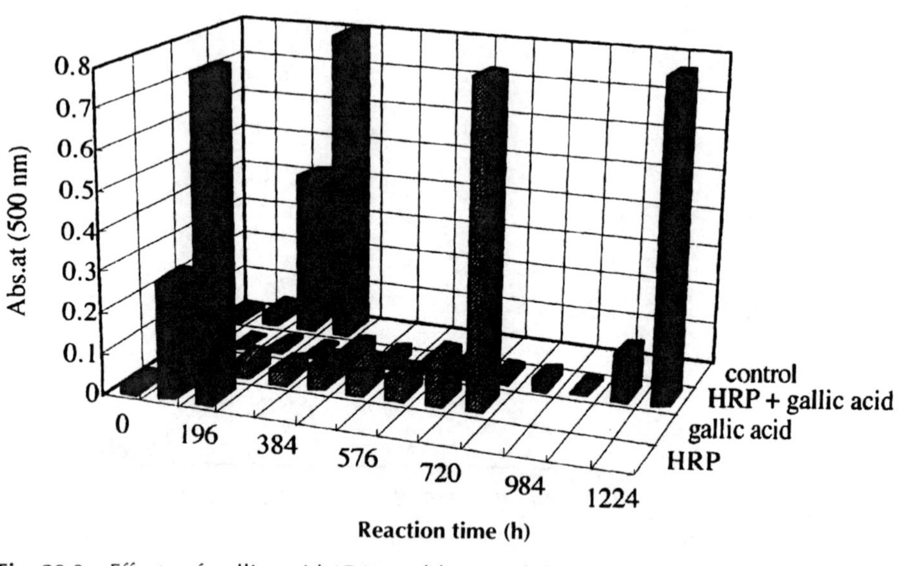

Fig. 29.3. Effects of gallic acid (GA) and horseradish peroxidase (HRP) on autoxidation of linoleic acid. The reaction mixture contained linoleic acid (2.51%) without (control) or with the indicated additives (1 mmol/L GA and 13.9 nmol/L HRP).

TABLE 29.2
Effects of Horseradish Peroxidase (HRP) and Gallic Acid (GA) on the Superoxide and Hydroxyl Radical–Scavenging Activities[a]

		$O_2^{-\bullet}$ inhibition (SOD unit/mL)		$HO^{\bullet}$ inhibition (%)	
		Without GA	0.5 mmol/L GA	Without GA	1.0 mmol/L GA
	(nmol/L)				
HRP	8.7			12.6	19.0
	17.4	0.43	1.24	11.2	32.0
	87.0			10.7	95.2
	174.0	1.54	2.46	6.7	97.3
	(mmol/L)				
	1.7			34.2	97.9
GA	0.5	1.41			
	1.0			9.3	

[a]The reaction mixture contained HRP and 0.5 or 1.0 mmol/L GA.

Figure 29.4 shows a possible mechanism for the XYZ system during autoxidation of linoleic acid, based on the results of lipid autoxidation and the ESR method. HRP has a hydrogen-abstracting activity in the autoxidation system of linoleic acid because HRP acts as a prooxidant (route 1). Although the prooxidative activity of

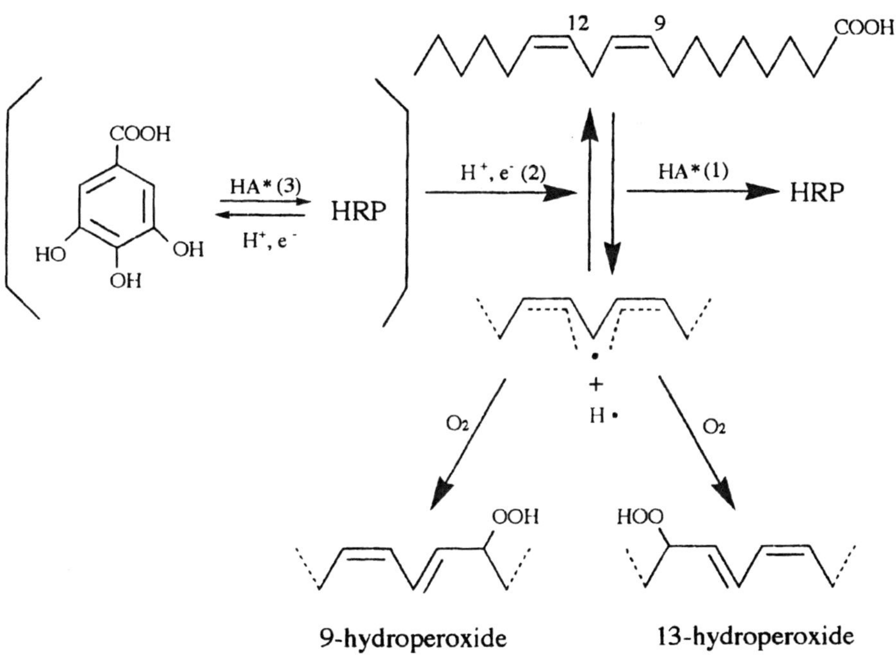

Fig. 29.4. Possible mechanism for gallic acid and horseradish peroxidase (HRP) during autoxidation of linoleic acid; *hydrogen abstraction.

HRP indicated that HRP had hydrogen-abstracting activity, the gallic acid and HRP mixture showed electron-donating activity. Furthermore, ESR analysis suggested that the synergistic antioxidative activity of the Y (gallic acid) and Z (HRP) mixture could be inferred by route 2; that is, the linoleic acid radical received hydrogen generated by hydrogen- or electron-translation between Y and Z (route 3). The same results, associated with the properties of hydrogen abstraction and electron donation, were obtained using heme-protein, which acted as Z in the XYZ CL system. It has been reported that the oxidation state ($Fe^{IV} = O$), after treating heme-protein with H_2O_2, is confirmed by the nuclear magnetic resonance (NMR) technique and by calculation of spin densities (La Mar *et al.* 1983). The crucial step in the formation of reactive oxidation state intermediates is the O-O bond cleavage of an iron-peroxo precursor (Fe^{III}-O_2^{2-}). The O-O bond scission controls the reaction rate by the axial ligand and positioning of the proton on the leaving group. Acid catalysis of O-O bond heterolysis was also evident, leading to the electronically excited carbonyl state (Groves and Han 1995). The present data of CL properties strongly suggested that CL in the H_2O_2/gallic acid/HRP system arose from Y and Z intermediates or a complex associated with the electron and hydrogen transfer mechanism.

Low-level CL in heme-proteins included in enzymes has remained elusive due to the inability to make a reliable assessment of its quantitative production. This study did examine the CL of the XYZ system to define the pathways involved in the process. Inhibition or scavenging of reactive oxygen species in the XYZ system might proceed *via* a hydrogen- or electron-transfer mechanism. This hypothesis is supported by the observation of the synergistic effects on antioxidative and hydroxyl radical–scavenging activities in the XYZ system.

References

Araiso, T., K. Miyashi, and I. Yamazaki, Mechanisms of Electron Transfer from Sulfite to Horseradish Peroxidase-Hydroperoxide Compounds, *Biochemistry 15*: 3059–3063 (1976).

Cadenas, E., A. Boveris, and B. Chance, Spectral Analysis of the Low Level Chemiluminescence of H_2O_2-Supplemented Ferricytochrome c, *FEBS Lett. 112*: 285–288 (1980).

Campa, A., L. Nassi, and G. Cilento, Triplet Energy Transfer to Chloroplasts from Peroxidase-Generated Excited Aliphatic Aldehydes, *Photochem. Photobiol. 40*: 127–131.

Groves, J.T., and Y. Han, Models and Mechanisms of Cytochrome P450 Action, in *Cytochrome P450*, edited by P.R.O. Montellano, Plenum Press, New York, 1995, pp. 3–48.

Hanson, L.K., C.K. Cahang, M.S. Davis, and J. Fajer, Electron Pathways in Catalase and Peroxidase Enzymic Catalysis, Metal and Macrocycle Oxidations of Iron Porphyrins and Chlorins, *J. Am. Chem. Soc. 103*: 663–670 (1981).

Inatani, R., N. Nakatani, and H. Fuwa, Antioxidative Effect of the Constituents of Rosemary (*Rosemarinus officianalis* L.) and Their Derivatives, *Agric. Biol. Chem. 47*: 521–528 (1983).

Kanofsky, R.J., Singlet Oxygen Production by Lactoperoxidase, *J. Biol. Chem. 258*: 5991–5993 (1983).

Kobayashi, S., K. Yamaguchi, and H. Inaba, Excitation of Indole Acetate in Myeloperoxidase-H_2O_2 System; Possible Formation of Indole Acetate Cation Radical, *Biochim. Biophys. Res. Commun. 93*: 967–973 (1980).

La Mar, G.N., J.S. de Ropp, L. Latos-Grazynski, A.L. Balch, R.B. Johnson, K.M. Smith, D.W. Parish, and R. Cheng, Proton NMR Characterization of the Ferryl Group in Model Heme Complexes and Hemoproteins; Evidence for the $Fe^{IV} = O$ Group in Ferryl Myoglobin and Compound II of Horseradish Peroxidase, *J. Am. Chem. Soc. 105*: 782–787 (1983).

Lissi, E.A., J. Escobar, C. Pascual, M. Castillo, T.H. Schmitt, and P. Mascio, Visible Chemiluminescence Associated with the Reaction Between Methemoglobin or Oxyhemoglobin with Hydrogen Peroxide, *Photochem. Photobiol. 60*: 405–411 (1994).

Mitsuta, K., Y. Mizuta, M. Kohno, M. Hiramatsu, and A. Mori, The Application of ESR Spin-Trapping Technique to the Evaluation of SOD-Like Activity of Biological Substances, *Bull. Chem. Soc. Jpn. 58*: 1971–1975 (1990).

Nassi, L., and G. Cilento, Energy Transfer from Enzyme-Generated Triplet Carbonyls to Thylkoid Membrane Fractions Enriched in Photosystems I and II, *Photochem. Photobiol. 41*: 195–201 (1985).

Slawlnska, D., and J. Slawlnski, Chemiluminescent Flow Method for Determination of Formaldehyde, *Anal. Chem. 47*: 2102–2109 (1975).

Yonetani, T., Studies on Cytochrome c Peroxidase, *J. Biol. Chem. 242*: 5008–5013 (1967).

Yoshiki, Y., T. Kahara, K. Okubo, K. Igarashi, and K. Yotsuhashi, Mechanism of Catechin Chemiluminescence in the Presence of Active Oxygen, *J. Biolumin. Chemilumin. 11*: 131–136 (1996).

Yoshiki, Y., K. Okubo, M. Onuma, and K. Igarashi, Chemiluminescence of Benzoic and Cinnamic Acids, and Flavonoids in the Presence of Aldehyderate and Hydrogen Peroxide or Hydroxyl Radical by Fenton Reaction, *Phytochemistry 39*: 225–229 (1995).

Chapter 30

Bioactive Metabolites from Yucatecan Medicinal Plants

Rocío de L. Borges-Argáez, Fabiola Escalante-Erosa, Filogonio May-Pat, Leticia Medina-Baizabál, Yazmín del C. Ojeda-Uc, Sergio R. Peraza-Sánchez, Matilde Pérez-Rodríguez, Norma E. Salazar-Aguilar, Paulino Simá-Polanco, and Luis M. Peña-Rodríguez*

Grupo de Química Orgánica, Unidad de Biotecnología, Centro de Investigación Científica de Yucatán, A.C., Mérida Yucatán, México 97310

Introduction

From the beginning, humans have learned to recognize and classify plants according to their uses and properties. On the one hand, they have come to know those plants that can be used as food and, on the other, plants that can be utilized for medicinal, insecticidal, and building purposes (Der Marderosian and Liberti 1988, Domínguez 1973). The practice of traditional medicine, based on the use of plants to alleviate common ailments, has never disappeared altogether; in remote rural zones or among ethnic minorities of modern industrial society, the use of plants for medicinal purposes still exists as a complement for the low income population or as an alternative to nonexistent medical assistance (Thompson 1980).

Recently, as a result of the search for new and better drugs, there has been a renewed interest on the part of the pharmaceutical industry, researchers, and physicians in plants used in traditional medicine as possible sources of biologically active metabolites with important pharmaceutical applications. Hundreds of plants have ben evaluated, and a significant number of bioactive metabolites have been obtained. However, it has been estimated that only 5–15% of all existing plants (considering a total number of $2.5–7.5 \times 10^5$) have been evaluated for their production of biologically active natural products; there are ~35,000 medicinal plants that can be considered as possible natural sources of new pharmaceuticals. To date, there exist in the market ~120 drug preparations derived from natural plant products (Der Marderosian and Liberti 1988), and it is expected that by the year 2000, as consumer preference moves toward naturally derived pharmaceuticals, the world market for drugs of plant origin will reach the 50 billion dollar mark. It is important to point out that of all the pharmaceutical products of plant origin on the market today, only 10 are prepared synthetically; the rest are still extracted directly from plants (Balandrin *et al.* 1985, Farnsworth and Soejarto 1985, Tyler 1986).

The study of bioactive natural plant products requires both accurate taxonomic information, in order to identify groups of plants as producers of biologically active metabolites (Del Amo and Anaya 1982), and the correct use of an appropriate bioassay procedure to ensure the detection of biological activity. In recent studies on the isolation of bioactive metabolites from plant sources, both crude extracts and purified fractions are evaluated in one or more bioassays, and only the bioactive fractions are pursued. This ensures that all isolation efforts are directed toward obtaining biologically active natural products (McLaughlin 1991, Trease 1988).

A good bioassay must be simple, inexpensive, reproducible, and it should require only small amounts of material (Alkofahi *et al.* 1988, McLaughlin and Anderson 1989). Three of the simplest (and cheapest) bioassays are the antimicrobial (Clark 1989, Stachel *et al.* 1985), brine shrimp lethality (Ferrigni *et al.* 1984, Hui *et al.* 1989, McLaughlin 1991, Meyer *et al.* 1982), and inhibition of crown gall tumor assays (Ferrigni *et al.* 1982, Galsky *et al.* 1980, Xiao-Hua and McLaughlin 1989). The antimicrobial assay detects metabolites with potential antibiotic activity by evaluating their ability to limit or inhibit the growth of pathogenic bacteria and fungi. On the other hand, the brine shrimp lethality assay is based on the fact that the biological activity of a number of metabolites is expressed as toxicity to *Artemia salina* larvae, and that this activity is often indicative of cytotoxic activity when the same metabolite is tested in the 9KB assay. Finally, it has also been demonstrated that those metabolites capable of inhibiting the growth of tumors on potato discs infected with *Agrobacterium tumefaciens* show antitumor activity when tested in the *in vivo* 3PS assay.

In Mexico, there are >3000 plant species reported to have medicinal properties. Of these, ~800 are reportedly used in the practice of Yucatecan traditional medicine. However, to date, <5% of these plants have been evaluated in terms of their production of bioactive metabolites. This clearly reflects the need to increase significantly the number of studies; this would result in an improved phytochemical knowledge of some of the most common Mexican medicinal plants.

To learn more about the production of bioactive metabolites from Yucatecan medicinal plants, our group has established a project directed toward the detection, isolation, and identification of bioactive metabolites produced by some of the most commonly used plant species in Yucatecan traditional medicine. The objectives of the project have been to establish simple bioassay techniques that can detect biological activity in plant crude extracts and then use them to guide the purification process, thus resulting in the isolation and identification of the corresponding biologically active natural products. We wish to report here some of the bioactive metabolites we have obtained during our evaluation of a number of Yucatecan medicinal plants using three different bioassay techniques.

Project Results

Our work in the project was initiated by carrying out literature and field surveys to identify those medicinal plants that were most commonly used by the people of the Yucatan peninsula. The plants were evaluated with consideration of the popular knowledge of their medicinal properties, the importance of the diseases they cured, their availability in the region, and the number of phytochemical studies to which they had been subjected. On the basis of these considerations, the following eight species were selected to be studied in terms of their production of biologically active metabolites: *Ocimum basilicum* L. (*Albahaca*) (*Labiateae*), whose root infusion is used to cure skin infections (Morton 1981); *Ocimum micranthum* Willd. (Xka-kaltún, *Albahaca silvestre*) (*Labiateae*), used as an infusion to treat dysentery; *Cnidoscolus aconitifolius* (Miller) I.M. Johnston (*Chaya silvestre*) (*Euphorbiaceae*), whose edible leaves are prepared as an infusion for the treatment of arthritis; *Bursera simaruba* (L.) Sarg. (Cha-ká) (*Burseraceae*), whose resin is used, diluted in water, against skin irritations; *Chiococca alba* (L.) Hitch. (Ka'anchakche') (*Rubiaceae*), whose root infusion is used to treat dysentery, as an anti-inflammatory, and against snake bites; *Thevetia gaumeri* Hemsley (K'aanloo) (*Apocynaceae*), reportedly used as an analgesic and for the treatment of ulcers; *Capraria biflora* L. (Apote de monte) (*Scrophulariaceae*), used as an infusion to alleviate intestinal fevers (Mendieta and Del Amo 1981); and *Urechites andrieuxii* Muell. Arg. (Contrayerba) (*Apocynaceae*), whose root infusion is used to treat lesions derived from leishmaniasis (Lezama, personal communication).

Each plant was prepared according to traditional recipes, and the resulting infusions were evaluated in the antimicrobial (Clark 1989), brine shrimp lethality (Hui *et al.* 1989), and inhibition of crown gall tumor assays (Galsky *et al.* 1980). Although no biological activity was detected in any of the infusions tested, biological activity was detected in at least one of the bioassays when the corresponding organic crude extracts, from each of the infusions and from the remaining plant materials, were evaluated. These results suggest the presence of bioactive metabolites in the infusion, but at a concentration that is too low for detection. At the same time, the presence of biological activity in the organic crude extract of the remaining plant material indicated that the medicinal properties of the various plants did not require a complete extraction of the biologically active natural products.

Ocimum basilicum *and* Ocimum micranthum

To date there are no reports about the type of metabolites produced by these two species because most of the phytochemical studies done on both *O. basilicum* and *O. micranthum* have been focused on the composition of its essential oils (Charles *et al.* 1990, Wagner and Bladt 1984). During our work, it was found that when the crude methanolic extract of the roots of both plants was partitioned between solvents of increasing polarity, the low polarity (hexane) fraction showed the presence of biological activity in both the antimicrobial [*Bacilus subtilis*; American Type Culture Collection (ATCC) 6633, Rockville, MD] and inhibition of crown gall tumor assays. A bioassay-guided purification resulted in the isolation of two bioactive fractions, each containing a single component on thin-layer chromatography (TLC). However, when the fraction with antimicrobial activity was subjected to gas chromatography/mass spectrometry (GC/MS) analysis, it was identified as an inseparable mixture of taraxasterol (**1**) and α-amyrin (**2**) (Fig. 30.1); in the same manner, a GC/MS analysis of the

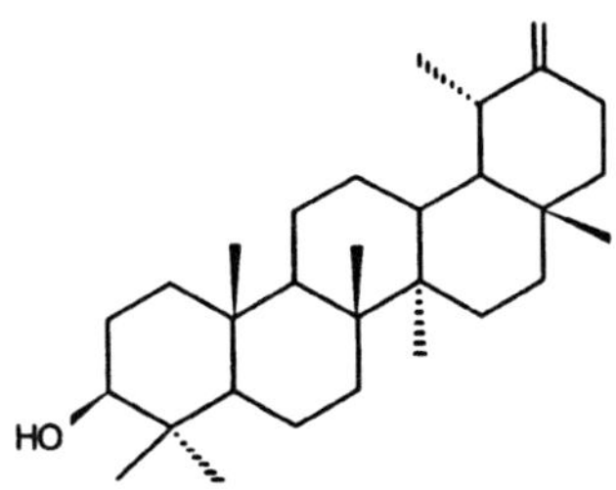

Taraxasterol (**1**)

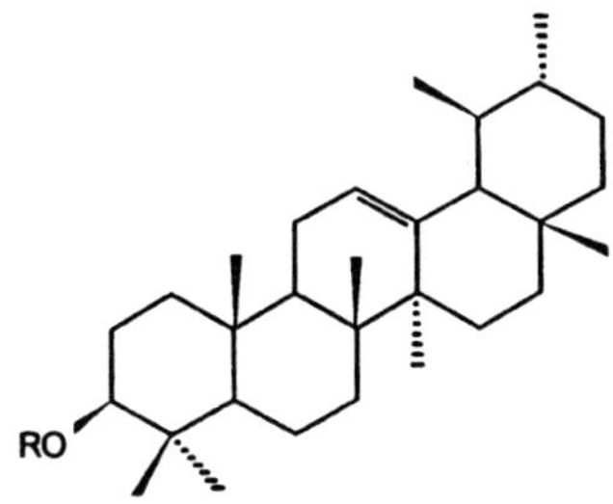

α-Amyrin (R = H, **2**)
α-Amyrin acetate (R = Ac, **3**)

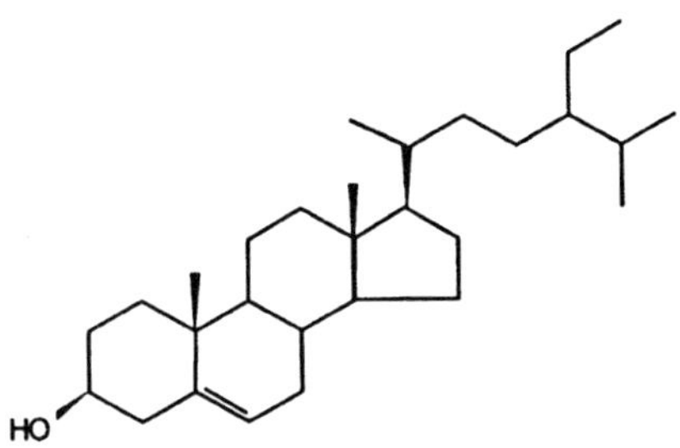

γ-Sitosterol (**4**)

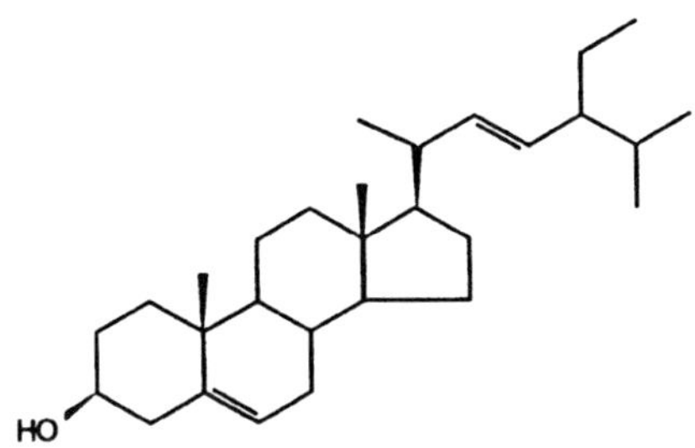

Stigmasterol (**5**)

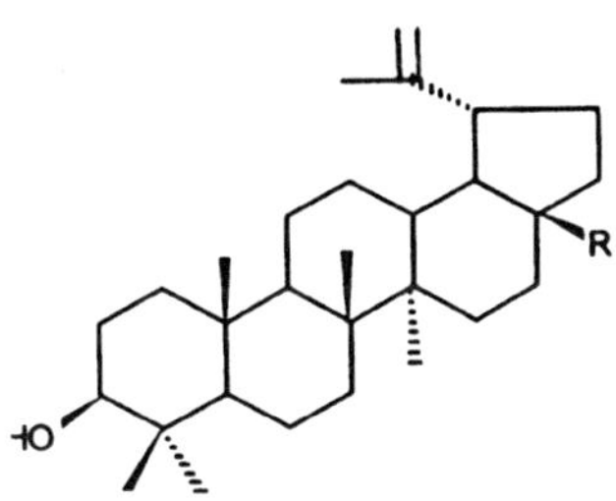

Betulin (R = CH_2OH, **6**)
Betulin acid (R = COOH, **7**)

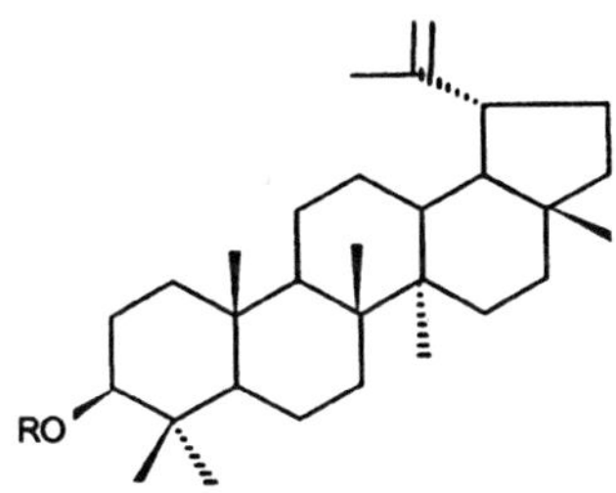

Lupeol (R = H, **8**)
Lupeol acetate (R = Ac, **9**)

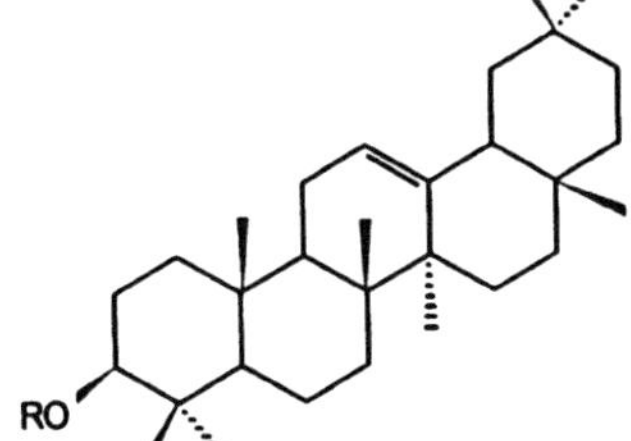

β-Amyrin (R = H, **10**)
β-Amyrin acetate (R = Ac, **11**)

Fig. 30.1. Triterpenes from Yucatecan medicinal plants.

fraction active in the inhibition of crown gall tumor assay identified it as a mixture of γ-sitosterol (**4**) and stigmasterol (**5**) (Fig. 30.1). Even though it appears that the phytochemical composition of both plants is very similar, it is interesting to point out that although the major triterpene of *O. basilicum* was identified as betulin (**6**) (Fig. 30.1), the presence of this metabolite was not detected in the extracts of *O. micranthum*, which, instead, yielded betulinic acid (**7**) (Fig. 30.1) as its major triterpene.

Cnidoscolus aconitifolius

In reviewing the literature, it was found that there are several reports on the phytochemistry of different *Cnidoscolus* species (Caballero *et al.* 1984), but no studies have been reported on *C. aconitifolius*. Although this is one of the most popular medicinal plants used by the people of the Yucatan peninsula, the organic crude extract from the leaves of this plant showed only a weak antimicrobial activity (*Staphylococcus aureus* ATCC 6538). Purification of the organic crude extract of the leaves of *C. aconitifolius* yielded three inactive fractions identified as the acetylated derivatives of lupeol (**9**), β-amyrin (**11**), and an inseparable mixture of α- and β-amyrin (**2** and **10**) (Fig. 30.1).

Bursera simaruba

Earlier investigations of the chemical constituents of different species from the genus *Bursera* have resulted in the isolation of triterpenes (Crowley 1964, Jolad *et al.* 1977a, Parsons *et al.* 1991, Syamasundar *et al.* 1991), bilignans (Hernandez *et al.* 1983), podophyllotoxin-like lignans (Bianchi *et al.* 1968, Jolad *et al.* 1977b, McDoniel and Cole 1972), and flavonoids (Souza *et al.* 1989). The single phytochemical study of metabolites produced by *B. simaruba* reported the isolation of elemicine and amyrenol from the resin extract (Pernet 1972). During our work, it was found that the resin chloroform extract of *B. simaruba* showed the presence of biological activity when tested using the brine shrimp lethality assay. A bioassay-guided purification of the resin crude extract resulted in the identification of the bioactive metabolite as picropolygamain (**14**) (Fig. 30.2), a lignan originally reported as a constituent of the resin of *Commiphora incisa* (*Burseraceae*) (Provan and Waterman 1985). An *in vitro* evaluation of picropolygamain against three human tumor cell lines indicated that this metabolite has cytotoxic activity comparable to that of adriamycin and merits further evaluation (Peraza-Sánchez and Peña-Rodríguez 1992). During the isolation of (**14**), a number of inactive triterpenes were isolated and identified, including the new natural product lup-20(29)-en-3β,23-diol (**15**) and the known metabolites lupeol (**8**), epilupeol (**13**), epiglutinol (**12**), α-amyrin (**2**), and β-amyrin (**10**) (Figs. 30.1 and 30.2) (Peraza-Sánchez *et al.* 1995).

Chiococca alba

A survey of the literature showed that little is known about the phytochemistry of *C. alba*; there are only three studies reporting the isolation of quinoline-type alkaloids (El

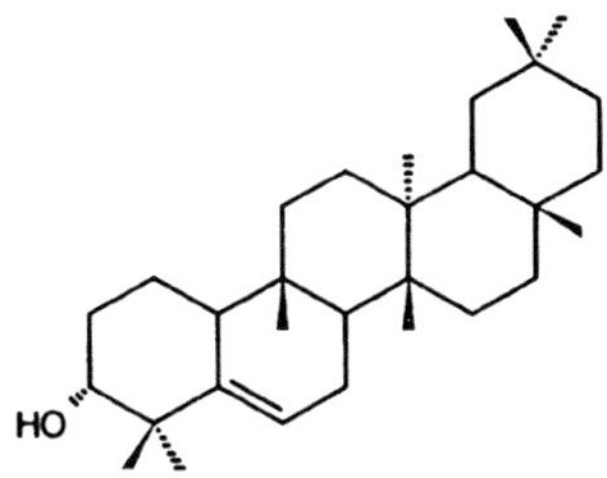

Epiglutinol (**12**)

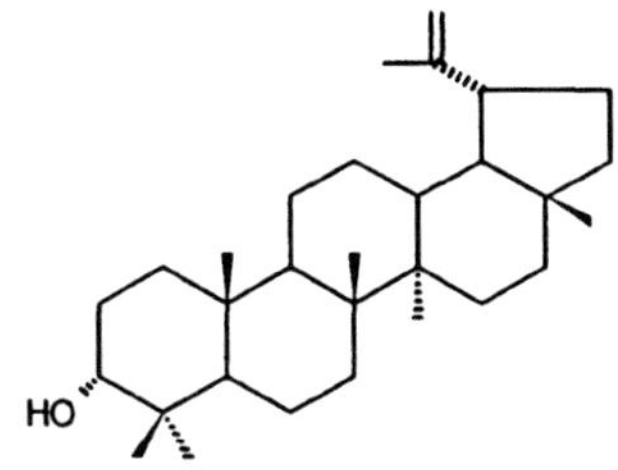

Epilupeol (**13**)

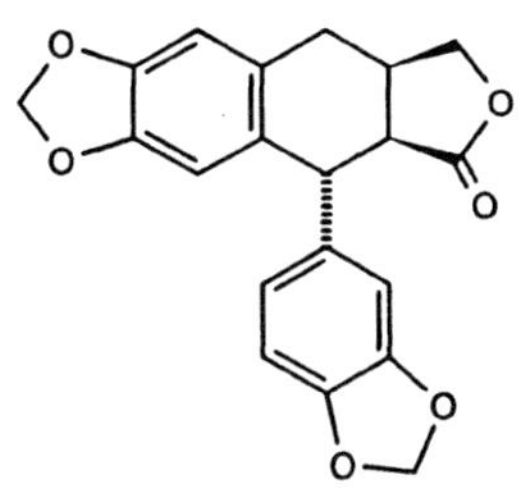

Picropolygamain (**14**)

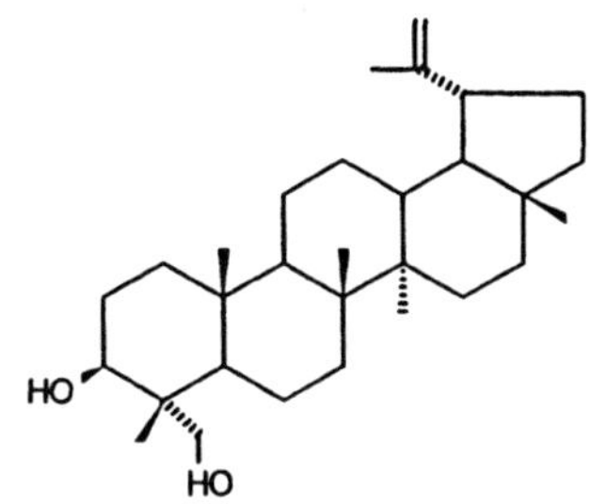

Lup-20(29)-en-3β,23-diol (**15**)

Fig. 30.2. Metabolites isolated from the resin of *Bursera simaruba*.

Abbadi *et al.* 1989), ketoalcohols (Abd El-Hafiz *et al.* 1991), and triterpenes (Bhattacharya and Cunha 1992) from the root extract of the plant. Our evaluation of the organic crude extract from the roots of *C. alba* showed the presence of biological activity in both the brine shrimp lethality and antimicrobial (*S. aureus*) assays. A bioassay-guided purification of the crude root extract resulted in the isolation of a new bioactive diterpene identified as *ent*-17-hydroxy-16α-kauran-3-one (**18**) (Fig. 30.3) (Borges-Argáez *et al.* 1997).

Thevetia gaumeri

Although reports exist in the literature on the chemical constituents of some *Thevetia* species (Sticher and Schmid 1969, Sticher 1970), to date there are no reports on the phytochemistry of *T. gaumeri*. We found that the organic crude extract of the roots of *T. gaumeri* showed the presence of a weak biological activity when tested in the brine shrimp lethality assay. Purification of the crude extract resulted in the isolation of a major, inactive component identified as theviridoside (**16**) (Fig. 30.3), an iridoid glycoside previously reported from the leaves of *T. peruviana* (Abe *et al.* 1995).

Capraria biflora

During our investigation of the chemical constituents of *C. biflora* we detected a strong biological activity in the root extract of this plant when tested in both the antimicrobial assay (*S. aureus*, *B. subtilis*, and *Candida albicans* ATCC 10231) and against *Mycobacterium tuberculosis* (Franzblau, personal communication). A

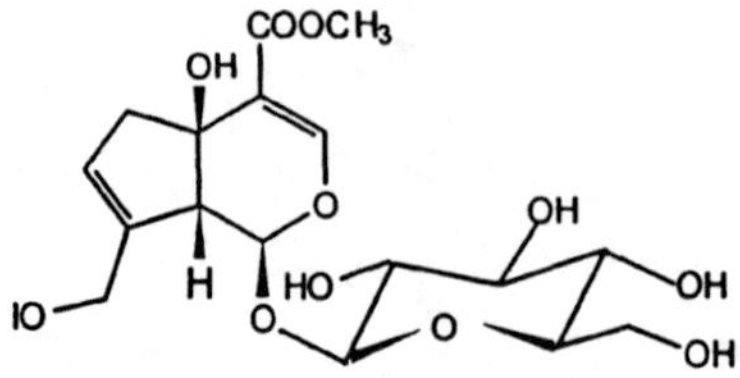

Theveridoside (**16**)

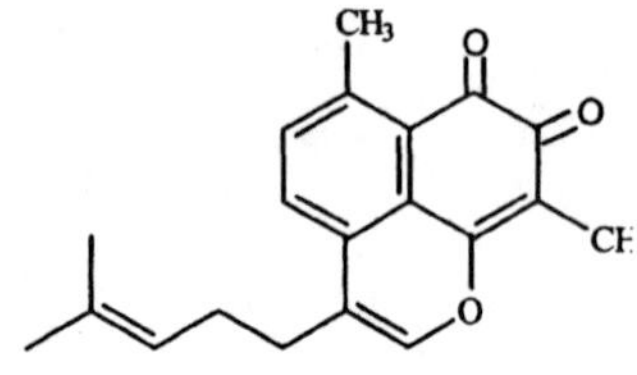

Biflorin (**17**)

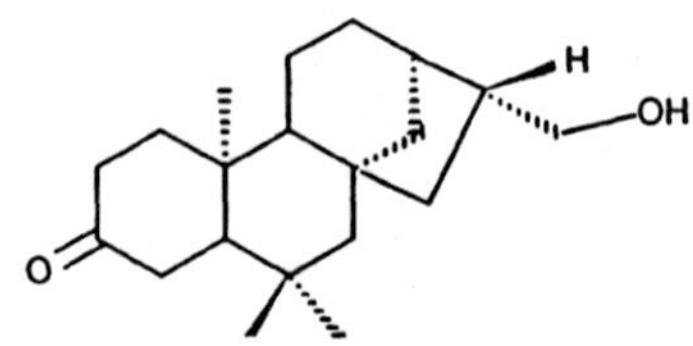

ent-17-Hydroxy-16α-kauran-3-one (**18**)

Fig. 30.3. Metabolites from *Chiococca alba, Thevetia gaumeri,* and *Capraria biflora.*

bioassay-guided purification of the organic crude extract resulted in the isolation of a single, biologically active metabolite identified as biflorin (**17**) (Fig. 30.3), a naphto[1,8-bc]pyran quinonoid originally reported from the same species (Goncalves de Lima *et al.* 1958).

Urechites andrieuxii

A survey of the literature showed that there are no reported studies of any kind on *U. andrieuxii.* Preliminary investigations carried out in our laboratories indicated that the root extract of this plant has a strong activity in both the brine shrimp lethality and leishmanicidal assays. At present, the bioassay-guided purification of the metabolites responsible for the biological activity is in progress.

Acknowledgments

We wish to thank Dr. Francisco Talamás Murra (Syntex, México), Dr. Guillermo Delgado Lamas (Instituto de Química UNAM, México), Prof. John A. Findlay (University of New Brunswick, Canada), Prof. William A. Ayer (University of Alberta, Canada), Dr. Lois M. Browne (University of Alberta, Canada), and Prof. Tom Goodwin (Hendrix College Arkansas, USA) for spectroscopic support; Dr. Javier Flores Abuxapqui (CIR Hideyo Noguchi, Mérida), Dr. Miguel Olvera (CINVESTAV, Mérida), Mr. Julian Coello Coello (CICY, Mérida), Prof. J.L. McLaughlin (Purdue University, USA), Purdue Cell Culture Laboratory (Purdue University, USA), Prof. Ricardo O. Guerrero (University of Puerto Rico, USA), Prof. Scott Franzblau (Louisiana State University, USA), and Dr. Fernando Andrade (CIR Hideyo Noguchi, Mérida) for their assistance in establishing or running bioassay evaluations; Ms. Karlina García Sosa, Ms. Fátima Alejos González, Mr. Irving Ramírez Erosa, Ms. Isela Flores Montenegro, Mr. Salvador Hernández, and Ms. Trinidad Pluma Angulo for technical assistance; Facultad de Química (UADY, Mérida) for initial large-scale extractions and Departamento de Recursos Naturales (CICY, Mérida) for their invaluable assistance in locating, identifying, and collecting plant material. This work was supported by grants F/1744-1 and F/1744-2 from the International Foundation for Science.

References

Abd El-Hafiz, M.A., B. Weniger, J.C. Quiron, and R. Anton, Ketoalcohols, Lignans and Coumarins from *Chiococca alba*, *Phytochemistry 30*: 2029–2031 (1991).

Abe, F., T. Yamauchi, S. Yahara, and T. Nohara, Minor Iridoids from *Thevetia peruviana*, *Phytochemistry 38*: 793–794 (1995).

Alkofahi, A., J.L. McLaughlin, and J.E. Anderson, Search for New Pesticides from Higher Plants, in *Proceedings from the Third Chemical Congress of North America*, Toronto, Ontario, Canada, 1988.

Balandrin, M.F., J.A. Klocke, E.S. Wurtele, and W.H. Bollinger, Natural Plant Chemicals: Sources of Industrial and Medicinal Materials, *Science 228*: 1154–1160 (1985).

Bhattacharya, J., and E.V.L. Cunha, A Triterpenoid from the Root-Bark of *Chiococca alba*, *Phytochemistry 31*: 2546–2547 (1992).

Bianchi, E., M.E. Caldwell, and J.R. Cole, Antitumor Agents from *Bursera microphylla* (*Burseraceae*) I, Isolation and Characterization of Deoxypodophyllotoxin, *J. Pharm. Sci. 57*: 696–697 (1968).

Borges-Argáez, R., L. Medina-Baizabál, F. May-Pat, and L.M. Peña-Rodríguez, A New *ent*-Kaurane from the Root Extract of *Chiococca alba*, *Can. J. Chem. 75*: 801–804 (1997).

Caballero, P., F.R. Fronczek, and N.H. Fischer, The Isolation and Molecular Structure of Lupeol Beta-Phenyl Propionate from *Cnidoscolus elasticus*, *J. Nat. Prod. 47*: 819–821 (1984).

Charles, D.J., J.E. Simon, and K.V. Wood, Essential Oil Constituents of *Ocimum micranthum* Willd., *J. Agric. Food Chem. 38*: 120–122 (1990).

Clark, A., Antimicrobial Screening, in *Workshop on Bioassays*, Proceedings of the 30th Annual Meeting of the American Society of Pharmacognosy, San Juan, Puerto Rico, 1989.

Crowley, K.J., Some Terpenic Constituents of *Bursera graveolens* (H.B.K.) Tr. et Pl. var. *villosula* Cuatr., *J. Chem. Soc.* 4254–4256 (1964).

Del Amo, S.R., and A.L. Anaya, Importancia de la Sistematización de la Información sobre Plantas Medicinales, *Biotica 7*: 2–6 (1982).

Der Marderosian, A., and L.E. Liberti, *Natural Product Medicine*, George F. Sticley Co. Philadelphia, 1988.

Domínguez, X.A., *Métodos de Investigación Fitoquímica*, Editorial Limusa, Monterrey, México, 1973.

El Abbadi, N., B. Weniger, A. Lobstein, J.C. Quiron, and R. Anton, New Alkaloids from *Chiococca alba*, *Planta Med. 55*: 603–604 (1989).

Farnsworth, N.R., and D.D. Soejarto, Potential Consequence of Plant Extinction in the United States on the Current and Future Availability of Prescription Drugs, *Econ. Bot. 39*: 231–240 (1985).

Ferrigni, N.R., J.L. McLaughlin, R.G. Powell, and C.R. Smith, Use of Potato Disc and Brine Shrimp Bioassays as the Antileukemic Principle from the Seeds of *Euphorbia lagascae*, *J. Nat. Prod. 47*: 347–352 (1984).

Ferrigni, N.R., J.E. Putnam, B. Anderson, L.B. Jacobsen, D.E. Nichols, D.S. Moore, and J.L. McLaughlin, Modification and Evaluation of the Potato Disc Assay and Antitumor Screening of *Euphorbiaceae* Seeds, *J. Nat. Prod. 45*: 679–686 (1982).

Galsky, A.G., J.P. Wilsey, and R.G. Powell, Crown Gall Tumor Disc Bioassay, a Possible Aid in the Detection of Compounds with Antitumor Activity, *Plant Physiol. 65*: 184–185 (1980).

Goncalves de Lima, O., W. Keller-Schierlein, and V. Prelog, Uber das Biflorin, *Helv. Chim. Acta 41*: 1386–1390 (1958).

Hernández, J.D., L.U. Román, J. Espiñeira, and P. Joseph-Nathan, Ariensin, a New Lignan from *Bursera ariensis*, *Planta Med. 47*: 215–217 (1983).

Hui, Y.H., H.K. Rupprecht, Y.M. Liu, J.E. Anderson, D.L. Smith, C.J. Chang, and J.L. McLaughlin, Bullatacin and Bullatacinone: Two Highly Potent Bioactive Acetogenins from *Annona bullata*, *J. Nat. Prod. 52*: 463–477 (1989).

Jolad, S.D., R.M. Wiedhopf, and J.R. Cole, Cytotoxic Agents from *Bursera klugii* (*Burseraceae*) I: Isolation of Sapelins A and B, *J. Pharm. Sci. 66*: 889–890 (1977a).

Jolad, S.D., R.M. Wiedhopf, and J.R. Cole, Cytotoxic Agents from *Bursera morelensis* (*Burseraceae*): Deoxypodophyllotoxin and a New Lignan, 5′-Desmethoxydeoxypodophyllotoxin, *J. Pharm. Sci. 66*: 892–893 (1977b).

McDoniel, P.B., and J.R. Cole, Antitumor Activity of *Bursera schlechtendalii* (*Burseraceae*): Isolation and Structure Determination of Two New Lignans, *J. Pharm. Sci. 61*: 1992–1994 (1972).

McLaughlin, J.L., Crown Gall Tumors on Potato Discs and Brine Shrimp Lethality: Two Simple Bioassays for Higher Plant Screening and Fractionation, in *Methods in Plant Biochemistry, Vol. 6, Assays for Activity,* edited by K. Hostettmann, Academic Press, San Diego, 1991, pp. 1–32.

McLaughlin, J.L., and E.J. Anderson, Brine Shrimp, Crown Gall Tumors on Potato Disc and *Lemna minor*, in *Workshop on Bioassays,* Proceedings of the 30th Annual Meeting of the American, Society of Pharmacognosy, San Juan, Puerto Rico, 1989.

Mendieta, R.M., and S.R. Del Amo, *Plantas Medicinales del Estado de Yucatán,* Editorial CECSA, México D.F., 1981.

Meyer, B.N., N.R. Ferrigni, J.E. Putnam, L.B. Jacobsen, D.E. Nichols, and J.L. McLaughlin, Brine Shrimp: A Convenient General Bioassay for Active Plant Constituents, *Planta Med. 45*: 31–34 (1982).

Morton, J.F., *Atlas of Medicinal Plants of Middle America,* Charles C. Thomas Publisher, New York, 1981.

Parsons, I.C., A.I. Gray, C. Lavaud, G. Massiot, and P.G. Waterman, Seco Ring-A Triterpene Acids from the Resin of *Dacroydes normandii*, *Phytochemistry 30*: 1221–1223 (1991).

Peraza-Sánchez, S.R., and L.M. Peña-Rodríguez, Isolation of Picropolygamain from the Resin of *Bursera simaruba*, *J. Nat. Prod. 55*: 1768–1771 (1992).

Peraza-Sánchez, S.R., N.E. Salazar-Aguilar, and L.M. Peña-Rodríguez, A New Triterpene from the Resin of *Bursera simaruba*, *J. Nat. Prod. 58*: 271–274 (1995).

Pernet, R., Phytochimie des *Burseracees*, *Lloydia 35*: 280–287 (1972).

Provan, G.J., and P.G. Waterman, Picropolygamain: A New Lignan from *Commiphora incisa* Resin, *Planta Med.* 271–272 (1985).

Souza, M.P., M.I.L. Machado, and R. Braz-Filho, Six Flavonoids from *Bursera leptophloeos*, *Phytochemistry 28*: 2467–2470 (1989).

Stachel, S.E., E. Messens, M. Van Montagu, and P. Zambryski, Identification of the Signal Molecules Produced by Wounded Plant Cells that Activate T-DNA Transfer in *Agrobacterium tumefaciens*, *Nature 318*: 624–629 (1985).

Sticher, O., Thevesid, ein Neues Iridoid Glucosid aus *Thevetia peruviana* (Pers.) K. Schum (*Apocynaceae*), *Tetrahedron Lett.* 3195–3196 (1970).

Sticher, O., and H. Scmid, Theviridosid, ein Iridoid-Glucosid aus *Thevetia peruviana* (Pers.) K. Schum (*Thevetia neriifolia* Juss.), *Helv. Chim. Acta 52*: 478–481 (1969).

Syamasundar, K.V., G.R. Mallavarapu, and E.M. Krishna, Triterpenoids of the Resin of *Bursera delpechiana*, *Phytochemistry 30*: 362–363 (1991).

Thompson, W., *Guía Práctica Ilustrada de las Plantas Medicinales,* Editorial Blume, Madrid, 1980.

Trease, E., *Tratado de Farmacognosia,* Décimosegunda Edición, Editorial Interamericana, Mexico, D.F. 1988.

Tyler, V.E., Plant Drugs in the Twenty-First Century, *Econ. Bot. 40*: 279–288 (1986).

Wagner, H., and S. Bladt, *Plant Drug Analysis,* Springer-Verlag, 1984.

Xiao-Hua, L., and J.L. McLaughlin, Bioactive Compounds from the Root of *Myrsine africana*, *J. Nat. Prod. 52*: 660–662 (1989).

Chapter 31

Antioxidative Fractions from the Leaves of Betel (*Piper betle* L.)

N. Andarwulan, T.J. Susanto, and R. Syarief

Department of Food Technology and Human Nutrition, Faculty of Agricultural Technology, Bogor Agriculture University, Bogor 16002, Indonesia

Introduction

Antioxidants, inhibitors of lipid peroxidation, are important not only for food protection, but also for the defense of living cells against oxidative damage. Although food industries have used effective synthetic antioxidants, consumers have come to prefer natural antioxidants to synthetic additives on the basis of the assumption that natural compounds are safe. Many investigators have found different types of antioxidants in various kinds of plants (Larson 1988). We have been interested in betel leaves as a new source of natural antioxidant because the leaves are traditionally used to maintain healthy teeth. The custom of betel chewing, which is common throughout India, the entire Malay archipelago, including some places in Indonesia, and in southern China, consists of chewing a betel leaf (*Sirih*) from *Piper betle* L. together with lime, some gambier (from *Uncaria gambir* Roxb.), a piece of areca, or pinang, or betel nut. The betel leaves owe their spicy and burning taste to a volatile oil, which is distilled repeatedly. The volatile oil and oleoresins have found no practical application thus far. In this chapter, the isolation of antioxidant fractions from green and yellow cultivars of betel leaves is reported.

Materials and Methods

Materials

Green and yellow (Fig. 31.1) betel leaves were freeze-dried and ground into a powder. The powders were then deodorized with the use of vapor distillation for 3 h and then freeze-dried.

Extraction

Freeze-dried green betel leaf powder was extracted using the method described by Hammerschmidt and Pratt (1978) with slight modification regarding the time of extraction (4 h) and substituting ethanol as the solvent. Total phenolic content (AOAC 1984) and antioxidant activity of betel leaf extracts (BLE) were then measured.

Fig. 31.1. Betel leaves (*Piper betle* L.).

Measurement of Antioxidant Activity

The antioxidant activity test was carried out using the thiocyanate method (Mitsuda *et al.* 1966). The induction period was used as the time needed to achieve an absorbance value of 0.30 (Chen *et al.* 1995); the protection factor defined to express antioxidant activity was determined as the ratio of the sample's induction period to that of the control.

Fractionation of BLE using Column Chromatography

The crude samples of BLE were fractionated using a silicic acid column (diameter 2.5 cm) as described by Tian and White (1994). The silicic acid (20 g) was activated overnight at 120°C; then a slurry of the silicic acid was made by adding 50 mL hexane. After the hexane slurry was packed into the column, 1 mL green BLE (GBLE) was applied onto the top of it. The column was eluted stepwise with 500 mL hexane and 500 mL methanol, respectively. The two separated fractions were then evaporated, using a rotary evaporator, to 1 mL each at 45°C and stored at –10°C until the time of the analysis. The antioxidant activity of the fractions was then measured.

Thin-Layer Chromatography (TLC) to Identify the Antioxidant Fraction

Green betel leaf extracts and the polar fraction of GBLE obtained after column chromatography were streaked onto a TLC plate (0.25 mm) and developed first in the solvent system chloroform/ethanol/acetic acid (98:2:2, by vol) and then in the

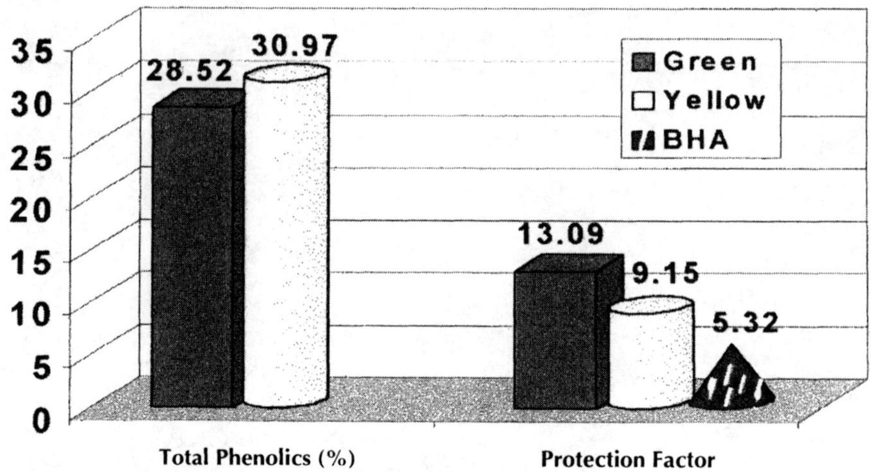

Fig. 31.2. Total phenolics and antioxidant activity of betel leaf extracts.

solvent system hexane/diethyl ether (3:7, vol/vol). The plate was viewed under ultraviolet (UV) light (200–280 nm); the separated bands were scraped from the plate and extracted in 2 mL ethanol. The antioxidant activity of the latter extracts was then tested.

Results and Discussion

Total phenol and antioxidant activities of green betel leaf extract (GBLE) and yellow betel leaf extract (YBLE) were measured (Fig. 31.2). Compared with YBLE

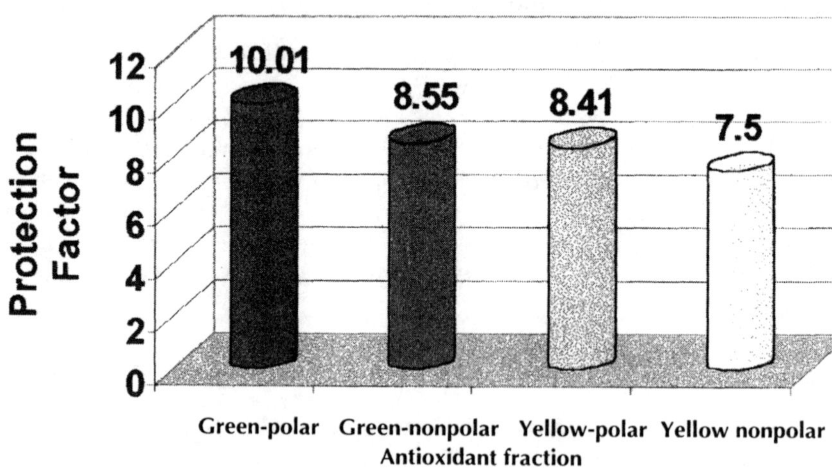

Fig. 31.3. Antioxidant activity of fraction after column chromatography.

and butylated hydroxianisole (BHA), GBLE showed a higher antioxidant activity, but had a lower content of total phenolics. It is suspected that GBLE had certain more active phenolic compounds and that the antioxidant activity of the extracts did not depend as much on the amount of total phenolics. The polar and nonpolar fractions of GBLE and YBLE had a high antioxidant activity (Fig. 31.3), whereas the polar fraction of GBLE contained the highest potential for antioxidant activity. GBLE and the polar fraction of GBLE were selected for further study. GBLE was further fractionated using a two-dimensional TLC yielding 57 spots (Fig. 31.4). Thirty-three spots were antioxidative, six of which (fractions 5, 25, 26, 32, 37, and 50) had similar activity to BHA (Fig. 31.5). Spectra of these fractions had λmax in

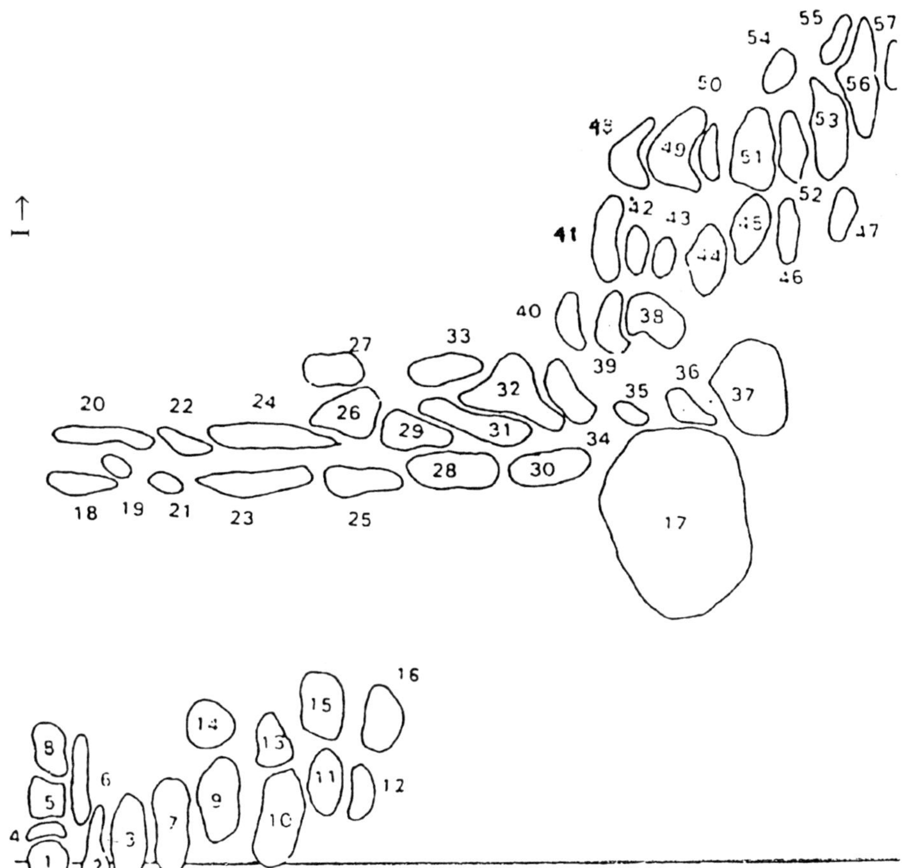

Fig. 31.4. Chromatogram of two-dimensional thin-layer chromatography of green betel leaf extracts. First and second solvent systems were chloroform/ethanol/acetic acid (98:2:2, by vol) and hexane/ethyl ether (3:7, vol/vol), respectively.

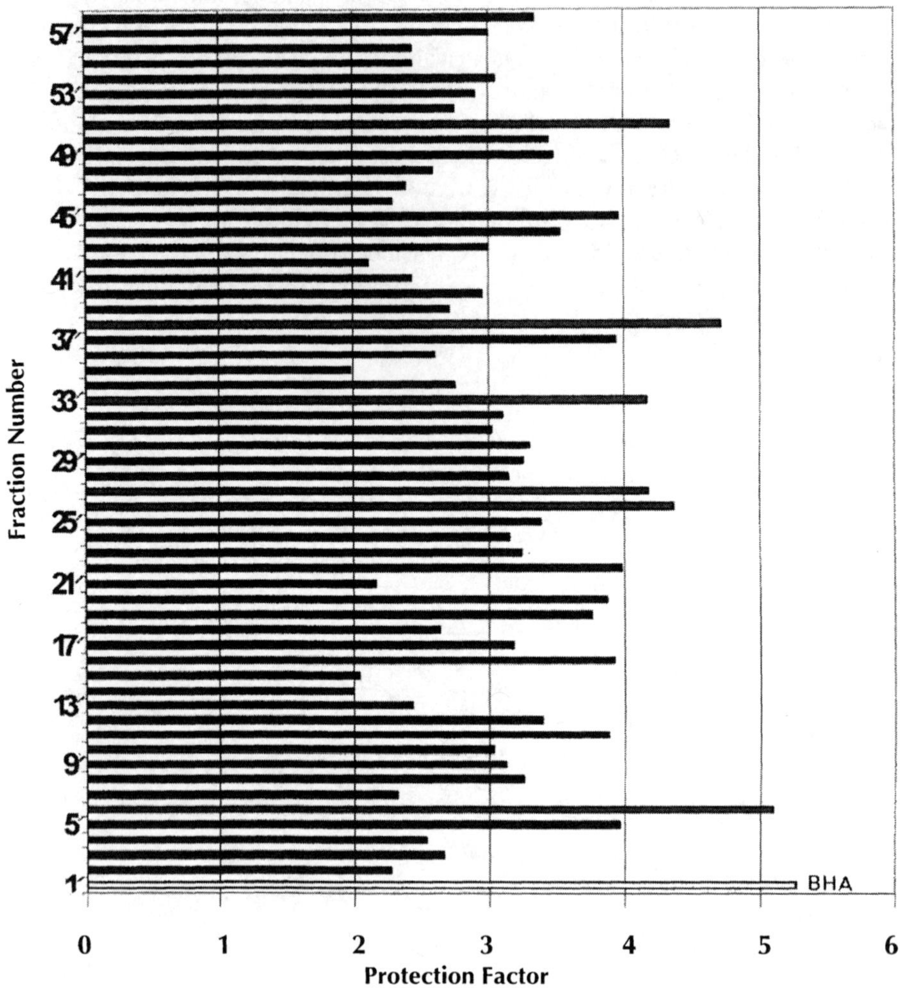

Fig. 31.5. Antioxidant activity of fraction from green betel leaf extract after thin-layer chromatography.

the UV-visible range (data not shown) except fraction 37 (λmax 226.8 and 279.6 nm). The polar fraction of GBLE after two-dimensional TLC yielded 23 spots (Fig. 31.6). Sixteen spots were antioxidative, two of which (fractions 2 and 8) had similar activity compared with BHA (Fig. 31.7). The spectrum of fraction 2 had λmax at 280 nm and that of fraction 8 at 567 nm. Fractionation using TLC showed that betel leaf extract contained compounds that could potentially be used as natural antioxidants. The identification and characterization of these antioxidants is being pursued.

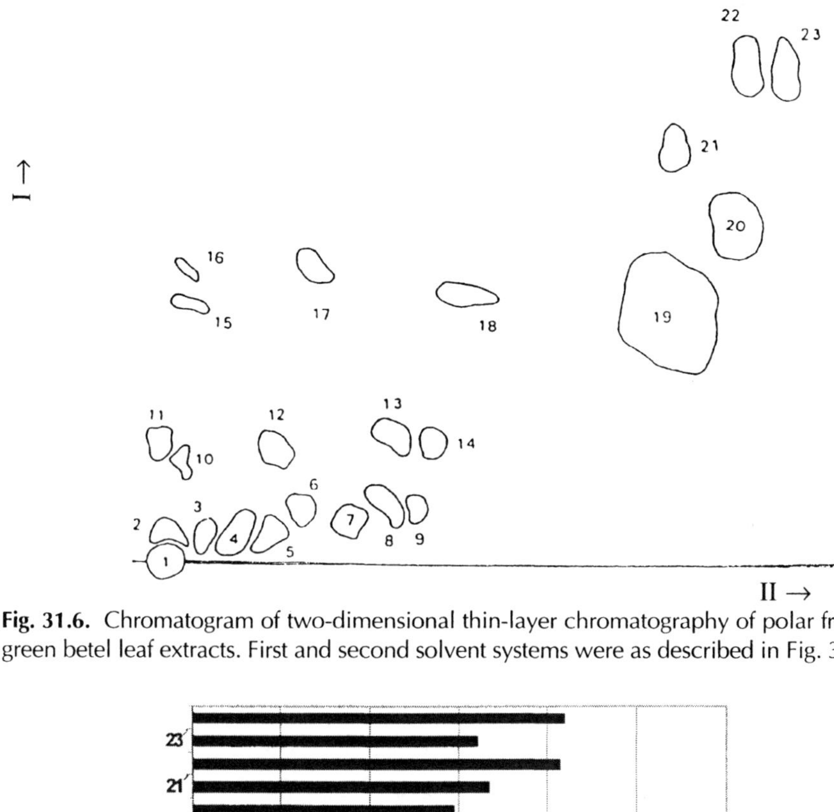

Fig. 31.6. Chromatogram of two-dimensional thin-layer chromatography of polar fraction green betel leaf extracts. First and second solvent systems were as described in Fig. 31.4.

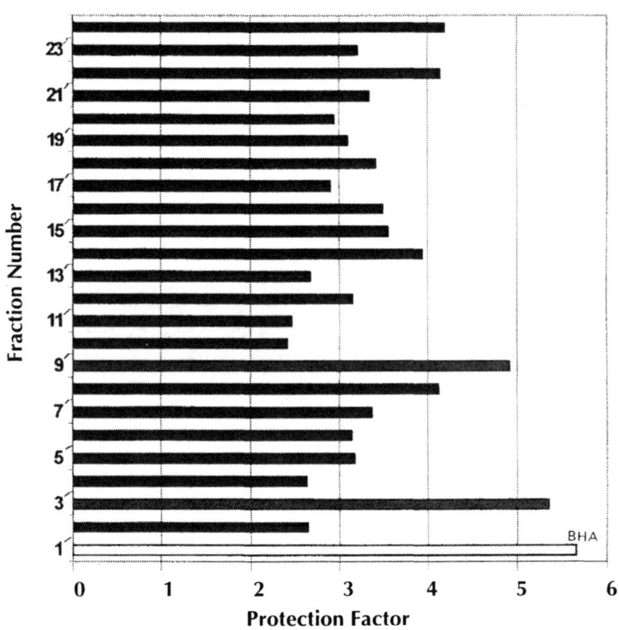

Fig. 31.7. Antioxidant activity of fraction from polar fraction green betel leaf extract after thin-layer chromatography.

References

Association of Official Analytical Chemists, *Official Methods of Analysis*, AOAC, Washington, DC, 1984.

Chen, H.M., K. Muramoto and F. Yamauchi, Structural Analysis of Antioxidative Peptides from Soybean-Conglycinin, *J. Agric. Food Chem. 43*: 574–578 (1995).

Hammerschmidt, P.A., and D.E. Pratt, Phenolic Antioxidants of Dried Soybeans, *J. Food Sci. 43*: 556–559 (1978).

Larson, R.A., The Antioxidants of Higher Plants, *Phytochemistry 27*: 969–978 (1988).

Mitsuda, H., K. Yasumoto, and K Iwami, Antioxidative Action of Indole Compounds During the Autoxidation of Linoleic Acid, *Eiyoto Shokuryo 19*: 210–214 (1966).

Tian, L.L., and P.J. White, Antioxidative Activity of Oat Extract in Soybean and Cottonseed Oils, *J. Am. Oil Chem. Soc. 71*: 1079–1086 (1994).

Chapter 32

Affinities of Dietary Phenolic Antioxidants for Lipid Bilayers

Tsutomu Nakayama

School of Food and Nutritional Sciences, University of Shizuoka, 52-1 Yada, Shizuoka 422-8526, Japan

Introduction: Affinities of Polyphenols for Lipid Bilayers Determine Their Biological Activities

The biological activities of numerous natural and synthetic antioxidants have been examined by various methods both *in vitro* and *in vivo* to elucidate their preventive effects on the diseases induced by oxidative stress such as cancer, aging, and AIDS. The activity evaluated in experiments *in vitro* using cultured mammalian cells or bacteria reflects mainly two factors. The first is the genuine activity, such as antioxidant activity, which should be considered under the condition that every sample be incorporated equally into the cells. The second is the amount incorporated into the cells during incubation. If the genuine activity is the same in the samples examined, only the incorporated amount determines the apparent biological activities. In other words, even if its concentration in the medium is low, a sample incorporated efficiently into the cells could show high activity. We have previously reported that phenolic antioxidants and ubiquinones suppress hydrogen peroxide (H_2O_2)-induced cytotoxicity toward mammalian and bacterial cells (Nakayama *et al.* 1992, 1993a, 1993b, 1994, 1996, and 1997). The structure-activity relationships obtained from these experiments, using a series of compounds with a common chemical moiety responsible for the antioxidant activity, suggest that the incorporated amounts of the compounds determine their dose dependencies.

Previous Methods to Evaluate Affinities of Polyphenols for Lipid Bilayers or Their Incorporated Amounts into Cells

The partition coefficients of polyphenols in *n*-octanol/aqueous buffer might be parameters reflecting their affinities for lipid bilayers. Because all compounds in some groups, such as curcuminoids, were partitioned in the alcohol layers, they cannot be distinguished according to these partition coefficients. In the pharmaceutical sciences, liposome-partitioning systems have been used to investigate the correlation between the biological activities of various compounds and their interaction with lipid bilayers; it has been concluded that liposome systems display more selective partitioning in models than in the *n*-octanol–buffer system (Betageri and Rogers 1987, Rogers and Choi 1993). In the studies of plant polyphenols, interaction of four flavonoids (quercetin, hesperetin, naringenin, and rutin) with liposomes has been investigated by differential scanning calorimetry (Saija *et al.* 1995). The authors described that aglycons such as quercetin, hesperetin, and naringenin interact with liposomes, but rutin, a glycoside of quercetin, does not. However, there have been no reports on the direct measurements of phenolic antioxidants incorporated into model membranes. Furthermore, measurements of the amounts of the compounds incorporated into the cells during a certain incubation period were not always successful because some compounds were metabolized in the cells immediately after addition to the medium. Thus, it is necessary to establish an accurate and general method to estimate the amounts of antioxidants incorporated into model membranes. The problem in measuring their amounts in lipid bilayers of liposomes resulted from the difficulty of separation of the liposomes and the surrounding aqueous medium after incubation of polyphenols. For example, gel-filtration chromatography was inadequate for this purpose because most polyphenols were abstracted from the liposomes and adsorbed into the gel.

A Novel Method to Evaluate the Affinities for Lipid Bilayers

We solved the above problem by separating the medium and the liposomes with a high-density internal aqueous phase by simple centrifugation (Nakayama *et al.* 1998). Figure 32.1 shows our method and its principle. Phosphatidylcholine (PC) from egg yolk (100 mg) in 300 mmol/L of aqueous glucose solution (10 mL) was sonicated in an ultrasonic cleaner. The resulting multilamellar vesicle (MLV) solution was placed in water in a 50-mL plastic centrifuge tube in the cavity of a cup-horn–type sonicator and sonicated for 10 min (step 1) to change MLV to small unilamellar vesicles (SUV). The liposomal solution was diluted 10 times with water (step 2). Untrapped glucose was easily removed by centrifugation at 130,000 × *g* for 5 min at 20°C (steps 3 and 4), because the density of the internal aqueous phase of the liposomes (SUV) was higher than that of the diluted external solution (Oku *et al.* 1994). The liposomes in the sediment were resuspended in phosphate-buffered saline (PBS, pH 7.4) to adjust the final concentration of egg PC in a liposomal solution to 1 mg/mL. In a typical experiment to measure an amount incorporated into the lipid bilayers of liposomes, a polyphenol (or a ubiquinone) solution in ethanol (0.1 mL) was added to 0.9 mL of the liposome solution in PBS and vortexed in a centrifuge tube (step 5). We confirmed that this amount of ethanol had no effect on the structure of the liposomes. After incubation for 20 min at 20°C (step 6), the mixture was centrifuged at 130,000 × *g* for 5 min (step 7). After the supernatant was discarded, the liposomes containing the polyphenol were suspended with PBS, centrifuged, and dissolved with 1 mL ethanol. The amount of polyphenol in the ethanol solution was measured using high-performance liquid chromatography (HPLC) (step 8). Then the amount of the polyphenol incorporated into the liposomes was calculated.

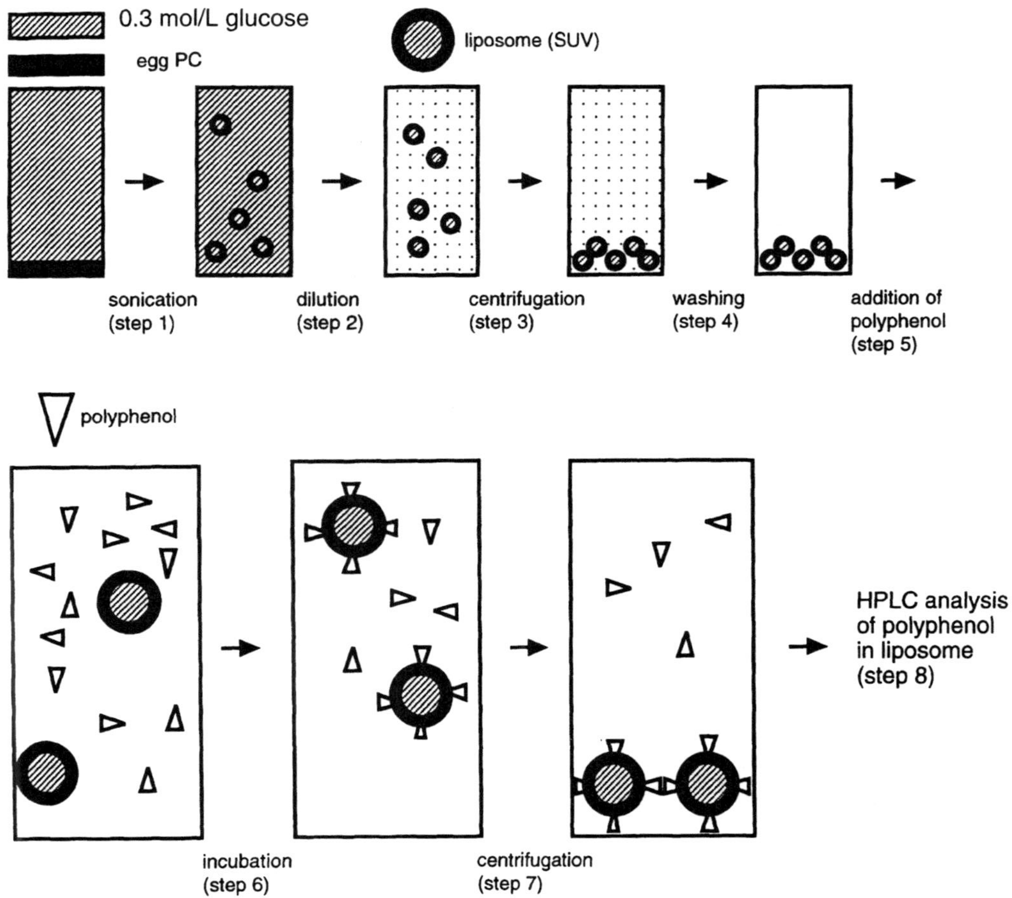

Fig. 32.1. Measurement of affinities of phenolic antioxidants for the lipid bilayers using high-density liposomes. Abbreviations: PC, phosphtidylcholine; SUV, small unilamellar vesicles; HPLC, high-performance liquid chromatography.

Affinities of Polyphenols for Lipid Bilayers Reflect the Dose Dependencies of Their Biological Activities in *in vitro* Experiments

Figure 32.2 shows typical dose effects of four groups of polyphenols and ubiquinones on the amounts incorporated into the liposomes. In most cases, the incorporated amounts were proportional to the added amounts for up to 10 nmol in the liposomal solutions. The relationships between these results with the dose dependencies of *in vitro* experiments on the respective groups are described below on the premise that antioxidants incorporated into the liposomes in large amounts should have high affinities for lipid bilayers.

Gallic Acid Esters

The amount of gallic acid lauryl ester (GALE) incorporated was highest, followed by gallic acid propyl ester (GAPE) (Fig. 32.2A). Neither gallic acid methyl ester (GAME) nor gallic acid was detected in the liposomes in the range of the added amounts. This result indicates that the affinity of a gallic acid ester for the lipid bilayers depends simply on the chain length of its ester moiety. We have already reported that 200 nmol/L GALE, 20 mol/L GAPE, and 80 mmol/L GAME were necessary for complete protection against H_2O_2-induced cytotoxicity (Nakayama *et al.* 1992), and that GALE was most effective among the gallic acid esters for lowering the SOS induction raised by H_2O_2 (Nakayama *et al.* 1993a). The previous results support our hypothesis that gallic acid esters with higher affinities exhibit biological activities in lower concentrations.

Caffeic Acid Esters

The incorporated amount of caffeic acid phenethyl ester (CAPE) was highest, followed by caffeic acid ethyl ester (CAEE) and caffeic acid methyl ester (CAME) (Fig. 32.2B). We have already reported that CAEE and CAME suppressed H_2O_2-induced cytotoxicity and H_2O_2-induced DNA single-strand breaks, but caffeic acid did not (Nakayama *et al.* 1992 and 1996). We compared dose dependencies of CAPE, CAEE, and CAME on H_2O_2-induced cytotoxicity by the experiment carried out at same time. Figure 32.3 shows that at 2 mmol/L, CAPE was most effective, followed by CAEE. This reflects "the higher affinity of CAPE than that of CAEE." We supposed that at 10 µmol/L, CAPE showed cytotoxic effects rather than preventive effects for H_2O_2-induced cytotoxicity. This is further evidence for the high affinity of CAPE for the cell membranes. CAPE (Fig. 32.4) is a natural caffeic acid ester found in propolis. Inhibitory effects of CAPE on azoxymethane-induced biochemical changes and aberrant crypt foci formation in rat colon, growth of human leukemia HL-60 cells, and activation of nuclear transcription factor NF-κB have been reported (Chen *et al.* 1996, Natarajan *et al.* 1996, Rao *et al.* 1993). If CAPE and other caffeic acid esters are hydrolyzed by esterase, the result-

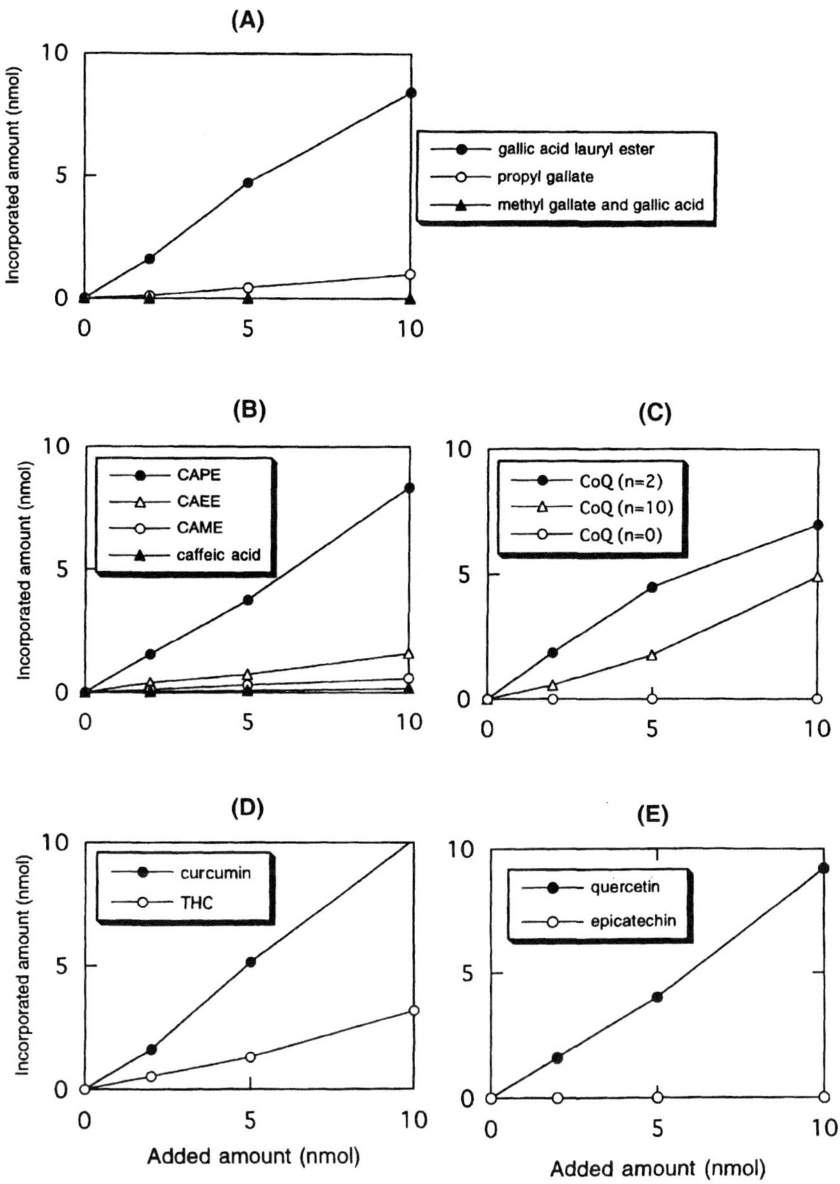

Fig. 32.2. Dose effects of dietary antioxidative polyphenols and ubiquinones on the amount incorporated into the liposomes. Each polyphenol or ubiquinone in 0.1 mL of ethanol was added to 0.9 mL of a liposomal solution and incubated for 20 min at 20°C. After centrifugation, the amount of the polyphenol incorporated into the liposomes was measured with high-performance liquid chromatography. Abbreviations: CAPE, caffeic acid phenethyl ester; CAEE, caffeic acid ethyl ester; CAME, caffeic acid methyl ester; THC, tetrahydrocurcumin; CoQ, coenzyme Q.

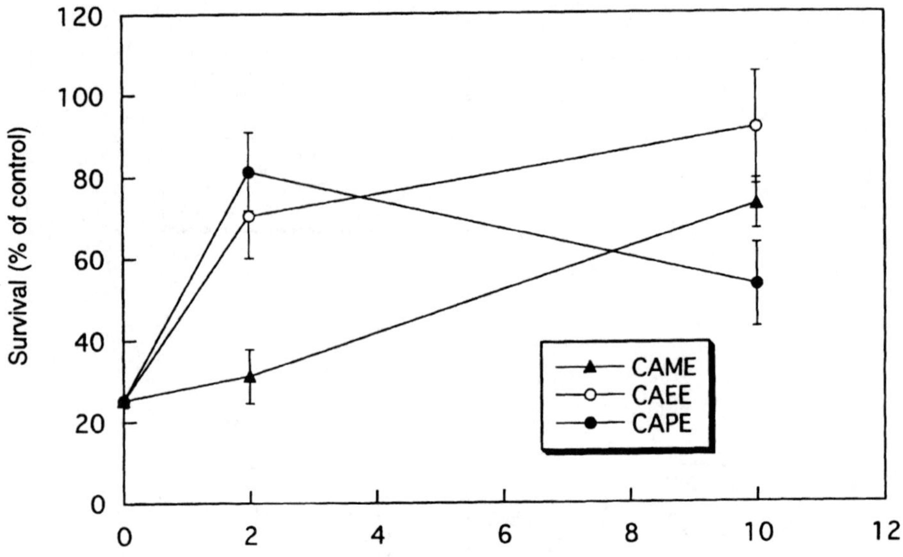

Fig. 32.3. Dose effects of caffeic acid esters on H_2O_2-induced cytotoxicity. Chinese hamster V79 cells were seeded in 60-mm petri dishes (200 cells/dish) and incubated in a minimum essential medium (MEM) supplemented with 10% heat-inactivated fetal bovine serum (FBS) in a humidified atmosphere of 5% CO_2 in air at 37°C. After the medium had been changed to MEM free of FBS, the cells were incubated with the indicated concentrations of the respective caffeic acid ester. After being washed with HEPES buffered saline (HBS, pH 7.3), the cells were incubated with 150 μmol/L of H_2O_2 in HBS. After 5 d of culturing, the number of the colonies stained with a Giemsa (azur eosin methylene-blue) solution was counted. The cell survival (% of control) was calculated by dividing the number of colonies of the cells treated with H_2O_2 and/or the respective caffeic acid ester by the number of colonies of untreated control cells. The results are expressed as the mean ± SD of four separately treated cultures. See Figure 32.2 legend for caffeic acid ester abbreviations.

ing caffeic acid should be the very compound exerting those inhibitory effects in the cells, i.e., their affinities for the cell membranes should determine their apparent biological activities. Actually, various activities of CAPE, such as inhibitory effects on lipoxygenase and cyclooxygenase, are higher than those of CAME (Rao *et al.* 1993).

Ubiquinones

In mammalian and bacterial cells, ubiquinones (Fig. 32.4) are reduced to ubiquinols, which act as antioxidants (Nakayama *et al.* 1994). Thus, it is important to study the interactions of ubiquinones with cell membranes. The affinities of ubiquinones (coenzyme Q) did not depend on the chain length of their isoprene units (Fig. 32.2C).

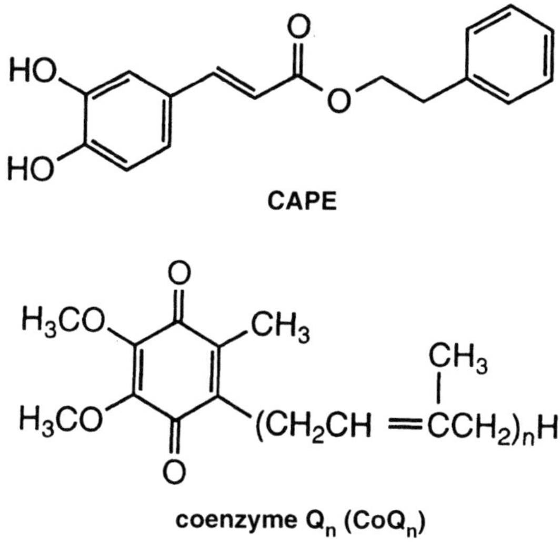

Fig. 32.4. Structure of caffeic acid phenethyl ester (CAPE) and coenzyme Q_n (CoQ_n).

Decylubiquinone (CoQ_2) has a higher affinity than that of CoQ_{10}. In the cell experiments, neither 2,3-dimethoxy-5-methyl-1,4-benzoquinone (CoQ_0) nor CoQ_{10} showed any suppressive effects for H_2O_2-induced cytotoxicity, whereas 1 mol/L of decylubiquinone (CoQ_2) did (Nakayama *et al.* 1994). We assume that the prenyl chain of CoQ_{10} would be too long to disperse uniformly in the liposomal solutions or in the medium of the cultured cells. Therefore, even in this case, the liposome experiment reflects the interaction of CoQ_{10} with the cell membranes.

Curcuminoids

Curcumin is the major yellow component of turmeric. Colorless tetrahydrocurcumin (THC), suggested to be a metabolite of curcumin, has been shown to be a more potent antioxidant than curcumin (Sugiyama *et al.* 1996). We attempted to obtain the partition coefficients of the polyphenols in an *n*-octanol/PBS or *n*-butanol/PBS system. Because all curcumin and THC were partitioned in the alcohol layers, the two compounds could not be distinguished with the partition coefficients. In this study, the affinity of THC was lower than that of curcumin (Fig. 32.2D), which exhibited cytotoxicity in the cultured cells at concentrations somewhat lower than did THC (Nakayama *et al.* 1997). We assume that the flexible and bulky structure of THC makes its incorporation into the lipid bilayers difficult. On the other hand, the high affinity for the lipid bilayers of curcumin should be partly ascribed to its rigid and planar structure due to the conjugated system (Fig. 32.5), which also causes its yellow color. It is natural that the three-dimensional interaction of these compounds with the phospholipid bilayers of the liposomes simulate well that with the cell membranes.

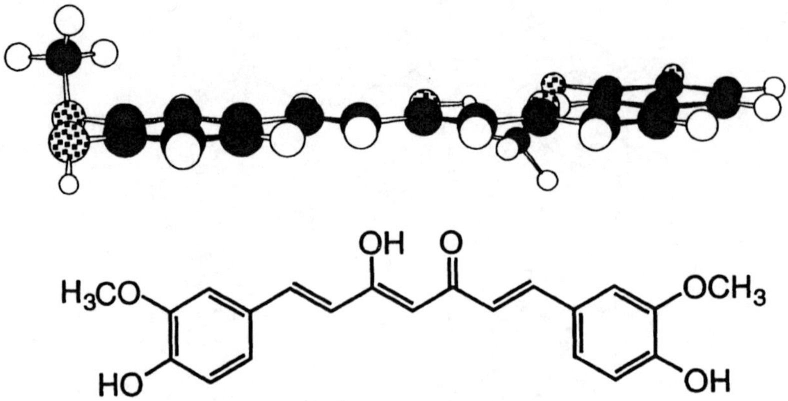

curcumin

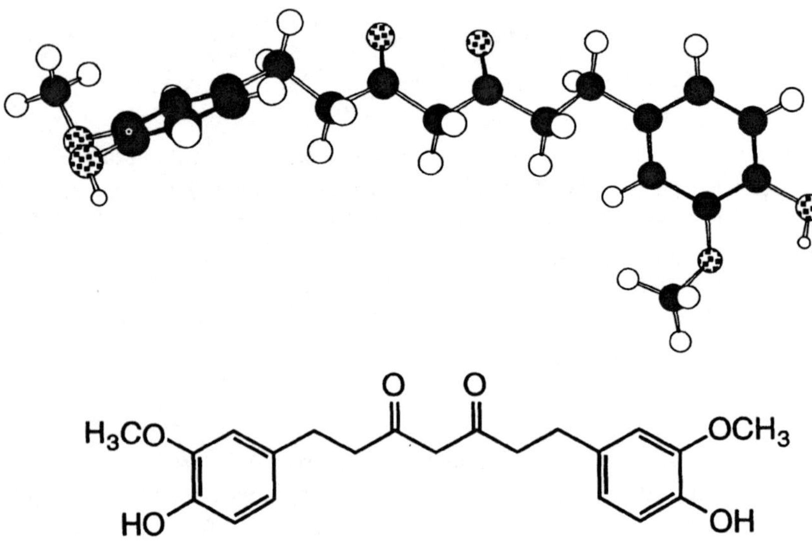

THC

Fig. 32.5. Two- and three-dimensional structures of curcumin and tetrahydrocurcumin (THC).

Catechin and Quercetin

At a glance, quercetin and epicatechin have similar chemical structures, but the two compounds have completely different affinities for lipid bilayers (Fig. 32.2E). The added quercetin was almost recovered in the liposomes. On the other hand, epicatechin was not incorporated into the liposomes. The three-dimensional structures of quercetin and epicatechin (Fig. 32.6) show that quercetin has a rigid and planar structure due to its conjugated systems. On the other hand, epicatechin has a flexible and bulky struc-

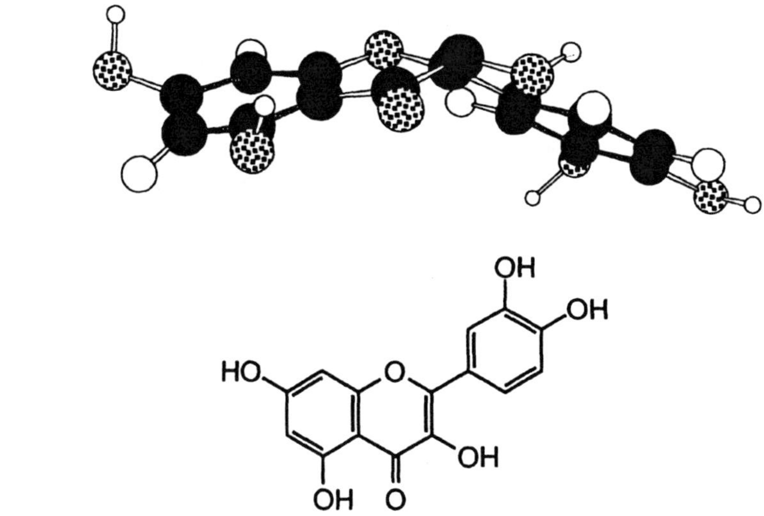

quercetin

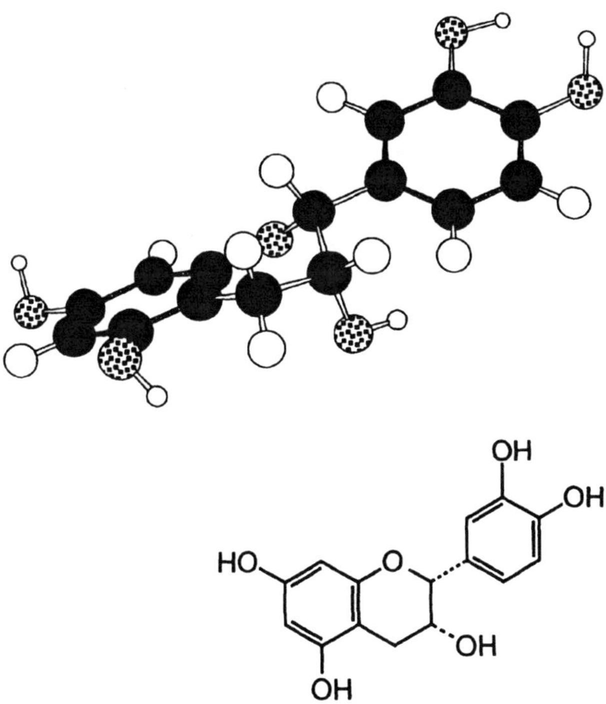

epicatechin

Fig. 32.6. Two- and three-dimensional structures of quercetin and epicatechin.

ture because of the rotation of the C-C bond between the B and C rings. The contrast between yellow quercetin and colorless epicatechin resembles the contrast between yellow curcumin and colorless THC. We suppose that their steric characters, in addition to their simple lipophilicities, determine the affinities for lipid bilayers. In the cell experiments, 1 mmol/L catechin was necessary to exhibit protective effects similar to those of 20 µmol/L quercetin against H_2O_2-induced cytotoxicity (Nakayama *et al.* 1993b). These results indicate the close relationship between the measured affinities of two compounds and their dose dependencies in the cell experiments. In conclusion, our new method is useful in evaluating the affinities of a series of phenolic antioxidants and ubiquinones for lipid bilayers and explaining the correlation between their affinities and the dose effects in cultured cells.

References

Betageri, C.V., and J.A. Rogers, Thermodynamics of Partitioning of β-Blockers and the *n*-Octanol-Buffer and Liposome Systems, *Int. J. Pharm. 36*: 165–173 (1987).

Chen, J.-H., Y. Shao, M.-T. Huang, C.-K. Chin, and C.-T. Ho, Inhibitory Effect of Caffeic Acid Phenethyl Ester on Human Leukemia HL-60 Cells, *Cancer Lett. 108*: 211–214 (1996).

Nakayama, T., I. Haraguchi, K. Hashimoto, Y. Sugiyama, and T. Osawa, Suppression of Hydrogen Peroxide-Induced Cytotoxicity Towards Chinese Hamster Lung Fibroblasts by Chemically Modified Curcumin, *Food Sci. Technol. Int. Tokyo 3*: 74–76 (1997).

Nakayama, T., M. Hiramitsu, T. Osawa, and S. Kawakishi, The Protective Role of Gallic Acid Esters in Bacterial Cytotoxicity and SOS Responses Induced by Hydrogen Peroxide, *Mutat. Res. 303*: 29–34 (1993a).

Nakayama, T., T. Niimi, T. Osawa, and S. Kawakishi, The Protective Role of Polyphenols in Cytotoxicity of Hydrogen Peroxide, *Mutat. Res. 281*: 77–80 (1992).

Nakayama, T., Y. Ogiso, T. Osawa, and S. Kawakishi, Suppression of Hydrogen Peroxide-Induced Cytotoxicity Toward Chinese Hamster Lung Fibroblasts by Decylubiquinone, a Coenzyme Q Homolog, *Biosci. Biotechnol. Biochem. 58*: 1702–1704 (1994).

Nakayama, T., K. Ono, and K. Hashimoto, Affinity of Antioxidative Polyphenols for Lipid Bilayers Evaluauted with a Liposome Systems, *Biosci. Biotechnol. Biochem. 62:* 1005–1007 (1998).

Nakayama, T., M. Yamada, T. Osawa, and S. Kawakishi, Suppression of Active Oxygen-Induced Mammalian Cytotoxicity by Flavonoids, *Biochem. Pharmacol. 45*: 265–267 (1993b).

Nakayama, T., M. Yamada, T. Osawa, and S. Kawakishi, Inhibitory Effects of Caffeic Acid Ethyl Ester on H_2O_2-Induced Cytotoxicity and DNA Single-Strand Breaks in Chinese Hamster V79 Cells, *Biosci. Biotechnol. Biochem. 60*: 316–318 (1996).

Natarajan, K., S. Singh, T.R. Burke, Jr., D. Grunberger, and B.B. Aggarwal, Caffeic Acid Phenethyl Ester Is a Potent and Specific Inhibitor of Activation of Nuclear Transcription Factor NF-κB, *Proc. Natl. Acad. Sci. USA 93*: 9090–9095 (1996).

Oku, N., R. Naruse, K. Doi, and S. Okada, Potential Usage of Thermosensitive Liposomes for Macromolecule Delivery, *Biochim. Biophys. Acta 1191*: 389–391 (1994).

Rao, C.V., D. Desai, B. Simi, N. Kulkarni, S. Amin, and B.S. Reddy, Inhibitory Effect of Caffeic Acid Esters on Azoxymethane-Induced Biochemical Changes and Aberrant Crypt Foci Formation in Rat Colon, *Cancer Res. 53*: 4182–4188 (1993).

Rogers, J.A., and Y.W. Choi, The Liposome Partitioning System for Correlating Biological Activities of Imidazolidine Derivatives, *Pharm. Res. 10*: 913–917 (1993).

Saija, A., M. Scalese, M. Lanza, D. Marzullo, F. Bonina, and F. Castelli, Flavonoids as Antioxidant Agents: Importance of Their Interaction with Biomembranes, *Free Radic. Biol. Med. 19*: 481–486 (1995).

Sugiyama, Y., S. Kawakishi, and T. Osawa, Involvement of the β-Diketone Moiety in the Antioxidative Mechanism of Tetrahydrocurcumin, *Biochem. Pharmacol. 52*: 519–525 (1996).

Chapter 33

CimiPure (*Cimicifuga racemosa*): A Standardized Black Cohosh Extract with Novel Triterpene Glycoside for Menopausal Women

Qun Y. Zheng, Kan He, Lynn Pilkington, Yu Shao, and Bolin Zheng

Department of Research and Development, Pure World Botanicals, S. Hackensack, NJ 07606

Introduction

Black cohosh, also called Cimicifuga, black snakeroot, and *Actaeae racemosae radix*, consists of the dried rhizome and roots of *Cimicifuga racemosa*. For a long time, the root of *Cimicifuga* was considered a remedy by Indians in North America and utilized for numerous indications, including internal use for relieving pain during delivery and external use against snake bites. The fluid extract prepared from black cohosh was used in Europe in the last century as an antirheumatic and for fever, chorea, vertigo, spermatorrhoea, dysmenorrhea, and amenorrhea (Jaspersen-Schib 1990). In Japanese oriental medicine, the rhizomes of *Cimicifuga* species are frequently used as antipyretic, analgesic, and anti-inflammatory drugs.

Pharmacologic studies in animals and humans have demonstrated that black cohosh contains substances with endocrine activity (Jarry and Harnischfeger 1985, Jarry *et al.* 1985) and that it is a safe and reasonably effective treatment for hot flashes and other menopausal symptoms in menopausal women. Its estrogenic effect is similar to that of estrogen replacement therapy (HRT), which has been used routinely for treatment of menopause; however, HRT has become increasingly controversial as a result of its numerous disadvantages, including links with endometrial and breast cancers. Therefore, black cohosh could be used as a natural alternative to HRT. The active components of black cohosh are listed in Table 33.1 and the structures are shown in Figure 33.1.

Chemistry and Analysis of the Active Principles

Liquid Chromatography/Mass Spectrometry (LC/MS) Analysis of CimiPure

The most common commercially available preparation of black cohosh is a powdered extract, which is standardized for its triterpene glycoside content (calculated as 27-deoxyactein). With the increasing demand for black cohosh preparations in the United States, it is necessary to develop a good analytical method to quantitate these triterpene glycosides. Currently, there is no analytical method available.

TABLE 33.1
Active Components of Black Cohosh (*Cimicifuga racemosa*)

Component	
Triterpene glycosides	Principally actein, 27-deoxyactein, cimicifugoside, and related aglycones
Isoflavones	Including formononetin (Jarry and Harnischfeger 1985)
Aromatic acids	Including isoferulic acid (IFA), ferulic acid (FA), and salicylic acid

These compounds have very weak absorption in the short wavelength range near 203 nm, making it difficult to detect them for quantification. It was our goal to obtain each pure compound to accurately identify and quantitate the total triterpene saponins with accuracy. In the reversed-phase high-performance liquid chromatography (HPLC) analysis of the extract, we observed that four compounds, actein, 27-deoxyactein, cimicifugoside, and the cimicifugoside isomer were the major saponin constituents of the alcoholic extract. LC/MS/MS methodology with atmospheric pressure chemical ionization (APCI) was used to identify a group of triterpene glycosides in the black cohosh extract CimiPure. As shown in Figure 33.2,

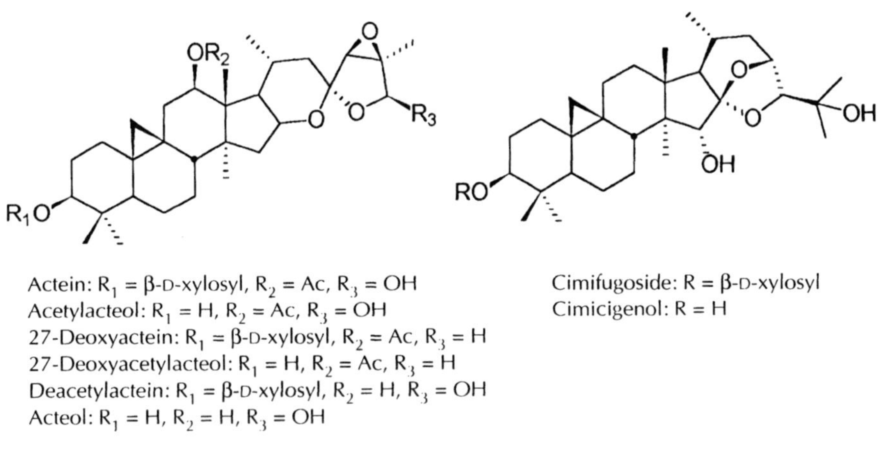

Actein: R_1 = β-D-xylosyl, R_2 = Ac, R_3 = OH
Acetylacteol: R_1 = H, R_2 = Ac, R_3 = OH
27-Deoxyactein: R_1 = β-D-xylosyl, R_2 = Ac, R_3 = H
27-Deoxyacetylacteol: R_1 = H, R_2 = Ac, R_3 = H
Deacetylactein: R_1 = β-D-xylosyl, R_2 = H, R_3 = OH
Acteol: R_1 = H, R_2 = H, R_3 = OH

Cimifugoside: R = β-D-xylosyl
Cimicigenol: R = H

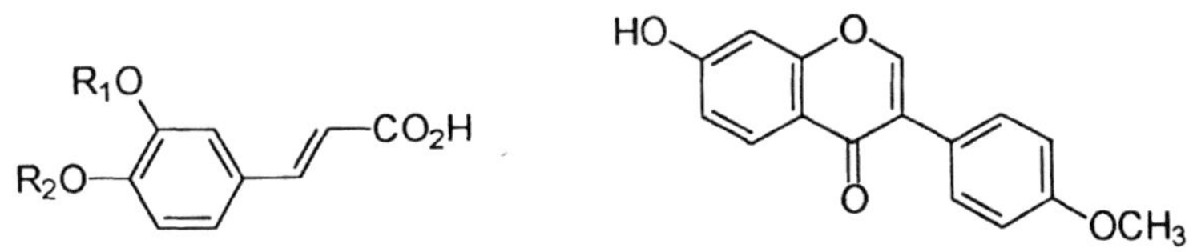

Ferulic acid: R_1 = CH_3, R_3 = H
Isoferulic acid: R_1 = H, R_2 = CH_3

Formononetin

Fig. 33.1. Active components of black cohosh, belonging to the family of triterpene glycosides, isoflavones, and aromatic acids.

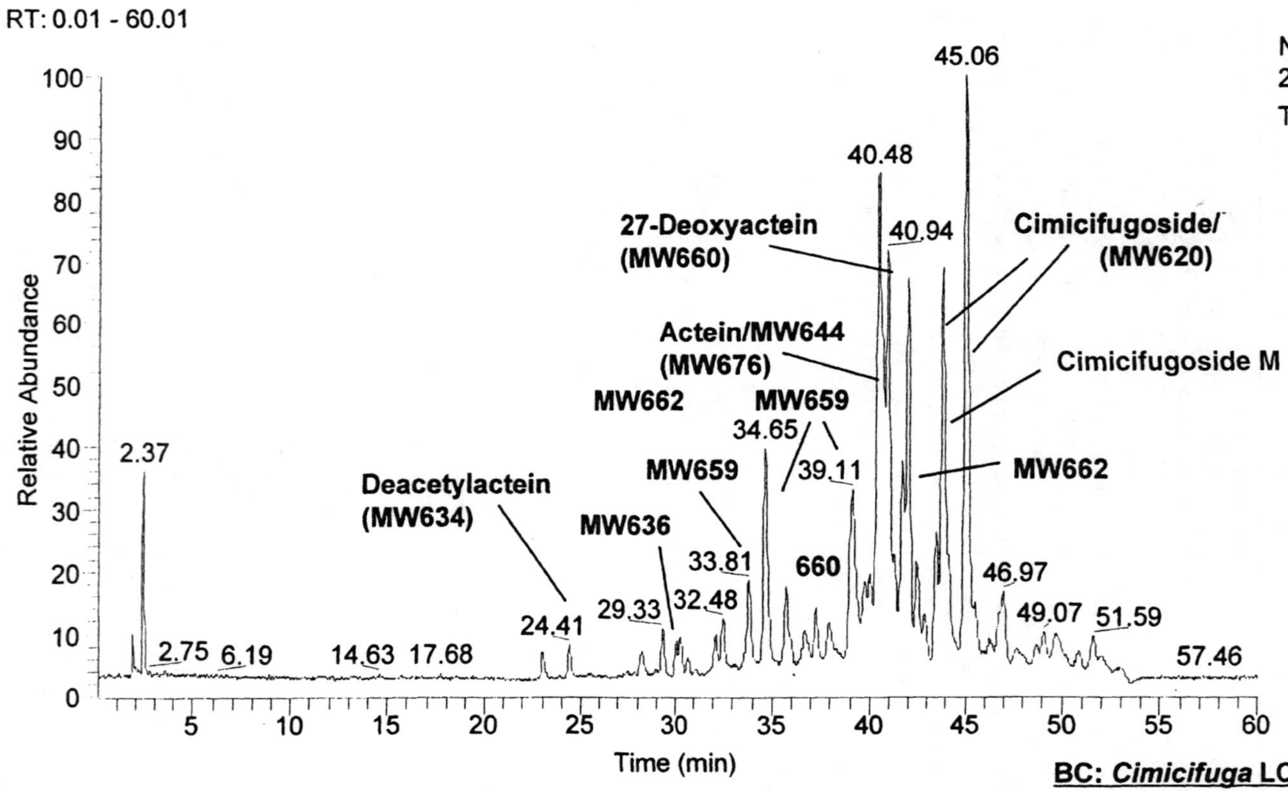

Fig. 33.2. Liquid chromatography-mass spectrometry (LC-MS) chromatogram total ion current (TIC) of *Cimicifuga racemosa* (CimiPure, standardized black cohosh extract, 2.5% total triterpene glycosides).

known triterpene glycosides such as cimicifugoside (620), 27-deoxyactein (MW 660), actein (MW 676), and deacetylactein (MW 634) were present. Some unknown peaks (MW 662, 659, 644, 662, and 636) have fragmentation patterns that are very similar to those of actein, 27-deoxyactein, and cimicifugoside, thus suggesting that these components may be triterpene glycosides as well.

In Figure 33.3, the LC/MS chromatogram and mass spectrum of 27-deoxyactein are presented. The peak at 40.94 min shows a mass spectrum with peaks at 661 $[M+1]^+$, 643 $[M\text{-}H_2O+1]^+$, 601 $[M\text{-AcOH}+1]^+$, 583 $[M\text{-AcOH-}H_2O+1]^+$, 529 $[M\text{-xylose}+1]^+$, 511 $[M\text{-xylose-}H_2O+1]^+$, 469 $[M\text{-AcOH-xylose}+1]^+$, 451 $[M\text{-AcOH-xylose-}H_2O+1]^+$, and 433 $[M\text{-AcOH-xylose-}2H_2O+1]^+$. A proposed fragmentation pattern of 27-deoxyactein is shown in Scheme 33.1. 27-Deoxyactein (MW 661) can lose an acetyl group to form a fragment at 601, which can in turn lose a water molecule to form a fragment at 583. After loss of an acetyl group and a water molecule, this fragment can continue to lose a xylosyl group to form the fragment at 469. The fragmentation can end when the 469 fragment loses a water molecule to give a fragment at 451. 27-Deoxyactein can also lose water directly to form the fragment at 643. The major fragments in this pattern are at 601 and 469, corresponding to the loss of an acetyl group and a xylosyl group, respectively. Loss of an acetyl group and a xylosyl group gives the characteristic fragmentation pattern of 27-deoxyactein.

The result of analysis of cimicifugoside is shown in Figure 33.4. The peak at 45.06 min shows a mass spectrum with peaks at 621 $[M+1]^+$, 603 $[M\text{-}H_2O+1]^+$, 585.4 $[M\text{-}2H_2O+1]^+$, 489.3 $[M\text{-xylose}+1]^+$, 471.4 $[M\text{-xylose-}H_2O+1]^+$, 453.5 $[M\text{-xylose-}2H_2O+1]^+$, and 435.5 $[M\text{-xylose-}3H_2O+1]^+$. A proposed fragmentation pattern of cimicifugoside is shown in Scheme 33.2. Cimicifugoside (MW 621) can lose a xylosyl group to form a fragment at 489. The 489 fragment can then lose four water molecules sequentially to form fragments at 471, 453, 435, and 417. Cimicifugoside can also lose two water molecules directly to form fragments at 603 and 585. Loss of a xylosyl group and a series of water molecules gives the characteristic fragmentation pattern of cimicifugoside. A peak appearing at 43.82 min shows a similar molecular weight ion and fragmentation pattern; therefore, this peak was identified as cimicifugoside M (He *et al.*).

Thin-Layer Chromatography and HPLC Analysis of CimiPure

Thin-layer chromatography (TLC) and HPLC were used to compare CimiPure and Remifemin, known products in the market. Figure 33.5 indicates a TLC spectrum of the 27-deoxyactein standard, CimiPure, and Remifemin. This TLC was run on a silica gel G/ultraviolet (UV) plate (Waterman®, Clifton, New Jersey) with a mobile phase of methylene chloride (CH_2Cl_2) and methanol (MeOH) 92:8 (vol/vol). The detection was done by spraying a solution of 1% $Ce(SO_4)_2$ in 10% H_2SO_4 on the TLC plate and heating it to 105°C for 5 min. For both the CimiPure and Remifemin samples, a concentration of 3 mg/mL of triterpene glycosides was used, and 3 μL of the solution was applied to the plate. As indicated by the TLC, CimiPure, which is very rich in triterpene glycosides, compares favorably to

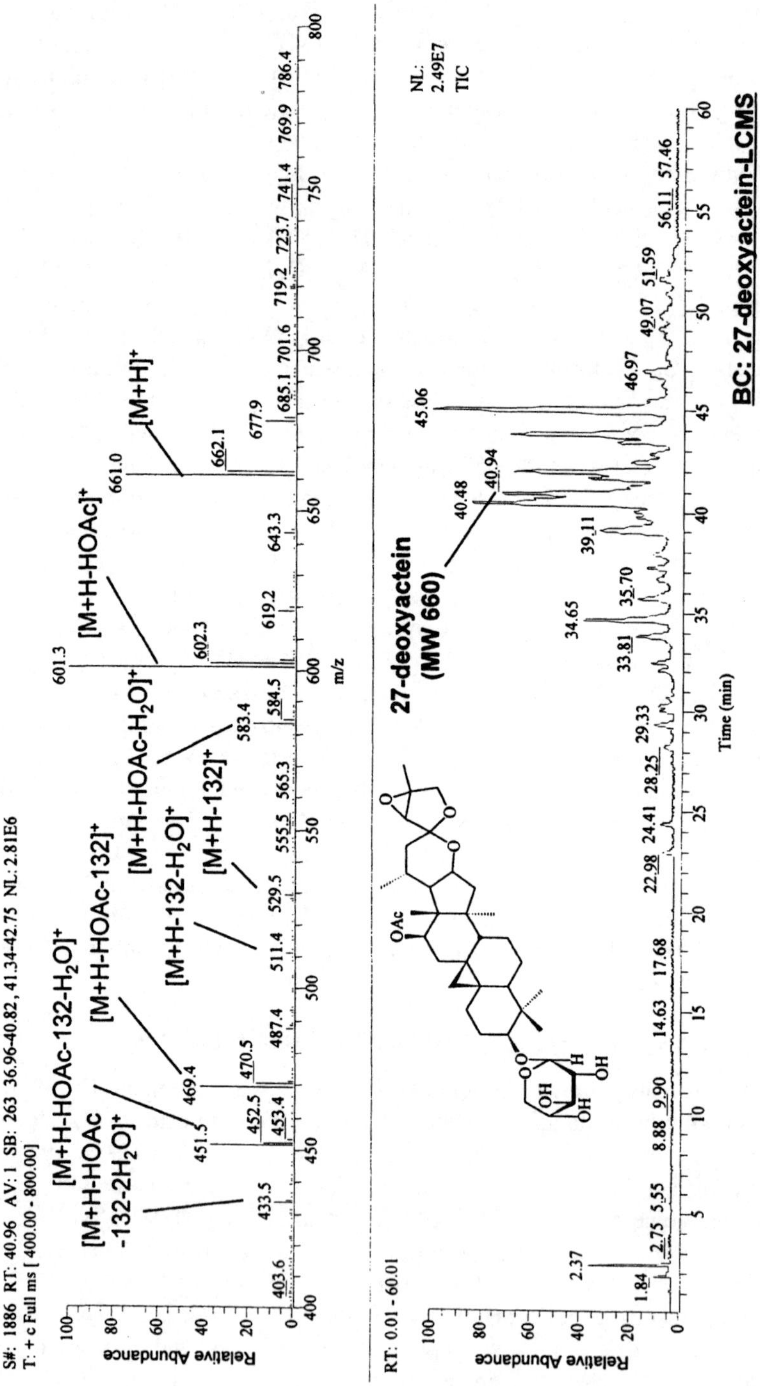

Fig. 33.3. Liquid chromatography-mass spectrometry (LC-MS) chromatogram and mass spectrum of 27-deoxyactein.

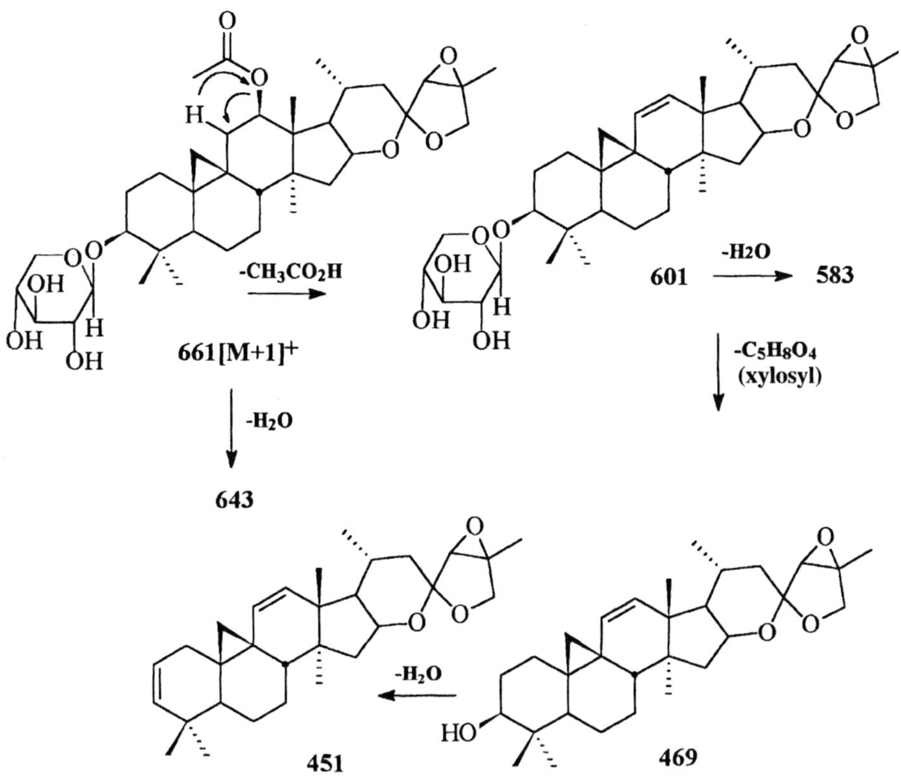

Scheme 33.1. Proposed fragmentation pattern of 27-deoxyacetein by liquid chromatography-mass spectrometry (LC-MS). Loss of an acetic acid (HOAc) group, a sugar, and a series of H_2O molecules gives the characteristic fragmentation pattern of 27-deoxyacetin by atmospheric pressure chemical ionization/MS.

Remifemin. CimiPure and Remifemin showed a very similar fingerprint when they were compared by HPLC as shown in Figure 33.6. The difference in this comparison is that the CimiPure extract has more components. CimiPure is a complete extract with a natural ratio of all the triterpene glycosides. The specifications for CimiPure are given in Table 33.2.

Mechanism of Action

Endocrine Activity

The extract from *Cimicifuga racemosa* reduces the serum concentration of pituitary luteinizing hormone (LH) in menopausal women and ovariectomized rats (Duker 1991). The LH suppressive effect is caused by at least three different synergistically acting compounds as follows: (i) compounds that are not ligands for the estrogen receptors but suppress LH release, (ii) compounds that bind to estrogen

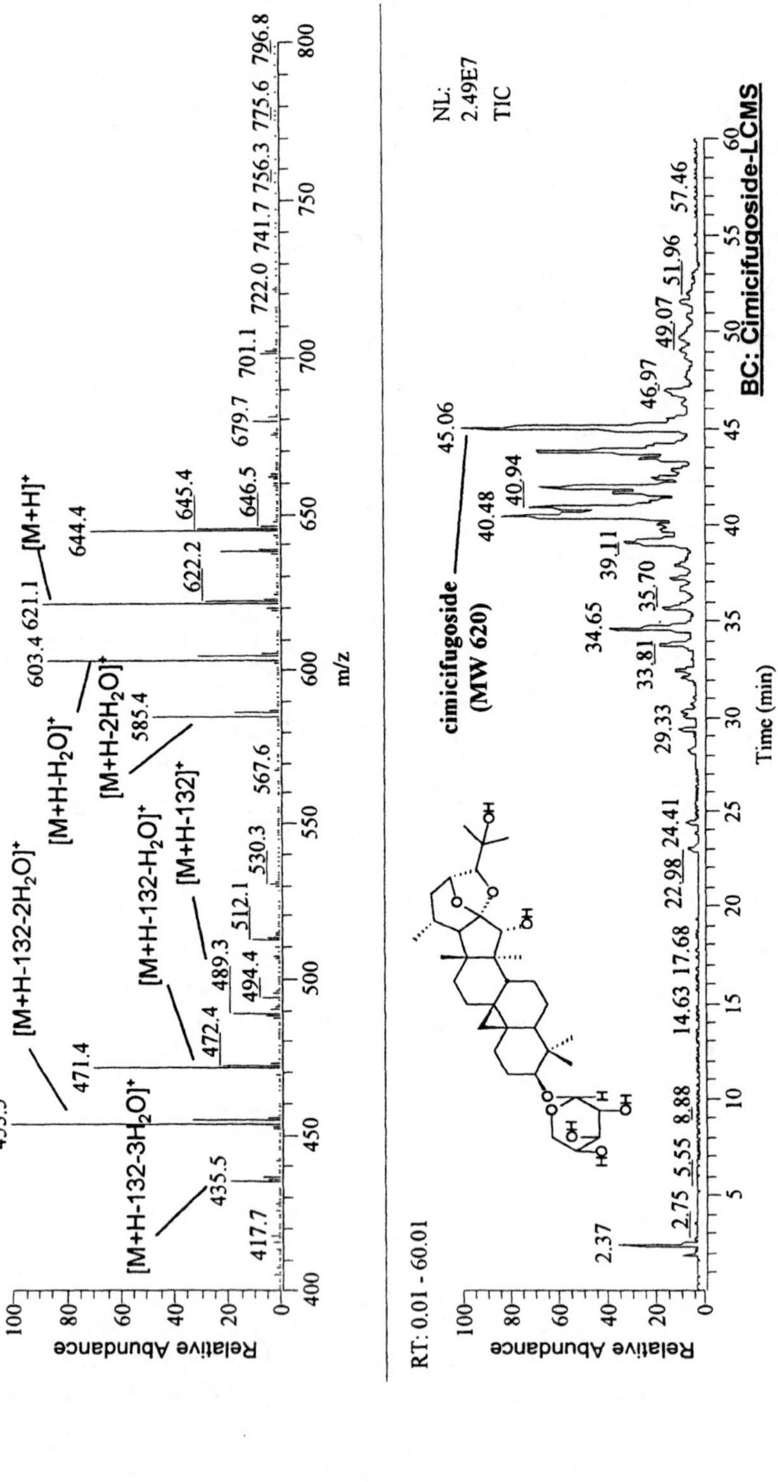

Fig. 33.4. Liquid chromatography-mass spectrometry (LC-MS) chromatogram and mass spectrum of cimicifugoside.

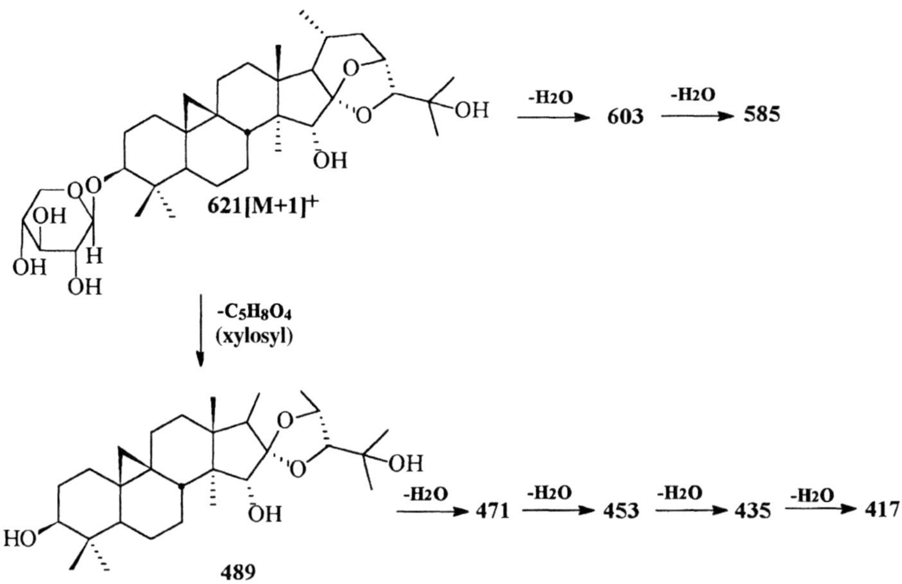

Scheme 33.2. Proposed fragments of cimicifugoside by liquid chromatography-mass spectrometry (LC-MS). Loss of a sugar and a series of H_2O molecules gives the characteristic fragmentation pattern of cimicifugoside by atmospheric pressure chemical ionization/MS.

receptors and also suppress LH release, and (iii) compounds that bind to the estrogen receptors but without any effect on LH release.

The extract exerts an estrogenic activity. The vaginal cytology of menopausal women is affected similarly by Remifemin, a commercially available extract of *Cimicifuga racemosa*, as it is by an estrogen treatment. LH is the major stimulant of gonadal steroid (including estrogens) production. On the other hand, LH secretion could be inhibited by estrogen through negative feedback. Some compounds in *C. racemosa* are able to bind to estrogen receptors, indicating that they are structurally related to estrogen. Thus, the suppressive effect of the extract of *C. racemosa* on LH release is at least in part an expression of its estrogen activity.

Reduced estrogen production and increased LH and follicle-stimulating hormone (FSH) secretions are characteristic endocrine changes of menopause. Consequently, osteoporosis, vasomotor symptoms (hot flashes), and menopausal syndrome (including insomnia and depression) occur in menopausal women. The hot flash is the hallmark of menopause, and it appears to coincide with a surge of LH, not FSH. Black cohosh selectively suppresses LH, but not FSH, release. This indicates that black cohosh is a suitable treatment for hot flashes.

Anti-Inflammatory Effect

The main components responsible for the anti-inflammatory effect of *Cimicifuga* are ferulic acid (FA) and its isomer, isoferulic acid (IFA). Murine interleukin 8 (IL-8)

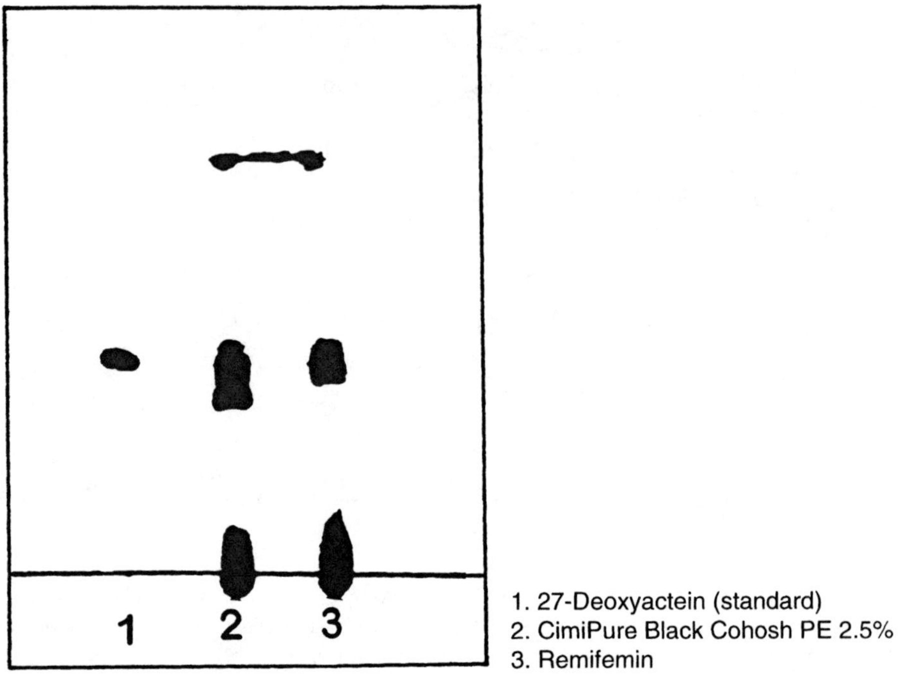

Fig. 33.5. Thin-layer chromatographic (TLC) analysis of black cohosh extracts (CimiPure vs. Remifemin). Conditions: TLC plate, silica gel G/ultraviolet (UV) plate (Waterman); mobile phase, CH_2Cl_2:MeOH (92:8); detection, 1% $Ce(SO_4)_2$ in 10% H_2SO_4 and heated at 105°C for 5 min; sample concentration, both CimiPure black cohosh powdered extract (PE) 2.5% and Remifemin solutions were prepared at 3 mg/mL of triterpene glycosides and a 3-μL aliquot was applied to the TLC plate.

production in response to influenza virus infections was inhibited by FA and IFA *in vitro* as well as *in vivo* (Hirabayashi *et al.* 1995). A similar phenomenon has also been observed with the extract of *Cimicifuga heracleifolia*. IL-8 is a potent chemotactic factor for neutrophils and T lymphocytes. An accumulation of neutrophils plays a proinflammatory role and modulates the inflammatory reaction (Watanabe *et al.* 1989). It should be noted that the FA and IFA exert their anti-inflammatory effect by reducing not only the IL-8 production but also neutrophil infiltration provoked by the influenza virus infection.

Therapeutic Uses

Menopausal Disorders, Particularly Hot Flashes and Depression

Black cohosh is the main component in three commercial European preparations for menopausal disorder. Numerous clinical studies have confirmed its efficacy (Lehmann-Willenbrock and Riedel 1988) and that the effect of *Cimicifuga* was as

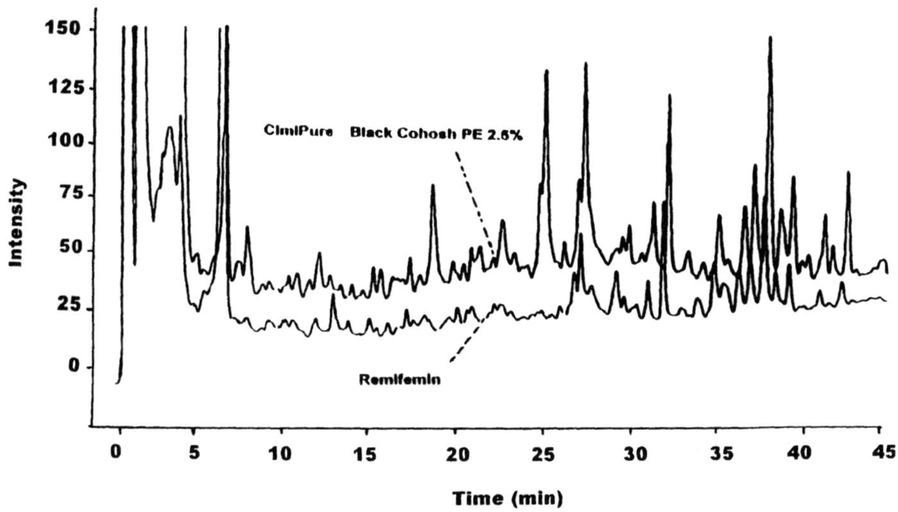

Fig. 33.6. Comparative high-performance liquid chromatography (HPLC) analysis of CimiPure and Reminfemin.

good as a dose of hormone for moderate menopausal symptoms. In 80% of subjects, the symptoms improved with treatment using *Cimicifuga* (Jaspersen-Schib 1990). Other disorders for which *Cimicifuga* is an effective treatment include premenstrual complaints, dysmenorrhoea, and uterine spasm; muscular rheumatism, rheumatoid arthritis, and tinnitus; and influenza virus infections (Shibata *et al.* 1975 and 1977).

TABLE 33.2
Specifications of CimiPure® PE 2.5%

Specification	
Botanical source	Black cohosh (*Cimicifuga racemosa*)
Part of plant used	Roots and rhizome
Description	Brownish powder with characteristic odor and taste
Standardization	2.5% Triterpene glycosides (27-deoxyactein, actein, cimicifugoside, and cimicifugoside M)
Assay method	High-performance liquid chromatography (27-deoxyactein as a standard)
Moisture	<5%
Excipient	Maltodextrin
Particle size	80 mesh (95% through)
Microbiology	
Total aerobic count	Not more than 5000 cfu/g
Escherichia coli	None detectable
Salmonella	None detectable

Dosage

According to the literature review, CimiPure (Pure World Botanical, Inc., South Hackensack, NJ) was designed to provide 2 mg of triterpene glycosides from 80 mg of the extract. A suggested dosage could be 160 mg/d (80 mg extract twice a day). For dried rhizome and root, the dosage would be 40–200 mg/d. Remifemin (Sharper & Brummer GmbH & Co., Germany) claims that each tablet contains 1 mg of triterpene glycosides from *Cimicifuga racemosa* extract. A dose of 4 mg triterpene glycosides per day is suggested (2 tablets each time, 2 times per day). Femicin (Amidro, Biel, Switzerland) contains 25 mg of dried extract of *Cimicifuga* per tablet. A suggested dose is 50 mg/d (1 tablet each time, 2 times per day). The average duration of treatment could be 2–6 mo.

Cimicifuga racemosa is not toxic in therapeutic doses; however, occasional gastrointestinal disturbances occur. Contraindications include pregnancy, lactation, and estrogen-dependent tumors, as well as heart disease.

References

Duker, E.M., L. Kopanski, H. Jarry, and W. Wuttke, Effects of Extract from *Cimicifuga racemosa* on Gonadotropin Release in Menopausal Women and Ovariectomized Rats, *Planta Med. 57:* 420–424 (1991).

Hirabayashi, T., H. Ochiai, S. Sakai, K. Nakajima, and K. Terasawa, Inhibitory Effect of Ferulic Acid and Isoferulic Acid on Murine Interleukin-8 Production in Response to Influenza Virus Infections *in vitro* and *in vivo*, *Planta Med. 61:* 221–226 (1995).

Jarry, H., and G. Harnischfeger, Studies on the Endocrine Effects of Contents of *Cimicifuga racemosa* 1. Influence on the Serum Concentration of Pituitary Hormones in Ovariectomized Rats, *Planta Med. 51:* 46–49 [German] (1985).

Jarry, H., G. Harnischfeger, and E. Duker, Studies on the Endocrine Effects of the Contents of *Cimicifuga racemosa* 2. *In vitro* Binding of Compounds to Estrogen Receptors, *Planta Med. 51:* 316–319 [German] (1985).

Jaspersen-Schib, R., Cimicifuga et Troubles de la Ménopause, *Schweiz. Apoth. Ztg. 128:* 155–157 (1990).

Lehmann-Willenbrock, E., and H.H. Riedel, Clinical and Endocrinologic Studies of the Treatment of Ovarian Insufficiency Manifestations Following Hysterectomy with Intact Adnexa, *Zentbl. Gynaekol. 110:* 611–618 [German] (1988).

Shibata, M., N. Sakurai, and M. Onoda, Pharmacological Studies on the Chinese Crude Drug "Shoma." II. Anti-Inflammatory Action of *Cimicifuga rhizoma*, *Cimicifuga simplex* wormsk, *Yakugaku Zasshi* (*J. Pharm. Soc. Jpn.*) *97:* 911–915 [Japanese] (1977).

Shibata, M., Y. Yamatake, Y. Amagaya, and M. Fukushima, Pharmacological Studies on the Chinese Crude Drug "Shoma." I. Acute Toxicity and Anti-Inflammatory Action of *Cimicifuga rhizome, Cimicifuga dahurica* maxim, Yakugaku Zasshi (*J. Pharm. Soc. Jpn.*) *5:* 539–546 [Japanese] (1975).

Watanabe, K., K. Konishi, and M. Fujioka, The Neutrophil Chemoattractant Produced by the Rat Kidney Epithelioid Cell Line NRK-52E Is a Protein Related to the KC/gro Protein, *J. Biol. Chem. 264:* 19559–19563 (1989).

Chapter 34

Natural Products of Agrochemical Interest from Mexican Labiatae

Baldomero Esquivel[a], Ana Adela Sánchez[a], and Eduardo Aranda[b]

[a]Instituto de Química de la Universidad Nacional Autónoma de México, Circuito Exterior de Cd. Universitaria Coyoacán 04510, México, D.F.

[b]Centro de Investigación en Biotecnología Universidad Autónoma del Estado de Morelos, Cuernavaca, Morelos, México

Introduction

The Labiatae (Lamiaceae) is a widespread and diversified family of plants with 224 genera and ~4000 species. The members of the family are found in the tropics, subtropics, and temperate parts of the world. The areas of high diversity in Lamiaceae are the Mediterranean, Central Asia, the Americas, the Pacific Islands, tropical Africa, and China. From the taxonomic point of view, the diversity of the family in the Mediterranean area and Central Asia suggests the possibility that the progenitors of the present-day Lamiaceae may have originated in these parts of the world during the early to mid-Cretaceous period (140–100 million years ago) (Hedge 1992).

One of the most distinguishing features of Lamiaceae is the terpene-containing glands found in several members of the family. These glands are responsible for their aromatic odor; as a consequence, some members of the family are used in folk medicine and in food preparation in several parts of the world, including Mexico, although the roots of some of these plants are also used frequently as traditional medicine (Ramamoorthy and Elliot 1993).

From the chemical point of view, the family Lamiaceae has been subjected to intensive studies focused on the isolation of several types of compounds. Plants of different genera, including *Ajuga, Rabdosia, Teucrium, Salvia, Scutellaria, Stachys, Leonorus, Ballota*, and *Coleus*, have been studied intensively. These studies have led to the isolation of a large number of secondary metabolites such as sterols, flavonoids, iridoids, sesquiterpenes, diterpenes, triterpenes, and sesterterpenes (Esquivel *et al.* 1996).

Some of the secondary metabolites isolated from Lamiaceae plants have shown interesting biological functions, including antispasmodic, antiviral, and antibacterial activities, for example. Antifeedant activity against economically important insects has been observed as well for some neo-clerodane diterpenoids isolated from several members of the family. Therefore, these plants are of medicinal and agrochemical interest (van Beek and de Groot 1986).

This chapter provides a brief summary of our recent chemical findings regarding the *Salvia* and *Scutellaria* genera. A survey of the indirect insecticidal activity (antifeedants, moulting disruptors, and oviposition inhibitors) found for some neo-clerodane diterpenoids isolated from these plant genera is presented.

Recent Advances in the Chemistry of Mexican Salvias

The genus *Salvia* L. (Tribe Mentheae) is the largest of the Lamiaceae with 900 species widespread throughout the world. These plants were organized by Bentham (1876) into four subgenera (Salvia, Sclarea, Leonia, and Calosphace); however, recent taxonomic and chemical data suggest a reconsideration of this classification. Table 34.1 shows a very acceptable organization of these plants based on recent data. This new classification includes the subgenus Audibertia (Harley and Heywood 1992), which comprises the Californian and Baja Californian Salvias. The name of the Calosphace subgenus for Jungia has been changed, based on historical reasons (Wood and Harley 1989). The correct distribution of the subgenera Sclarea and Leonia is also discussed. Subgenus Sclarea is present not only in Asia and Europe, but also in Mexico; Leonia is found in Europe, Asia, and Africa and not only in North America as was erroneously stated in various earlier phytochemical publications (Rodriguez-Hahn *et al.* 1994, and references cited therein).

Phytochemical studies of several *Salvia* species have established an interesting chemical profile for the genus. Sterols, such as β-sitosterol and stigmasterol, and flavonoids are frequently found in the aerial parts of these plants. Triterpenes derived from lupane, oleanane, dammarane, and ursane skeletons and some phenolic compounds have also been isolated. The most diversified and perhaps the most characteristic secondary metabolites of the genus are the diterpenoids. Over 200 abietane and rearranged abietane diterpenoids have been described for species belonging to all of the subgenera in both aerial parts and roots. The major concentration and diversity of these types of metabolites are represented in the subgenera Salvia and Sclarea. The diterpenoids isolated from the *Salvia* species from Turkey, which belong to subgenera Salvia, Sclarea, and Leonia, are good representatives of this structural diversity. To

TABLE 34.1
Classification and Distribution of Genus *Salvia* at the Subgenus Level

Subgenus	Distribution	Reference
Audibertia	California (USA) and Baja California (Mexico)	Epling 1938
Jungia (formerly Calosphace)	Mexico, Central and South America	Epling 1939 Wood and Harley 1989
Leonia	Asia, Europe, Africa, and North America	Bentham 1876
Salvia	Asia and Europe	Bentham 1876
Sclarea	Asia, Europe, and Mexico	Bentham 1876

date, >100 neo-clerodane diterpenoids and related compounds have been isolated, mainly from species belonging to the subgenus Jungia (Calosphace) (Hanson 1999).

One of the most distinguishing features of the *Salvia* genus is the number of rearranged diterpene skeletons isolated from these plants. To date, 28 rearranged diterpene skeletons have been described, one derived from a probable pimarane precursor and remainder almost equally of clerodanic and abietanic origin.

An interesting relationship among the diterpenoid content, geographical distribution, and botanical classification of these plants has been established regarding subgenera, following Bentham classification. With the consideration of the subgenus Audibertia and the distribution shown in Table 34.1, a more natural relation could be proposed. As a result of the phytochemical studies of several Mexican *Salvia* spp. (subgenus Jungia) a close relationship between the diterpenoid content of species and their taxonomic section was proposed in an earlier stage of these studies. However, with the present data, it becomes clear that botanical sectional differentiation does not correlate with chemical composition.

In spite of the contribution of several groups throughout the world, only ~15% of the Mexican *Salvia* spp. have been studied. The following paragraphs describe our recent phytochemical results obtained from the study of some *Salvia* species, as well as their probable chemosystematic and evolutionary significance.

From a population of *Salvia mexicana* var. *mexicana* (sect. Briquetia) (Epling 1939), we isolated a new languidulane-type diterpenoid (**1**), whose structure was established on the basis of spectroscopic methods and confirmed by X-ray analysis. Compound **1**, named salvimexicanolide, possesses an A/B *cis* ring fusion and is the first example of a *cis*-languidulane derivative. *Trans*-languidulane derivatives are frequently found in species of the botanically unrelated sections Angulatae (Sánchez *et al.* 1995) and Polystachyae (Esquivel *et al.* 1988).

The chemical analysis of *Salvia xalapensis* Benth. (sect. Angulatae) (Epling 1939) (Scheme 34.1) led to the isolation of the languidulane derivatives, salvixalapenolide (**2**) and salvisousolide (**3**). Although compound **2** is new, salvisousolide (**3**) had been isolated previously from *Salvia sousae* Ramam. (Jungia sect. Polystachyae) (Esquivel *et al.* 1988). The salvigenane-type diterpenoids, **4** and **5**, were also isolated from this species. Salvigenolide (**4**), previously known from *Salvia fulgens* Cav. (sect. Fulgentes) (Esquivel *et al.* 1985), is the first salvigenane diterpenoid isolated. The structure of the new derivative (**5**) was established by spectroscopic methods and comparison with **4**. Other salvigenane diterpenoids are also known from *Salvia leucantha* Cav. (sect. Albolanatae) and *S. lavanduloides* (sect. Lavanduloideae) (Esquivel *et al.* 1994).

Subgenus Audibertia is represented in Mexico by 12 species, growing in the Baja Californian peninsula (Epling 1938). Recently, some species belonging to this subgenus, such as *Salvia munzii* (Luis *et al.* 1993 and references cited therein), *S. apiana* (Luis *et al.* 1996), and *S. mellifera* (Luis and San Andrés 1993) have been analyzed extensively; the chemical composition that was determined indicated a close relationship with some Asiatic species. Our interest in other Mexican representatives of this subgenus prompted us to undertake the analysis of *Salvia pachy-*

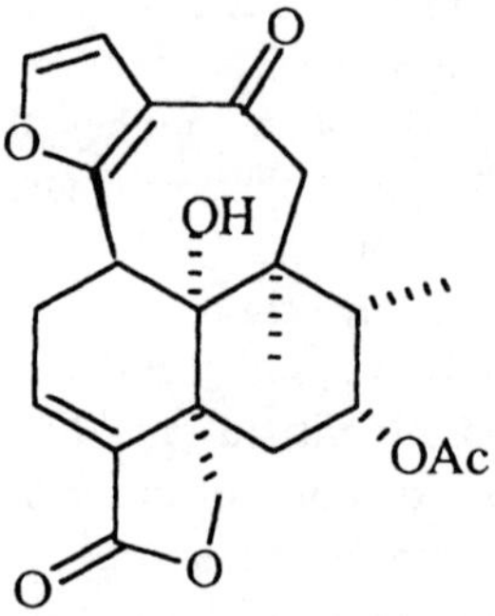

(**1**) Salvimexicanolide

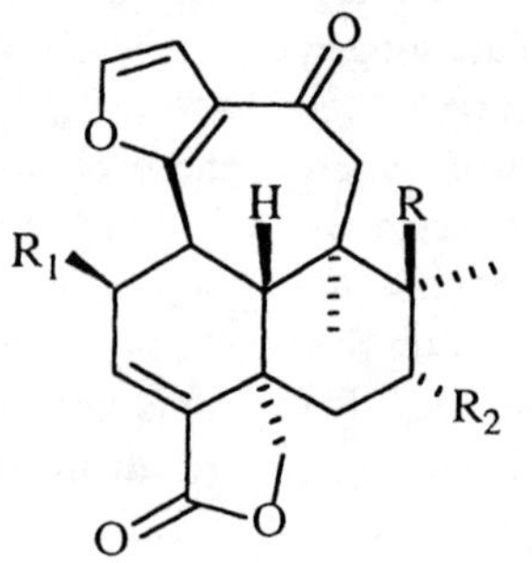

(**2**) R = OH, $R_1 = R_2$ = H Salvixalapenolide
(**3**) R = H, R_1 = OH, R_2 = OAc, Salvisousolide

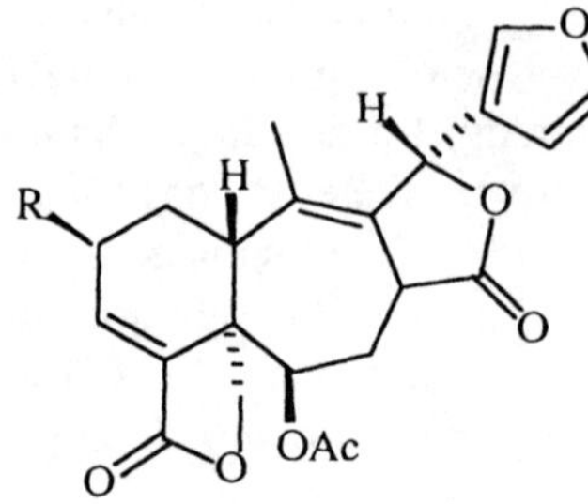

(**4**) R = H, Salvigenolide
(**5**) R = OH, 2β-hydroxy-salvigenolide

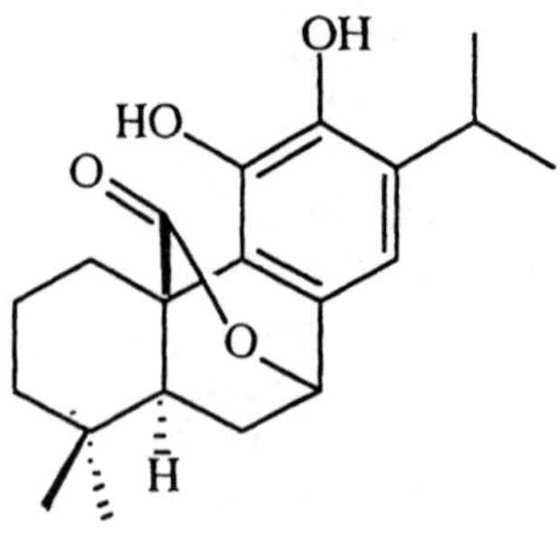

(**6**) Carnosol

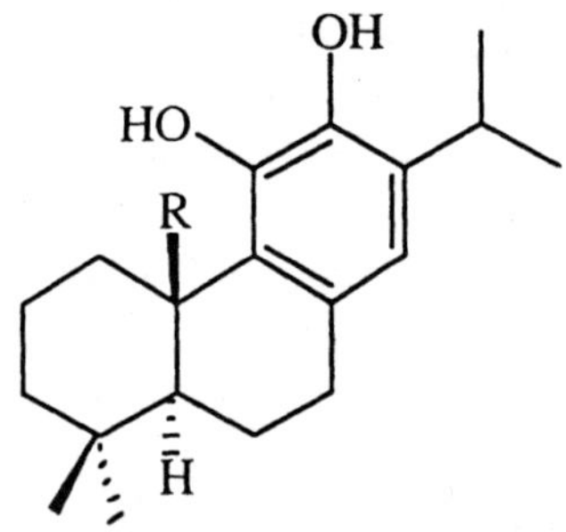

(**7**) R = CO_2H; Carnosic acid

Scheme 34.1. Diterpenoid constituents of *Salvia xalapensis* and *S. pachyphylla.*

phylla Epl. (sect. Jepsonia). From the aerial parts of this species, we isolated carnosol (**6**) and carnosic acid (**7**). These compounds have been isolated previously from other species of the same subgenera and from *Salvia canariensis* (subgenus Salvia, sect. Hymenosphace) (González *et al.* 1991 and references cited therein).

A very interesting chemical finding is the presence of cryptotanshinone (**8**) in both aerial parts and roots of *Salvia axillaris*, which belongs to section Axillares. This section is the most primitive of the Jungia subgenus and is closer to Audibertian plants (Esquivel *et al.* 1997a). Compound **8** is a C-20 *nor*-abietane derivative and was isolated previously from *Salvia miltiorrhiza* Bunge (subgenus Salvia, sect. Drymosphace), a Chinese medicinal plant (Chang *et al.* 1990 and references cited therein), and recently from some species classified in subgenus Audibertia, such as *Salvia apiana* (González *et al.* 1992), *S. mellifera* (Luis *et al.* 1993), and *S. munzii* (Luis and Grillo 1993). This geographical distribution and the presence of **8** in these plants could be considered as chemical evidence of a probable evolutionary link between Asian, Californian, and Mexican Salvias.

Salvia anastomosans Ramam. is an endemic species from the south of Mexico and was recently classified in section Tomentellae. From the roots of this species, we isolated several abietane derivatives (**9–12**). Cariocal (**9**) is the first example of a 6,7-seco-abietane diterpenoid from Mexican Salvia and is also present in the roots of the Mexican endemic species *Salvia aspera* M. *et* G., which belongs to section Conzattiana (Esquivel *et al.*, unpublished results). Other seco-abietane derivatives (**13** and **14**) have also been isolated from *Salvia miltiorrhiza* Bunge (Haro *et al.* 1991) and recently from several plants of the Audibertia subgenus (Luis *et al.* 1994). The geographical distribution found for this type of compound is the same as that observed for cryptotanshinone; this fact reinforces the evolutionary link previously proposed. From the roots of *S. anastomosans* we also isolated arucadiol (miltiodiol) (**12**). This C-20 *nor*-abietane derivative was also present in *Salvia militiorrhiza* and in Californian plants as in the case of cryptotanshinone and the 6,7-seco-derivatives (Scheme 34.2).

These findings are of phytogeographical and evolutionary significance and indicate a close chemical relation between some Chinese and American Salvias. Although some botanical similarities between some areas of vegetation in Mexico and Asia (China) are well documented (Raven and Axelrod 1974), to the best of our knowledge, this is the first time that this relationship is chemically supported for *Salvia* genus. The same chemical relation has been found in some Mexican *Scutellaria* spp., such as *Scutellaria drummondii* (Esquivel *et al.* 1995), *S. caerulea, S. seleriana* (Esquivel *et al.* 1998), and *S. guatemalensis* (Esquivel *et al.* 1996). These findings also suggest that some elements of the Mexican Labiatae could have migrated in the past from the Asiatic continent through North America *via* the Bering Strait and then to Mexico and South America. The recent isolation of ferulic acid depsides, commonly found in Chinese species, from the Mexican endemic species *Salvia adenophora* reinforces this chemical relation (Esquivel *et al.* 1996).

Neo-Clerodane Diterpenoids in Mexican Scutellaria Plants

Scutellaria L. is a large subcosmopolitan genus of the Labiatae with ~360 currently recognized species (Paton 1990a). Recently, plants belonging to this genus have attracted much attention due to the interesting biological activities found for some of the neo-clerodane diterpenoids isolated from them, in particular as insect antifeedants (Anderson *et al.* 1989) and against plant pathogenic fungi (Cole *et al.* 1991). In Mexico, *Scutellaria* is represented by ~32 species, most of which grow in the mountains near the center of the country (Epling 1942). Recently, we reported the isolation of several neo-clerodane diterpenoids from *Scutellaria drummondii* (Esquivel *et al.* 1995). As part of our ongoing search for diterpenoids from plants of the Labiatae with potential antifeedant activity, we report on our recent chemical findings in the analysis and the structure of additional neo-clerodane diterpenoids isolated from *Scutellaria caerulea* Moc. *et* Sessé, *S. guatemalensis* Leon, and *S. seleriana* Loesen.

From *Scutellaria caerulea*, we isolated five new neo-clerodane diterpenoids (**15–19**) with an oxidation pattern closely related to that found in the diterpenoids

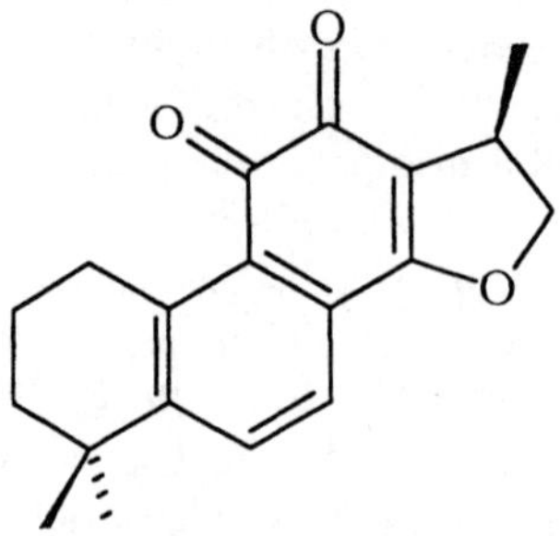

(**8**) Cryptotanshinone

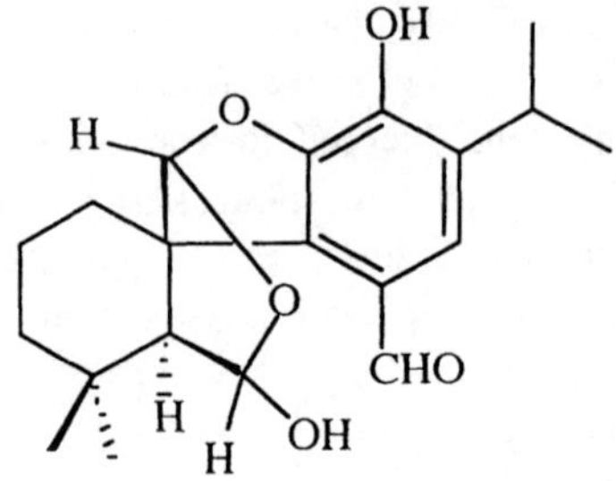

(**9**) Cariocal

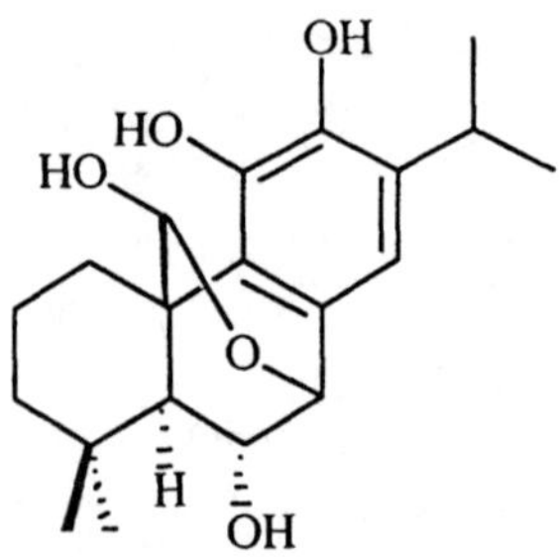

(**10**)

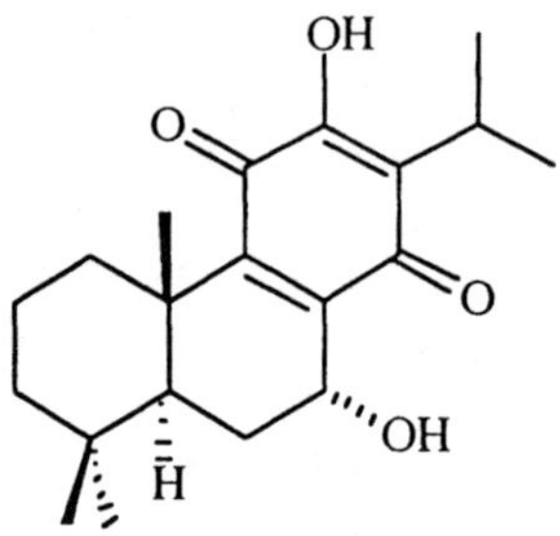

(**11**)

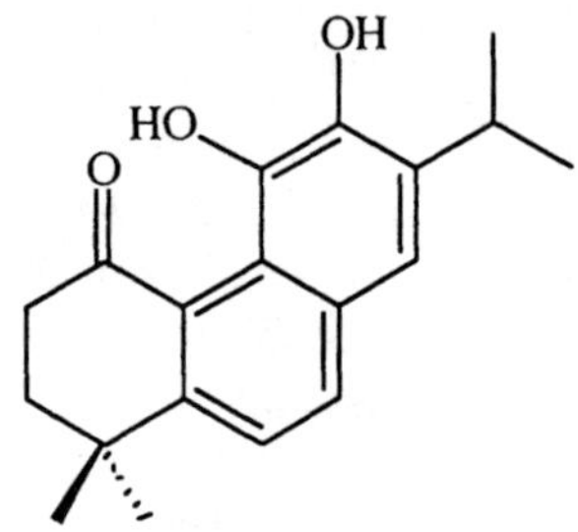

(**12**) Arucadiol

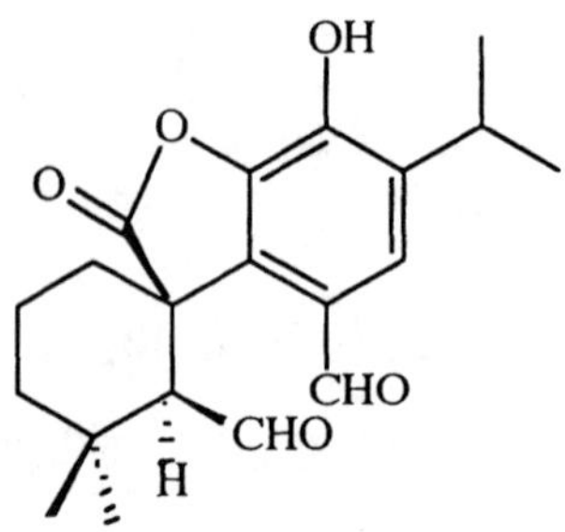

(**13**)

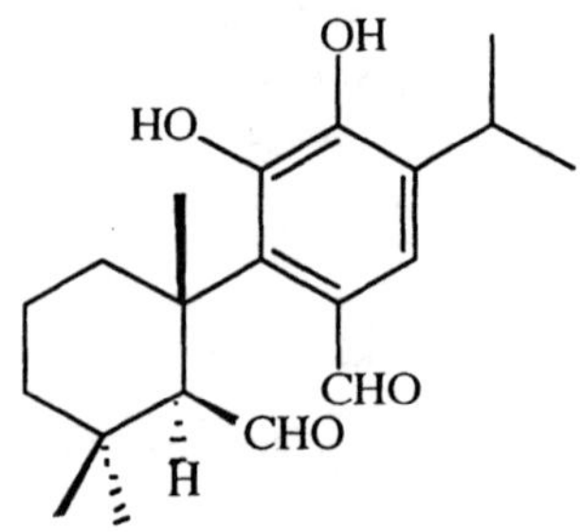

(**14**)

Scheme 34.2. Abietane and rearranged abietane diterpenoids of evolutionary significance from *Salvia* spp.

isolated from the Chinese species *Scutellaria rivularis*. From *Scutellaria guatemalensis* Leon, the scuteguatemalines (**20–23**) were isolated. The structures of these compounds were established by spectroscopic methods, including heteronuclear multiple-bond correlation (HMBC), heteronuclear multiple-quantum correlation (HMQC), and nuclear Overhauser effect spectroscopy (NOESY) spectra. A common feature of compounds **20–23** is the boat conformation of the A-ring. This arrangement was firmly established by the NOESY spectrum and is due to the spatial proximity of the substituents located at C-1 and C-11 (Esquivel *et al.*

1997b). The oxidation pattern of scuteguatemalines is similar to that found for **15–19**. Scuteselerin (**24**), was the sole neo-clerodane diterpenoid isolated from *Scutellaria seleriana* Loesen (Scheme 34.3). The A-ring of this compound adopts a boat conformation as in the case of **20–23**. This fact is illustrated in Figure 34.1.

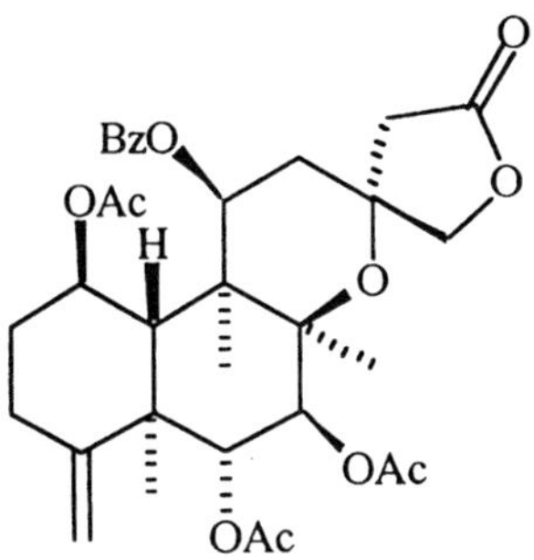

(**15**) Scuterulein A

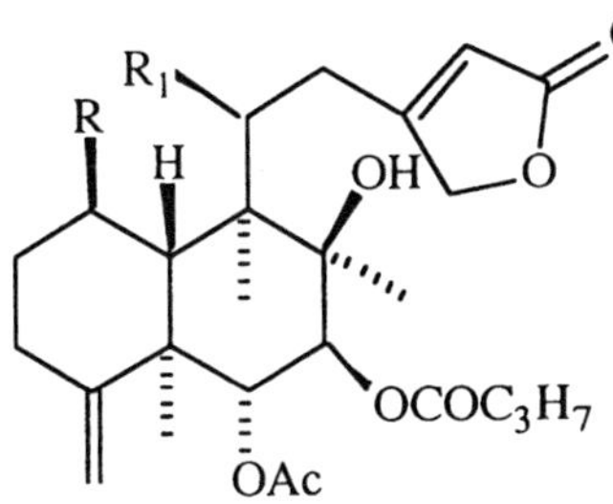

(**16**) Scuterulein B, R = R_1 = OAc
(**17**) Desacetilscuterulein B, R = H, R_1 = OAc
(**18**) Scuterulein C, R = OH, R_1 = OisoBut

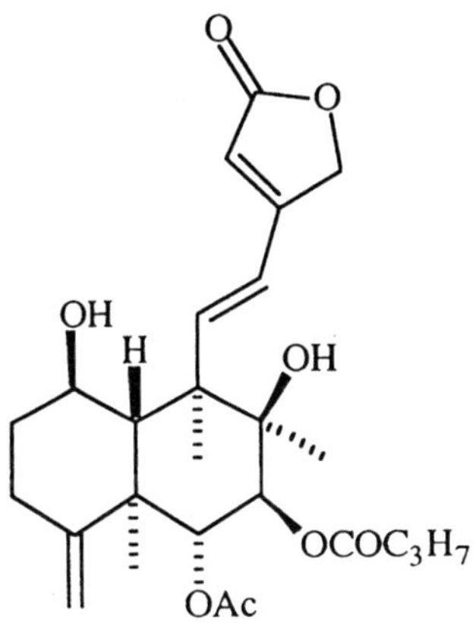

(**19**) Scuterulein D

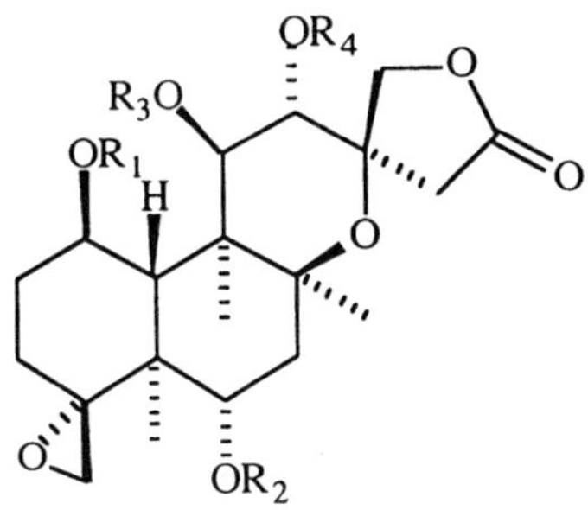

(**20**) R_1 = isoBut, R_2 = Ac, R_3 = Ac, R_4 = isoBut
(**21**) R_1 = isoBut, R_2 = Ac, R_3 = Ac, R_4 = Bz
(**22**) R_1 = Bz, R_2 = R_3 = R_4 = Ac
(**23**) R_1 = isoBut, R_2 = R_3 = R_4 = Ac

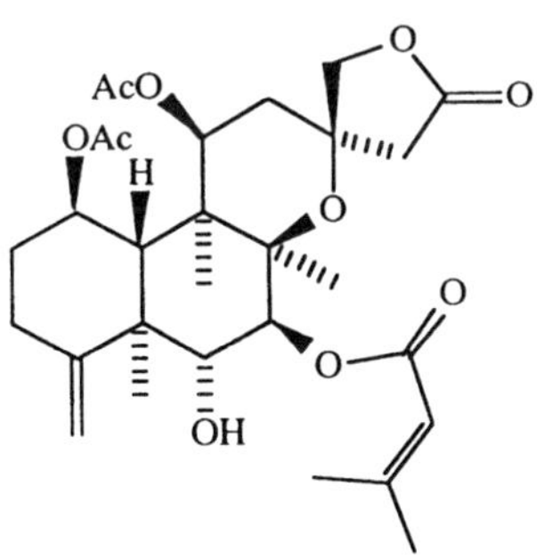

(**24**) Scuteselerin

Scheme 34.3. Neo-clerodane diterpenoids from Mexican *Scutellaria* spp.

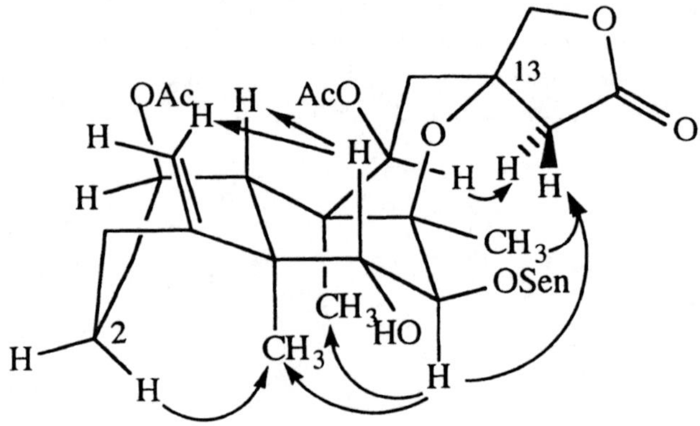

Fig. 34.1. Nuclear Overhauser effect spectroscopy (NOESY) correlations for Scuteselerin.

The oxidation pattern of the neo-clerodane diterpenoids isolated from Mexican Scutellarias in comparison with that found for some compounds from Chinese representatives of this genus, indicates a phytogeographical and perhaps an evolutionary relation between some Asiatic and American *Scutellaria* species as proposed before for the *Salvia* genus. This relation has been proposed previously, based on both botanical considerations and the global distribution of the genus (Paton 1990b). The evaluation of the biological activity of these diterpenes as insect antifeedants is in progress.

Antifeedant Activity of Neo-Clerodane Diterpenoids from Mexican Salvia Species

Several neo-clerodane diterpenoids, isolated from Labiatae plants, have shown antifeedant activity against economically important insects including *Spodoptera littoralis*, *S. exempta*, *S. litura,* and *Bombix mori*. Ajugarin I (**25**), from the African plant *Ajuga remota* (Labiatae), and Jodrellin B (**26**) a constituent of *Scutellaria woronowii*, are two examples (Rodríguez-Hahn *et al.* 1994) (Scheme 34.4). Compound **26** is considered the most potent antifeedant neo-clerodane diterpenoid yet found against *Spodoptera littoralis* (Anderson *et al.* 1989).

Recently, some neo-clerodane diterpenoids isolated from Mexican Salvias have been evaluated as insect antifeedants against *Spodoptera littoralis*. Several diterpenes from *Salvia lineata* (**27**), *S. ryacophylla* (**28** and **29**), *S. melissodora* (**30**), and *S. keerlii* (**31**) exhibited a potent activity with 1(10)-dehydrosalviarin (**27**) demonstrating high activity even at 10 ppm (see Scheme 34.5) (Simmonds *et al.* 1996).

Compound **27** deserves special mention. This compound is also very active against the adult western corn root worm, *Diabrotica virgifera virgifera* Le Conte, at very low concentrations. In this bioassay 1(10)-dehydrosalviarin (**27**) approaches azadirachtin (**32**) in antifeedant potency as illustrated in Scheme 34.6 (Mullin 1994).

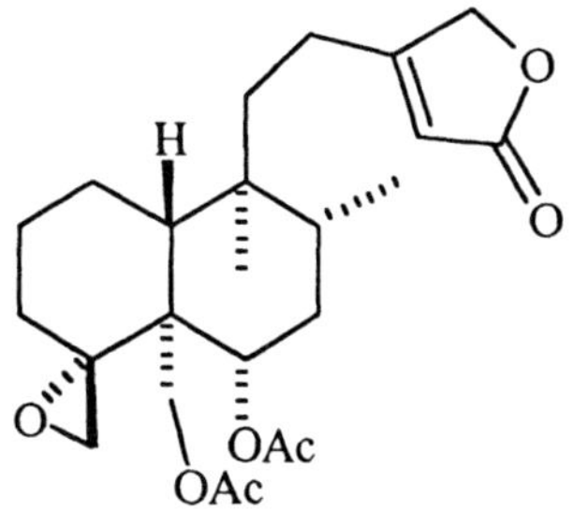

(**25**) Ajugarin I

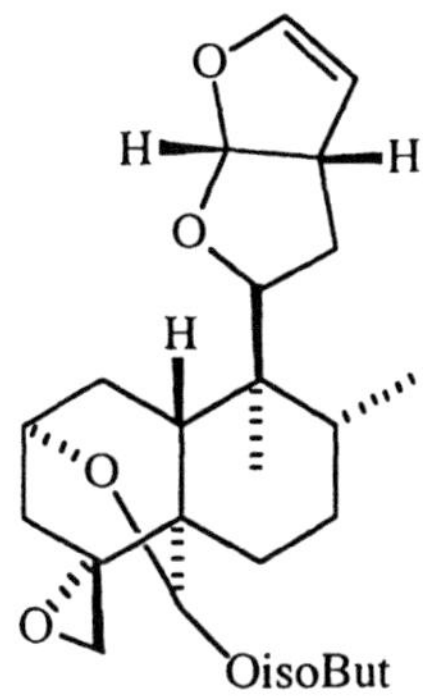

(**26**) Jodrellin B

Scheme 34.4. Antifeedant neo-clerodane diterpenoids from Labiatae plants.

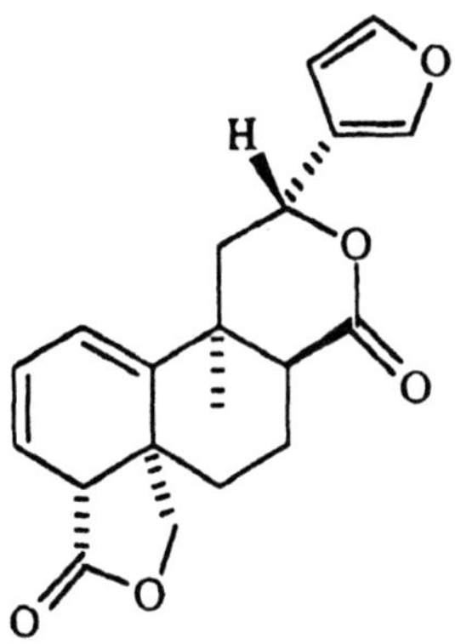

(**27**) 1(10)-dehidrosalviarin,
M = 66 *Salvia lineata*

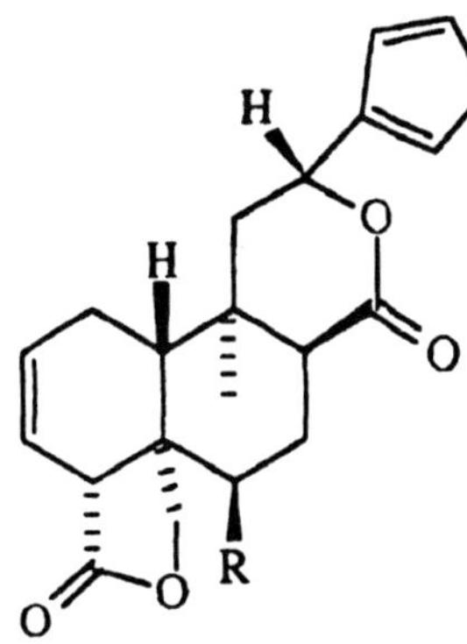

(**28**) R = H Salviarin, M = 59
(**29**) R = OH 6β-hydroxysalviarin,
M = 66 *Salvia ryacophilla*

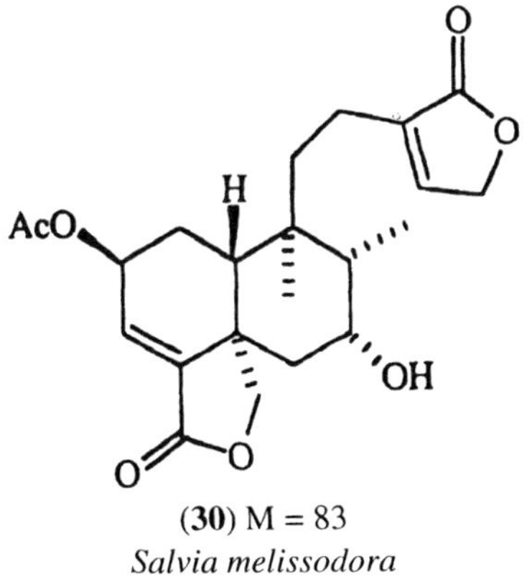

(**30**) M = 83
Salvia melissodora

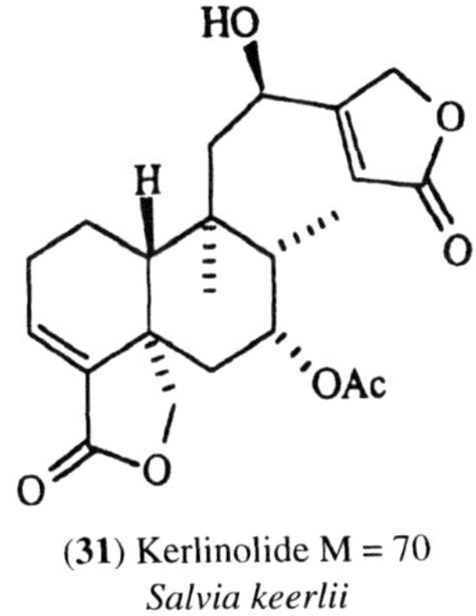

(**31**) Kerlinolide M = 70
Salvia keerlii

Scheme 34.5. Antifeedant activity of some neo-clerodane diterpenoids from Mexican Salvia species at 100 ppm choice bioassay. M = mean antifeedant index [(C-T)/(C+T)] · 100 at 100 ppm. A value >50 indicates potent activity.

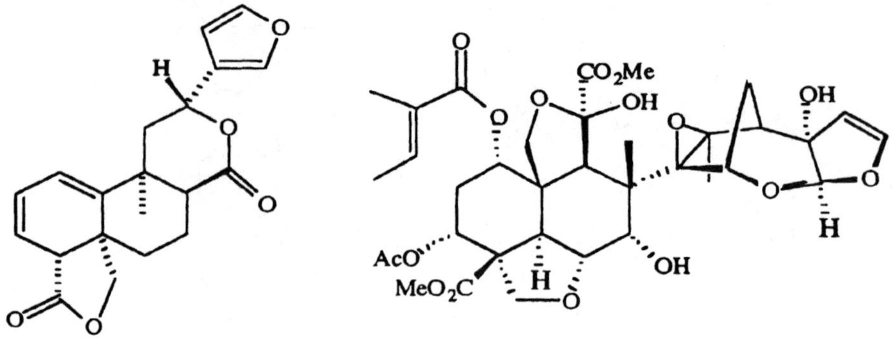

Scheme 34.6. Antifeedant activity of 1 (10)-dehydrosalviarin against adult western corn rootworm *Diabrotica virgifera virgifera* Le Conte. For (**27**) the 50% effective dose (ED_{50}) is 1.1 nmol/disk; for (**32**), the ED_{50} is 0.39 nmol/disk.

Although compound **32** is almost three times more active than **27**, the difference in structural complexity makes 1(10)-dehydrosalviarin an interesting prototype for the development of new agents useful for the control of *Diabrotica virgifera virgifera.*

Although these results are very interesting, neither *Spodoptera littoralis* nor *Diabrotica virgifera virgifera* are important plagues in Mexico. One of the most important pests in Mexico and some Latin American countries is *Spodoptera frugiperda* (fall army worm) (Hernández 1988). This insect attacks corn plants and 87 commercially important crops including ornamental plants (Brown and Dewhurst 1975). In 1995, this insect, together with *Diatrea grandiosella* and *Rhopalosiphum maidis*, was responsible for the loss of almost 80% of the corn crops in Mexico. In some parts of the country, some toxic insecticides that are commercially available are still used to control this insect. In Mexico, as in other third world countries, it is imperative to develop new, less toxic, and ecologically safe crop protective agents. On these bases, we started the evaluation of the neo-clerodane diterpenoids that we isolated from *Salvia* spp. against *Spodoptera frugiperda.*

Antifeedant activity is, of course, very important as a part of an integrated pest management program and has been established for a plethora of different types of compounds, but is not the only alternative. If we consider the life cycle of a Lepidopteran species (Fig. 34.2), it is clear that there are alternative ways of interfering with the development of the insect without killing the entire population.

The normal cycle of a Lepidopteran insect requires adequate larval growth, high emergence ratio of adults from the pupal stage, and a ratio of female to male adults of 3–4 to mate and secure the reproduction of the species. If we could alter some of these stages, we could keep the insect population under control without having to kill all of the insects. This is important from an ecological point of view. In Figure 34.3, some of these alternatives are illustrated. In addition to antifeedant activity, it is important to evaluate the biological activity of one compound in the complete cycle, searching for a moulting disruptor or an oviposition inhibitor, for example.

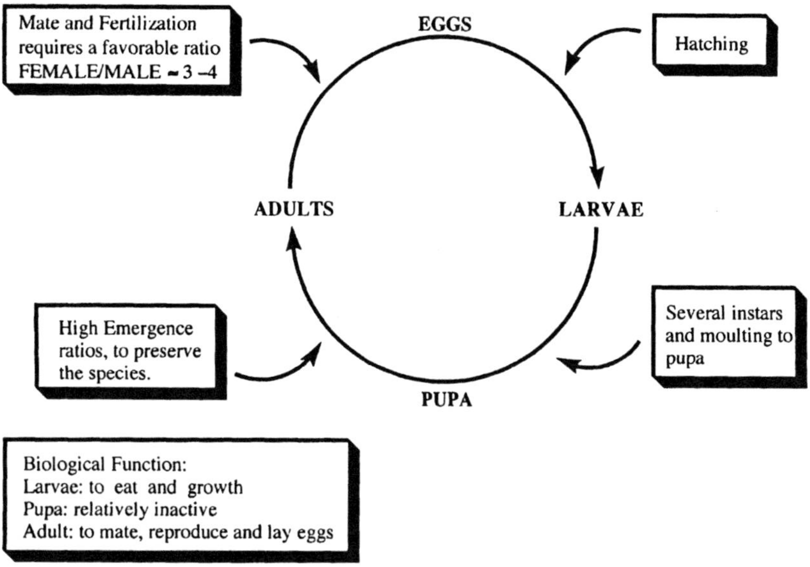

Fig. 34.2. Lepidopteran life cycle.

With this idea, we evaluated, very carefully, the influence of several neo-clerodane diterpenoids from Mexican Salvias in the complete cycle of *Spodoptera frugiperda*; a summary of the most relevant results of these experiments is provided (Scheme 34.7).

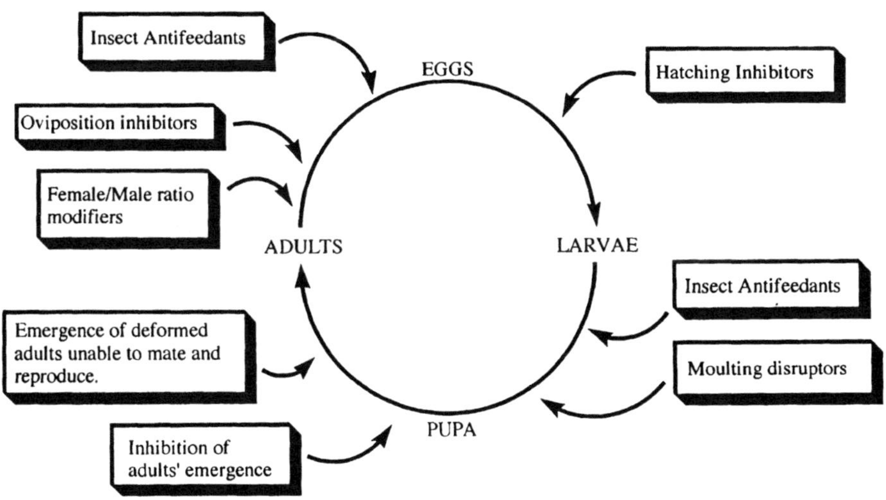

Fig. 34.3. Alternative effects of secondary metabolites of potential utility in a pest control program.

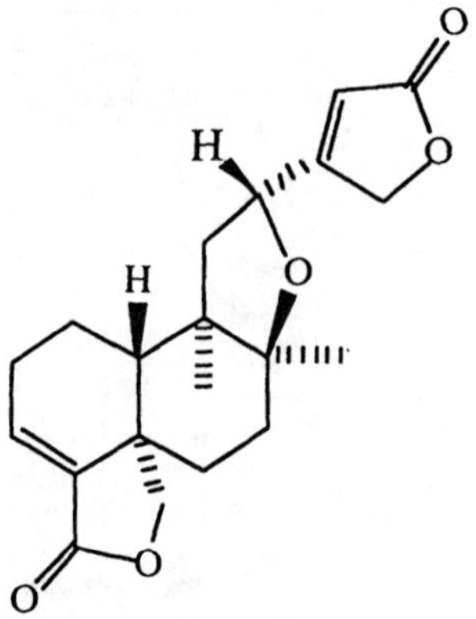

(**33**) Kerline

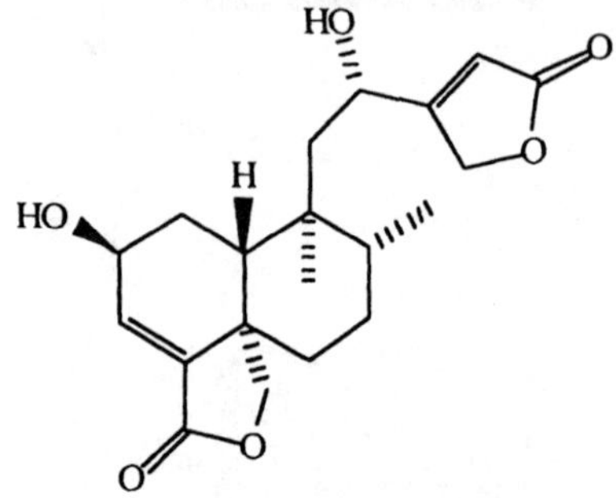

(**34**) Semiatrin

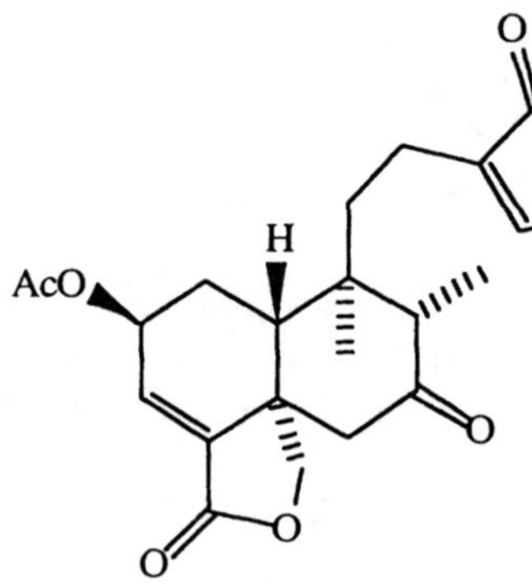

(**35**)

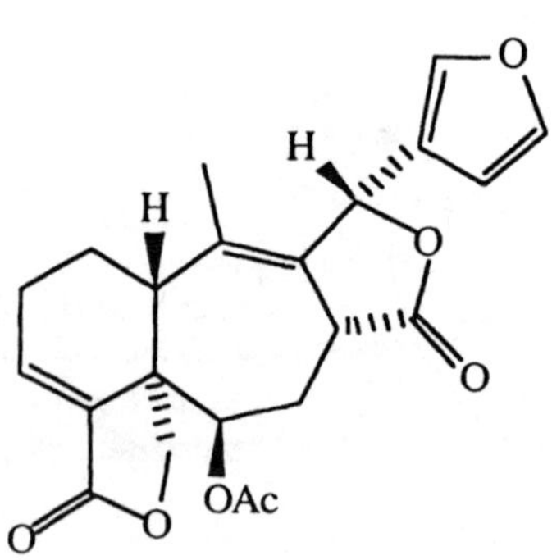

(**36**) Salvigenolide

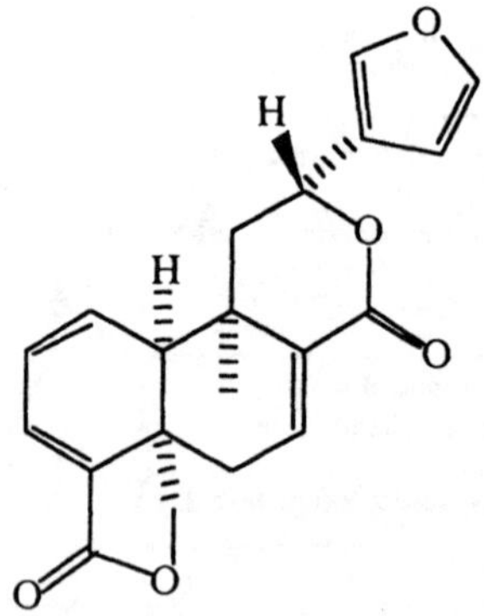

(**37**) Linearolactone

Scheme 34.7. Bioactive neo-clerodane diterpenoids from Mexican *Salvia* spp., against *Spodoptera frugiperda* and *Trichoplusia ni*.

Compounds **33–35**, isolated from *Salvia keerlii* Benth., *S. semiatrata* Kunth., and *S. melissodora* Lag., respectively, exert a significant moulting disruptor activity in the transit from larva to pupa at low concentration. Although these compounds exhibited a moderate antifeedant activity, they could be useful to keep the pest population below economically damaging levels (Aranda *et al.,* unpublished results).

Salvigenolide (**36**), a rearranged diterpenoid, was isolated previously from *Salvia fulgens* Cav. and recently from *Salvia xalapensis* Benth. This product causes a drastic reduction of the percentage of emergence (63% at 1.2 nmol/L) of adult insects from the pupal stage and a significant reduction of the female to male ratio to 1:2 in comparison with the control (3:1). This compound can alter the normal development of the insect. With fewer adults and more males than females, reproduction is limited.

A very interesting result was obtained for linearolactone (**37**). This product was isolated from *Salvia lineata* Cav. (Esquivel *et al.* 1986) and was inactive against *Spodoptera littoralis* Boisduval (Simmonds *et al.* 1996). However, it was shown to be very toxic to *Spodoptera frugiperda*. With a 75% larval mortality and a complete inhibition of emergence of adults (both at 1.2 nmol/L), together with a moulting-delaying activity (from larvae to pupa), this compound is the most interesting one of the series tested to date (Aranda *et al.*, unpublished results).

The evaluation of this type of compound in the complete cycle of the cabbage looper, *Trichlopusia ni*, a pest of Cruciferae plants such as broccoli, cauliflower, and cabbage (Hill 1987), indicates that compound **37** was also a very promising prototype for the development of new strategies for the control of this insect. Linearolactone (**37**) exerts a complete inhibition of oviposition (at 0.3 and 0.6 nmol/L) and reduces the female to male ratio to 1:1 in relation to the control (4:1). With a low antifeedant index of 27 at 1.9 nmol/L, this product could be very useful in the biological control of this pest (Aranda *et al.*, unpublished results).

These results demonstrate the potential of the neo-clerodane diterpenoids from Mexican Salvias not only as insect antifeedants, but also as moulting disruptors and oviposition inhibitors. These activities could be very useful as a part of an integrated control of pest insects such as *Spodoptera frugiperda* and *Trichoplusia ni*. The evaluation of these compounds with other commercially important insects such as *Epilachna varivestis* is in progress. The systematic chemical analysis of further *Salvia* and *Scutellaria* species is proceeding.

Acknowledgments

We are in debt to Julieta Flores-Hidalgo, Isabel Chávez, Beatriz Quiroz, Rocio Patiño, Luis Velasco, and Javier Pérez for technical assistance. We also thank Drs. Francisco González Medrano, Fernando Chiang, and Miguel Angel Martínez (Instituto de Biología UNAM) for their helpful commentaries on the biogeography of Mexican plants. The contributions of my colleagues and students named in the references are deeply appreciated. This work was supported, in part, by DGAPA-UNAM (Grant PAPIID IN 212594).

References

Anderson, J.C., W.M. Blaney, M.D. Cole, L.L. Fellows, S.V. Ley, R.N. Sheppard, and M.S.J. Simmonds, The Structure of Two New Clerodane Diterpenoid Potent Insect Antifeedants from *Scutellaria woronowii* (JUZ); Jodrellin A and B, *Tetrahedron Lett. 30:* 4737 (1989).

Bentham, G., Labiatae, in *Genera Plantarum* , edited by G. Bentham, and J.D. Hooker, vol. 2 , Reeve & Co., London, 1876, pp. 1160–1223.

Brown, E., and C. Dewhurst, The Genus *Spodoptera* (Lepidoptera=Noctuidae) in Africa and the Near East, *Bull. Entomol. Res. 65:* 221–262 (1975).

Chang, H.M., K.P. Cheng, T.F. Choang, H.F. Chow, K.Y. Chui, P.M. Hon, F.W.L. Tan, Y. Yang, Z.P. Zhong, C.M. Lee, H.L. Sham, C.F. Chan, Y.X. Cui, and H.N.C. Wong, Structure Elucidation and Total Synthesis of New Tanshinones Isolated from *Salvia miltiorrhiza* bunge (Danshen), *J. Org. Chem. 55:* 3537–3543 (1990).

Cole, M.D., P.D. Bridge, J.E. Dellar, L.E. Fellows, M.C. Cornish, and J.C. Anderson, Antifungal Activity of Neo-Clerodane Diterpenoids from *Scutellaria, Phytochemistry 30:* 1125–1127 (1991).

Epling, C., The Californian Salvias: A Review of *Salvia*, Section Audibertia, *Ann. Mo. Bot. Gard. 25:* 95–152 (1938).

Epling, C., A Revision of *Salvia* Subgenus *Calosphace, Feddes Repert. Specierum. Nov. Regni. Beih. 110:* 1–383 (1939).

Epling, C., American Species of *Scutellaria, Univ. Cal. Publ. Bot. 20:* 1–146 (1942).

Esquivel, B., J.S. Calderón, and E. Flores, A Neoclerodane Diterpenoid from Scutellaria-Seleriana, *Phytochemistry 47:* 135–137 (1998).

Esquivel, B., J.S. Calderón, E. Flores, A.A. Sánchez, and R. Rosas-Rivera, Abietane and Icetexane Diterpenoids from *Salvia ballotaeflora* and *Salvia axillaris, Phytochemistry 46:* 531–534 (1997a).

Esquivel, B., J.S. Calderón, A.A. Sánchez, T.P. Ramamoorthy, E.A. Flores, and R.M. Domínguez, Recent Advances in Phytochemistry and Biological Activity of Mexican Labiatae, *Rev. Latinoam. Quim 24:* 44–64 (1996).

Esquivel, B., J. Cardenas, T.P. Ramamoorthy, and L. Rodriguez-Hahn, Clerodane Diterpenoids of *Salvia lineata, Phytochemistry 25:* 2381–2384 (1986).

Esquivel, B., J. Cardenas, A. Toscano, M. Soriano-Garcia, and L. Rodriguez-Hahn, Structure of Salvigenolide, a Novel Diterpenoid with a Rearranged Neo-Clerodane Skeleton from *Salvia fulgens, Tetrahedron 41:* 3213–3217 (1985).

Esquivel, B., R.M. Domínguez, S. Hernández-Ortega, R.A. Toscano, and L. Rodríguez-Hahn, Salvigenane and Isosalvipuberulan Diterpenoids from *Salvia leucantha, Tetrahedron 40:* 11593–11600 (1994).

Esquivel, B., R.M. Domínguez, M.A. Martínez, and G. Espinosa-Pérez, A Neoclerodane Diterpenoid for *Scutellaria-Guatemalensis, Heterocycles 45:* 2247–2252 (1997b).

Esquivel, B., E. Flores, S. Hernández-Ortega, and R.A. Toscano, Neo-Clerodane Diterpenoids from *Scutellaria drummondii, Phytochemistry 38:* 175–179 (1995).

Esquivel, B., J. Ochoa, J. Cardenas, T.P. Ramamoorthy, L. Rodriguez-Hahn, Clerodane-Type Diterpenoids from *Salvia sousae*, *Phytochemistry 27:* 483–486 (1988).

González, A.G., Z.E. Aguiar, T.A. Grillo, and J.G. Luis, Diterpenes and Diterpene Quinones from the Roots of *Salvia apiana, Phytochemistry 31:* 1691–1695 (1992).

González, A.G., L. San Andrés, J.G. Luis, I. Brito, and M.L. Rodríguez, Diterpenes from *Salvia mellifera, Phytochemistry 30:* 4067–4070 (1991).

Hanson, J.R., Diterpenoids, *Nat. Prod. Reports 16:* 209–219 (1999).

Harley, R.M., and C.A. Heywood, Chromosome Numbers in Tropical American Labiatae, in *Advances in Labiatae Science*, edited by R.M. Harley, and T. Reynolds, Royal Botanic Gardens, Kew, UK, 1992, pp. 211–246.

Haro, G., M. Mori, M.O. Ishitsuka, T. Kusumi, Y. Inouye, and H. Kakisawa, Structures of Diterpenoid Tropolones, Salviolone and Miltipolone from the Root of *Salvia miltiorrhiza* Bunge, *Bull. Chem. Soc. Jpn. 64:* 3422–3426 (1991).

Hedge, I.C., A Global Survey of the Biogeography of the Labiatae, in *Advances in Labiatae Science*, edited by R.M. Harley, and T. Reynolds, Royal Botanical Gardens, Kew, UK, 1992, pp. 7–17.

Hernández, J., Evaluation de la Toxicité de *Bacillus thuringiensis* sur *Spodoptera frugiperda, Entomophaga 33:* 163–171 (1988).

Hill, D., *Agricultural Insect Pests of the Tropics and Their Control*, 2nd edn., Cambridge University Press, Cambridge, 1987.

Luis, J.G., and L. San Andrés, C-16 Hydroxylated Abietane Diterpenes from *Salvia mellifera, Phytochemistry 33:* 635–638 (1993).

Luis, J.G., L. San Andrés, and A. Perales, C-16 Hydroxylated Abietane Diterpenes from *Salvia mellifera:* Absolute Configuration and Biogenetic Implications, *Tetrahedron 49:* 4993–5000 (1993).

Luis, J.G., and T.A. Grillo, New Diterpenes from *Salvia munzii:* Chemical and Biogenetic Aspects, *Tetrahedron 49:* 6277–6284 (1993).

Luis, J.G., E.H. Lahlou, L. San Andrés, G.H.N. Sood, and M.M. Ripoll, Apiananes: Terpenoids with a New Type of Skeleton from *Salvia apiana, Tetrahedron Lett. 37:* 4213–4216 (1996).

Luis, J.G., W. Quiñones, T.A. Grillo, and M.P. Kishi, Diterpenes from the Aerial Part of *Salvia columbariae, Phytochemistry 35:* 1373–1374 (1994).

Mullin, C., *Insect Taste Perception Mechanism for Plant Antifeedants,* Plenary Lecture in the II Simposio Internacional Química de Productos Naturales y Sus Aplicaciones, Concepción, Chile, December 1994.

Paton, A., A Global Taxonomic Investigation of *Scutellaria* (Labiatae), *Kew Bull. 45:* 399–450 (1990a).

Paton, A., The Phytogeography of *Scutellaria* L., *Notes R. Bot. Gard. Edinb. 46:* 345–359 (1990b).

Ramamoorthy, T.P., and M. Elliot, Mexican Lamiaceae: Evolution, Distribution and Endemism, in *Biological Diversity of Mexico: Origins and Distribution*, edited by T.P. Ramamoorthy, R. Bye, A. Lot, and J. Fa, Oxford University Press, New York, 1993.

Raven, P.H., and D.I. Axelrod, Angiosperm Biogeography and Past Continental Movements, *Ann. Mo. Bot. Gard. 61:* 539–673 (1974).

Rodríguez-Hahn, L., B. Esquivel, and J. Cardenas, Clerodane Diterpenes in the Labiatae, *Prog. Chem. Nat. Prod. 63:* 107–196 (1994).

Sánchez, A.A., B. Esquivel, T.P. Ramamoorthy, and L. Rodríguez-Hahn, Clerodane Diterpenoids from *Salvia urolepis, Phytochemistry 38:* 171–174 (1995).

Simmonds, M.S.J., W.M. Blaney, B. Esquivel, and L. Rodríguez-Hahn, Effect of Clerodane-Type Diterpenoids Isolated from *Salvia* spp. on the Feeding Behavior of *Spodoptera littoralis, Pestic. Sci. 47:* 17–23 (1996).

van Beek, T.A., and A.E. de Groot, Terpenoid Antifeedants, Part I: An Overview of Terpenoid Antifeedants of Natural Origin, *Recl. Trav. Chim. Pays-Bas 105:* 513–527 (1986).

Wood, J.R.I., and R.M. Harley, The Genus *Salvia* (Labiatae) in Colombia, *Kew Bull. 44:* 211–278 (1989).

Chapter 35

Synthetic Study of a Highly Antitumorigenic Marine Phytochemical, Halichondrin B

Kiyoshi Horita[a], Sansei Nishibe[a], and Osamu Yonemitsu[b]

[a]Faculty of Pharmaceutical Sciences, Health Sciences, University of Hokkaido, Japan
[b]Faculty of Sciences, Okayama University of Sciences, Japan

Introduction

Halichondrins, eight antitumor derivatives, were isolated from a marine sponge, *Halichondria okadai* Kadota, in 4×10^{-7} to 5×10^{-6}% yields by Uemura and Hirata (1985). Halichondrin B (**1**), the most representative compound, exhibited remarkably powerful antitumor activity *in vitro* and *in vivo*. A few doses of 10 µg/kg indicated effective antitumor activities against B-16 melanoma and P-388 leukemia such that test group/control group (T/C) exceeded 200 and 300%, respectively, in mice. However, the very limited supply of **1** from natural sources has prevented further development of their potential clinical application. At this stage, synthetic preparation of **1** seemed to be the most likely possible approach.

Actually, the structure of **1** is quite complex. The 32 asymmetric carbons in **1** allow >4×10^9 stereoisomers. Therefore, a practical and efficient total synthesis must be highly stereoselective. Furthermore, these asymmetric carbons are located in several regions of the molecule that are apparently isolated from one another. Additionally, many kinds of functional groups, as represented in 2,6,9-trioxatricyclo[3.3.2.$0^{3,7}$]decane system, are included in the structure of **1**. Therefore, it would be most difficult to generate a desired configuration within one region under a stereocontrolling influence of chirality already present in one of the other sectors. For the past decade, many efforts (Burke *et al.* 1994 and 1995, Horita *et al.* 1994a, 1994b, 1994c, 1994d, and 1994e, 1996 and 1997a, 1996b, and 1997c, Kishi and Aicher 1987, Kishi *et al.* 1992b, Kishi and Duan 1993, Kishi and Stamos 1996, Salomon and Kim 1989, Salomon and Cooper 1990, Salomon *et al.* 1993a and 1993b toward the total synthesis of **1** have been made because of its highly complex structure and antitumor activity; the first total synthesis was completed by Kishi *et al.* (1992a). In connection with our continuing research program on the antitumor active site of **1**, we established a flexible and efficient synthetic pathway of the macrolactone part (**2**) and the polyether part (**3**) from cheap and commercially available starting materials such as dimethyl L-tartarte, methyl (2*S*)-3-hydroxy-2-methylpropionate and D-glucose.

Synthetic Plan

Our synthetic plan is summarized in Scheme 35.1. It seems that the macrolactone part (**2**) and the polyether part (**3**) structurally overlap at the F-ring portion in Halichondrin B (**1**). Therefore, we considered that Halichondrin B (**1**) could be divided synthetically into the two fragments, the macrolactone part (**2**) and the polyether part (**3**) at the F-ring portion. Furthermore, **2** was expected to be synthesized from the C_1-C_{15} fragment (**4**), the C-C fragment (**5**), and the C_{27}-C_{36} fragment (**6**). On the other hand, the synthesis of **3** must be achieved by coupling at the C_{36}- and C_{37}-position of **6** and the C_{37}-C_{54} fragment (**7**), respectively, followed by

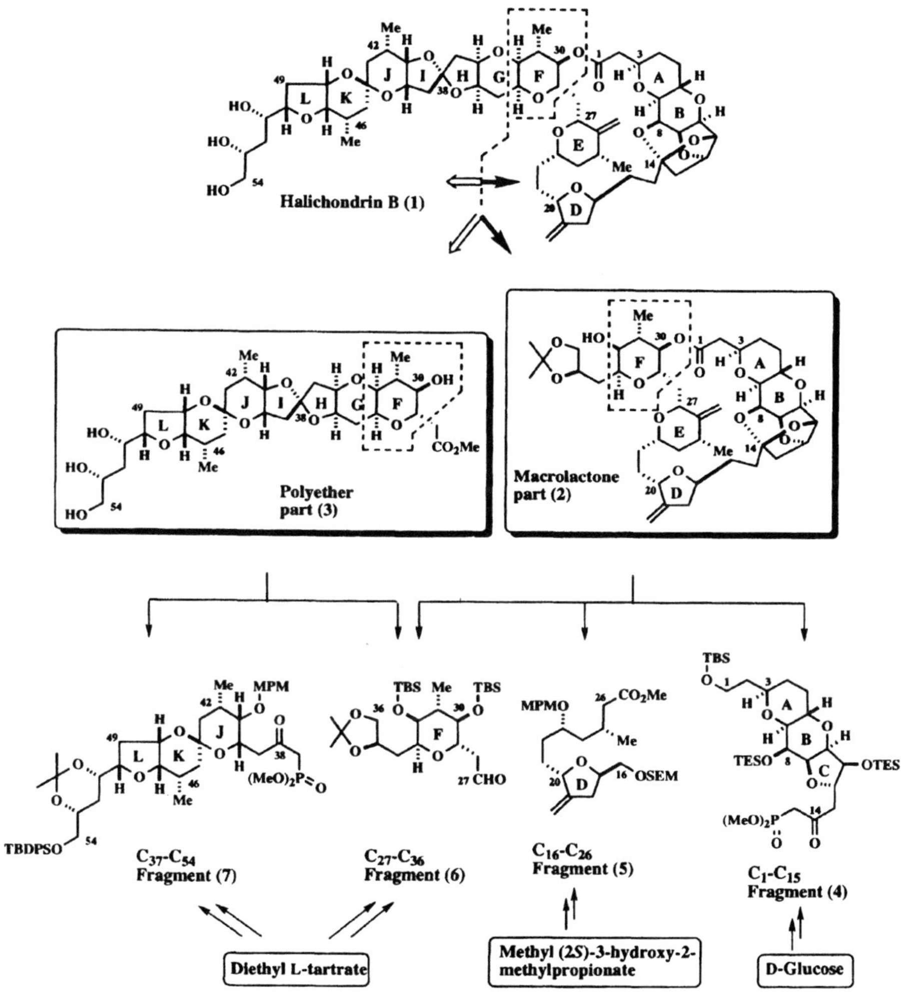

Scheme 35.1.

the H- and I-ring construction. The program of synthesis for the four fragments was completed in our previous studies (Horita *et al.* 1994a, 1994b, 1994c, 1994d, and 1994e). Thus, **4**, **5**, **6** and **7** were stereoselectively synthesized from D-glucose, methyl 2(*S*)-3-hydroxy-2 methylpropionate, and dimethyl L-tartrate in gram-scale amounts.

Synthesis of the Macrolactone Part (2)

Synthesis of 12 *via Coupling Reaction of* 5 *with* 6 *and E-Ring Construction*

For the synthesis of **12**, it would be most difficult to construct the E-ring stereoselectively. Accordingly, a newly generated hydroxyl group at the C_{27}-position, after the aldol reaction of the ester enolate of **5** with the aldehyde (**6**), was expected to play an important role in the ring closure between the C_{23}- and C_{27}-positions. Indeed, the aldol reaction of **5** and **6** proceeded smoothly and stereoselectively as follows. When the lithium enolate of **5** was generated with lithium diisopropyl amide in tetrahydrofuran at –78°C and then reacted with the aldehyde (**6**), a completely stereoselective reaction proceeded and was completed within 10 min to give a single aldol adduct (**8**). It was clear that the newly generated hydroxyl group at the C_{27}-position had the most suitable configuration as a leaving group in the case of the E-ring construction by using an SN-2-type reaction from the C_{23} side to the C_{27} side. However, all attempts to convert the C_{27} hydroxy group into the corresponding leaving groups, such as the mesyloxy group, were unsuccessful.

The efficient and stereoselective C_{23}-C_{27} E-ring construction was successfully achieved by using the reductive deoxygenation reaction of the δ-*p*-methoxybenzyloxycarbonyl portion of **9** with Et_3SiH in the presence of $BF_3 \cdot OEt_2$ in acetonitrile as follows. Dess-Martin oxidation (Martin and Dess 1983) of **8** followed by the exchange of the isopropylidene group at the C_{35} and C_{36} positions to diacetyl derivative gave **9** (a 1:1 mixture at the C_{26}-position). When **9** was treated with Et_3SiH and $BF_3 \cdot OEt_2$ in acetonitrile at 0°C, a quite rapid reaction occurred to give the desired **10** as a sole product in 87% yield. After the diacetyl groups at the C_{35}- and the C_{36}-positions were replaced again by the isopropylidene group, the C_{26} methoxycarbonyl group was converted to an *exo*-methylene group to give **11**, which was converted to the important synthetic intermediate (**12**) *via* four steps. The overall yield for the 16 steps from **6** to **2** was 20% (Scheme 35.2).

Synthesis of 2 via *Coupling Reaction of* 12 *with* 4 *and Macrolactone Ring Formation*

The Horner-Emmons coupling reaction of **12** with the lithium salt of **4** proceeded smoothly to give the enone, which was treated with a copper (I) hydride cluster under Stryker's conditions (Stryker and Mahoney 1988, 1989). A selective reduction of the C_{15}-C_{16} double bond portion occurred to give **13**. To construct the intramolecular acetal ring at the C_8 and C_{11} hydroxyl groups with a C_{14} carbonyl

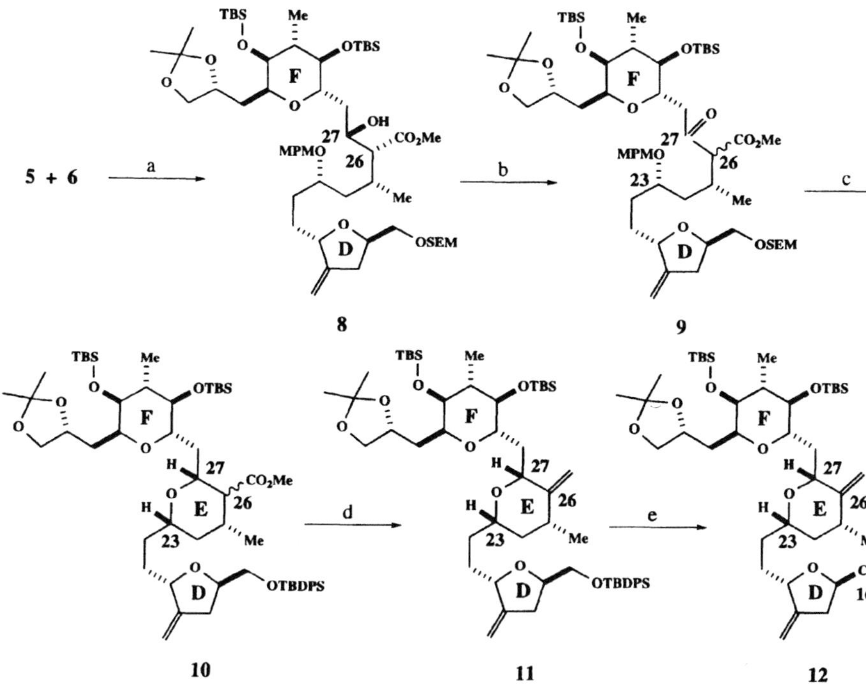

Scheme 35.2. Conditions: (a) LDA, THF, –78°C, 76%. (b) (1) Dess-Martin oxidation, 100%; (2i) PPTS, MeOH; Ac_2O, TEA, DMAP, CH_2Cl_2, 77%. (c) Et_3SiH, BF3 · OEt_2, MeCN, 0°C, 5 min; TBDPSCl, imidazole, CH_2Cl_2, 87%. (d) (1) K_2CO_3, MeOH; $Me_2C(OMe)_2$, CSA, benzene, 92%; (2) TBSOTf, 2,6-lutidine, CH_2Cl_2, 89%; (3) DIBAH, toluene, 86%; (4) TsCl, TEA, DMAP, 100%; (5) NaI, $MaHCO_3$, DBU, THF, 83%; (e) (1) TBAF, THF, BzCl, pyridine, CH_2Cl_2, 97%; (2) TBSOTf, 2,6-lutidine, CH_2Cl_2, 91%; (3) K_2CO_3, MeOH, 93%; (4) Swern oxidation, 83%.

group, two C_8 and C_{11} triethylsilyl (TES) groups and a C_1 primary tert-butyldimethylsilyl (TBS) group of **13** were carefully and selectively deprotected by treatment with 1 N HCl in tetrahydrofuran at room temperature. Under these conditions, two C_{28} and C_{30} TBS groups remained. When the triol obtained was treated with pyridinium *p*-toluenesulfonate in tetrahydrofuran/methanol at room temperature, the intramolecular acetal ring was constructed smoothly to give **14** in 80% yield *via* two steps. Dess-Martin oxidation (Martin and Dess 1983) of **14** followed by $NaClO_2$ oxidation and deprotection of TBS groups by treatment with tetrabutylammonium fluoride gave the seco-acid (**15**). The seco-acid (**15**) was converted to an activated ester by treatment with 2,4,6-trichlorobenzoyl chloride and triethylamine and then exposed in a large excess of 4-(*N*,*N*-dimethyl)aminopyridine in benzene at room temperature to give the title compound (**2**), the macrolide part of halichondrin B (**1**). However, the yield is still unsatisfactory as shown in Scheme 35.3.

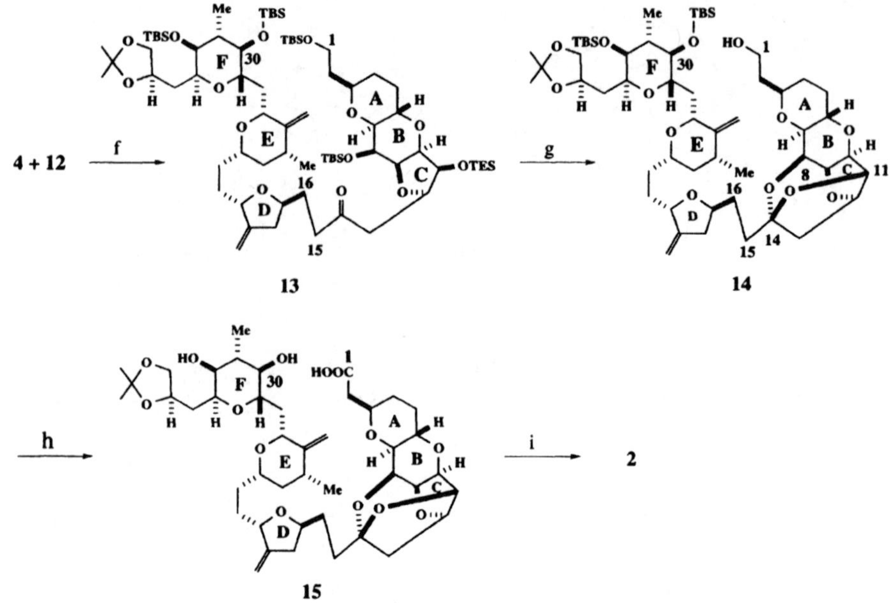

Scheme 35.3. Conditions: (f) (1) nBuLi THF, room temperature, 74%; (2) Ph_3CuH, benzene-H_2O, room temperature, 90%. (g) 1 N HCl, THF, room temperature; PPTS, THF-MeOH, 80%. (h) Dess-Martin oxidation; $NaClO_2$, NaH_2PO_4, $Me_2C{=}CMe_2$; TBAF, THF, 70%. (i) $Cl_3C_6H_2COCl$, TEA, THF; DMAP, benzene, room temperature, 35%.

Synthesis of the Polyether Part (3)

Construction of the G-Ring

We originally planned to synthesize the polyether part (**3**) of Halichondrin B (**1**) *via* coupling the aldehyde (**16a**) and C_{37}-C_{54} β-ketophosphonate (**7**) followed by two thermodynamically controlled reactions, namely, an intramolecular Michael reaction of the C_{32} hydroxyl group to the C_{36} double bond portion to form the G-ring and then a subsequent spiroketalization of the two C_{35} and C_{41} hydroxyl groups to the C_{38} ketone carbonyl group by acid treatment, because the natural halichondrin (**1**) is expected to have the most stable configuration. Therefore, even if stereoisomeric compounds with respect to the C_{32}- and the C_{38}-positions are produced under kinetic conditions, they must be converted to the desired compound with correct stereochemistry by exposure under thermodynamic conditions (Kishi *et al.* 1992a). The Horner-Emmons coupling reaction of **16a** with the lithium salt of **7** gave the enone (**17a**). Although the G-ring was stereoselectively constructed to give the ketone (**18**) as a sole product by the intramolecular Michael reaction of **17a** on treatment with tetrabutylammonium fluoride and the subsequent deprotection reaction with 2,3-dichloro-5,6-dicyano-1,4-benzoquinone (DDQ) oxidation, no spiroketalization by acid treatment to give **19** proceeded entirely. This disap-

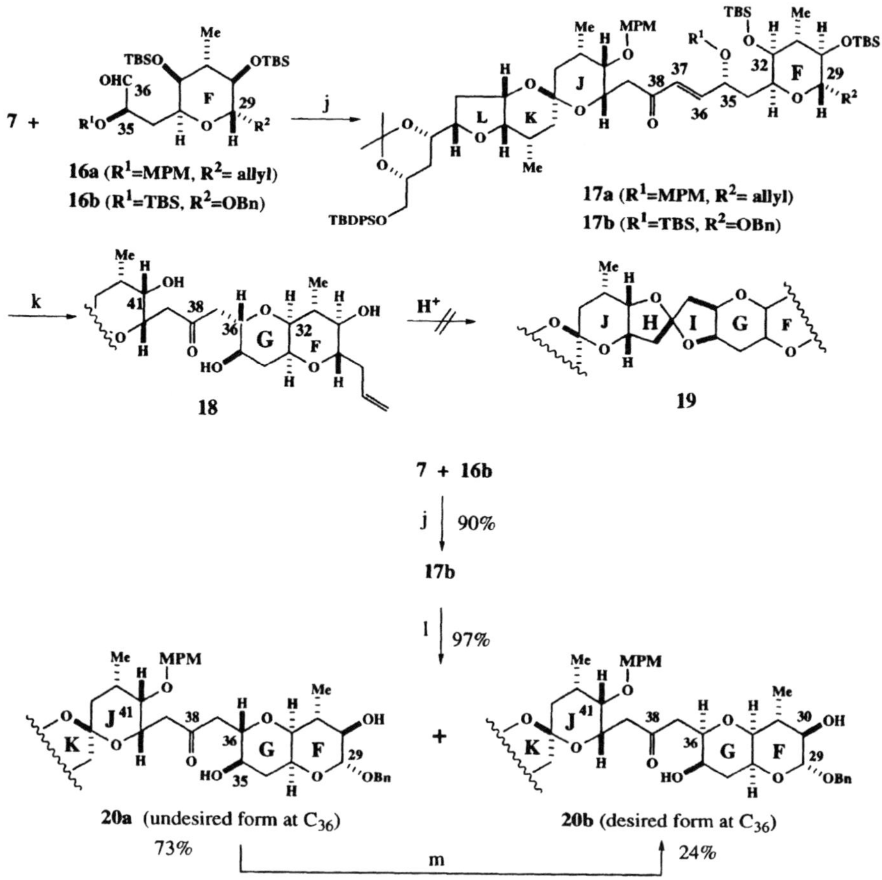

Scheme 35.4. Conditions: (j) nBuLi THF, room temperature. (k) (1) TBAF, THF, room temperature, 65%; (2) DDQ, CH_2Cl_2/H_2O (10:1), room temperature, 66%. (l)TBAF (5 eq), AcOH (1.25 eq), THF, room temperature, 97%. (m) K_2CO_3, MeOH, 80°C, 82%.

pointing result can be explained in terms of the wrong configuration at the C_{36}-position of **18**. This critical problem was solved by using the aldehyde (**16b**) instead of **16a** as follows. The Horner-Emmons coupling reaction of **16b** with lithium salt, generated with *n*-butyllithium in tetrahydrofuran at –78°C, proceeded easily and smoothly to give the enone (**17b**) in 90% yield as described above. The two F-rings of **16a** and **16b** are chair forms but have quite opposite configurations from each other. When **17b** was treated with tetrabutylammonium fluoride in tetrahydrofuran at room temperature, the intramolecular cyclization proceeded smoothly to give a 3:1 mixture of **20a** and **20b** in 97% yield. Although the undesired **20a** at the C_{36}-position was still rich in this mixture, **20a** was easily and completely converted into the desired compound (**20b**) by treatment with K_2CO_3 in methanol at

80°C in 82% yield. Thus, the stereoselective construction of the G-ring with the correct stereochemistry was achieved as shown in Scheme 35.4.

Construction of the H- and I-Rings and Synthesis of the Polyether (3)

Conversion of **20b** to the polyether part (**3**) is illustrated in Scheme 35.5. The p-methoxyphenylmethyl (MPM) group of **20b** at the C_{41}-position was removed by DDQ oxidation. When the resulting triol was treated with *dl*-camphorsulfonic acid (CSA) in dichloromethane at room temperature, an intramolecular spiroketalization took place smoothly to construct the H- and I-rings as a 1:1 mixture with respect to the C_{38}-position in excellent yield this time. The mixture was converted into the tetraacetates, **21a** and **21b,** respectively. For the purpose of elongating two carbon units from the C_{29}-position, deprotection of the C_{29} benzyl group by hydrogenolysis with H_2-$Pd(OH)_2$ in ether and the subsequent Wittig reaction of the resulting hemiacetal derivative were successfully performed, although the stereo center at the C_{38}-position was still a mixture. However, the undesired isomer was easily and smoothly converted into the desired compound (**22**) by treatment with

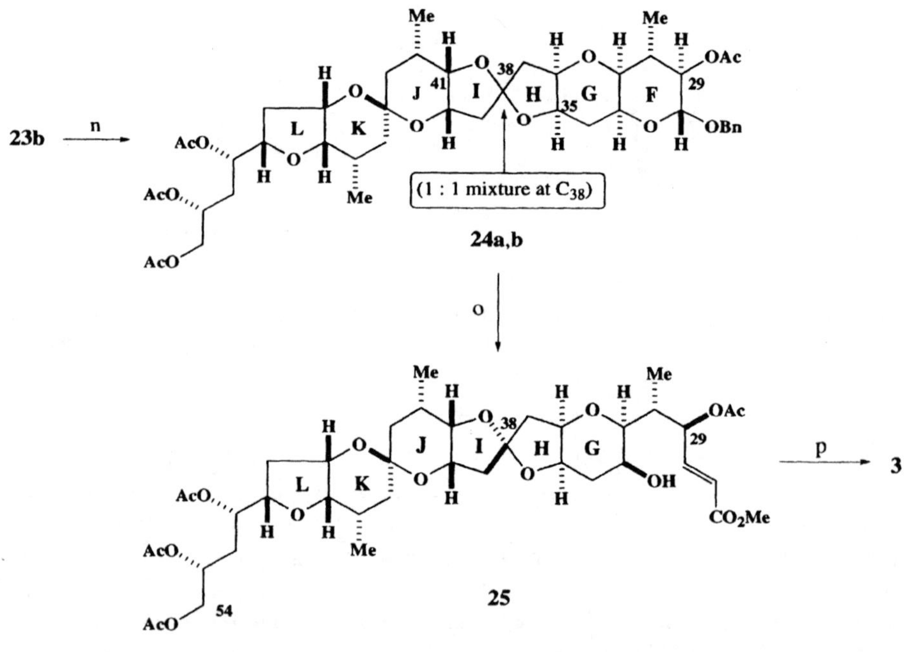

Scheme 35.5. (n) (1) DDQ, CH_2Cl_2/H_2O (20:1), room temperature; (2) CSA, CH_2Cl_2, room temperature, 93% *via* two steps; (3) Ac_2O, DMAP, Et_3N, CH_2Cl_2, room temperature, 90%; (o) (1) H_2, $Pd(OH)_2$, ether, room temperature; (2) $Ph_3P{=}CHCO_2Me$, CH_3CN, 66% *via* two steps; (3) CSA, CH_2Cl_2, room temperature, 86%. (p) (1) TBAF (5 eq), H_2O (10 eq), THF, room temperature, 61%; (2) K_2CO_3, MeOH, room temperature, 99%.

CSA in dichloromethane at room temperature. Finally, synthesis of the polyether part (**3**) was achieved by removal of acetyl groups.

Preparation of the Macrolactone Part (2) and the Polyether Part (3)

Macrolactone Part (2). The alcohol (**14**) (7.5 mg, 0.0073 mmol) was treated with Dess-Martin reagent (31 mg, 0.073 mmol) in anhydrous CH_2Cl_2 (3.0 mL) for 5 min. Saturated aqueous sodium thiosulfate solution and ether (30 mL) were added into the resultant until it became a clear solution. The aqueous mixture was extracted with ether (20 mL × 3) and then the organic layer was washed with brine. After the organic layer was dried over anhydrous Na_2SO_4, the organic solvent was evaporated under reduced pressure. The residual crude oil was oxidized with $NaClO_2$ (13 mg, 0.15 mmol) in the presence of NaH_2PO_4 (23 mg, 0.15 mmol) and dimethyl-2-butene (100 μL) in *t*-butanol (3.0 mL) for 10 min at room temperature without purification. The mixture was extracted with ether (20 mL × 3) and then dried over anhydrous Na_2SO_4. After removal of the solvent *in vacuo*, the residue was purified on a silica gel column using ethyl acetate/*n*-hexane (2:1) as an eluate to give **15** as an oil (5.6 mg, 73%). The hydroxy carboxylic acid (**15**) (2.4 mg, 0.0029 mmol) was treated with 2,4,6-trichlorobenzoyl chloride (36 mg, 0.15 mmol) in the presence of triethylamine (30 mg, 0.29 mmol) in an anhydrous tetrahydrofuran (0.5 mL) for 24 h. The resulting solution was slowly added into anhydrous benzene solution (5.0 mL) of 4-(dimethylamino)pyridine (70 mg, 0.59 mmol) over the period of 1 h. After the mixture was stirred for 8 h at room temperature, methanol (1.0 mL) was added into the reaction mixture. H_2O was added into the resultant and then the aqueous mixture was extracted with ether (20 mL × 3). The combined extracts were washed with brine and then dried over anhydrous Na_2SO_4. After removal of the solvent *in vacuo*, the residue was purified on a silica gel column using ethyl acetate as an eluate to give the macrolactone part (**2**) (0.8 mg, 35%).

$[\alpha]_D{}^{25}$ + 3.8° (c = 0.14, $CHCl_3$) infrared (IR) (neat) cm^{-1}: 2960, 2920, 2840, 1720, 1690, 1645. ^1H nuclear magnetic resonance (NMR) (600 MHz, $CDCl_3$) δ: 0.70–1.10 (*m*, 3H), 0.87 (*d*, J = 6.5 Hz, 3H), 0.91 (*d*, J = 6.5 Hz, 3H), 0.99 (*s*, 3H), 1.01 (*s*, 3H), 1.13–1.95 (*m*, 12H), 2.42 (*ddd*, J = 5.0, 10.0, 15.0 Hz, 1H), 2.45–2.60 (*m*, 3H), 3.35–3.43 (*m*, 1H), 3.44–3.48 (*m*, 1H), 3.52–3.58 (*m*, 1H), 3.64 (*t*, J = 7.5 Hz, 1H), 3.78 (*dd*, J = 4.0, 6.5 Hz, 1H), 3.94–3.99 (*m*, 3H), 3.98 (*dd*, J = 6.0, 8.0 Hz, 1H), 4.03 (*dd*, J = 6.0, 8.0 Hz, 1H), 4.15–4.22 (*m*, 4H), 4.23–4.31 (*m*, 2H), 4.32–4.43 (*m*, 2H), 4.51–4.53 (*m*, 1H), 4.53–4.59 (*m*, 2H), 4.67 (*m*, 1H), 4.83 (*d*, J = 2.5 Hz, 1H), 4.88 (*dt*, J = 2.0, 2.5 Hz, 1H), 4.98 (*s*, 2H), 5.26 (*dd*, J = 5.5, 8.5 Hz). Fast atom bombardment mass spectrometry (FAB-MS) *m/z*: 787 (MH^+, 0.15), 391 (10), 307 (10), 137 (100). High-resolution (HR)-MS (FAB) calculated for $C_{43}H_{63}O_{13}$ (MH^+): 787.4273. Found: 787.4297.

Polyether Part (3). Hydrogenolysis of **21a,b** (59 mg, 0.070 mmol) was carried out in the presence of $Pd(OH)_2$ (20 mg) in diethyl ether (1.5 mL) for 2 h. After removal of the insoluble catalyst by filtration, the filtrate was evaporated under reduced pressure. The crude residue was treated with methyl (triphenylphosphoranylidene)acetate (123 mg, 0.35 mmol) in acetonitrile (1.0 mL) at 60°C for 15 h. After removal of the solvent, the residue was passed through a silica gel column using *n*-hexane/ethyl acetate (1:3) as an eluate to give a colorless oil (29 mg, 55%) as two diasteomeric mixtures at the C_{38}-position. This oily mixture (10 mg, 0.013 mmol) was treated with *dl*-10-camphorsulfonic acid (3.0 mg, 0.375 mmol) in anhydrous CH_2Cl_2 at room temperature for 2 h. After neutralization of the mixture with triethylamine, solvent was removed under reduced pressure. The residue was purified on a silica gel colomun using *n*-hexane/ethyl acetate (1:3) as an eluate to give an oil (8.8 mg, 86%). This oily compound (8.8 mg, 0.01 mmol) was treated with 1 mol/L tetrabutylammonium fluoride in THF (54 μL, 0.05 mmol) in H_2O (1.9 μL) and THF (0.5 mL) at room temperature for 2 h. After dilution of the mixture with H_2O, the aqueous mixture was extracted with ether (10 mL × 3). The combined extracts were washed with brine and dried over anhydrous Na_2SO_4. After evaporation of the solvent *in vacuo*, the residue was purified on a silica gel column using *n*-hexane/ethyl acetate (1:3) as an eluate to give **22** as a colorless oil (5.4 mg, 61%).

$[\alpha]_D^{25}$ –70° (*c* = 0.19, CHCl). IR (neat) cm^{-1}: 2930, 2875, 1740, 1435, 1370, 1230, 1110, 1030. ^1H NMR (400 MHz, C_6D_6)δ: 1.07 (*d*, 3H, *J* = 7.0 Hz), 1.15 (*d*, 3H, *J* = 7.0 Hz), 1.17 (*d*, 3H, *J* = 6.5 Hz), 1.30–1.44 (*m*, 2H), 1.64 (*s*, 3H), 1.52–1.66 (*m*, 2H), 1.68 (*m*,1H), 1.70 (*s*, 3H), 1.78 (*s*, 3H), 1.76–1.78 (*m*, 2H), 1.84 (*s*, 3H), 1.86–1.97 (*m*, 2H), 2.02 (*dd*, 1H, *J* = 7.5, 14.0 Hz), 2.14 (*d*, 1H, *J* = 14.0 Hz), 2.17 (*dt*, 1H, *J* = 11.5, 5.0 Hz), 2.24–2.32 (*m*, 3H), 2.54 (q, 1H, *J* = 11.5 Hz), 2.61 (*dd*, 1H, *J* = 5.0, 15.0 Hz), 2.65 (*dd*, 1H, *J* = 8.0, 15.0 Hz), 2.90 (*d*, 1H, *J* = 14.0 Hz), 2.90 (*dd*, 1H, *J* = 5.0, 14.0 Hz), 2.94 (*dd*, 1H, *J* = 6.5, 8.5 Hz), 3.30 (*dd*, 1H, *J* = 2.0, 3.0 Hz), 3.39 (*s*, 3H), 3.56 (*ddd*, 1H, *J* = 5.5, 6.5, 11.5 Hz), 3.66 (*ddd*, 1H, *J* = 0.5, 6.0, 7.5 Hz), 3.79 (*dd*, 1H, *J* = 2.0, 4.5 Hz), 3.84 (*dd*, 1H, *J* = 2.0, 5.0 Hz), 3.89 (*dd*, 1H, *J* = 2.0, 3.0 Hz), 3.90–3.98 (*m*, 2H), 4.19 (*dd*, 1H, *J* = 6.0, 12.0 Hz), 4.33 (*dt*, 1H, *J* = 5.0, 8.5 Hz), 4.40 (*dd*, 1H, *J* = 3.5, 12.0 Hz), 4.64 (*dd*, 1H, *J* = 8.5, 10.5 Hz), 5.31 (*m*, 1H), 5.42 (*m*, 1H). ^{13}C NMR (500 MHz, CD_3OD)δ: 16.2, 19.0, 27.7, 27.8, 39.2, 46.1, 46.5, 69.7, 73.2, 73.9, 74.5, 76.8, 78.4, 78.7, 79.6, 80.5, 81.0, 82.1, 98.9, 116.7. FAB-MS *m/z* (%): 811 (MH^+, 11.4), 810 (3.31), 793 (12.9), 751 (3.08), 441 (5.89), 411 (11.0), 154 (100). HR-MS (FAB) calculated for $C_{40}H_{59}O_{17}$ (MH^+): 811.3753. Found: 811.3729.

The tetraacetate (20 mg, 0.025 mmol) was hydrolyzed with potassium carbonate (68 mg, 0.512 mmol) in methanol (1.0 mL) at room temperature for 1 h. The resulting mixture was diluted with ether (30 mL, which was washed with H_2O and brine. The ether layer was dried over anhydrous Na_2SO_4 and then evaporated under reduced pressure. The residue was passed through a silica gel column using ethyl acetate/methanol (10:1) as an eluate to give **3** (16 mg, 99%).

$[\alpha]_D^{25}$–56° (*c* = 0.63, $CHCl_3$). IR (neat) cm^{-1}: 3450, 2930, 2855, 1735, 1610, 1515, 1460, 1250. ^1H NMR (500 MHz, CD_3OD)δ: 1.01 (*d*, 3H, *J* = 7.0 Hz), 1.02 (*d*, 3H, *J* = 7.0 Hz), 1.14 (*d*, 3H, *J* = 6.5 Hz), 1.35 (*m*, 1H), 1.36 (*m*, 1H), 1.43 (*m*,

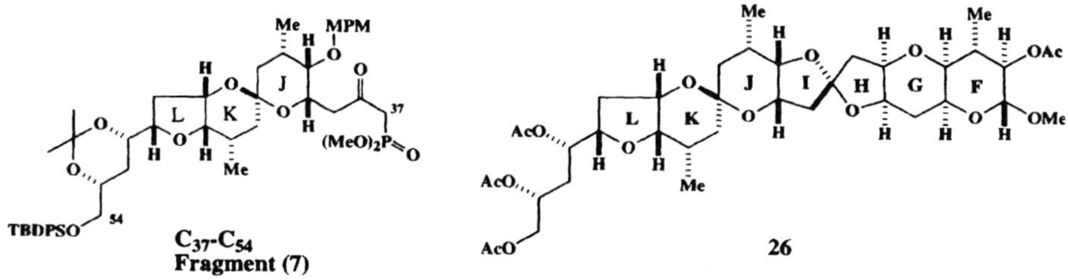

Scheme 35.4. Conditions: (j) nBuLi THF, room temperature. (k) (1) TBAF, THF, room temperature, 65%; (2) DDQ, CH_2Cl_2/H_2O (10:1), room temperature, 66%. (l)TBAF (5 eq), AcOH (1.25 eq), THF, room temperature, 97%. (m) K_2CO_3, MeOH, 80°C, 82%.

1H), 1.50 (*m*, 1H), 1.58 (*dt*, 1H, *J* = 14.5, 8.0 Hz), 1.73 (*dt*, 1H, *J* = 14.5, 4.5 Hz), 1.82 (*m*, 1H), 1.90 (*ddd*, 1H, *J* = 4.5, 6.0, 12.0 Hz), 2.07 (*d*, 1H, *J* = 14.0 Hz), 2.12 (*dt*, 1H, *J* = 8.5, 12.0 Hz), 2.21 (*m*, 1H), 2.23 (*m*, 1H), 2.29 (*m*, 1H), 2.30 (*m*, 1H), 2.35 (*dd*, 1H, *J* = 4.5, 13.5 Hz), 2.98 (*dd*, 1H, *J* = 8.0, 9.0 Hz), 3.26 (*dd*, 1H, *J* = 6.0, 8.0 Hz), 3.46 (*dd*, 1H, *J* = 6.0, 11.0 Hz), 3.52 (*dd*, 1H, *J* = 4.5, 11.0 Hz), 3.58 (*t*, 1H, *J* = 2.5 Hz), 3.77 (*dt*, 1H, *J* = 9.0, 4.5 Hz), 3.81 (*ddd*, 1H, *J* = 4.0, 8.0, 8.5 Hz), 3.85 (*m*, 1H), 3.86 (*m*, 1H), 3.87 (*m*, 1H), 3.97 (*dt*, 1H, *J* = 9.0, 4.5 Hz), 4.05 (*m*, 1H), 4.06 (*m*, 1H), 4.07 (*m*, 1H), 4.11 (*m*, 1H). ^{13}C NMR (400 MHz, CD_3OD)δ: 15.4, 18.2, 18.2, 26.9, 30.8, 36.2, 37.5, 38.2, 38.4, 40.2, 45.5, 45.9, 67.2, 68.1, 71.7, 73.2, 73.7, 75.4, 76.5, 77.3, 78.1, 78.4, 78.9, 80.2, 81.1, 81.3, 98.2, 116.0. FAB-MS *m/z* (%): 643 (MH^+, 0.2), 460 (0.4), 391 (0.5), 307 (2.0), 289 (1.3), 155 (100). HR-MS (FAB) calculated for $C_{32}H_{51}O_{13}$ (MH^+): 643.3330. Found: 643.3348.

Cytotoxicity Against KB and L1210 Cells. As described above, we succeeded in establishing a flexible and efficient synthetic pathway of the macrolactone part (**2**) and the polyether part (**3**) of the highly antitumorigenic polyether macrolide, halichondrin B (**1**), from cheap and commercially available starting materials. Two compounds (**7**) and (**23**) in our synthetic intermediates were compared by cytotoxicity against KB and L1210 *in vitro* as summarized in the legend of Scheme 35.6. It was found, for the first time, that **26**, corresponding to the polyether unit (**3**), had relatively strong cytotoxicity, although the activities of the two compounds examined were not as high as those of natural halichondrin B (**1**). Unfortunately, we could not examine the cytotoxicity of the macrolactone part (**3**) because of the very low yields of macrolactonization from **15** to **2**. Further investigation is now in progress.

References

Burke, S.D., K.Y. Jung, J.R. Phillips, and R.E. Perri, An Expeditious Synthesis of the C(1)-C(14) Subunit of Halichondrin B, *Tetrahedron Lett. 35*: 703–706 (1994).

Burke, S.D., G. Zang, and J.L. Buchanan, Enantioselective Synthesis of a Halichondrin B C(20)-C(36) Precursor, *Tetrahedron Lett. 36:* 7023–7026 (1995).

Horita, K., S. Hachiya, M. Nagasawa, M. Hikota, and O. Yonemitsu, Synthetic Studies of Halichondrin B, an Antitumor Polyether Macrolide Isolated from a Marine Sponge, 1, *Synlett* 38–40 (1994a).

Horita, K., S. Hachiya, K. Ogihara, Y. Yoshida, M. Nagasawa, and O. Yonemitsu, Synthetic Studies of Halichondrin B, an Antitumor Polyether Macrolide Isolated from a Marine Sponge, 5, *Heterocycles 42*: 99–104 (1996).

Horita, K., S. Hachiya, T. Yamazaki, T. Naitou, J. Uenishi, and O. Yonemitsu, Synthetic Studies of Halichondrin B, an Antitumor Polyether Macrolide Isolated from a Marine Sponge, 6, *Chem. Pharm. Bull. 45*: 1265–1281 (1997a).

Horita, K., M. Nagasawa, S. Hachiya, Y. Sakurai, T. Yamazaki, J. Uenishi, and O. Yonemitsu, Synthetic Studies of Halichondrin B, an Antitumor Polyether Macrolide Isolated from a Marine Sponge, 8, *Tetrahedron Lett. 38*: 8965–8968 (1997b).

Horita, K., M. Nagasawa, S. Hachiya, and O. Yonemitsu, Synthetic Studies of Halichondrin B, an Antitumor Polyether Macrolide Isolated from a Marine Sponge, 2, *Synlett* 40–43 (1994b).

Horita, K., Y. Sakurai, S. Hachiya, M. Nagasawa, and O. Yonemitsu, Large Laboratory Scale Synthesis of (2*S*, 3*S*)-2-(4-Methoxybenzyloxy)-3,4-pentylidenedioxybutan-1-ol, a Versatile Chiral Building Block in Natural Product Synthesis, *Chem. Pharm. Bull. 42*: 683–685 (1994c).

Horita, K., Y. Sakurai, M. Nagasawa, S. Hachiya, and O. Yonemitsu, Synthetic Studies of Halichondrin B, an Antitumor Polyether Macrolide Isolated from a Marine Sponge, 3, *Synlett* 43–45 (1994d).

Horita, K., Y. Sakurai, M. Nagasawa, K. Maeno, S. Hachiya, and O. Yonemitsu, Synthetic Studies of Halichondrin B, an Antitumor Polyether Macrolide Isolated from a Marine Sponge, 4, *Synlett* 46–48 (1994e).

Horita, K., Y. Sakurai, M. Nagasawa, and O. Yonemitsu, Synthetic Studies of Halichondrin B, an Antitumor Polyether Macrolide Isolated from a Marine Sponge, 7, *Chem. Pharm. Bull. 45*: 1558–1572 (1997c).

Kishi, Y., and T.D. Aicher, Synthetic Studies Towards Halichondrin, *Tetrahedron Lett. 28*: 3463–3466 (1987).

Kishi, Y., T.D. Aicher, K.R. Buszek, F.G. Fang, C.J. Forsyth, S.H. Jung, M.C. Matelich, P. M. Scola, D.M. Spero, and S.K. Toon, Total Synthesis of Halichondrin B and *nor*-Halichondrin B, *J. Am. Chem. Soc. 114*: 3162–3164 (1992a).

Kishi, Y., T.D. Aicher, K.R. Buszek, F.G. Fang, C.J. Forsyth, S.H. Jung, and P.M. Scola, Synthetic Studies Towards Halichondrin: Synthesis of the C.27-C.38 segment, *Tetrahedron Lett. 33*: 1549–1552 (1992b).

Kishi, Y., and J.J.W. Duan, Synthetic Studies on Halichondrin: A New Practical Synthesis of the C.1-C.12 Segment, *Tetrahedron Lett. 34*: 7541–7544 (1993).

Kishi, Y., M.D. Lewis, and J.K. Cha, Highly Stereoselective Approaches to α- and β-*C*-Glycopyranosides, *J. Am. Chem. Soc. 104*: 4976–4978 (1982).

Kishi, Y., and D.P. Stamos, Synthetic Studies on Halichondrins: A New Practical Synthesis of C.1-C.13 Segment, *Tetrahedron Lett. 37*: 8643–8646 (1996).

Martin, J.C., and D.B. Dess, Readily Accessible 12-I-5 Oxidant for the Conversion of Primary and Secondary Alcohols to Aldehydes and Ketones, *J. Org. Chem. 48*: 4155–4156 (1983).

Salomon, R.G., and A.J. Cooper, Total Synthesis of Halichondrins: Enantioselective Construction of a Homochiral Pentacyclic C1-C15 Intermediate from D-Ribose, *Tetrahedron Lett. 31*: 3813– 3816 (1990).

Salomon, R.G., A.J. Cooper, and W. Pan, Total Synthesis of Halichondrin B from Common Sugars: An F-Ring Intermediate from D-Glucose and Efficient Construction of the C1 to C21 Segment, *Tetrahedron Lett. 34*: 8193–8196 (1993a).

Salomon, R.G., E. DiFranco, and V.T. Ravikumar, Total Synthesis of Halichondrins: Enantioselective Construction of a Homochiral Tetracyclic KLMN-Ring Intermediate from D-Mannitol, *Tetrahedron Lett. 34*: 3247–3250 (1993b).

Salomon, R.G., and S. Kim, Total Synthesis of Halichondrins: Highly Stereoselective Construction of a Homochiral Penta-Substituted H-Ring Pyran Intermediate from a D-Glucose, *Tetrahedron Lett. 30*: 6279–6282 (1989).

Stryker, J.M., and W.S. Mahoney, Hydride-Mediated Homogeneous Catalysis, Catalytic Reduction of, β-Unsaturated Ketones Using $[(Ph_3P)CuH]_6$ and H_2, *J. Am. Chem. Soc. 111*: 8818–8823 (1989).

Stryker, J.M., W.S. Mahoney, and D.M. Brestensky, Selective Hydride-Mediated Conjugate, Reduction of, β-Unsaturated Carboxy Compound Using $[(Ph_3P)CuH]_6$, *J. Am. Chem. Soc. 110*: 291–293 (1988).

Uemura, D., K. Takahashi, and T. Yamamoto, Norhalicondrin A: An Antitumor Polyether Macrolide from a Marine Sponge, *J. Am. Chem. Soc. 107*: 4796–4798 (1985).

Uemura, D., and Y. Hirata, Halichondrins—Antitumor Polyether Macrolides from a Marine Sponge, *Pure Appl. Chem. 58*: 701–710 (1986).

Chapter 36

Phenolics from Deciduous Leaves and Coniferous Needles as Sources of Novel Control Agents for Lepidopteran Forest Pests

Mamdouh M. Abou-Zaid*, Gary G. Grant, Blair V. Helson, Clifford W. Beninger, and Peter de Groot

Natural Resources Canada, Canadian Forest Service, Great Lakes Forestry Centre, Sault Ste. Marie, ON, Canada P6A 5M7

Introduction

Environmental and public health concerns over the widespread use of conventional pesticides in forestry have provided strong incentives to develop alternative methods of pest control. There has been considerable interest in natural products as new pest control agents, but little evaluation of indigenous North American trees as potential sources of active compounds has been carried out. Evidence is accumulating, however, that an array of secondary metabolites present in these trees provides a degree of natural defense against forest insects. Many of these compounds are phenolic in nature and are found in foliage in various combinations and concentrations. Phenolics (including phenolic acids, phenylpropanoids, and flavonoids) are hydrophilic substances; their common origin is the aromatic precursor shikimic acid. Classes of phenolic compounds found in deciduous and coniferous species include the following (Fig. 36.1): (i) *phenolic acids* such as protocatechuic acid, gallic acid, and ellagic acid; (ii) *phenylpropanoids* such as ρ-coumaric acid, caffeic acid, and ferulic acid; (iii) *flavan-3-ols* such as (+)-catechin, (-)-epicatechin, and (+)-gallocatechin; (iv) *flavanones* such as pinocembrin, naringenin, and eriodictyol; (v) *dihydroflavonols* such as taxifolin and taxifolin glycosides; (vi) *flavones* such as apigenin; and (vii) *flavonol glycosides* such as kaempferol and quercetin glycosides (Abou-Zaid and Nozzolillo 1998 and in press, Abou-Zaid *et al.*, unpublished data).

We have begun a series of investigations to increase our understanding of the role that phenolics play in plant-insect interactions in forestry, particularly those involving lepidopteran defoliators of maples (*Acer* spp.), pines (*Pinus* spp.) and balsam fir (*Abies balsamae* L. Mill.), and to assess the potential of these compounds as environmentally benign insect control agents and as host resistance factors. Phenolics from conifers and maples have not been reported previously as stimulants or deterrents affecting feeding, development, or oviposition of forest lepidopteran pests. However, phenolics from trembling aspen (*Populus tremuloides* Michx.) have been shown to affect the performance of several important tree-feeding lepidopteran lar-

a) Phenolic acids:

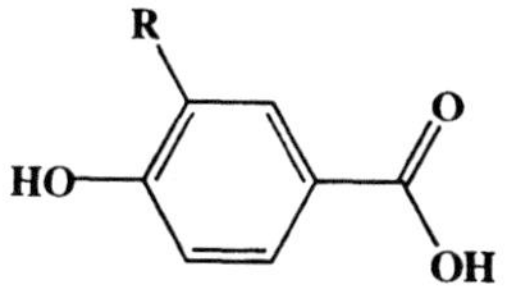

R = OH Protocatechuic acid

R = OCH_3 Vanillic acid

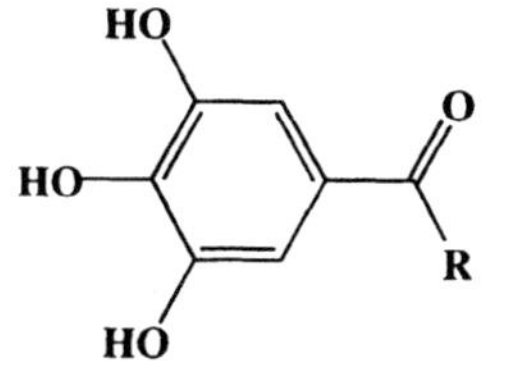

R = OH Gallic acid

R = OCH_3 Gallic acid methyl ester

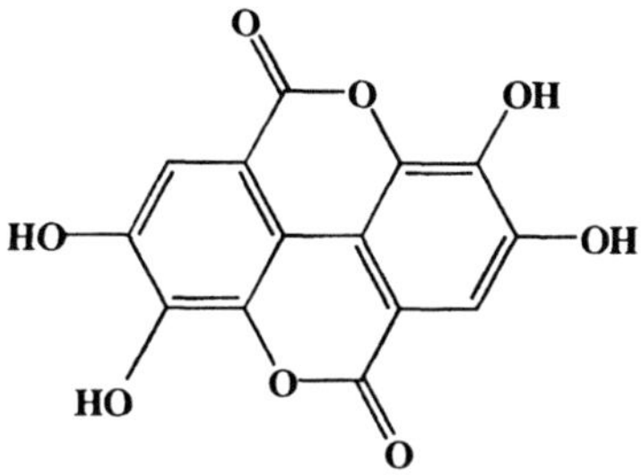

Ellagic acid

b) Phenylpropanoids:

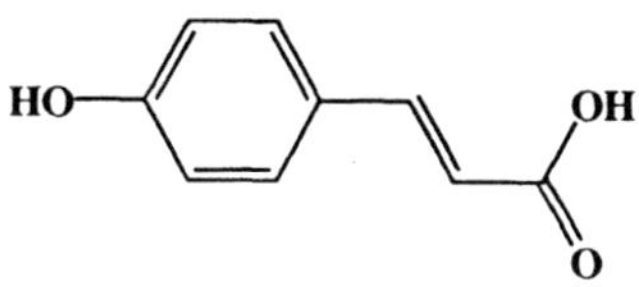

p-Coumaric acid

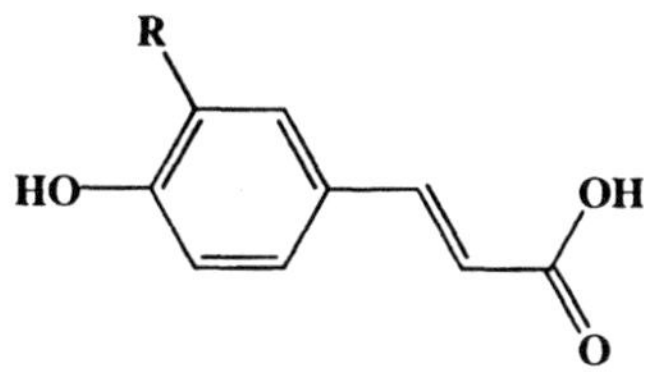

R = OH Caffeic acid

R = OCH_3 Ferulic acid

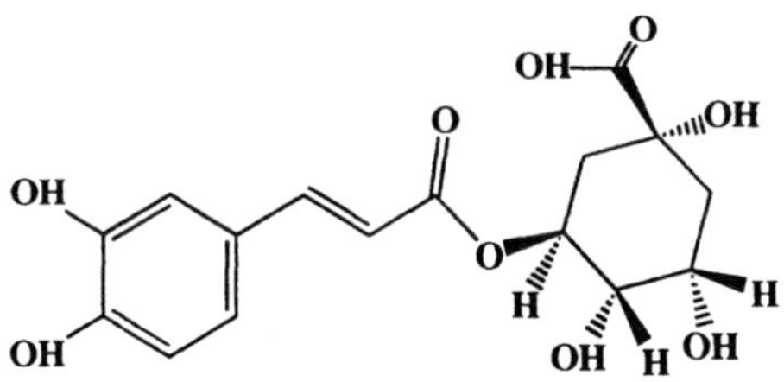

Fig. 36.1. Structures of various classes of phenolic compounds that have been identified in ethanolic extracts of deciduous and coniferous tree species from North America.

vae including the forest tent caterpillar, *Malacosoma disstria* Hbn. (Lindroth and Bloomer 1991); the gypsy moth, *Lymantria dispar* (L.) (Lindroth and Hemming, 1990); and the large aspen tortrix, *Choristoneura conflictana* (Wlk.) (Clausen *et al.* 1989). In addition, there are many examples in the literature that demonstrate the

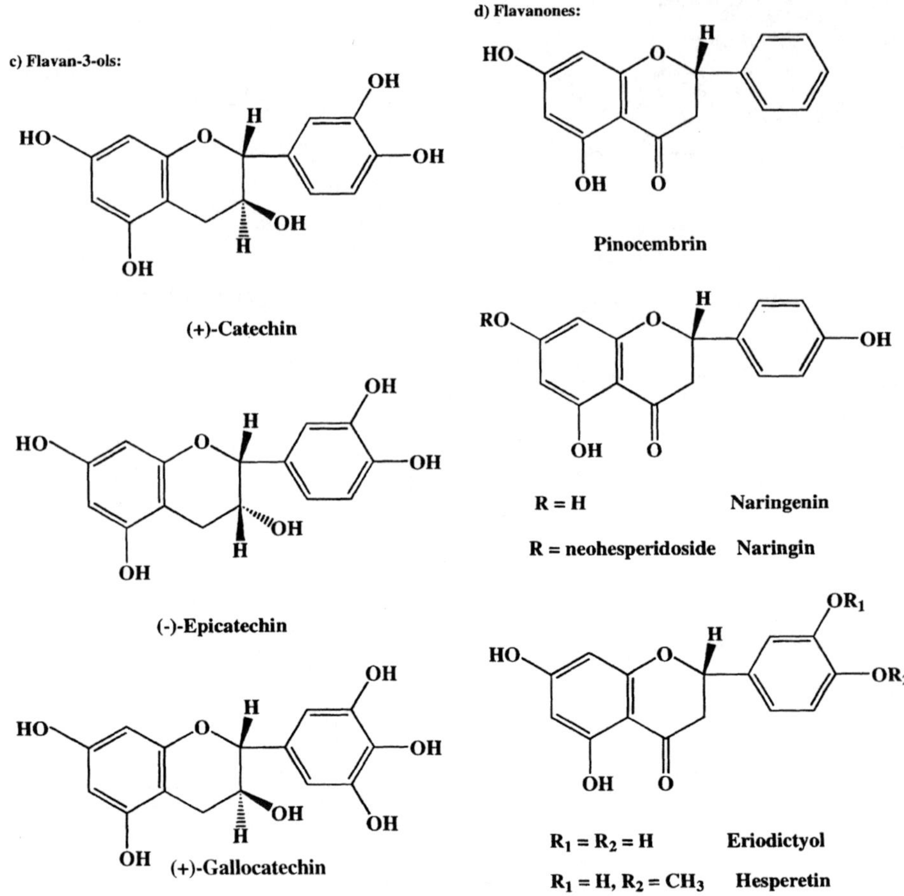

Fig. 36.1. (*Continued*).

importance of phenolics as cues regulating feeding and oviposition in other lepidopteran species (Abou-Zaid *et al.* 1993, Elliger *et al.* 1980, Harborne 1994, Harborne and Grayer 1994, Haribal and Renwick 1996, Isman and Duffey 1982, Lindroth and Peterson 1988, Renwick and Huang 1995, Shaver and Lukefahr 1969, Summers and Felton 1994). Initially, we focused on three key forest insect pests, i.e., the gypsy moth, the forest tent caterpillar, and the spruce budworm, *Choristoneura fumiferana* (Clem.), because these species are major defoliators of North American forests and they are continuously available in large numbers from our insect-rearing facility. In this contribution, we outline our procedures for collecting, analyzing, and identifying phenolics; we describe our bioassay techniques and discuss how phenolics affect insect feeding, development, and oviposition; and we consider their potential role in host plant resistance for pest management. Our long-term objectives are to develop new pest control agents and strategies for use in forest protection.

e) Dihydroflavonols:

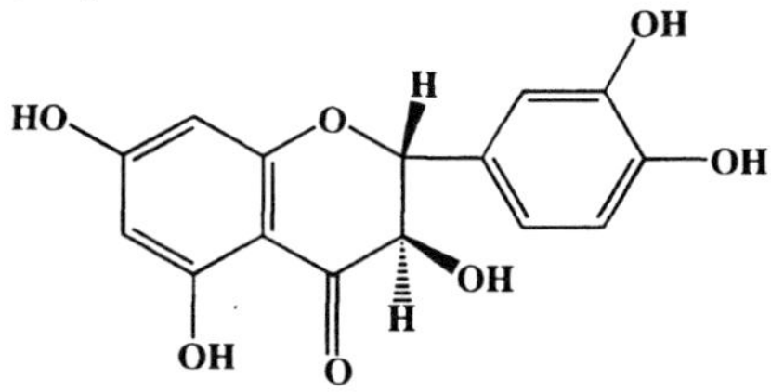

Taxifolin

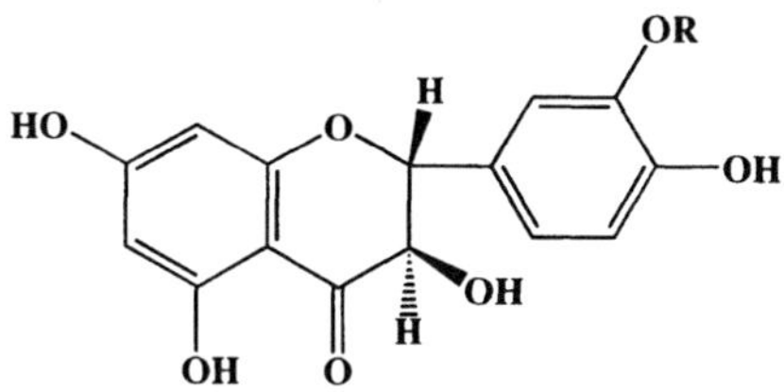

R = glucosyl Taxifolin 3'-*O*-glucoside

f) Flavone glycosides:

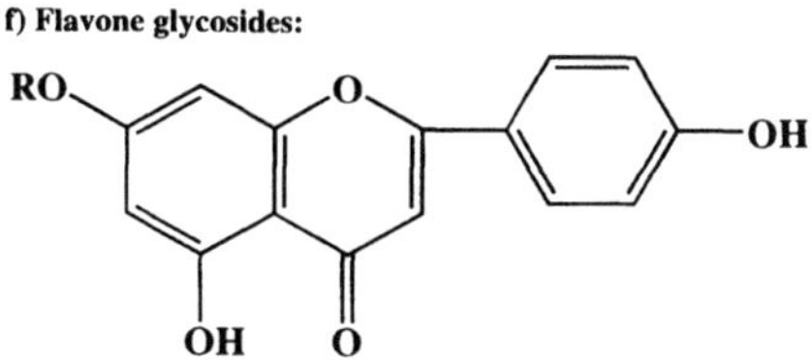

R = H	**Apigenin**
R = glucosyl	**Apigenin 7-*O*-glucoside**

g) Flavonol glycosides:

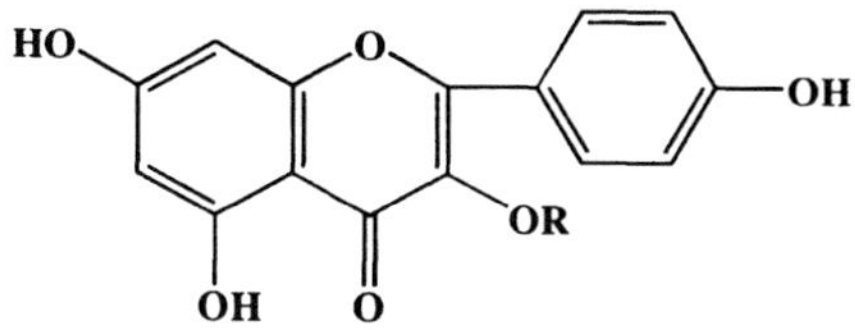

R=glucosyl	**Kaempferol 3-*O*-glucoside**
R=galactosyl	**Kaempferol 3-*O*-galactoside**
R=rhamnosyl	**Kaempferol 3-*O*-rhamnoside**
R=rhamnoglucosyl	**Kaempferol 3-*O*-rutinoside**

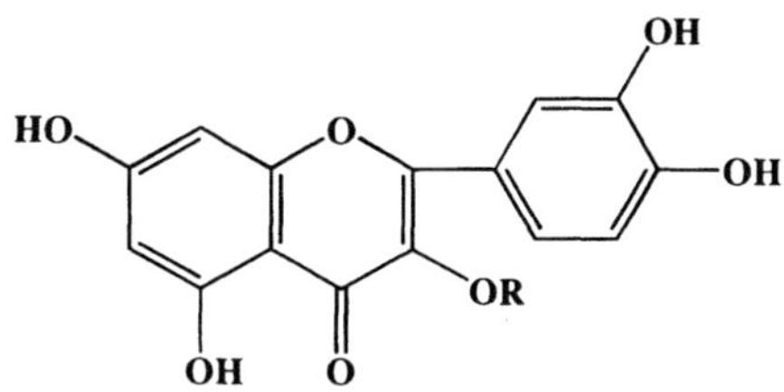

R=glucosyl	**Quercetin 3-*O*-glucoside**
R=rhamnosyl	**Quercetin 3-*O*-rhamnoside**
R=rhamnoglucosyl	**Quercetin 3-*O*-rutinoside (Rutin)**

Fig. 36.1. (*Continued*).

Identification and Assay of Phenolics

The procedure used for examining plant extracts for behavioral activity is summarized in the following.

Plant Identification

Target tree species were sampled and the appropriate tree parts collected. Taxonomic identification of the selected trees was authenticated by an acknowledged authority; voucher specimens were mounted and labeled for permanent accession. Information on each of the plants, including location, a description of habitat, and notes on growth and development parameters, were recorded.

Extraction of Active Principles

The choices of solvents for extracting phenolics (Fig. 36.2) were cold or hot methanol, ethanol, or acetone in combination with water (Harborne 1989, Mabry *et al.* 1970,

Markham 1982). Fresh plant materials (~2 Kg fresh weight) were extracted at room temperature in two steps as follows: (i) steeping for 24 h in 100% EtOH (4 L), followed by chopping in a commercial Waring blender (USA) and decanting the solvent; and (ii) steeping for 24 h with 4 L EtOH/H_2O (1:1). The combined extracts were evaporated under reduced pressure until most of the EtOH had been removed. The residue was freeze-dried and weighed.

Fractionation

The ethanolic freeze-dried extracts were adsorbed onto polyvinylpoly-pyrrolidone (PVPP) powder (Sigma Chemical, USA) packed in a Buchner (USA) funnel (2 L). Elution was carried out at a slow rate, initially with water, followed by aliquots of increasing concentrations (0, 20, 50, 70, and 100%) of ethanol (Fig. 36.3). Fractions were selected for further analysis on the basis of their activity in a bioassay (see below). To isolate compounds, the active fraction was concentrated under

Fresh Foliage

|

Steep with 100% EtOH for 24 h and then grind in blender

|

Drain and steep residue with EtOH:H_2O 1:1 for 24 h

|

Combine ethanolic extracts and concentrate at 50°C under reduced pressure to remove ethanol

|

Fractionate on PVPP (100% H_2O--->100% EtOH) (5 Fractions)

|

Reduce volume of fractions to 10 to 50-mL volumes and freeze dry

Fig. 36.2. Procedure for the preparation of foliage extracts of maple (*Acer*) and pine (*Pinus*) species.

Fractionation on PVPP Column
(Left on the column high MW condensed tannins and gallotannins)

100% H_2O — **50% EtOH** — **100% EtOH**

Sugars Free amino acids and other water soluble nonphenolics

Simple Phenolics

Flavonoid Glycosides

Methylated Aglycones and Chlorophyll

Further Fractionation and Purification

Preparative Column Chromatography
(PVPP, Silica Gel, Sephadex LH-20)

TLC and Paper Chromatography and HPLC Analysis of Purified Compounds

Identification of isolated compounds is achieved by the following: (i) Chemical analyses: Acid hydrolysis; enzymatic hydrolysis; hydrogen peroxide oxidation; (ii) physical analyses: UV, ^{1}H-NMR, ^{13}C-NMR, and FAB-MS.

Fig. 36.3. Outline of the fractionation scheme and identification procedures in the analysis of phenolic compounds found in maple (*Acer* spp.) leaves and pine (*Pinus* spp.) needles. (Highly glycosylated flavonols appear in the 20–50% fractions, and methylated phenolics elute only with higher concentrations of alcohol). Abbreviations: PVPP, polyvinyl-pyrrolidone; MW, molecular weight; TLC, thin-layer chromatography; HPLC, high-performance liquid chromatography; UV, ultraviolet; NMR, nuclear magnetic resonance; FAB-MS, fast atom bombardment mass spectrometry.

vacuum and chromatographed (one-dimensional) on Whatman (England) No.1 chromatography paper using *n*-butanol/acetic acid/water (BAW, 4:1:5, upper phase), water, or acetic acid/water (15:85). Bands detected by absorbance/fluorescence under a short-wave light (254 nm) and long-wave light (366 nm) were eluted on a PVPP column with the following solvent systems: (i) CH_2Cl_2/EtOH/MeCOEt/Me_2CO (1:1:1:1), (ii) EtOH/MeCOEt/Me_2CO/H_2O (1:1:1:1), and (iii) EtOH/H_2O (1:1). Purification was achieved with the aid of a low-pressure liquid chromatograph (Chemco low-prep pump, (Japan) model 9 1-M-8R, with 6-port valve, max. 80 mL/min). Final cleanup of the compounds was achieved on a Sephadex LH-20 column (Sweden) (1 cm × 50 cm), using methanol as the eluting solvent, a step essential to obtaining good spectra of purified compounds.

Identification of Purified Compounds

Structural elucidation (Fig. 36.3) was achieved by the following analyses: (i) chemical analyses: acid hydrolysis in 2 and 0.1 M HCl (mild hydrolysis) at 100°C for 60 min; enzymatic hydrolysis with β-glucosidase using an acetate buffer (pH 5); hydrogen peroxide oxidation; and (ii) physical analyses: ultraviolet (UV) spectroscopy; ^{1}H nuclear magnetic resonance (NMR); ^{13}C NMR and (positive and negative) fast atom bombardment (FAB)-mass spectroscopy (Harborne 1989, Mabry *et al.* 1970, Markham 1982). An important step is confirmation of the structure by comparison with authentic samples when available.

High-Performance Liquid Chromatography

A dilute solution of freeze-dried crude extract (10 mg/mL) was filtered through an Acrodisc 4 CR PTFE, 0.45-μm filter (USA), and 10 μL was injected onto an HPLC column with or without spiking with a standard. A Waters Delta Prep 4000 Liquid Chromatography system (USA) equipped with computer-operated Millennium 2010 software, (USA) an autoscan photodiode array spectrophotometric detector 996, and a Waters reverse-phase analytical column (3.9 × 150 mm i.d.) packed with Nova-Pak (USA) C18 60° 4 m were used. Modified methods of Van Sumere *et al.* (1994) and Slacanin *et al.* (1991) were followed. A gradient chromatographic technique was applied at room temperature as follows: solvent A = 5% aqueous HCOOH; solvent B = CH_3CN/MeOH (5:95); the flow rate is set at 0.9 mL/min; elution program: 0–3 min, isocratic (85% A:15% B); 3–8 min, gradient (85% A:15% B–76%A:24%B); 8–11 min, isocratic (76% A:24% B); 11–18 min, gradient (76% A:24% B–66% A:34% B); 18–28 min, gradient (66% A:34% B–54% A:46% B); 28–39 min, gradient (54% A:46% B–5% A:95% B); 39–42 min, isocratic (5% A:95% B); 42–47 min, gradient (5% A:95% B–85% A:15% B); 47–55 min, isocratic (85% A:15% B). Two fixed detection wavelengths at 280 and 350 nm were used, and resolved peaks were scanned by the photodiode array detector from 220 to 390 nm. Compounds were identified on the basis of retention time and UV spectrum. Peak height, measured as absorbance at 280 and 350 nm, gave an indication of abundance. Such HPLC analyses were performed in triplicate.

Bioassay-Guided Isolation of Active Constituents

A small portion of the crude extract of each fraction from the PVPP column was removed for testing on insects. Active extracts were fractionated and bioassayed. At present, we are using several feeding and oviposition bioassays to assess activity.

Leaf Disk Feeding Preference Assay

In a feeding choice test, 2-cm diameter disks were cut with a #13 cork borer from freshly collected leaves of a species susceptible to the test insect and placed in 9-cm petri plates. The disks were treated alternately with acetone/water only (control) or with acetone/water solutions of the extract or fraction being tested. The solutions (50 μL) were applied to both sides of a leaf disk and allowed to dry completely. The disks were pinned in place through small holes in the bottom of the plate. Five newly molted larvae of the test insect were placed in each plate. After 24 h, the leaf disks were photocopied and the percentage of each disk remaining was determined with an Image Analyzer (Artek, Image Editor and Video Memory, Model 940 (USA). The antifeedant index (AI) for each plate was calculated as $\%AI = (C-T)/(C + T) \times 100$, where C is the percentage of two control disks eaten and T is the percentage of two treated disks eaten. For tests with individual compounds, a modified procedure was developed to permit testing of small quantities (10–20 mg) of each compound at 1% concentrations. Smaller, 1.5-cm diameter leaf disks were cut with a #8 cork borer and placed in petri dishes as above. Four applications of 12.5 μL (50 μL total) were applied to the upper surface only, to avoid loss of material. Larvae were added and the percentage of each disk eaten was visually estimated at 1- to 2-h intervals until 35–65% of the disks were consumed in a plate. Each compound was tested in 10 plates, each with two treated disks and two control disks.

Diet Incorporation: Larval Development Assay

Whole extracts, extract fractions, or individual compounds were dissolved in acetone and incorporated into a warm artificial diet. After thorough mixing in a Waring blender, the diet was poured into 9-cm plastic petri plates. The plates were sealed with parafilm and kept in the dark at 4°C until needed. A plug of this diet was cut with a #5 cork borer and placed in a petri plate with 10 preweighed first or second instar larvae. The diet plug was changed every 2 d. Typically six plates per treatment plus six controls with diet treated with acetone only were set up for each experiment. The larvae were kept in an environmental chamber at 21°C, 80% relative humidity with 16 h light:8 h dark, and weighed at regular intervals until pupation. The mean larval weights in each dish were used for statistical analysis.

Nutritional indices have also been measured in diet-incorporation feeding assays (Abou-Zaid and Nozzolillo 1998, Abou-Zaid *et al.*, unpublished data). Naive fourth instar larvae were placed individually in vials containing treated diet. After 4 d of feeding, each larva, uneaten diet, and frass were oven dried at 60°C for 3 d and weighed separately. A pretest group of larvae and artificial diet was also dried to estimate initial dry weight. From these weights, several indices were calculated according to Waldbauer (1968), including relative consumption rate, relative growth rate, approximate digestibility, efficiency of conversion of ingested food, and efficiency of conversion of digested food. The values of these indices compared with those of controls can segregate behavioral and physiologic dietary effects and indicate mechanisms of action (Nicol *et al.* 1997).

Oviposition Preference Assay

A simple, dual-choice bioassay was developed to screen foliage extracts and phenolic compounds for ovipositional activity against the spruce budworm (Grant and

Langevin 1994). Briefly, individual pairs of mating insects were placed in small screen cages, lined top and bottom with filter paper, which served as the oviposition substrate. Suitable concentrations of foliage extract (5–15 mg/mL) or test chemical (10–100 mmol/L) were pipetted (0.25 mL) evenly over one half (31.8 cm^2) of each filter paper, and the solvent used to dissolve the test material was applied to the other half. After two nights, the number of egg masses on treated and control surfaces were counted. Stimulatory or deterrent activity was expressed either as a ratio of egg masses or eggs on treated and control surfaces or by an oviposition index (Grant and Langevin 1995) that is analogous to the feeding index described above. By counting the number of eggs per egg mass, the effects on clutch size could also be evaluated (Grant and Langevin 1995). Deterrent material was evaluated further in laboratory assays using host foliage as the treated substrate.

Phenolics in Coniferous and Broadleaf Trees

Phenolics and polyphenols are major constituents of coniferous shoots and needles (Fig. 36.1) (Abou-Zaid and Nozzolillo 1991 and 1998, Beninger and Abou-Zaid 1997, Kraus and Spiteller 1997a and 1997b, Strack *et al.* 1989). Abou-Zaid and Nozzolillo (1991) have identified flavonol glycosides in jack pine (*Pinus banksiana* Lamb.). The presence of these compounds may be responsible for jack pine's resistance to attack by gypsy moth, whereas their absence in white pine (*P. strobus* L.) may account for its susceptibility to this insect. Subsequently, Beninger and Abou-Zaid (1997) described the inhibition of early instar gypsy moth development by flavonol glycosides from four pine species. In the case of red maple (*Acer rubrum* L.), phenolic constituents in leaves play a role in the resistance to feeding by forest tent caterpillar larvae.

HPLC separations of phenolic compounds in crude ethanolic extracts of *Acer saccharum* Marsh (sugar maple), *A. rubrum*, *P. resinosa* Ait. (red pine), *P. strobus*, *P. banksiana*, *P. sylvestris* L. (scotch pine) (Pinaceae), as shown by absorbance at 280 nm and 350 nm (Figs. 36.4–36.6), demonstrate the wide array of compounds produced and the characteristic "fingerprint" of the mixtures found in each species. Phenolic acids, phenylpropanoids, catechins, flavanones, flavones, and flavonols have been identified in both *Pinus* needles and *Acer* leaves (Fig. 36.1) (Abou-Zaid and Nozzolillo 1998 and in press, Abou-Zaid *et al.*, unpublished data, Delendick, 1984). Typically, the elution order (in blocked time periods) of various classes of phenolic compounds found in dilute ethanolic extracts (10 mg freeze-dried extract/mL) of maple (*Acer* spp.) leaves and pine (*Pinus* spp.) needles is as follows: 0–10 min, phenolic acid aglycones; 10–25 min, phenolic acid *O*-glycosides, phenylpropanoid aglycones, and their glycosides, and flavan-3-ols; 20–35 min, the *O*-glycosides of flavone and flavonol and their *C*-glycosides; 30–45 min, the aglycones and glycosides of flavanone and dihydroflavanol, the aglycones of flavone and flavonol and their acylated glycosides; and 45–55 min, proanthocyanidins (Figs. 36.4–36.6). The identification of phenolic compounds in balsam fir (*A. bal-*

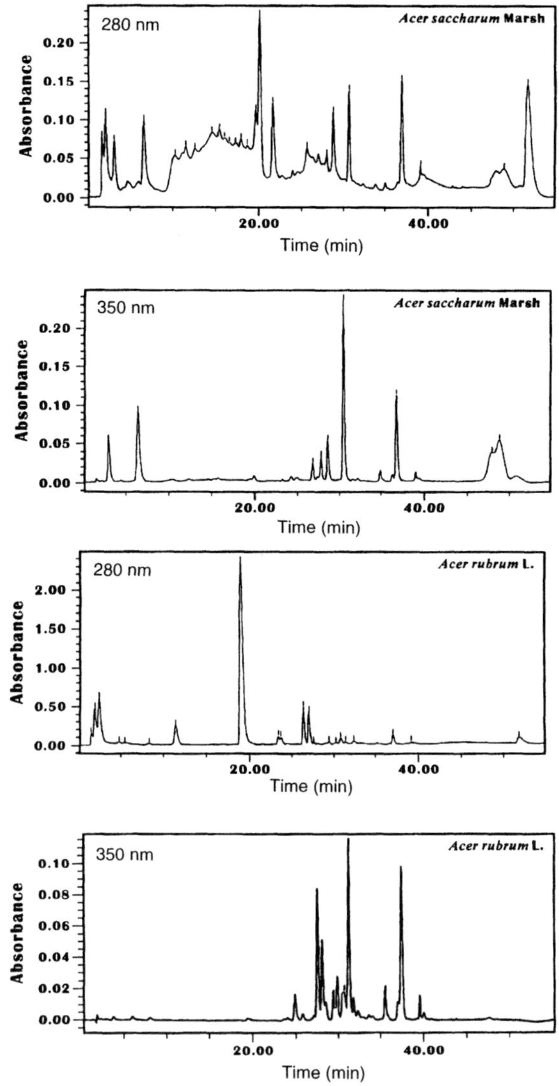

Fig. 36.4. High-performance liquid chromatography (HPLC) separation of phenolic compounds in dilute ethanolic extracts (10 mg freeze-dried extract/mL) of *Acer saccharum* Marsh (sugar maple) and *Acer rubrum* L. (red maple) leaves as shown by absorbance at 280 and 350 nm. The elution order of various classes of phenolic compounds found in dilute ethanolic extracts of maple leaves (Fig. 36.4) and pine needles (Figs. 36.5 and 36.6) is characterized as follows: 0–10 min, phenolic acid aglycones; 10–25 min, phenolic acid *O*-glycosides, phenylpropanoid aglycones, and their glycosides, and flavan-3-ols; 20–35 min, the *O*-glycosides of flavone and flavonol and their *C*-glycosides; 30–45 min, the aglycones and glycosides of flavanone and dihydroflavanol, the aglycones of flavone and flavonol, and their acylated glycosides; and 45–55 min, proanthocyanidins.

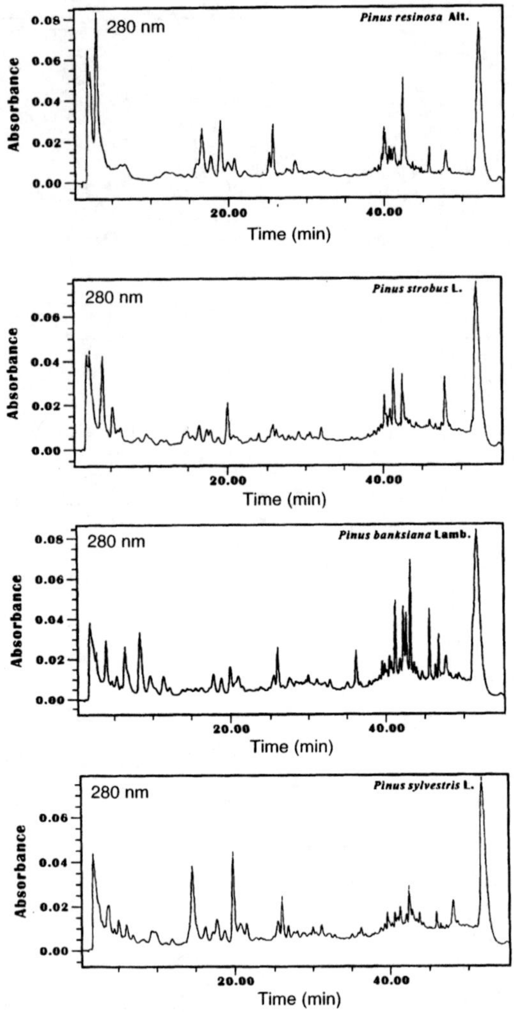

Fig. 36.5. High-performance liquid chromatography (HPLC) separation of phenolic compounds in dilute ethanolic needle extracts (10 mg freeze-dried extract/mL) of *Pinus resinosa* Ait. (red pine); *P. strobus* L. (white pine); *P. banksiana* Lamb. (jack pine); and *P. sylvestris* L. (scotch pine) as shown by absorbance at 280 nm.

samea) is currently in progress. These compounds will be assessed initially against the oviposition behavior of the spruce budworm.

Phenolics and Insect Feeding and Growth

Forest tent caterpillar larvae will not feed on leaves of red maple but do feed readily on sugar maple leaves (Nicol *et al.* 1997). An ethanolic extract of red maple

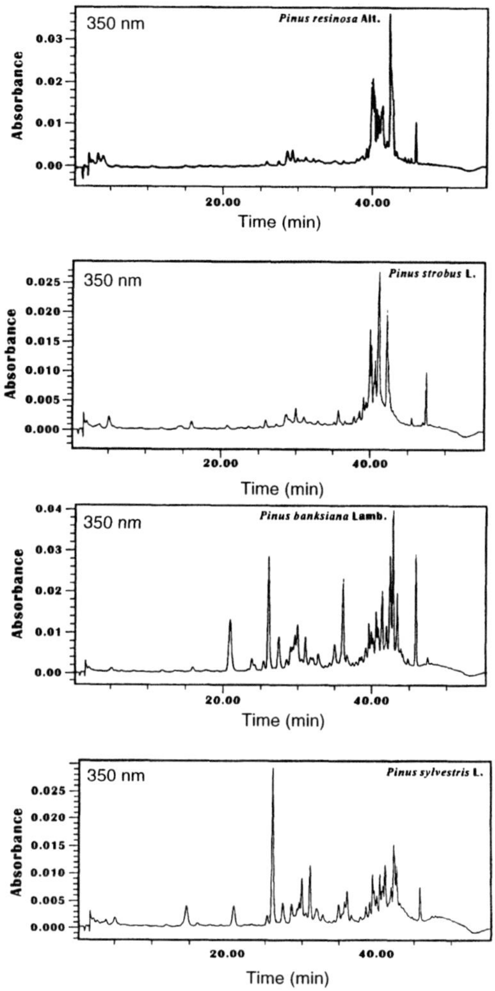

Fig. 36.6. High-performance liquid chromatography (HPLC) separation of phenolic compounds in dilute ethanolic needles extracts (10 mg freeze-dried extract/mL) of *Pinus resinosa* Ait. (red pine); *P. strobus* L. (white pine); *P. banksiana* Lamb. (jack pine); and *P. sylvestris* L. (scotch pine) as shown by absorbance at 350 nm.

leaves applied to aspen leaf disks deterred forest tent caterpillar larvae from feeding (Abou-Zaid *et al.*, unpublished data). A fraction of this extract containing phenolic compounds was highly deterrent in these choice feeding assays. Similarly, both the extract and phenolic fraction in the diet resulted in significant reductions in growth and survival of larvae, suggesting that phenolic compounds in red maple could be involved in its resistance to feeding by forest tent caterpillar. When gallic acid and gallic acid methyl ester, two common phenolic acids found in red maple, were tested on aspen leaf disks, they acted as deterrents to a significant degree (Fig. 36.7). At low concentrations, gallic acid methyl ester was a stronger deterrent than gallic acid. Additional phenolic compounds, all flavonol glycosides, from red maple extract did not act as deterrents on aspen leaf disks (Fig. 36.7). Interestingly, kaempferol 3-O-rhamnoglucoside significantly stimulated feeding (Fig. 36.7). In nutritional indices tests, gallic acid methyl ester and ellagic acid incorporated into an artificial diet also exhibited antifeedant activity, whereas the effects of gallic acid and rutin were insignificant (Abou-Zaid *et al.*, unpublished data). Ellagic acid reduced the efficiency of conversion of digested food, suggesting that it may also cause postdigestive growth reduction. Although some individual compounds found in red maple deter feeding, the overall mechanism of resistance will likely be the product of a complex interaction among the >20 phenolic compounds in red maple leaves. Investigations are continuing on the levels and types of bioactivity of these red maple compounds for the forest tent caterpillar to gain a better understanding of this resistance mechanism.

By contrast, forest tent caterpillar larvae will feed on sugar maple, and extracts

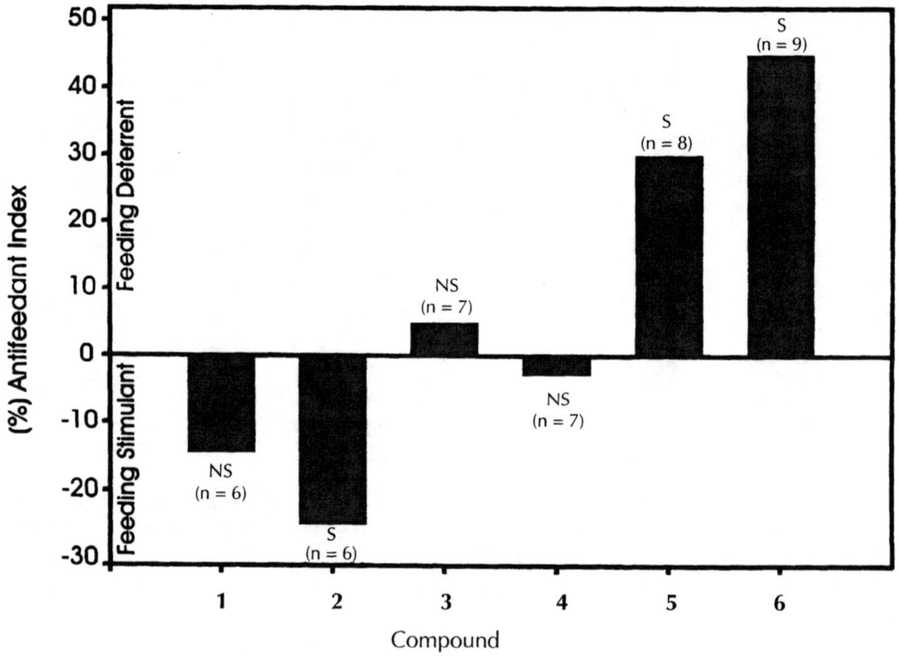

Fig. 36.7. Effects of phenolic compounds (1%) isolated from red maple on feeding by forest tent caterpillar larvae in leaf disk feeding preference assay: (**1**) kaempferol 3-*O*-β-D-glucoside (astragalin); (**2**) kaempferol-3-*O*-β-rhamnoglucoside (nicotoflorin); (**3**) quercetin 3-*O*-β-D-glucoside (isoquercetin); (**4**) quercetin 3-*O*-rhamnoglucoside (rutin); (**5**) 3,4,5-trihydroxybenzoic acid (gallic acid); and (**6**) methyl 3,4,5-trihydroxybenzoate (gallic acid methyl ester). S, significant antifeedant or phagostimulant ($P < 0.05$, one-sample *t*-test).

of sugar maple leaves do not possess antifeedant activity when applied to aspen leaf disks. However, feeding on sugar maple leaves is detrimental to larval growth, and sugar maple extracts in diet cause severe growth inhibition (Nicol 1996). Nutritional indices tests indicated that extracts caused postdigestive toxicity to larvae. Further investigations will attempt to identify the cause of this effect and specifically the role of phenolic compounds present in sugar maple.

Phenolics and Insect Oviposition

Although phenolics (typically flavonol glycosides) are important host stimuli, eliciting oviposition from several papilionid (Harborne and Grayer 1994) and danaid (Haribal and Renwick 1996) butterfly species, we are not aware of any reports that flavonoids or other phenolic compounds serve as oviposition stimulants for moth species. However, in empirical tests, xanthoxylin deterred oviposition of the angoumois grain moth, *Sitotroga cerealella* (Olivier) on treated paper substrate (Ge and Weston 1995). In addition, rutin applied to host foliage of the diamond-backed moth, *Plutella xylostella* (L.), (Tabashnik 1985), and naringin, applied to the host foliage of the tobacco budworm, *Heliothis virescens* (F.), (Ramaswamy *et al.* 1992) were effective oviposition deterrents.

Studies on the effects of phenolic stimuli on oviposition behavior of the spruce budworm, and several other *Choristoneura* species have been initiated. Polar extract fractions of host (*Abies, Picea*) and nonhost (*Pinus*) foliage deterred oviposition of the spruce budworm in laboratory bioassays (Grant and Langevin 1995). Further fractionation of active extracts of *A. balsamea* by column chromatography produced phenolic-containing fractions that were generally deterrent, although a few were stimulatory. Active compounds in these fractions are currently being identified. In empirical tests, representative phenolics have shown significant deterrent activity; none so far was stimulatory (Table 36.1). Some of these compounds have been evaluated on host foliage, but only naringin has deterred oviposition. Single-cell electrophysiology recording from contact chemoreceptors (van Loon 1990) will be incorporated into these studies to screen fractions or identified compounds rapidly (van Loon and van Meer 1991) in an attempt to simplify behavioral bioassays.

Some lepidopteran contact chemoreceptor neurons respond specifically to phenolic compounds (van Loon 1990). However, flavonol glycosides, such as rutin, have a pale yellow color potentially visible to insects, and it is possible that their behavioral effects on diurnally active adult insects (e.g., spruce budworm) could be due to visual rather than chemosensory responses or both. In experiments with the diamond-backed moth, Tabashnik (1985) found that less oviposition occurred on rutin-treated host plants held under continuous light conditions than on similarly treated plants held under continuous dark conditions. He suggested that under dark conditions, the deterrent activity of rutin could be mediated by female chemoreceptors or mechanoreceptors, whereas under light conditions, vision played an addi-

TABLE 36.1
Oviposition Response in a Laboratory Bioassay of Spruce Budworm, *Choristoneura fumiferana,* to Representative Phenolic Compounds Applied to Filter Paper Substrate

Compound (100 mmol/L)	Females (n)	Ratio of egg masses Treated/Control	*P*-value[a]
Phenolic acids			
Vanillic acid	44	0.43	0.001
Phenylpropanoids			
Caffeic acid	42	0.56	0.01
Chlorogenic acid	44	0.50	0.005
Flavan-3-ols			
(+)-Catechin	46	0.91	NS
Flavanones			
Naringenin	46	1.05	NS
Naringin	42	0.48	0.006
Flavonols			
Quercetin	40	0.81	NS
Rutin	43	0.67	0.02

[a]Wilcoxon matched pairs, signed-rank test; NS, not significant, $P > 0.05$.

tional role in the deterrent activity of rutin. The notion that flavonoids and other compounds can deter oviposition because of the visual signals they impart is not widely appreciated; hence, this mode of action is usually not considered in oviposition experiments with these phytochemicals.

Host Plant Resistance and Variation of Phenolics

Knowledge of the variation in phenolic compounds among plant species has been an effective tool for solving taxonomic and phylogenetic questions (Niemann 1988). Within species, the quantity and presence of phenolics can vary with organs, cells, and age of the individual plant, its response to damage, as well as the season and geographic location (Baldwin *et al.* 1987, Niemann 1988). Genetic variation of phenolic compounds is also known and forms the basis, in part, for developing plants that are resistant or partially resistant to insects. Yet, despite the many documented cases showing significant intra- and interspecific variation in the susceptibility of *Pinus* to insects, breeding for insect resistance is rare (de Groot and Turgeon 1998). Nonetheless, the study of intraspecific phenolic variation in forest trees is an emerging field and will help identify resistant or partially resistant genotypes and further our understanding of the complex dynamics of insect-plant interactions.

Of the four species of pine studied in our laboratory, variation in phenolic compounds is best known for *P. sylvestris*. There is considerable seasonal variation in total phenolic concentrations of *P. sylvestris* seedlings in Europe, with the highest concentrations occurring in the spring (Nerg *et al.* 1994). Two flavanoid chemomorphs of *P. sylvestris*, one rich in prodelphinidin and lacking taxifolin, and the other with taxifolin and exhibiting more *o*-dihydroxyflavonol, including quercetin have been found (Laracine-Pittet and Lebreton 1988). The first chemomorph occurs in high-latitude or high-altitude populations; this taxifolin dimorphism has been confirmed (Lundgren and Theander 1988). One simple locus may be responsible for most of the differences in the inheritance pattern of these chemomorphs (Yazdani and Lebreton 1991).

P. banksiana is another widely distributed species, and it would not be surprising to find phytochemical heterogeneity as in *P. sylvestris*; however, Abou-Zaid and Nozzolillo (1991) found that fully developed needles of 4-mo-old, 16-mo-old, and mature plants appeared to be qualitatively identical in their flavonol profiles, but quantitative differences were not assessed. Currently, we are examining the phenolic and terpenoid profiles of 10 families of *P. banksiana* to determine seasonal and genetic variation and their role on insect herbivory. Flavonol glycosides from *P. resinosa* and *P. strobus* have also been identified recently (Beninger and Abou-Zaid 1997), but we are not aware of any studies that have examined the seasonal, geographic, or genetic variation of phenolics for these trees.

Little work has been done on phenolic variation in maples (*Acer* spp.). Baldwin *et al.* (1987) found that the within-tree, individual, and seasonal variations in leaf tannins of *A. saccharum* were too great to identify variation arising from family, population, or even geographic resources. Our recent identification of several phenolics and flavonol glycosides of *A. saccharum* and *A. rubrum* (Abou-Zaid *et al.* unpubished data) will provide a useful reference point for examining within-species variation of these compounds.

Conclusions

It is anticipated that our studies will increase the understanding of how phenolic compounds affect the preference, performance, and fitness of forest insects. These data may lead to the development of new pest control agents with a primarily behavioral mode of action (e.g., feeding and oviposition deterrents). This knowledge may also be useful in predicting the vulnerability and susceptibility (risk factors) of forest stands and help identify the range of host plant resistance to insect pests. We agree with van Emden (1991) (see also Harborne and Grayer 1994) that a prudent ‹approach will be to develop cultivars or clones with partial resistance to reduce selection pressure on pests and to maintain a natural enemy complex. There is clear evidence from our studies with red maple that phenolics are resistance factors that deter larval feeding and development of forest tent caterpillar larvae. We have found that some oviposition deterrents have interesting additional effects of significant value to their potential use as control agents. In dual-choice laboratory bioassays involving cotton gauze substrates, newly emerged spruce budworm larvae ‹were deterred from spinning up in treated gauze and diapausing. Consequently, these larvae were forced to wander in search of more suitable (i.e., untreated) hibernation sites. Control agents with such dual effects are of considerable practical value because the effects are obtained with a single treatment application. Phenolics in North American forest trees are potential sources of new control agents that can be applied in novel control strategies.

Acknowledgments

We acknowledge the financial support of the Natural Resources Canada, Pest Management Methods-Natural Products and Semiochemicals Network, Science and Technology Opportunities Fund and Integrated Forest Pest Management, Green Plan Initiative. We thank Dr. C. Nozzolillo, University of Ottawa, for reviewing early drafts of this manuscript.

References

Abou-Zaid, M.M., and C. Nozzolillo, 1-*O*-Galloyl-α-L-rhamnose from *Acer rubrum. Phytochemistry.* In Press.

Abou-Zaid, M.M., C.W. Beninger, J.T. Arnason, and C. Nozzolillo, The Effect of One Flavanone, Two Catechins and Four Flavonols on Mortality and Growth of the European Corn Borer (*Ostrinia nubilalis* Hubner), *Biochem. Syst. Ecol. 21:* 415–420 (1993).

Abou-Zaid, M.M., and C. Nozzolillo, Flavonol Glycosides from the Needles of *Pinus banksiana, Biochem. Syst. Ecol. 19:* 237–240 (1991).

Abou-Zaid, M.M., and M. Nozzolillo, Phenolic Composition of Needles of Four Pine Species Growing in the Great Lakes Forest Region of Canada, *Polyphenol Commun. 2:* 599–600 (1998).

Baldwin, I.T., J.C. Schultz, and D. Ward, Patterns and Sources of Leaf Tannin Variation in Yellow Birch (*Betula allegheniensis*) and Sugar Maple (*Acer saccharum*), *J. Chem. Ecol. 13:* 1069–1079 (1987).

Beninger, C.W., and M.M. Abou-Zaid, Flavonoids from Four Pine Species That Inhibit Larval Gypsy Moth (Lepidoptera: Lymantriidae) Development, *Biochem. Syst. Ecol. 25:* 505–512 (1997).

Clausen, T.P., P.B. Reichardt, J.P. Bryant, R.A. Werner, K. Post, and K. Frisby, Chemical Model for Short-Term in Quaking Aspen (*Populus tremuloides*) Foliage Against Herbivores, *J. Chem. Ecol. 15:* 2335 (1989).

de Groot, P. and J.J. Turgeon, Insect-Pine Interactions, in *Ecology and Biogeography of Pinus*, edited by D.M. Richardson, Cambridge University Press, Cambridge, 1998, pp. 354–380.

Delendick, T.J., Reconsideration of Two Infraspecific Taxa of the Fullmoon Maple *Acer japonicum* (Aceraceae), *Brittonia 36:* 49–58 (1984).

Elliger, C.A., B.C. Chan, and A.C. Waiss, Jr., Flavonoids as Larval Growth‹ Inhibitors—Structural Factors Governing Toxicity, *Naturwissenschaften 67:* 358–360 (1980).

Ge, X., and P.A. Weston, Ovipositional and Feeding Deterrent from Chinese Prickly Ash Against Angoumois Grain Moth (Lepidoptera: Galechiidae), *J. Econ. Entomol. 88:* 1771–1775 (1995).

Grant, G.G., and D. Langevin, Oviposition Response of Four *Choristoneura* Species to Chemical and Physical Stimuli of Host and Non-Host Trees, *Environ. Entomol. 23:* 447–456 (1994).

Grant, G.G., and D. Langevin, Oviposition Deterrence, Stimulation, and Effect on Clutch Size of *Choristoneura* (Lepidoptera: Tortricidae) Species by Extract Fractions of Host and Non-Host Foliage, *Environ. Entomol. 24:* 1656–1663 (1995).

Harborne, J.B., *Methods in Plant Biochemistry*, Vol. 1, Academic Press, London, 1989.

Harborne, J.B., *Flavonoid Pigments, in Herbivores*, Their Interaction with Secondary Plant Metabolites, Vol I, The Chemical Participants, 2nd edn., edited by G.A. Rosenthal, and M.R. Berenbaum, Academic Press, San Diego, 1991, pp. 389–429.

Harborne, J.B., *Introduction to Ecological Biochemistry*, 4th edn., Academic Press, New York, 1994.

Harborne, J.B., and R.J. Grayer, Flavonoids and Insects, in *The Flavonoids, Advances in Research Since 1986*, edited by J.B. Harborne, Chapman and Hall, London, 1994, pp. 589–618.

Haribal, M., and J.A.A. Renwick, Oviposition Stimulants for the Monarch Butterfly: Flavonol Glycosides from *Asclepias curassavica, Phytochemistry 41:* 139–144 (1996).

Isman, M.B., and S.S. Duffey, Toxicity of Tomato Phenolic Compounds to the Fruitworm, *Heliothis zea, Entomol. Exp. Appl. 31:* 370–376 (1982).

Krauss, C., and G. Spiteller, Comparison of Phenolic Compounds from Galls and Shoots of *Picea glauca, Phytochemistry 44:* 59–67 (1997a).

Kraus, C., and G. Spiteller, Phenolic Compounds from Ageing Shoots of *Picea abies, Z. Naturforsch. Sect. C, J. Biosci. 52:* 308–312 (1997b).

Laracine-Pittet, C., and P. Lebreton, Flavonoid Variability Within *Pinus sylvestris, Phytochemistry 27:* 2663–2666 (1988).

Lindroth, R.L., and M.S. Bloomer, Biochemical Ecology of the Forest Tent Caterpillar: Responses to Dietary Protein and Phenolic Glycosides, *Oecologia 86:* 408–413 (1991).

Lindroth, R.L., and J.D.C. Hemming, Responses of the Gypsy Moth (Lepidoptera: Lymantriidae) to Tremulacin, an Aspen Phenolic Glycoside, *Environ. Entomol. 19:* 842–847 (1990).

Lindroth, R.L., and S.S. Peterson, Effects of Plant Phenols on Performance of Southern Armyworm Larvae, *Oecologia 75:* 185–189 (1988).

Lundgren, D., and O. Theander, *Cis-* and *trans-*Dihydroquercetin Glucosides from Needles of *Pinus sylvestris, Phytochemistry 27:* 829–832 (1988).

Mabry, T.J., K.R. Markham, and M.B. Thomas, *The Systematic Identification of Flavonoids,* Springer-Verlag, Berlin, 1970.

Markham, K.R., *Techniques of Flavonoid Identification*, Academic Press, London, 1982.

Nerg, A., P. Kainulainen, M. Vuorinen, M. Hanso, J.K. Holopainen, and T. Kurkela, Seasonal and Geographic Variation of Terpenes, Resins Acids and Total Phenolics in Nursery Grown Seedlings of Scots Pine (*Pinus sylvestris* L.), *New Phytol. 128:* 703–713 (1994).

Niemann, G.J., Distribution and Evolution of the Flavonoids in Gymnosperms, in *The Flavonoids*, edited by J.B. Harborne, Chapman and Hall, London, 1988, pp. 469–478.

Nicol, R.W., The Activity of the Phytochemical Defenses of Red Maple (*Acer rubrum* L.) and Sugar Maple (*Acer saccharum* Marsh.) Against the Forest Tent Caterpillar (*Malacosoma disstria* Hübner), M.Sc. Thesis, University of Ottawa, Ottawa, Canada, 1996.

Nicol, R.W., J.T. Arnason, B. Helson, and M.M. Abou-Zaid, Effect of Host and Non-Host Trees on the Growth and Development of the Forest Tent Caterpillar, *Malacosoma disstria* Hubner (Lepidoptera: Lasiocampidae), *Can. Entomol. 129:* 995–1003 (1997).

Ramaswamy, S.B., N.E. Cohen, and F.E. Hanson, Deterrence of Feeding and Oviposition Responses of Adult *Heliothis virescens* by Some Compounds Bitter-Tasting to Humans, *Entomol. Exp. Appl. 65:* 81–93 (1992).

Renwick, J.A.A., and X.P. Huang, Rejection of Host Plant by Larvae of Cabbage Butterfly: Diet-Dependent Sensitivity to an Antifeedant, *J. Chem. Ecol. 21:* 465–475 (1995).

Shaver, T.N., and M.J. Lukefahr, Effect of Flavonoid Pigments and Gossypol on Growth and Development of the Bollworm, Tobacco Budworm, and Pink Bollworm, *J. Econ. Entomol. 62:* 643–646 (1969).

Slacanin, I., A. Marston, K. Hostettmann, N. Delabays, and C. Darabellay, Isolation and Determination of Flavonol Glycosides from *Epilobium* Species, *J. Chromatogr. 557:* 391–394 (1991).

Strack, D., J. Heilemann, V. Wray, and H. Dirks, Structures and Accumulation Patterns of Soluble and Insoluble Phenolics from Norway Spruce Needles, *Phytochemistry 28:* 2071–2078 (1989).

Summers, C.B., and G.W. Felton, Prooxidant Effects of Phenolic Acids on Generalist Herbivore *Helicoverpa zea* (Lepidoptera: Noctuidae): Potential Mode of Action for Phenolic Compounds in Plant Anti-Herbivore Chemistry, *Insect Biochem. Mol. Biol. 24:* 943–953 (1994).

Tabashnik, B.E., Deterrence of Diamondback Moth (Lepidoptera: Plutellidae) Oviposition by Plant Compounds, *Environ. Entomol. 14:* 575–587 (1985).

Van Emden, H.F., The Role of Host Plant Resistance in Insect Pest Mismanagement, *Bull. Entomol. Res. 81:* 123–126 (1991).

Van Loon, J.J.A., Chemoreception of Phenolic Acids and Flavonoids in Larvae of Two Species of *Pieris, J. Comp. Physiol. 166:* 889–899 (1990).

Van Loon, J.J.A., and M.M.M. Van Meer, Chemosensory Perception of Surface Chemicals by Ovipositing *Pieris brassicae* L. Butterflies, *Proc. Exp. Appl. Entomol.,* N.E.V. Amsterdam 2: 56–61 (1991).

Van Sumere, C., P. Fache, K.V. Casteele, L. De Cooman, E. Everaert, R. De Loose, and W. Hutsebaut, Improved Extraction and Reversed Phase High Performance Liquid Chromatographic Separation of Flavonoids and Identification of *Rosa* Cultivars, *Phytochem. Anal. 4:* 279–292 (1994).

Waldbauer, G.P., The Consumption and Utilization of Food by Insects, *Adv. Insect Physiol. 5:* 229–288 (1968).

Yazdani, R., and P. Lebreton, Inheritance Pattern of the Flavonic Compounds in Scots Pine (*Pinus sylvestris* L.), *Silvae Genet. 40:* 57–59 (1991).

Chapter 37

Pharmacologic Aspects and Analysis of Potato Glycoalkaloids

Peter Sporns, Darcy C. Abell, and Darcy R. Driedger

Department of Agricultural, Food and Nutritional Science, University of Alberta, Edmonton, AB, Canada, T6G 2P5

Introduction

Glycoalkaloids (GA) are toxic secondary metabolites found in the *Solanaceae* family. This family includes some very important crops, i.e., potatoes, tomatoes, tomatillos, eggplants, sweet peppers, and tobacco. For most of these crops, it is the ripe fruit that is eaten, and the GA content in this plant part is low. Potatoes, however, are the fourth most important world crop in terms of amount of food produced and always contain significant amounts of GA. For some time, commercial potatoes (*Solanum tuberosum*) have been controlled by various methods so as not to exceed 200 mg GA/kg fresh weight potato.

Recently, there has been renewed interest in potato GA and numerous review articles have appeared in the literature. The article by Jadhav *et al.* (1997) and a very comprehensive review by Friedman and McDonald (1997) discuss all aspects of potato GA. Three other reviews (Krasowski *et al.* 1997, Plhak and Sporns 1997, Smith *et al.* 1996) concentrate on biological activity and medical concerns associated with potato GA.

Pharmacologic Aspects of Glycoalkaloids

It is well known that potatoes near the GA tolerance level of 200 mg/kg have a small safety margin (as little as a fourfold safety factor) before toxic effects can be demonstrated in humans. Although potato tubers represent a portion of the plant with relatively low GA levels, young growing portions of the plant can contain very high GA levels (Table 37.1).

GA toxicity seems to involve two major mechanisms, membrane destabilization and cholinesterase inhibition. It seems unlikely that GA are a concern as carcinogens because they are not mutagenic in the Ames test (Friedman and Henika 1992). Although some researchers have claimed teratogenic effects for GA, most scientists seem to agree that oral administration of GA does not pose a teratogenic hazard. Potato GA are very stable and remain largely unaffected by baking or frying.

Renewed interest in GA has occurred because of a number of different factors. High amounts of GA can be found in potato varieties due to genetic as well as

TABLE 37.1
Average Glycoalkaloid (GA) Levels in Different Parts of the Potato Plant[a]

Plant part	Average GA[b]
Leaves	600
Flowers	3000
Berries	400
Sprouts	3000
Skin	500
Peel	200
Flesh	50

[a]*Source:* Adapted from Friedman and McDonald (1997).
[b]Values are expressed as mg/kg fresh weight.

environmental factors, such as any stress condition. Although newly introduced potato varieties are now monitored for GA levels, there have been at least two documented cases of commercial varieties having to be withdrawn near or after commercial production because of high GA levels (Hellenäs *et al.* 1995, Zitnak and Johnston 1970). There are likely many other undocumented cases in which potato varieties have been rejected at various points in breeding programs because of failing to meet safe GA levels. However, because many of these high-GA potato varieties have other good agronomic properties, such as early maturation and pest resistance, they continue to be used in breeding programs. Moreover, wild potato varieties are also used in breeding programs to improve the properties of commercial potatoes. Wild potato varieties are known to have higher GA levels; they frequently contain unique new GA not found in commercial potato varieties. With faster, more efficient breeding programs based on new genetic engineering and tissue culture methods, analysis for GA as early as possible in breeding programs becomes more of a necessity.

Another factor in the increased interest in potato GA is the surprising fact that little is known about human metabolism of GA. Although GA analysis of potatoes is difficult, analysis of blood and tissue for GA and their metabolites is even more difficult because of the much smaller GA concentrations and potential variety of metabolites. It is known that GA have biological half-lives in humans from 10 to 20 h after consumption (Hellenäs *et al.* 1992) and that the sugar hydrolysis product, solanidine, and GA can be found in relatively large quantities in the livers of human cadavers (Friedman and McDonald 1997). Furthermore, it is clear that potato consumption is high in many countries and continues to rise. Many scientists have cautioned that the 200 mg GA/kg potato control level, based on potential toxic effects, is too high for chronic exposure.

Another factor in renewed potato GA interest is their cholinesterase inhibition effects. The two closely related enzymes, acetylcholinesterase and butyrylcholinesterase, are important in the processing of the neurotransmitter acetylcholine and in the metabolism of anesthetic drugs such as succinylcholine, mivacurium, and cocaine. GA inhibit both enzymes and can interfere with enzymatic action in clearing drugs from the body. To further complicate this interaction, atypical butyrylcholinesterases with different substrate specificity have developed in some people. It has been postulated that these atypical butyrylcholinesterases may have evolved as protective systems in populations exposed to high GA levels because these atypical enzymes are much less sensitive to GA (Krasowski *et al.* 1997).

Analysis of Glycoalkaloids

The two GA found in commercial potato varieties are α-chaconine and α-solanine (Fig. 37.1), typically in a 60:40% ratio. Although the carbohydrate portions of these GA are different, they both have the same alkaloid portion, solanidine. Problems with the analysis of these compounds are obvious from their structure.

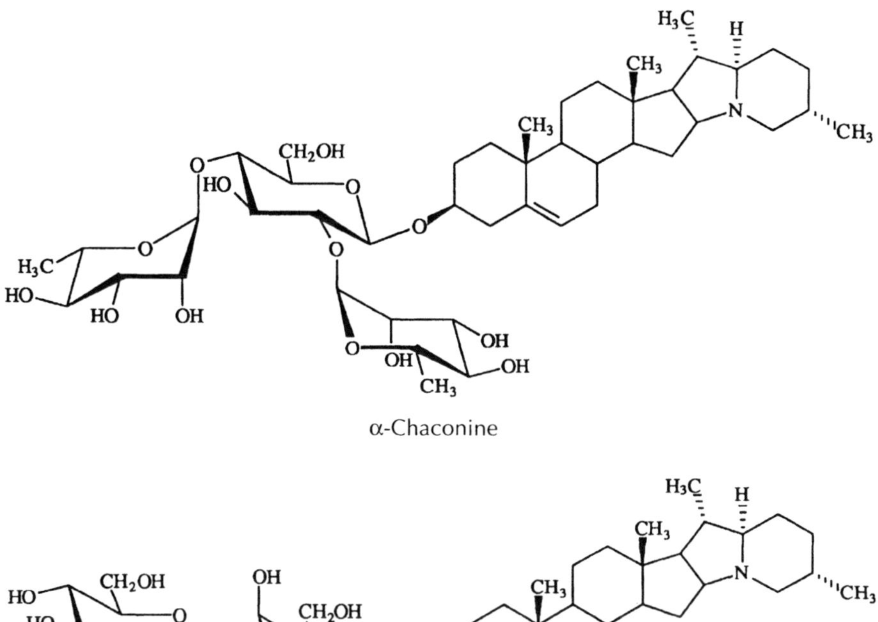

α-Chaconine

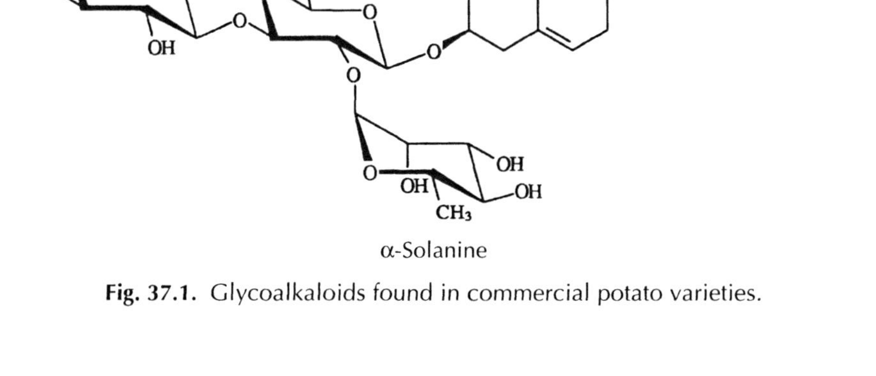

α-Solanine

Fig. 37.1. Glycoalkaloids found in commercial potato varieties.

They have soap-like properties (an explanation of their saponin-like membrane destabilization properties), making them difficult to extract and concentrate in any single solvent. In addition, the alkaloid has only a single double bond to absorb ultraviolet (UV) light (~205 nm). This means that with the use of the presently preferred analysis method, high-performance liquid chromatography (HPLC) with UV detection (Friedman and McDonald 1997), GA require extensive purification to allow their isolation from impurities that might be present in relatively small amounts, but with larger molar absorptivities at wavelengths ~205 nm.

Recently, we have had great success with a very simple GA analysis procedure that uses a new analytical technique, i.e., matrix-assisted laser desorption/ionization mass spectrometry (MALDI-MS; Abell and Sporns 1996). Other researchers have used modern MS techniques to aid in the determination of GA structural features (Claeys *et al.* 1996), but MALDI-MS offers some real advantages for food analysts with problems such as GA quantitation.

This unique ionization, MALDI, was first reported in 1988. The technique uses an excess of matrix material (500–50,000 molar excess) co-crystallized with the sample along with a later short nanosecond laser pulse to sublime matrix and sample in the mass spectrometer. This ionization technique is most often coupled with the simplest form of mass spectrometry, time-of-flight (TOF). The advantages of MALDI-MS were quickly recognized and commercial instruments began to appear in 1991. Although food applications are just now beginning to appear in the literature, two recent review articles have highlighted why this mass spectrometry technique should enjoy rapid acceptance by food analysts (Sporns and Abell 1996 and 1997). Advantages and disadvantages of MALDI-MS are given in Table 37.2.

TABLE 37.2
Advantages and Disadvantages of Matrix-Assisted Laser Desorption/Ionization Mass Spectrometry (MALDI-MS)

Advantages	Disadvantages
1) Time-of-flight has no theoretical mass limit, but the matrix crystallization excess limits mass to <1500 kDa.	1) Instruments are now very expensive. Entry level instruments cost ~U.S.$70,000.
2) The mass can be determined, on modern instruments, down to ppm accuracy.	2) Lack of research available on many food problems.
3) Crude samples are tolerated.	3) Matrix improvement to expand molecular scope of technique (to lipids, for example) is still being researched.
4) Fast analysis—typically, spectra—are generated in 1–2 min.	
5) Quantitation with an internal standard is possible.	
6) Typical sample size ≤100 fmol.	
7) Easy to learn to use.	

TABLE 37.3
Comparison of Glycoalkaloid (GA) Levels Found Using High-Performance Liquid Chromatography (HPLC) and [Matrix-Assisted Laser Desorption/Ionization Mass Spectrometry (MALDI-MS)][a]

Sample	α-Chaconine	α-Solanine	Total GA
A	133 [132]	119 [85]	252 [217]
B	194 [234]	147 [129]	340 [364]
C	81 [60]	50 [32]	131 [92]
D	66 [46]	41 [30]	107 [76]
E	9 [9]	4 [8]	12 [16]
F	18 [20]	4 [10]	22 [30]

[a]Values are expressed as mg/kg fresh weight.

The major advantages for food analysis are the ability to identify the mass of even very large molecules, and to quantify and deal with crude sample mixtures. The major disadvantage is the cost of the instrument.

The currently preferred method for GA analysis in potatoes, HPLC, is compared with the developed MALDI-MS method in Tables 37.3 [adapted from Abell and Sporns (1996)] and 37.4. Although there is some variation in the amounts of GA (Table 37.3) using the two methods, it is not clear which method offers the greater accuracy. When the analysis methods themselves are compared, the MALDI-MS method has several advantages. With the MALDI-MS method, complete GA extraction is not important. The only requirement is that GA equilibration (both potato GA and the internal standard, tomatine) occur in the solvent. This equilibrated sample can be examined repeatedly if desired because only a small (0.5 μL) amount of the

TABLE 37.4
Comparison of High-Performance Liquid Chromatography (HPLC) and Matrix-Assisted Laser Desorption/Ionization Mass Spectrometry (MALDI-MS) Methods

HPLC	MALDI-MS
1) Glycoalkaloids (GA) from 1-g sample extracted with methanol/water	1) GA in 200-mg sample equilibrated with tomatine standard in methanol/water
2) Filtering and washing of sample with more methanol/water to complete extraction	2) Centrifuge to obtain supernatant
3) Sep-Pak column purification	
4) Concentration of GA sample	
5) HPLC analysis: ~1 h/sample (analysis from triplicate samplings)	5) MALDI-MS analysis: ~6 min/sample (analysis from triplicate samplings)
6) GA assignment dependent on HPLC chromatography behavior	6) Accurate mass identification
7) Not applicable to all GA or metabolites (e.g., solanidine)	7) Applicable to all GA (no double bond required) and metabolites

methanol/water (1:1) solvent (10 mL total) is used for each MALDI-MS matrix spot. Washing, an extra column (Sep-Pak), (Waters Corp., Milford, MA) and concentration steps required with the HPLC method are not required for the MALDI-MS method. Moreover, the analysis itself is an order of magnitude faster (6 vs. 60 min). Finally, although the MALDI-MS analyses were performed using 200 mg of potato sample, even smaller amounts of potato could be sampled. This might make the technique applicable to the small amounts of potato sample available early in breeding programs.

Our future plans are to expand the utility of GA analysis using MALDI-MS in a number of ways. We want to demonstrate that the analysis method works with many different potato varieties and different parts of the potato plant. We want to reduce the detection limit of GA from micromolar to nanomolar concentrations, if possible. This lower detection limit would allow for analysis of blood and tissue samples. One way that this lower detection limit can be obtained is to use antibodies to capture, purify,

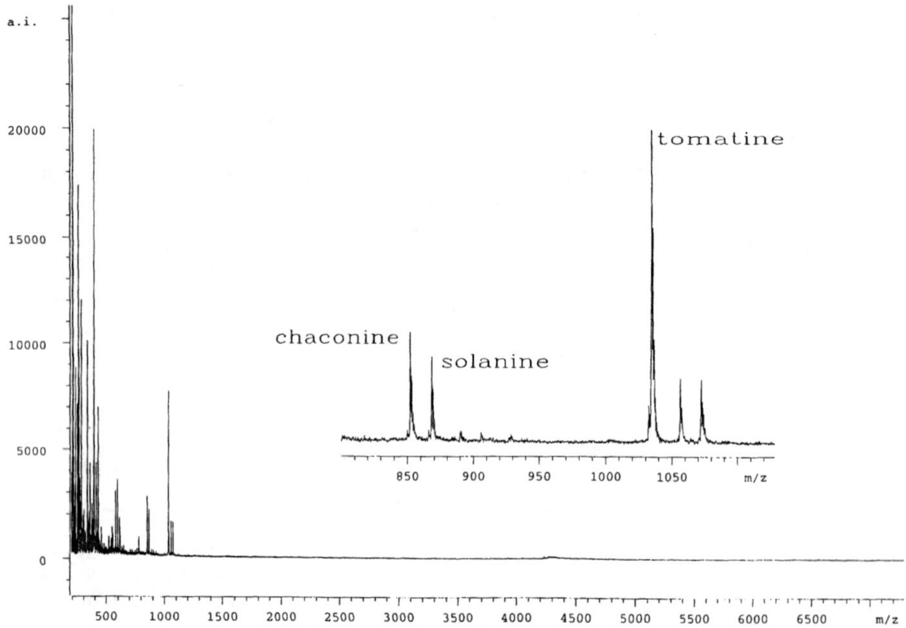

Fig. 37.2. Matrix-assisted laser desorption/ionization mass spectrometry (MALDI-MS) spectrum of potato equilibrated solution. The sample preparation was a slight modification of that given in Abell and Sporns (1996), using sample E [peeled commercial potato variety containing 16 mg total glycoalkaloid (GA)/kg fresh potato by earlier MALDI-MS analysis (see also Table 37.3)]. Freeze-dried potato (400-mg sample) was shaken for 1 h with 10 mL 1:1, methanol/water containing a total of 0.108 mg tomatine standard and centrifuged. Saturated trihydroxyacetophenone in 0.4 µL acetone was air dried on the probe followed by 0.5 µL prepared potato solution, also air dried. The spectrum was taken on a Bruker Proflex instrument (Bruker, Billerica, MA) fitted with pulsed ion extraction using 100 shots at a laser power ~20% above threshold levels. Insert is an expansion of the 800–1100 mass region.

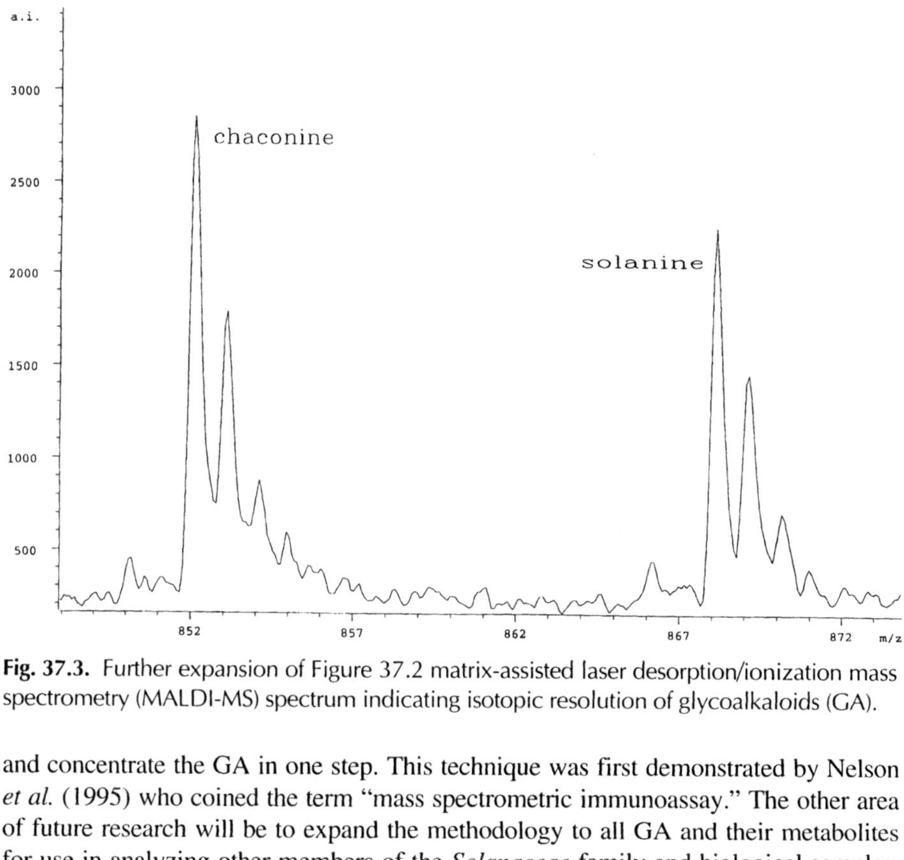

Fig. 37.3. Further expansion of Figure 37.2 matrix-assisted laser desorption/ionization mass spectrometry (MALDI-MS) spectrum indicating isotopic resolution of glycoalkaloids (GA).

and concentrate the GA in one step. This technique was first demonstrated by Nelson *et al.* (1995) who coined the term "mass spectrometric immunoassay." The other area of future research will be to expand the methodology to all GA and their metabolites for use in analyzing other members of the *Solanaceae* family and biological samples. A requirement for this expansion will be to prepare another internal standard because the standard that is presently in use, tomatine, is found in some *Solanaceae* members such as tomatoes. The new internal standard will require synthetic modification of a readily available GA to increase or decrease its mass.

Finally, MALDI-MS improvements have been dramatic in the last few years. Figs. 37.2 and 37.3 show the remarkable resolution available now with even simple instruments.

Acknowledgment

Funding for this research came from the Natural Sciences and Engineering Research Council (NSERC) of Canada.

References

Abell, D.C., and P. Sporns, Rapid Quantitation of Potato Glycoalkaloids by Matrix-Assisted Laser Desorption/Ionization Time-of-Flight Mass Spectrometry, *J. Agric. Food Chem. 44:* 2292–2296 (1996).

Claeys, M., H. Vandenheuvel, S. Chen, P.J. Derrick, F.A. Mellon, and K.R. Price, Comparison of High and Low Energy Collision Induced: Dissociation Tandem Mass Spectrometry in the Analysis of Glycoalkaloids and Their Aglycons, *J. Am. Soc. Mass Spectrom. 7:* 173–181 (1996).

Friedman, M., and P.R. Henika, Absence of Genotoxicity of Potato Glycoalkaloids α-Chaconine, α-Solanine and Solanidine in the Ames Salmonella and Adult and Foetal Erythrocyte Micronucleus Assays, *Food Chem. Toxicol. 30:* 689–694 (1992).

Friedman, M., and G.M. McDonald, Potato Glycoalkaloids: Chemistry, Analysis, Safety, and Plant Physiology, *Crit. Rev. Plant Sci. 16:* 55–132 (1997).

Hellenäs, K.-E., C. Branzell, H. Johnsson, and P. Slanina, Glycoalkaloid Content of Early Potato Varieties, *J. Sci. Food Agric. 68:* 249–255 (1995).

Hellenäs, K.-E., A. Nyman, P. Slanina, L. Lööf, and J. Gabrielsson, Determination of Potato Glycoalkaloids and Their Aglycon in Blood Serum by High-Performance Liquid Chromatography. Application to Pharmacokinetic Studies in Humans, *J. Chromatogr. 573:* 69–78 (1992).

Jadhav, S.J., S.E. Lutz, G. Mazza, and D.K. Salunkhe, Potato Glycoalkaloids: Chemical, Analytical, and Biochemical Perspectives, in *Antinutrients and Phytochemicals in Food*, edited by F. Shahidi, ACS Symposium Series 662, American Chemical Society, Washington, DC, 1997, pp. 94–114.

Krasowski, M.D., D.S. McGehee, and J. Moss, Natural Inhibitors of Cholinesterases: Implications for Adverse Drug Reactions, *Can. J. Anaesth. 44:* 525–534 (1997).

Nelson, R.W., J.R. Krone, A.L. Bieber, and P. Williams, Mass Spectrometric Immunoassay, *Anal. Chem. 67:* 4581–4585 (1995).

Plhak, L.C., and P. Sporns, Biological Activities of Potato Glycoalkaloids, in *Antinutrients and Phytochemicals in Food*, edited by F. Shahidi, ACS Symposium Series 662, American Chemical Society, Washington, DC, 1997, pp. 115–126.

Smith, D.B., J.G. Roddick, and L.J. Jones, Potato Glycoalkaloids: Some Unanswered Questions, *Trends Food Sci. Technol. 7:* 126–131 (1996).

Sporns, P., and D.C. Abell, MALDI Mass Spectrometry for Food Analysis, *Trends Food Sci. Technol. 7:* 187–190 (1996).

Sporns, P., and D.C. Abell, MALDI-MS for Food Analysis, in *New Techniques in the Analysis of Foods*, edited by M.H. Tunick, S.A. Palumbo, and P.M. Fratamico, Plenum Press, New York, 1999, pp. 181–192.

Zitnak, A., and G.R. Johnston, Glycoalkaloid Content of B5141-6 Potatoes, *Am. Potato J. 47:* 256–260 (1970).

Index

A

Adzuki beans, superoxide-scavenging compound from, 261–269
Aged garlic extract (AGE)
  antioxidant activity of, 65–67
  health benefits of, 68–74
  mice stress and, 64–65
  organosulfur compounds in, 67
  and peripheral microcirculation, 67, 72–74
  pharmacological effects of, 62–63
Allicin, 61, 67
Alliinase, inactivation of, 67
Anthocyanins, cancer prevention and, 154–161
Antifeedants
  citrus limonoid aglycones, 91
  forest foliage phenolics, 398–406, 408–414
  Salvia neo-clerodane diterpenoids, 375–382
Antioxidant(s)
  activity of (-)-epicatechin in plasma, 202–206, 208, 210–211
  adzuki bean, 261–269
  aged garlic extract, 65–67, 69–72
  anthocyanin and flavonol, 154–161
  assessment with hypochlorous acid, 39–40
  assessment using peroxynitrite, 40–41
  betel leaf extract, 342–346
  cancer prevention by, 239, 242–247
  characterization of, 33–34, 36
  evening primrose, 278–284, 286–291
  highbush blueberry, 271–276
  methods for assessing prooxidant actions, 38–39
  in plant extracts, 32
  polyphenols, *see* Polyphenol antioxidants
  properties of dietary, 34, 36
  reserves and cardiovascular disease, 146–153
  rosemary phenolic diterpene, 296–298
  synergism, 185
  carotenoid-tocopherol interaction, 185–190
  α-tocopherol in liposomes, 132–136, 138–143
AP-1 activation, inhibition by tea polyphenols, 198
Apigenin, anticancer potential of, 242
Apoptosis
  curcumin induced, 244–245
  effect of flavonoids on, 157–160
  reactive oxygen species and, 246, 247
  tea polyphenols and, 198

B

Bactericidal activity, of tea polyphenols, 214, 216–218, 220–221
Betel, antioxidant activity of, 342–346
Biological screening
  for antiaging and memory-enhancing compounds, 53
  for antibacterial compounds, 53
  for anticancer activity, 52
  for antifungal compounds, 53
Black cohosh
  anti-inflammatory effect of, 367–368
  chemistry and analysis of active principles, 360–361, 363, 365
  endocrine activity of, 365, 367
  therapeutic uses, 368–370
Black tea. *See also* Green Tea; Tea
  bioavailability and relevance to humans, 198–200
  inhibition of tumors by, 195–198
  polyphenols in, 192
Blood circulation
  antioxidant reserves and strokes, 146–153
  effects of aged garlic extract on, 67, 72–74
Blueberry, antioxidative and cytotoxic compounds, 271–276
Boswellia serrata gum exudate
  antitumor activity of, 310–313
  major constituents of, 310
Bursera simaruba, 336

C

Caffeic acid esters, affinity for lipid bilayers, 352

Cancer chemopreventive agents, 1–2, 239, 242–247
apigen, 242
citrus limonoids, 4–5, 79–80, 82–83, 85, 95–103
curcumin, 242
apoptosis induction by, 244–245
flavonols and anthocyanins, 154–161
flaxseed lignans, 5–6
flaxseed oil, 108
garlic compounds, 6
aged garlic extract, 62, 71–72
generic Paclitaxel, 56–57
2″-*O*-glycolsyl-isovitexin, 243
isovitexin, 243
possible mechanisms of, 2
antioxidant activity, 239
screening for, 52
soy isoflavones, 2–3
genistein, 242–243
resveratrol, 243
rosemary, 299–300, 302, 305
tea constituents, 6–7, 192–200
polyphenols, 243–246
tomato lycopene, 3–4
Capraria biflora, 337–338
Carnosic acid, 296
Carnosol, 296
β-Carotene, 3
Carotenoids
antioxidant properties in membranes, 175–181, 183
cancer and lycopene, 3–4
β-carotene and lung cancer, 3
interaction with tocopherols, 185–190
Catechin(s)
activities of Adzuki bean, 265–269
affinity for lipid bilayers, 356, 358
effect on fecal flora, 218, 220–221
free radical-scavenging by, 222–236
aroxyl radical-scavenging rates with, 226–229
structure-activity relationship of, 229–235
vitamin E regeneration, 225–226, 235–236
H. pylori inhibition by, 216–217
tea, *see* Tea catechins
Cell proliferation
effects of tea on, 195–198
flavonols and anthocyanins and, 154, 157–161
α-Chaconine, 419–421
Chemiluminescence
of Glc-catechin in adzuki beans, 261, 265–266
of a hydrogen peroxide/gallic acid/horseradish peroxidase system, 321–330
in Maillard reaction, 252–260
Chiococca alba, 336–337
Cholesterol hydroperoxides
inhibition by (-)-epicatechins, 208, 210–211
oxidation by copper ions, 206-208
Cholinesterase inhibition, and potato glycoalkaloids, 417–419
Cimicifuga racemosa. See Black cohosh
CimiPure. *See* Black cohosh
Citrus limonoids. *See* Limonoids
Cnidoscolus aconitifolius, 336
Coniferous needles
phenolics of, 398–404, 406, 408
as pest control agents, 408–413
Consumer(s)
interest in nutrition, 22–24
and health-promoting foods, 24–26
new product benefits to, 26–29
genetically modified sources, 29–30
Curcumin
apoptosis induction by, 244–245
mechanisms of action of, 242
Curcuminoids, affinity for lipid bilayers, 355
Cyanidin, and cancer prevention, 154–161
Cyanogenic compounds, in flaxseed, 114–115

D

Daidzein, 2–3
in Japanese diet, 164–165, 168–173
Deciduous leaves
phenolics of, 398–404, 406, 408
as pest control agents, 408–413

Delayed bitterness, and limonoids, 82, 89
Diet, and cancer prevention, 1–7
Dietary Supplement Health and Education Act (DSHEA), 13
Disease prevention. *See* Health
Diterpenes, antioxidant activity of rosemary phenolic, 296–297
Drugs. *See* Phytopharmaceuticals

E

Enterodiol, 5
Enterolactone, 5
(-)-Epicatechin
  plasma antioxidative activity of, 202–206, 208, 210–211
    absorption profile and metabolites, 204–206
    consumption during plasma oxidation, 208–210
(-)-Epigallocatechin-3-gallate (EGCC), anticancer effects of, 245–246
Epirosmanol, 296
Estrogenic phytochemicals
  black cohosh, *see* Black cohosh
  genestein, 242–243
  lignans, 5
Evening primrose, antioxidants of, 278–284, 286–291

F

Flavones, in rosemary, 298
Flavonoids, 32, 61
  antioxidant activity in plasma, 202–206, 208, 210–211
  antioxidant activity of rosemary, 298
  apigenin, 242
  bioavailability in soybean milk, 172
  catechins, *see* Catechins; Tea Catechins
  effects on cell, 154
    proliferation and apoptosis, 157–160
  free radical-scavenging action of, 222–223
  in Japanese diet, 164–173
    bioactivity study, 168, 170
    nutritional study, 164–165, 168–169
    pharmacokinetic study, 165, 168, 169
Flavonols, 26, 27
  and cancer prevention, 154–161
Flavonones, in rosemary, 298
Flaxseed, 5
  components and health benefits, 106–112, 114–116
    cyanogenic compounds, 114–115
    dietary proteins, 108–110
    flavonoids, 112
    lignans, 5–6, 115
      medicinal, 122–123, 125–126, 128–129
    α-linolenic acid, 106–108
    lipoxygenase, 114
    oil, 106–108
    phenolic acids, 111–112
    phytic acid, 112
    polysaccharides, 110–111
    tocopherols, 114
Free radical-scavenging, by catechins, 222–236
Functional foods, regulation and research incentives, 13–20

G

Gallic acid esters, affinity for lipid bilayers, 352, 354
Garlic
  aged extract, *see* Aged garlic extract
  anticancer activity of, 6
  preparations and effects, 62–74
    antioxidation, 65–67
    antistress, 64–65
    on peripheral microcirculation, 67, 72–74
Genetically modified foods, benefits of, 29–30
Genistein, 2–3, 26
  anticancer potential of, 2–3, 242–243
  estrogenic activity, 242–243
  in Japanese diet, 164–165, 168–173
Glutathione-S-transferase, effect of limonoids on, 103
Glycoalkaloids, potato
  analysis of, 419–423
  pharmacologic aspects, 417–419
2″-*O*-Glycosyl-isovitexin, antioxidant activity of, 243
Green tea. *See also* Black tea; Tea

polyphenols in, 192
tumorigenesis inhibition by, 193–196, 197–198

H

Halichondrin B, synthesis of antitumorigenic, 386–395
Health
antioxidant supplementation and strokes, 146–153
claims, 13–20
consumer interest in, 22–24
foods, 24–26
genetic modification of, 29–30
new product development of, 26–29
plant components and, 1–7
Health claims, regulation and industry research, 13–20
Health-promoting foods, 24–26
Helicobacter pylori, inhibition by tea, 214, 216–217
Hesperitin, 298
Horseradish peroxidase, chemiluminescence of, 321–325, 326, 328–330

I

Intellectual property, and new ingredients, 27–28
Isoflavones. *See* Soy isoflavones
Isoflavonoids. *See also* Daidzein; Genistein
in Japanese diet, 164–173
Isorosmanol, 296
Isovitexin, antioxidant activity of, 243

K

Kaempferol, and cancer prevention, 154–161
Keystone National Policy Dialogue on Food, Nutrition and Health, 14, 15

L

Labiatae, insect control by, 371–383
Lamiaceae. *See also* Labiatae
Salvia diterpenoids, 372–375
antifeedant activity of, 375–383
Scutellaria diterpenoids, 375
Lepidopteran species
life cycle of, 380–381
management using forest foliage phenolics, 398–406, 408–414
Lignan glycoside, in rosemary, 299
Lignans, flaxseed
anticancer activity of, 5–6, 115
extraction of medicinal, 123–129
and purification of, 128–129
nature of, 5, 115, 122
d-Limonene, antitumor activity and metabolites of, 4–5
Limonin
anticancer activity of, 80, 82–83, 85, 95
and delayed bitterness, 89
Limonoid aglycones
anticancer activity of, 80, 82–83, 85
antifeedant activity of, 91
in debittering process, 89
Limonoid glucosides
anticancer activity of, 83
oral carcinogenesis and, 95–103
average intake of, 102
bulk isolation of, 85–88
in citrus, 79, 86–88, 102
degradation of, 90–91
genetic engineering of, 88–90
toxic evaluation of, 85
utilization of, 87–88
Limonoid glucosyl transferase, 89–90
Limonoids. *See also* Limonoid aglycones; Limonoid glucosides
anticancer activity of citrus, 4–5
α-Linolenic acid, health benefits of, 106–108
Lipids, antioxidants and, 32, 278
Liposomes, carotenoids and α-tocopherol in, 175–183
Lycopene
cancer risk and tomato, 3–4
interaction with carotenoids, 188–190

M

Maillard reaction substances (MRS), radical-scavenging activity of, 252–260
Medicinal plants
black cohosh, 368–370
evaluation for new pharmaceuticals, 332
screening for medicines, 48–53
from Yucatecan, 332–338

Menopause treatment, with black cohosh, 368–370
Miltirone, 297

N
Nitric oxide, toxicity of, 146, 149
Nomilin
  anticancer activity, 80, 82–83, 85, 95
  glucoside of, 86
  in limonoid biosynthesis, 88
Nutraceuticals. *See* Functional foods
Nutrition
  categories of health-promoting foods, 24–26
  consumer concerns for, 22–24
  new product opportunities, 26–29
    genetically modified, 29–30
  research policy initiatives, 13–20

O
Ocimum basilicum, metabolites of, 334–336
Ocimum micranthum, metabolites of, 334–336
Organosulfur compounds, and cancer, 6
Oxidative defense, regeneration of tocopherols and, 185–190
Oxygen toxicity, 146–147
  antioxidant supplementation and, 146–153
    evaluating serum, 152–153

P
Paclitaxel, generic version of, 56–58
Phenolic(s)
  antioxidants, *see* Polyphenol antioxidants
  forest foliage extracted, 398–406, 408–414
    in coniferous and broadleaf trees, 406, 408
    host plant resistance and variation of, 412–413
    identification and assay of, 401–406
    and insect feeding and growth, 408–411
    and insect oviposition, 411–412
Phospholipids, peroxidation and α-tocopherol, 133
Phytic acid, health benefits of, 112
Phytochemicals
  antiestrogenic, 5
  definition of, 2
  market segmentation for, 25–26
  research incentives, 14–17
  role in health, 2–7
  sources of, 24–25
Phytopharmaceuticals
  Xechem's strategy for discovering, 47–60
    drug development rationale, 47–48
    screening process, 48–56
    specific illustrations of, 56–58
Phytopolyphenols. *See* Polyphenol antioxidants
Piper betle L.. *See* Betel
Polyphenol antioxidants
  affinities for lipid bilayers, 349–350, 352, 354–356, 358
  antibacterial actions of tea, 214, 216–218, 220–221
  with anticancer potential, 239, 242–247
Polysaccharides, flaxseed, 110–111
Potato glycoalkaloids, *see* Glycoalkaloids
Prooxidant activity, of dietary antioxidants, 34, 36, 38–40
Prooxidative mitotic signaling, and phytopolyphenols, 246–247

Q
Quercetin
  affinity for lipid bilayers of, 356, 358
  cancer prevention by, 154–161

R
Radical-scavenging activity
  of catechin and related compounds, 222–236
  of Maillard reaction substances, 252
Reactive oxygen-scavenging compounds, from adzuki beans, 261–269
Reactive oxygen species (ROS), 246–247
Resveratrol, cancer prevention and, 243, 314–319
Rosemanol, 296
Rosmarinoside A, 299
Rosemary (*Rosmarinus officinalis* L.), 302, 305
  anticancer activity of, 299–300, 302, 305

antioxidant activity of, 296–299
Rutin, cancer prevention and, 154–161

S

Salvia, agrochemical potential of, 371–383
Secoisolariciresinol diglucoside, extraction of, 122–123, 125–126, 128–129
Signal transduction, and phytopolyphenols, 246–247
Singlet oxygen, antioxidants and membrane damage by, 175–181, 183
α-Solanine, 419–421
Soy, 2–3, 26, 27
Soy hypocotyl tea, antioxidant activity of, 164, 168, 172–173
Soy isoflavones, 2–3, 164–173
Stress, effects of aged garlic extract on, 64–65, 71
Stroke disease, 149
Superoxide, toxicity of, 146–147
Superoxide dismutase (SOD)-like activity, in adzuki beans, 261–269

T

Tea, 6–7. *See also* Black tea; Green tea
  bowel improvement by, 217–218, 220–221
  carcinogenesis inhibition by, 192–200
    induced lung tumorigenesis, 194–195
    spontaneous lung tumors and rhabdomyosarcomas, 197
  health effects of, 192
  inhibition of foodborne pathogens by, 214
Tea catechins, 6–7. *See also* Catechin(s); Tea polyphenols
Tea polyphenols
  antibacterial applications in humans, 214, 216–218, 220–221
    fecal composition and odor, 217–218, 220
    foodborne pathogens, 214
    H. pylori in stomach, 216–217
  carcinogenesis inhibition by, 192–200, 243–244
  investigations of action mechanisms, 245–246
Tetrahydrocurcumin, antioxidant activity of, 355
Theaflavins
  antibacterial action of, 214
  anticancer effects of, 192–200
  bioavailability of, 199
Thearubigins
  anticancer effects of, 192–200
  bioavailability of, 199
Thevetia gaumeri, 337
α-Tocopherol
  antioxidant activity of, 132–136, 138–143
    singlet oxygen and, 175–181, 183
  and formation of α-tocopherones, 136, 138
  interaction with carotenoids, 185–190
  serum levels and cardiovascular disease, 147–149
  supplementation in stroke patients, 149–150
Tocopherols
  and carotenoid interaction, 185–190
  flaxseed, 114
  regeneration, 225–226, 235–236
Tocotrienols, supplementation effect, 149–150
Tomato lycopene. *See* Lycopene
Triterpene glycosides, in black cohosh, 360–361, 363, 365
Tumors
  inhibition by
    Boswellin serrata extract, 310–313
    halichondrin B, 386–395
    resveratrol and related compounds, 314–319
    rosemary, 299–300, 302, 305
    tea and tea constituents, 192–200

U

Ubiquinones, affinity for lipid bilayers of, 354–355
Urechites andrieuxii, 338
Ursolic acid, 299

V

Vaccinium corymbosum L.. *See* Blueberry

Vigna angularis. See Adzuki beans
Vitamin C, and regeneration of tocopherol, 185, 235–236
Vitamin E. *See also* Tocopherols
  regeneration rates, 235–236

X

Xechem
  phytopharmaceutical research by, 47–60
    drug development rationale, 47–48
    screening process, 48–56
    specific illustration of, 56–58

Y

Yucatecan medicinal plants, bioactive metabolites of, 332–335, 336–338

CPSIA information can be obtained at www.ICGtesting.com
Printed in the USA
LVOW092252120412

277428LV00002B/3/P

9 781893 997059